WAVES, PARTICLES, AND STORMS IN GEOSPACE

Waves, Particles, and Storms in Geospace

A Complex Interplay

Edited by

Georgios Balasis

National Observatory of Athens, Greece

Ioannis A. Daglis

National and Kapodistrian University of Athens, Greece

Ian R. Mann

University of Alberta, Canada

OXFORD

UNIVERSITY PRESS

OXFORD
UNIVERSITY PRESS

Great Clarendon Street, Oxford, OX2 6DP,
United Kingdom

Oxford University Press is a department of the University of Oxford.
It furthers the University's objective of excellence in research, scholarship,
and education by publishing worldwide. Oxford is a registered trade mark of
Oxford University Press in the UK and in certain other countries

First Edition published in 2016

Published in the United States of America by Oxford University Press
198 Madison Avenue, New York, NY 10016, United States of America

British Library Cataloguing in Publication Data
Data available

Library of Congress Control Number: 2016932196

ISBN 978–0–19–870524–6

Preface

This book addresses the topic of Waves, Particles, and Storms in Geospace. In particular it provides an overview of the current state-of-the-art in understanding how waves and particles in geospace interact during magnetic storms, and how these processes combine to explain the dynamics of the space radiation observed in the Earth's Van Allen radiation belts. The book covers some of the very significant advances in our knowledge, which have been developed over the last decade or more, driven especially and most recently by new discoveries arising from the four-spacecraft Cluster mission of the European Space Agency (ESA), the National Aeronautics and Space Administration (NASA) five-spacecraft THEMIS mission, and most recently the twin spacecraft NASA Van Allen Probes[1] mission launched on August 30, 2012. Carrying a suite of instruments designed to measure fields, waves and particles in geospace with unprecedented levels of coverage, energy resolution, and cleanliness, the Van Allen Probes in particular have opened a new window on radiation belt science and this forms a significant focus of this book. Although these missions are all currently exploring connections between waves, particles, and storms, in geospace more than 50 years after the initial discovery of the radiation belts by James Van Allen in 1958, many significant questions and scientific puzzles remain.

This book is intended to capture the state of knowledge on the topic of waves, particles, and storms in geospace at the present time, up to and including the results from the Van Allen Probes mission until the end of its two year prime mission. After an introduction, the book comprises 17 chapters, which collectively reflect the current state-of-the-art in a range of topics related to the dynamics of the Earth's radiation belts and inner magnetosphere. It is written with a target audience of active researchers working in this field in mind, but it will also serve as a reference for graduate students and even senior undergraduate students with interests in the solar-terrestrial connection, and in space plasma physics, space radiation and space weather.

After the Introduction [Mann], the book begins with three chapters that address the historical overview of wave and particle measurements from missions in the Earth's neighbourhood [Baker], offer an introduction to wave-particle interactions and their impacts on energetic particles in geospace [Takahashi and Miyoshi], and examine the variable effects of magnetic storms on radiation belt dynamics [Reeves and Daglis]. The next eight chapters address the characteristics of various wave modes, their influence on

[1] Note that following their successful launch, and completion of two months of commissioning and early operations, NASA renamed the satellites in honour of the late James Van Allen from their previous name of the "Radiation Belt Storm Probes" to the "Van Allen Probes."

the radiation belts, including related modelling efforts. The first and second of these address current and open questions relating to the role of ultra-low frequency (ULF) waves in radiation belts [Elkington and Sarris] and modelling of energetic particles of the inner magnetosphere [Bourdarie et al.], before the next six chapters address the monitoring of ULF waves from low Earth orbit [Balasis et al.] and from the ground [Waters and Menk], the nature of chorus waves in geospace and their influence in radiation belt dynamics [Bortnik et al.], the processes of wave-driven diffusion in the radiation belts [Horne et al.], recent advances in understanding the role of electromagnetic ion cyclotron (EMIC) waves in radiation belt and ring current dynamics [Usanova and Mann], and multi-dimensional analysis of whistler-mode waves in the radiation belt region [Santolik]. The next two chapters address the extreme variability of radiation belt electrons [Turner and Angelopoulos], and radiation belt relativistic electron flux enhancements associated with substorms [Miyoshi et al.], before the next two chapters of the book turn to addressing the linkages between the radiation belt and the polar atmosphere and climate [Clilverd et al.] and the dynamics of waves and particles in the radiation belts of the outer planets [Krupp et al.]. Finally, the book concludes with two chapters addressing the recent advances in our understanding of the impacts of waves and particles on radiation belt dynamics [Kletzing], and an overview of the early results from the energetic particle, composition, and thermal plasma (ECT) suite [Spence et al.], of the Van Allen Probes mission.

Each of the 17 chapters described above was refereed by a leading expert in the field, as well as by one of the Editors. We wish to offer our heartfelt thanks to all of the referees for taking the time to contribute to the successful completion of this book: Vassilis Angelopoulos, Jacob Bortnik, Christopher Cully, Scot Elkington, Mark Engebretson, Mary Hudson, Xinlin Li, Barry Mauk, Nigel Meredith, Kyle Murphy, Yoshiharu Omura, Ondrej Santolik, Danny Summers, Kazue Takahashi, Richard Thorne and Maria Usanova. We also express our thanks to Dr Stavros Dimitrakoudis, University of Athens (now at the University of Alberta), for processing assistance associated with the preparation of the final version of the figures in the book and for compiling the Index, to Dr Andy Kale, University of Alberta, for the preparation of the cover art, to Sonke Adlung, Ania Wronski, and the team at Oxford University Press, for their help with the preparation of this book. Last but not least, we wish to thank our families for their support throughout this project: Maria, Thodoris, Olia, Anna, Alexandros, Thanasis, Dimitris, and Sonya.

Dr Georgios Balasis, National Observatory of Athens, Greece

Prof. Ioannis A. Daglis, University of Athens, Greece

Prof. Ian R. Mann, University of Alberta, Canada

January 2016

Contents

List of Contributors

Monika Andersson
Finnish Meteorological Institute, Helsinki, Finland, Monika.Andersson@fmi.fi

Vassilis Angelopoulos
Department of Earth, Planetary, and Space Sciences, University of California, Los Angeles, California, USA, vassilis@ucla.edu

Daniel N. Baker
Laboratory for Atmospheric and Space Physics, University of Colorado at Boulder, Boulder, CO, USA, daniel.baker@lasp.colorado.edu

Georgios Balasis
Institute for Astronomy, Astrophysics, Space Applications and Remote Sensing, National Observatory of Athens, Athens, Greece, gbalasis@noa.gr

Jacob Bortnik
Department of Atmospheric and Oceanic Sciences, University of California at Los Angeles, Los Angeles, California, United States of America, jbortnik@atmos.ucla.edu

Sebastien Bourdarie
DESP, ONERA The French Aerospace Lab, Toulouse, France, Sebastien.Bourdarie@onera.fr

Mark Clilverd
British Antarctic Survey (NERC), Cambridge, UK, macl@bas.ac.uk

Ioannis A. Daglis
Department of Physics, National and Kapodistrian University of Athens, Athens, Greece, iadaglis@phys.uoa.gr

Yusuke Ebihara
Research Institute for Sustainable Humanosphere, Kyoto University, Uji, Japan, ebihara@rish.kyoto-u.ac.jp

Scot Elkington
Laboratory for Atmospheric and Spacy Physics, University of Colorado, Boulder, USA scot.elkington@lasp.colorado.edu

Sarah Glauert
British Antarctic Survey, Cambridge, UK, sagl@bas.ac.uk

Richard Horne
British Antarctic Survey, Cambridge, UK, rh@bas.ac.uk

George Hospodarsky
Department of Physics and Astronomy, University of Iowa, Iowa City, IA, USA, george-hospodarsky@uiowa.edu

Vania K. Jordanova
Space Science and Applications Group, Los Alamos National Laboratory, Los Alamos, NM, USA, vania@lanl.gov

Ryuho Kataoka
National Institute of Polar Research, Tachikawa, Japan, kataoka.ryuho@nipr.ac.jp

Tobias Kersten
British Antarctic Survey, Cambridge, UK, tobker@bas.ac.uk

Ramona Kessel
Heliophysics Division of the Science Mission Directorate, NASA HQ, Washington, DC, USA, mona.kessel@nasa.gov

Craig Kletzing
Department of Physics and Astronomy, The University of Iowa, Iowa City, IA, USA, craig-kletzing@uiowa.edu

Norbert Krupp
Max Planck Institute for Solar System Research, Göttingen, Germany, krupp@mps.mpg.de

Wen Li
Department of Atmospheric and Oceanic Sciences, University of California at Los Angeles, Los Angeles, California, United States of America, moonli@atmos.ucla.edu

Michael Liemohn
University of Michigan, Department of Climate and Space Sciences and Engineering, Ann Arbor, USA, liemohn@umich.edu

Ian R. Mann
Department of Physics, University of Alberta, Edmonton, Canada, imann@ualberta.ca

Frederick Menk
Centre for Space Physics, School of Mathematical and Physical Sciences, University of Newcastle, New South Wales, Australia, fred.menk@newcastle.edu.au

Nigel Meredith
British Antarctic Survey, Cambridge, UK, nmer@bas.ac.uk

Yoshizumi Miyoshi
Institute for Space-Earth Environment Research, Nagoya University, Nagoya, Japan, miyoshi@isee.nagoya-u.ac.jp

Thomas Paul O'Brien
Space Sciences Department/Chantilly, The Aerospace Corporation, Chantilly VA, USA, paul.obrien@aero.org

Constantinos Papadimitriou
Institute for Astronomy, Astrophysics, Space Applications and Remote Sensing, National Observatory of Athens, Athens, Greece; Department of Physics, National and Kapodistrian University of Athens, Athens, Greece, constantinos@noa.gr

Chris Paranicas
The Johns Hopkins University Applied Physics Laboratory, Laurel, MD, USA, chris.paranicas@jhuapl.edu

Viacheslav Pilipenko
Space Research Institute, Moscow, Russia, space.soliton@gmail.com

Geoffrey D. Reeves
Space Science and Applications Group, Los Alamos National Laboratory, Los Alamos, New Mexico, USA; The New Mexico Consortium, Los Alamos, New Mexico, USA, geoff@reevesresearch.org

Craig Rodger
Department of Physics, University of Otago, Dunedin, New Zealand, craig.rodger@otago.ac.nz

Elias Roussos
Max Planck Institute for Solar System Research, Göttingen, Germany, roussos@mps.mpg.de

Ondrej Santolik
Institute of Atmospheric Physics ASCR, Prague, Czech Republic Faculty of Mathematics and Physics, Charles University, Prague, Czech Republic, os@ufa.cas.cz

Theodore Sarris
Department of Electrical and Computer Engineering, Democritus University of Thrace, Xanthi, Greece, tsarris@ee.duth.gr

Annika Seppälä
Finnish Meteorological Institute, Helsinki, Finland, Annika.Seppala@fmi.fi

Yuri Y. Shprits
Section 2.3: Earth's Magnetic Field, GFZ German Research Centre for Geosciences, Potsdam, Germany; Department of Earth, Planetary, and Space Sciences, University of California, Los Angeles, CA, USA, yuri.shprits@gfz-potsdam.de

Angelica Sicard
Onera, Toulouse, France, angelica.sicard@onera.fr

Harlan E. Spence
Institute for the Study of Earth, Oceans, and Space, University of New Hampshire, Durham, NH, USA, harlan.spence@unh.edu

Kazue Takahashi
The Johns Hopkins University Applied Physics Laboratory, Laurel, MD, USA, kazue.takahashi@jhuapl.edu

Xin Tao
Department of Geophysics and Planetary Sciences, University of Science and Technology of China, Hefei, Anhui, PR China, xtao@ustc.edu.cn

Richard Thorne
Department of Atmospheric and Oceanic Sciences, University of California at Los Angeles, Los Angeles, California, United States of America, rmt@atmos.ucla.edu

Drew Turner
Space Sciences Department, The Aerospace Corporation, El Segundo, California, USA, drew.lawson.turner@gmail.com

Maria Usanova
Laboratory for Atmospheric and Space Physics, University of Colorado at Boulder, Boulder, CO, USA, maria.usanova@lasp.colorado.edu

Pekka Verronen
Finnish Meteorological Institute, Helsinki, Finland, Pekka.Verronen@fmi.fi

Colin Waters
Centre for Space Physics, School of Mathematical and Physical Sciences, University of Newcastle, New South Wales, Australia, colin.waters@newcastle.edu.au

Eftyhia Zesta
Heliophysics Science Division, NASA Goddard Space Flight Center, Greenbelt, MD, USA, eftyhia.zesta@nasa.gov

Waves, Particles, and Storms in Geospace: An Introduction

Ian R. Mann

"My God, space is radioactive!"

Attributed to Ernest Ray, a member of the team led by James Van Allen which discovered the Earth's radiation belts in 1958 (see, e.g., Stern, 1996 and references therein).

■ Abstract

This chapter provides a brief introduction to the topic of waves, particles, and storms in geospace, and the unexpected acceleration of particles to relativistic energies to form the Van Allen radiation belts. A coordinated international research effort, combining unprecedented multiple satellite observations with theory and modeling, has led to remarkable new discoveries which are unlocking the secrets of geospace dynamics. In this complex system, particle populations whose energies span more than six orders of magnitude in energy, from eV to MeV, are coupled through the intermediary of a range of plasma wave modes. Coordinated international efforts are addressing the

Waves, Particles, and Storms in Geospace. First Edition. Georgios Balasis, Ioannis A. Daglis, Ian R. Mann.

fundamental science as well as the impacts of severe, as well as more moderate, space weather on modern twenty-first century technological infrastructure. We provide a brief overview of this heliophysics research, and thereby an introductory context for the later chapters of this book which examine in detail the state of the art of the complex interplay between waves, particles, and storms in geospace.

Geospace storms and radiation belt dynamics

More than 50 years ago, an unexpected population of highly energetic particles which we now call the Earth's radiation belts was discovered by the flight of the Explorer 1 and Explorer 3 spacecraft, launched in 1958 at the beginning of the space race[1] (e.g., Van Allen and Frank, 1959). Taking the name of the scientist credited with their discovery, James Van Allen from the University of Iowa, this could be considered to represent the beginnings of the modern era of solar-terrestrial and magnetospheric physics research. Even now, if one looks up toward the sky directly above, it's still hard to imagine that there should be a population of locally accelerated electrons and protons which move so fast that they have speeds approaching the speed of light—yet this was the unexpected discovery revealed by Van Allen and his team. Quasi-stably trapped in the Earth's magnetic field, and filling near-Earth space at the altitudes where Earth-orbiting satellites fly, we now know that these particles gain such huge energies as a result of wave-particle interaction processes which operate locally in geospace. It is these processes, which act on a planetary scale but remain rather poorly understood, that turn geospace into a giant particle accelerator. It is the nature of these waves and particles, and their interactions and role in storms in geospace that are the focus of this book.

The Van Allen radiation belts comprise energetic electrons and ions whose trajectories follow the structure of the near-Earth magnetic field. The radiation belts can span distances from altitudes above 100 km from the Earth's surface out to beyond geosynchronous orbit at distances greater than 42,000 km from the centre of the Earth. Typically comprising two (and sometimes more) distinct outer and inner belt regions, and despite early thinking that the belts were relatively stable, it is now known that the flux of this particle radiation in space is very dynamic. For example, the flux of electrons in the outer radiation belt can change by many orders of magnitude on a range of timescales ranging from years, to months, weeks, days, hours, and even down to minutes. The time-varying solar wind, originating from the solar atmosphere which is so hot that its expansion into space escapes the gravitational attraction of the Sun and so highly conducting that it carries magnetic field along with it, is responsible for shaping the morphology and ultimately controlling the dynamics of the Earth's radiation belts. However, how geospace and the Earth's radiation belts respond to the incident solar wind are an active area of research and this forms the focus for this book.

[1] The Soviet spacecraft Sputnik 2 (launched before Explorer 1 in 1957) and Sputnik 3 also contributed to early radiation belt research, with Sputnik 3 confirming the existence of the radiation belts as already discovered and reported by Van Allen—see, for example, Chapter 1.

More than 50 years after their discovery a detailed understanding of the acceleration, transport, and loss processes which underpin the dynamics of the Van Allen belts remains largely elusive. The nature of many of the potentially important physical processes is by now quite well understood in theory, and the results from satellite missions and supporting ground-based measurements are providing the ability to differentiate to some degree between proposed interaction mechanisms. However, understanding which of these processes are dominant, and under what conditions, how they are coupled, and how they superpose to produce the observed radiation belt response remains a major challenge in space science. In particular the goal of advancing this understanding to the point of predictability, essential for the production of accurate space weather and space radiation forecasts, has not yet been achieved. Simply put a greater overall physical understanding is required before this knowledge can be transitioned into accurate operational space radiation and space weather forecasts. Nonetheless, there has been an accelerated rate of discovery in understanding the plasma-physical processes which morph the radiation belts with especially fast and significant progress in recent years. Many of these recent research results are reviewed here in this book.

A new era of radiation belt research

The launch of the twin spacecraft NASA Van Allen Probes[2] mission on August 30, 2012, in particular, heralded a new era in radiation belt research. Carrying a suite of instruments designed to measure the fields, waves, and particles in geospace, numerous important discoveries have been made using data from this mission—a selection of which are reviewed in detail in this book. In particular the unprecedented energy resolution and cleanliness of the particle data from the Van Allen Probes payload have opened a new window on radiation belt science—see in particular Kletzing (Chapter 16) and Spence et al. (Chapter 17). In addition to the Van Allen Probes, a number of other international Earth-orbiting satellite missions such as the NASA Time History of Events and Macroscale Interactions during Substorms (THEMIS) mission, the ESA four spacecraft Cluster mission, and the Japanese Geotail mission (to name a few) have contributed to the recent development of an unprecedented level of understanding about the dynamics of the inner magnetosphere. When data from such geospace missions are combined with monitoring of the upstream driving solar wind and integrated with the signatures of the solar-terrestrial interaction in data from networks of ground-based instruments, such as magnetometers, radars, imagers, riometers, etc., we have a truly unique and global scale infrastructure with which to examine the physical processes which control the dynamics of this highly coupled system. Baker (Chapter 1) presents a historical overview of wave and particle measurements from missions in the Earth's neighborhood, Takahashi and Miyoshi (Chapter 2) provide an introduction to the wave-particle interactions active in

[2] Note that after commissioning and early operations, NASA renamed the satellites in honour of the late James Van Allen from their previous name of the "Radiation Belt Storm Probes" to the "Van Allen Probes."

geospace, and Reeves and Daglis (Chapter 3) provide an overview of the effects of geospace magnetic storms on radiation belt dynamics. Recent work addressing the extreme variability of the flux of radiation belt electrons in geospace is also examined by Turner and Angelopoulos (Chapter 12), while the connection between radiation belt relativistic electron flux enhancements and substorms is analyzed by Miyoshi et al. (Chapter 13).

Wave-particle interactions and their impacts on the radiation belts

Although the overall response of the electron radiation belts has long been understood to result from a balance between the action and superposition of different acceleration and loss processes (e.g., Friedel et al., 2002; Reeves et al., 2003), the detailed nature of individual wave-particle interaction processes is only now being revealed. For example, data from the Van Allen Probes have been instrumental in driving significant progress in understanding the processes of radiation belt acceleration. For example Reeves et al. (2013) and Thorne et al. (2013) have provided evidence for the rapid local acceleration of MeV energy electrons in the outer radiation belt, lower band chorus waves in the very low frequency (VLF) band having been identified as the most likely cause of the observed acceleration which is associated with local peaks in phase space density as a function of L-shell. Similarly, Baker et al. (2013a) presented observations of the creation of a third radiation belt. The characteristics of chorus waves in geospace are reviewed by Bortnik et al. (Chapter 8), and aspects of the response of radiation belt electrons to such waves within the diffusive paradigm are discussed in detail by Horne et al. (Chapter 9). Recent advances in the multi-dimensional analysis of whistler-mode waves in the radiation belt region are also presented by Santolík (Chapter 11).

In addition to local acceleration, longer period ultra-low frequency (ULF) waves can be excited by the solar wind with a scale size of the whole magnetosphere and these waves can also have a strong impact on charged particle dynamics in the radiation belts. For example, it is now well established not only that there is a strong correlation between solar wind speed and MeV energy electron flux at geosynchronous orbit (e.g., Paulikas and Blake, 1979) but also that there is a similar correlation between long-period Pc5 ULF wave power and MeV electron flux enhancements in the Van Allen belts (e.g., Rostoker et al., 1998; Mathie and Mann, 2000; Mann et al., 2012). ULF wave interactions with MeV electrons can occur through drift resonance, with electrons with energies up to around 5 MeV across the outer zone being able to resonantly interact with ULF wave frequencies ~0.1–100 mHz. Recent work has shown evidence for both strong coherent interactions between individual packets of ULF waves and MeV energy electrons (e.g., Claudepierre et al., 2013; Mann et al. 2013), as well as stochastic ULF wave-driven transport due to interactions with multiple wave packets which can be described using a radial diffusion paradigm (e.g., Fälthammar, 1965; Bratuigam and Alberta, 2000; Ozeke et al., 2012a, 2012b, 2014).

Such interactions typically conserve the first and second adiabatic invariants, while violating the third, and depending on the local gradient of electron phase space density the transport from radial diffusion can be either inward or outward. Hence in addition to causing the acceleration of particles during inward transport from sources in the plasma sheet, ULF wave radial diffusion can redistribute new relativistic flux created, for example, by local VLF chorus acceleration across L-shells, or act to transport particles outward to a depleted magnetopause through enhanced outward radial diffusion and magnetopause shadowing (e.g., Turner et al., 2012). In all such studies, amplitudes of the ULF pulsations and hence the rates of radial diffusion are a critical component determining the resulting impact on belt dynamics. Studies of the excitation, propagation and global characteristics of ULF waves, as well as their impact on MeV electron dynamics hence remain an active area of research. In this book, Balasis et al. (Chapter 6) address the monitoring of ULF waves from low Earth orbit, while Waters and Menk (Chapter 7) examine ULF wave characterization from the ground. Elkington and Sarris (Chapter 4) examine the effects of ultra-low frequency (ULF) waves on particle dynamics in the radiation belts, and Bourdarie et al. (Chapter 5) present modeling of some of the impacts of ULF waves on energetic particle populations in the inner magnetosphere.

In addition to these energization and transport processes, there are competing effects from wave-particle interactions which result in the loss of particles from the radiation belts. In addition to magnetopause shadowing described above, there are a number of other plasma wave-particle interaction processes which have been implicated in scattering of energetic particles into the loss cone and therefore causing their precipitation into the atmosphere. In comparison to particle acceleration processes, the physics of the dominant loss processes remains comparatively poorly understood. For example, electromagnetic ion cyclotron (EMIC) waves have been identified as a potentially dominant loss process for MeV electrons from the outer zone radiation belt through a Doppler-shifted cyclotron resonance and scattering into the loss cone (e.g., Horne and Thorne, 1998; Summers and Thorne, 2003). EMIC waves themselves are excited by gyro-resonance with energetic ions and occur most frequently and are most intense during magnetic storms—models suggesting prevalence along the high-density dusk-side plasmapause (Horne and Thorne, 1993; Kozyra et al., 1997; Jordanova et al., 2001). However, observational evidence for EMIC MeV electron loss is rather limited. There is some evidence from balloons (e.g., Foat et al., 1998; Millan et al., 2002) and recent observations and modeling (Usanova et al., 2010; Silin et al., 2011; Turner et al. 2013) have examined the role of EMIC wave loss on both ring current and radiation belt dynamics—but more work on these poorly understood loss processes is urgently needed.

In recent observations using the NASA Van Allen Probes, Usanova et al. (2014) did report the discovery of the first direct evidence for the scattering of large pitch angle ultra-relativistic electrons into the loss cone by EMIC waves. However, since the loss cone is small at equatorial altitudes the details of the wave-particle interaction could not be resolved. Since EMIC waves can be observed not only in the equatorial plane, but also along the field line and even on the ground (e.g., Usanova et al., 2010 and references therein), the combination of satellite and ground EMIC wave measurements has the potential to be used to determine both the spatial nature of the wave modes and their role in radiation belt loss (cf., for example, Mann et al., 2014). In this book, Usanova and Mann (Chapter 10) provide an overview of recent studies of the impact of EMIC waves on the ring current and radiation belt.

Higher frequency waves such as lower band chorus and plasmaspheric hiss may also play a role in loss by scattering relativistic electrons into the atmosphere (see, e.g., the review by Millan and Thorne, 2007). However, the relative efficiency of different plasma waves in driving energy input to the atmosphere hence remains relatively poorly understood. Interestingly, during the Van Allen Probes prime mission NASA also funded two campaigns comprising the flights of arrays of high altitude balloons from Antarctica during the southern hemisphere summers in the early months of 2013 and 2014 as part of the NASA Balloon Array for Radiation-belt Relativistic Electron Losses (BARREL) program. BARREL was focused on examining the wave-particle interaction processes causing radiation belt loss, and flew payloads measuring the Bremsstrahlung X-rays produced when electrons from the radiation belts precipitate into the upper atmosphere. The results from BARREL have revealed some evidence in support of the action of wave-particle loss, including for example from EMIC waves (e.g., Blum et al., 2015) and by plasmaspheric hiss by Breneman et al. (2015). However, one of the obvious major conclusions from BARREL is that multiple relatively proximal balloons often observe very different precipitation signatures. As a consequence, understanding which wave modes are responsible for scattering the radiation belt electrons into the atmosphere, when, and with what efficiency remains unresolved. Satellite missions with the capacity to provide full pitch angle resolution at lower altitudes, where the loss cone is larger and can be more easily resolved, promise an alternative and highly desirable future approach which might deliver a more detailed understanding of the nature and efficiency of radiation belt losses caused by plasma wave scattering into the atmosphere.

Overall, from a physics point of view, understanding the dynamics of the coupled geospace system is also especially challenging since not only are the plasmas collisionless, such that interactions have to occur through the intermediary of various plasma waves modes, but also the excitation and propagation of the relevant wave modes can be controlled by particles with a very wide range of energies. For example, the lowest energy particles in the magnetosphere (energies of ~eV, such as within the plasmasphere) have a strong influence on both the growth rates and propagation speeds of plasma waves. At the same time, electrons and proton distributions in the ring current (with energies ~1–400 keV) control the amount of free energy which is available to drive the growth of electromagnetic waves such as the whistler mode chorus waves and electromagnetic ion cyclotron (EMIC) waves. Finally, radiation belt particle resonances with these waves can introduce energy transport for the higher energy (~1–10+ MeV) particle populations in the radiation belts. This illustrates the important concept of cross-energy coupling in the inner magnetosphere[3], which can span up to six orders or more in energy and is manifest through the action of the intermediary of various plasma waves. This cross-energy coupling acts in parallel with other more well-known cross-scale coupling effects,

[3] The explicit paradigm of cross-energy coupling in this context was initially considered in the context of the scientific goals for the proposed Canadian Outer Radiation Belt Injection, Transport, Acceleration and Loss Satellite (ORBITALS) mission (Mann et al., 2006).

such as the global scale impacts arising from kinetic scale processes such as magnetic reconnection. Although the latter is the basis of a more familiar paradigm, the physical nature of the kinetic scale magnetic reconnection process is also still not fully understood and it forms the scientific target for the four-spacecraft NASA Magnetospheric Multi-Scale (MMS) mission which was launched on March 12, 2015.

At the heart of the challenge of comprehending, and ultimately predicting, radiation belt dynamics is the requirement to understand when, and under what driving solar wind and geomagnetic conditions, the radiation belt responds most strongly to which physical processes. For example, from among the multitude of known wave-particle interactions, determining which might dominate the overall response is highly desirable from a radiation belt forecasting perspective. It is also becoming increasingly clear that the initial and boundary conditions within the system can also have a very significant impact on the subsequent belt response, such that there may also be important pre-conditioning effects which influence the overall impact on the system. Certainly a number of important challenges remain before we can confidently state that we fully understand the strongly coupled ionosphere-plasmasphere-ring current-radiation belt system in the inner magnetosphere.

Coupling between the radiation belts and the neutral atmosphere: potential impact on climate

There is growing evidence that radiation belt precipitation is one of the most significant links in the space-atmosphere chain (e.g., Daglis, 2001), having distinct impacts on the dynamics of the neutral atmosphere. Recent studies have suggested this radiation belt-neutral atmosphere connection might extend all the way to the troposphere and even to ground level with potential impacts on surface air temperature having now also been reported (see, for example, Chapter 14 by Clilverd et al.). Therefore it is possible that the precipitation of radiation from the Van Allen belts can have an impact on terrestrial climate as well. For example, as discussed, for example, by Turunen et al. (2009), there is a pathway connecting energetic particle precipitation to surface temperature changes due to the catalytic effect of odd NOx and HOx populations on ozone chemistry in the stratosphere, possibly resulting in observable changes to the climate system.

Evidence for a direct linkage between geomagnetic activity and observed changes in surface temperature was presented by Seppälä et al. (2009), with a pathway linking NOx to energetic electron precipitation being presented by Rozanov et al. (2005). In the modeling presented by Turunen et al. (2009), auroral electron precipitation, relativistic electron precipitation from the radiation belts, as well as solar proton events can all potentially contribute to climatic effects. In contrast to higher energy radiation belt precipitation, the lower energy auroral precipitation also requires the action of transport processes to lower altitudes in order to cause impacts in the stratosphere. Very recently, Andersson et al. (2014) were able to quantify the effect arising from energetic electron precipitation (EEP) from the radiation belts, concluding not only that "EEP events strongly affect ozone at 60–80 km, leading to extremely large (up to 90%) short-term ozone depletion," but also that the effect is "comparable to that of large, but much less frequent, solar proton events." Therefore, Andersson et al. (2014) conclude, "it is reasonable to suspect that EEP could be an important part of solar influence on the atmosphere and climate system." There is little doubt that this will be an expanding and active area of future research.

Space plasma physics in the heliosphere

The dynamics of the Earth's geospace and radiation belts, and their impacts on the thermosphere and potentially even Earth's climate, represent one example of the outcomes arising from the influence of the Sun on the plasmas in the solar system. However there can be impacts on other planets as well. All of the planets, including the Earth, exist within the atmosphere of the Sun as it expands in the form of the solar wind and which fills the entire heliosphere. The solar wind expands until it reaches the heliopause, separating plasmas of solar origin from the interstellar medium. The dynamical coupling between magnetic structures in the solar atmosphere, the solar wind, and that in the ionized upper atmospheres of both magnetized and non-magnetized planets, out to the heliopause, has become known as heliophysics. In recent decades there has been a co-ordinated and concentrated international effort to operate a flotilla of spacecraft and satellites throughout the heliosphere, not just in local geospace around the Earth, as well as in monitoring and measuring the driving outputs from the Sun. For example multiple space agencies have recently coordinated their activities within the context of the International Living with a Star (ILWS) program.

Future international coordination in the launch and operation of new scientific missions, and the continuing operation of existing satellites, in the form of a great Heliophysics System Observatory (HSO) is the basis of ongoing efforts to further understand the fundamental plasma physics and active space weather in the heliosphere. Particular attention has been focused on the drivers at the Sun as well as dynamical coupling at the magnetized planets—creating energy and momentum transfer into their magnetospheres. In addition to the dynamics of the Earth's geospace environment described earlier, in this book the dynamics of waves and particles in the radiation belts of the outer planets are examined in Chapter 15 by Krupp et al.

The study of heliophysics is also fairly unique from the perspective of astrophysical space plasma physics because of the proximity and relative ease of access to detailed observational characterizations. For example, in general in the heliosphere the target astrophysical objects are sufficiently close to home that they can at least in principle be visited by in situ satellites and spacecraft and the plasmaphysical processes probed directly. This can be contrasted with more remote astrophysical objects where information is usually restricted to that which can be derived indirectly from the electromagnetic emissions from these distant plasma systems. As a result, although controlled laboratory type experiments still cannot be performed, heliophysics data are often sufficiently detailed and rich that specific scientific hypotheses can be directly tested and definitive scientific conclusions reached with a very high level of confidence. Of course, the lessons learned from the study of the proximal plasma systems in the solar system can then also be subsequently applied to studies of the more distant astrophysical plasma systems as well. Examples of such advances in fundamental plasma physics appear throughout this book, and with time these discoveries are being applied in the wider fields of not only astrophysical plasmas but also laboratory plasma research.

Applied heliophysics research: space weather

There is also an important applied element to heliophysics research relating to its application to understanding its space weather impacts on terrestrial and space-based infrastructure (e.g., Daglis, 2004). For example, the intense electrical currents which are driven in geospace during magnetic storms can also drive unwanted geomagnetically induced currents (GICs) in terrestrial power grids. These can cause premature aging, as well as in some cases the catastrophic loss, of transformers as well as instabilities in the phasing of electrical power, resulting in potential electrical power interruptions such as occurred in Canada during the HydroQuebec power outage in March 1989. Indeed, recent studies have assessed the economic impacts of such space weather to be potentially as large as $2 trillion dollars with a 4–10 year recovery in the U.S. alone following a severe magnetic storm (Baker et al., 2008). In relation to space radiation, the so-called satellite "killer" electrons with MeV energies in the Van Allen radiation belts can penetrate spacecraft shielding, generate a build-up of static charge within electrical components such as dielectrics, resulting in rogue commands, subsystem damage, and ultimately can even cause the total loss of Earth-orbiting satellites. The ability to accurately forecast the levels of space radiation to enable preventative action to protect sensitive systems, and to develop improved radiation belt specification models and establish worst case scenarios, are all key to future critical infrastructure protection activities. Such approaches also require advances in understanding the fundamental science, which is the focus of this book, as well as improved understanding and modeling of the pathways (see, e.g., Schrijver et al., 2015) and nature of the disturbances which have the capacity to cause the most damage arising from space weather in various technological systems.

The importance of the potential threat from space weather both in terms of the risk assessment of impact and likelihood is now being increasingly recognized internationally. For example, the United Nations Committee on the Peaceful Uses of Outer Space has established an Expert Group on Space Weather, and is also currently considering Guidelines and Best Practices for the Long-term Sustainability of Outer Space Activities. While the overall impact from a severe space storm might not be as large as that which could arise from the impact of an asteroid or near-Earth object, the likelihood of a severe space weather event is much higher. For example, a recent coronal mass ejection which was ejected from the surface of the Sun on July 23, 2012, missed the Earth by only a week or so, but was observed off the Sun–Earth line at 1 AU by the NASA Stereo-A spacecraft. Simulations based on the characteristics of the solar wind observed by Stereo predicted that it would have had a severe impact had it hit the Earth—perhaps as large, or larger, than the Carrington superstorm event of 1859 (Baker et al., 2013b). Although there were no satellites or large power grid infrastructures in 1859, there are contemporary accounts of telegraph wires being set on fire as a result of the GICs driven in the telegraph wires during the storm. In assessing the occurrence probability of an event of similar size to the Carrington event, Riley (2012) concluded that the probability of such an event happening in the next decade might be as high as ~12%. Even if such occurrence probability estimates turn out to be somewhat high, it seems likely that there could be very severe impacts arising from a severe space storm requiring appropriate critical infrastructure protection in the relatively near future. Although we have yet not experienced an event of the size of the Carrington 1859 event during the space age, the impacts of such an event would likely be very large—prompting a number of countries to include geospace storms in the list of hazards on the risk registers of their civil protection agencies. In our view, it is becoming increasingly important for all nations to undertake space weather risk and socio-economic impact assessments, and develop appropriate approaches and strategies to minimize and mitigate the detrimental effects from adverse space weather.

Conclusions and future perspective

Looking to the future, despite the huge recent strides in the advancement of the understanding of the radiation belts, there is still much to learn about the role of waves in controlling energetic particle dynamics during magnetic storms in geospace. For example, the Van Allen Probe spacecraft are estimated to be carrying sufficient fuel to operate at least the end of 2018, continuing instrument and spacecraft health permitting. In calendar year 2016 the Japanese Energy and Radiation in Geospace (ERG) mission is scheduled for launch, while the Russian Federation plans the RESONANCE mission which will examine long-duration wave-particle interactions through an orbit designed to allow the spacecraft to orbit along a fixed magnetic field line for long periods of up to ~ 3 hours. In combination with the extended mission phase operations of the Van Allen Probes and other international missions in the Great Heliophysics Observatory, such an inner magnetosphere constellation would represent a unique infrastructure through which to drive further discoveries uncovering the mysteries of the dynamics of the Earth's radiation belts. There certainly is no doubt that this is a truly golden age of discovery in radiation belt and inner magnetospheric physics. The continuing operation of an unprecedented number of scientific satellite missions dedicated to understanding the dynamics of this region will be sure to reveal not only a greater understanding of the dominant physical processes, but in all likelihood will reveal further new and unexpected results as well.

Ultimately, at the end of this new age of scientific discovery in geospace and the radiation belts, we expect to also be able to look forward to a reality of more accurate operational space weather radiation forecasts. For example, improvements in monitoring and understanding promise improved and more accurate climatological space radiation specification models. In turn, these can be used to define the expected radiation dose, and the related level of uncertainty, expected to be received by Earth-orbiting satellites operating in the radiation belts during specific epochs and phases of the solar cycle. Such information is required not only for protecting the longevity of operational satellites against space weather effects, but also for the design of mass-efficient and reliable space systems—thereby contributing directly to the protection of the critical space infrastructure upon which our twenty-first century technological society increasingly relies. Wernher von Braun is quoted in the article "Space: Reach for the Stars," which reported on the launch of Explorer I and the upcoming launch of Explorer II in *Time* magazine on February 17, 1958, as saying: "Everything in space obeys the laws of physics. If you know these laws, and obey them, then space will treat you kindly. And don't tell me man doesn't belong out there. Man belongs wherever he wants to go—and he'll do plenty well when he gets there." As we continue to expand our use of space to serve the needs of our highly inter-connected technological society, and in parallel probe the nature of waves, particles, and storms in geospace in our quest to comprehend how the laws of physics control radiation belt dynamics, we hope that space will indeed continue to treat us kindly.

REFERENCES

Andersson, M. E., P. T. Verronen, C. J. Rodger, M. A. Clilverd, and A. Seppälä (2014), Missing driver in the Sun–Earth connection from energetic electron precipitation impacts mesospheric ozone, *Nature Comms.*, **5**, doi:10.1038/ncomms6197.

Baker, D.N., Wave and particle measurements in Earth's neighborhood: a historical mission overview [Chapter 1 of this volume].

Baker, D. N., et al. (2013a), A long-lived relativistic electron storage ring embedded in Earth's outer Van Allen belt, *Science*, **340**, 186–190.

Baker, D. N., et al. (2013b), A major solar eruptive event in July 2012: defining extreme space weather scenarios, *Space Weather*, **11**, 585–591.

Baker, D. N., et al. (2008), Severe Space Weather Events—Understanding Societal and Economic Impacts: A Workshop Report. Space Studies Board. National Academies Press.

Balasis, G., P. Constantinos, E. Zesta, and V. Pilipenko, Monitoring ULF waves from low Earth orbit satellites [Chapter 6 of this volume].

Blum, L. W., et al. (2015), Observations of coincident EMIC wave activity and duskside energetic electron precipitation on 18–19 January, 2013, *Geophys. Res. Lett.*, **42**, 5727–5735.

Bortnik, J., R. M. Thorne, W. Li, and X. Tao, Chorus waves in geospace and their influence on radiation belt dynamics [Chapter 8 of this volume].

Bourdarie, S., V. Jordanova, M. Liemohn, and T. P, O'Brien, Modeling the energetic particles of the inner magnetosphere [Chapter 5 of this volume].

Brautigam, D. H. and J. M. Albert (2000), Radial diffusion analysis of outer radiation belt electrons during the October 9, 1990, magnetic storm, *J. Geophys. Res.: Space Phys.*, **105**(A1), 291–309, doi:10.1029/1999JA900344.

Breneman, A. W., et al. (2015), Global-scale coherence modulation of radiation-belt electron loss from plasmaspheric hiss, *Nature*, **523**, 193–195, doi:10.1038/nature14515.

Claudepierre, S. G., et al. (2013), Van Allen Probes observation of localized drift resonance between poloidal mode ultra-low frequency waves and 60 keV electrons. *Geophys. Res. Lett.*, **40**, 4491–4497.

Clilverd, M., C. Rodger, M. Andersson, A. Seppälä, and P. Verronen, Linkages between the radiation belts, polar atmosphere and climate: electron precipitation through wave particle interactions [Chapter 14 of this volume].

Daglis, I.A. (2001), Space storms, ring current and space-atmosphere coupling, in *Space Storms and Space Weather Hazards*, ed. by I.A. Daglis, pp. 1–42, Springer.

Daglis, I.A. (2004) (ed.), *Effects of Space Weather on Technology Infrastructure*, Springer.

Elkington, S. and T. Sarris, The role of Pc-5 ULF waves in the radiation belts: current understanding and open questions [Chapter 4 of this volume].

Fälthammar, C. G. (1965), Effects of time-dependent electric fields on geomagnetically trapped radiation, *J. Geophys. Res.*, **70**, 2503, doi:10.1029/JZ070i011p02503.

Foat, J. E., et al. (1998), First detection of a terrestrial MeV X-ray burst, *Geophys. Res. Lett.*, **25**, 4109–4112, doi:10.1029/1998GL900134.

Friedel, R.H.W., Reeves, G.D., and Obara, T. (2002), Relativistic electron dynamics in the inner magnetosphere—a review. *J. Atmos. Sol. Terr. Phys.*, **64**, 265–282.

Horne, R.B., N. P. Meredith, S. A. Glauert, and T. Kersten, Wave driven diffusion in radiation belt dynamics [Chapter 9 of this volume].

Horne, R. B. and R. M. Thorne (1998), Potential waves for relativistic electron scattering and stochastic acceleration during magnetic storms, *Geophys. Res. Lett.*, **25**, 3011–3014, doi:10.1029/98GL01002.

Horne, R. B. and R. M. Thorne (1993), On the preferred source location for the convective amplification of ion-cyclotron waves, *J. Geophys. Res.*, **98**, 9233–9247, doi:10.1029/92JA02972.

Jordanova, V. K., C. J. Farrugia, R. M. Thorne, G. V. Khazanov, G. D. Reeves, and M. F. Thomsen (2001), Modeling ring current proton precipitation by electromagnetic ion cyclotron waves during the May 14–16, 1997, storm, *J. Geophys. Res.*, **106**, 7–22, doi:10.1029/2000JA002008.

Kletzing, C., Fields and waves influencing radiation belt dynamics - Results from the Van Allen Probes mission [Chapter 16 of this volume].

Kozyra, J. U., A. F. Nagy, and D. W. Slater (1997), High-altitude energy source(s) for stable auroral red arcs, *Rev. Geophys.*, **35**, 155–190, doi:10.1029/96RG03194.

Krupp, N., E. Roussos, C. Paranicas, A. Sicard, G. Hospodarsky, and Y. Y. Shprits, Energetic particles and waves in the outer planet radiation belts [Chapter 15 of this volume].

Mann, I. R., et al. (2012), The role of ultralow frequency waves in radiation belt dynamics, in *Dynamics of the Earth's Radiation Belts and Inner Magnetosphere* (eds. D. Summers, I. R. Mann, D. N. Baker, and M. Schulz), American Geophysical Union, Washington, D.C. doi:10.1029/2012GM001349.

Mann, I. R., et al. (2006), The outer radiation belt injection, transport, acceleration and loss satellite (ORBITALS): a Canadian small satellite mission for ILWS, *Adv. Space Res.* **38** (8), 1838–1860.

Mann, I. R., et al. (2013), The Earth's Van Allen radiation belts: a geophysical synchrotron, *Nature comms.*, **4**, doi:10.1038/ncomms3795.

Mann, I. R., et al. (2014), Spatial localization and ducting of EMIC waves: Van Allen Probes and ground-based observations, *Geophys. Res. Lett.*, **41**, 785–792, doi:10.1002/2013GL058581.

Mathie, R. A. and I. R. Mann (2000), A correlation between extended intervals of ULF wave power and storm-time geosynchronous relativistic electron flux enhancements, *Geophys. Res. Lett.*, **27** (20), 3261–3264, doi:10.1029/2000GL003822.

Millan, R. M., R. P. Lin, D. M. Smith, K. R. Lorentzen, and M. P. McCarthy (2002), X-ray observations of MeV electron precipitation with a balloon-borne germanium spectrometer, *Geophys. Res. Lett.*, **29**, 2194, doi:10.1029/2002GL015922.

Millan, R. M. and R. M. Thorne (2007), Review of radiation belt relativistic electron losses, *J. Atmos. Sol. Terr. Phys.*, **69**, 362–377.

Miyoshi, Y., R. Kataoka and Y. Ebihara, Flux enhancement of relativistic electrons associated with substorms [Chapter 13 of this volume].

Ozeke, L. G., et al. (2012a), ULF wave derived radiation belt radial diffusion coefficients, *J. Geophys. Res.*, **117**, doi:10.1029/2011JA017463.

Ozeke, L. G., I. R. Mann, K. R. Murphy, I. J. Rae, and A. A. Chan (2012b), ULF wave-driven radial diffusion simulations of the outer radiation belt, *in Dynamics of the Earth's Radiation Belts and Inner Magnetosphere* (eds. D. Summers, I. R. Mann, D. N. Baker, and M. Schulz), American Geophysical Union, Washington, D. C. doi:10.1029/2012GM001332.

Ozeke, L. G., et al. (2014), Analytic expressions for ULF wave radiation belt radial diffusion coefficients, *J. Geophys. Res.*, **119**, 1587–1605.

Paulikas, G.A. and Blake, J.B. (1979), Effects of the solar wind on magnetospheric dynamics: energetic electrons at the synchronous orbit, in *Quantitative Modeling of Magnetospheric Processes* (ed. W. P. Olson), American Geophysical Union, Washington, D. C. doi:10.1029/GM021p0180.

Reeves, G. and I. A. Daglis, Geospace magnetic storms and the Van Allen radiation belts [Chapter 3 of this volume].

Reeves, G.D., et al. (2013), Electron acceleration in the heart of the Van Allen radiation belts, *Science*, **341**, 991–994.

Reeves, G. D., K. L. McAdams, R. H. W. Friedel, and T. P. O'Brien (2003), Acceleration and loss of relativistic electrons during geomagnetic storms, *Geophys. Res. Lett.*, **30** (10), 1529.

Riley, P. (2012), On the probability of occurrence of extreme space weather events, *Space Weather*, **10**, doi: 10.1029/2011SW000734.

Rostoker, G., S. Skone, and D. N. Baker (1998), On the origin of relativistic electrons in the magnetosphere associated with some geomagnetic storms, *Geophys. Res. Lett.*, **25**, 3701–3704, doi:10.1029/98GL02801.

Rozanov, E., L. Callis, M. Schlesinger, F. Yang, N. Andronova, and V. Zubov (2005), Atmospheric response to NOy source due to energetic electron precipitation, *Geophys. Res. Lett.*, **32**, L14811, doi:10.1029/2005GL023041.

Santolík, O., Multi-dimensional analysis of whistler-mode waves in the radiation belt region [Chapter 11 of this volume].

Schrijver, C. J., et al. (2015), Understanding space weather to shield society: a global road map for 2015–2025 commissioned by COSPAR and ILWS, *Adv. Space Res.*, **55**, 2745–2807.

Seppälä, A., et al. (2009), Geomagnetic activity and polar surface air temperature variability, *J. Geophys. Res.*, **114**, doi:10.1029/2008JA014029.

Silin, I., I. R. Mann, R. D. Sydora, D. Summers, and R. L. Mace (2011), Warm plasma effects on electromagnetic ion cyclotron wave MeV electron interactions in the magnetosphere, *J. Geophys. Res.*, **116**, doi:10.1029/2010JA016398.

Spence, H. E., G. D. Reeves, and R. Kessel, An overview of early results from the radiation belt storm probes energetic particle, composition, and thermal plasma suite on NASA's Van Allen Probes mission [Chapter 17 of this volume].

Stern, D. P. (1996), A brief history of magnetospheric physics during the space age, *Rev. Geophys.*, **34**, 1–31.

Summers, D. and R. M. Thorne (2003), Relativistic electron pitch-angle scattering by electromagnetic ion cyclotron waves during geomagnetic storms, *J. Geophys. Res.*, **108**, 1143, doi:10.1029/2002JA009489.

Takahashi, K. and Y. Miyoshi, Introduction to wave-particle interactions and their impact on energetic particles in geospace [Chapter 2 of this volume].

Thorne, R. M., et al. (2013), Rapid local acceleration of relativistic radiation-belt electrons by magnetospheric chorus. *Nature*, **504**, 411–414.

Turner D. and V. Angelopoulos, Extreme variability of relativistic electrons in Earth's outer radiation belt: an overview and recent revelations [Chapter 12 of this volume].

Turner, D. L., Y. Shprits, M. Hartinger, and V. Angelopoulos (2012), Explaining sudden losses of outer radiation belt electrons during geomagnetic storms, *Nature Phys.*, **8**, 208–212, doi:10.1038/NPHYS2185.

Turner, D. L., et al. (2013), On the storm-time evolution of relativistic electron phase space density in Earth's outer radiation belt, *J. Geophys. Res.*, **118**, doi:10.1002/jgra.50151.

Turunen, E., et al. (2009), Impact of different energies of precipitating particles on NOx generation in the middle and upper atmosphere during geomagnetic storms, *J. Atmos. Sol. Terr. Phys.*, **71**, 1176–1189.

Usanova, M. and I. R. Mann, Understanding the role of EMIC waves in radiation belt and ring current dynamics: recent advances [Chapter 10 of this volume].

Usanova, M. E., et al. (2014), Effect of EMIC waves on relativistic and ultrarelativistic electron populations: ground-based and Van Allen Probes observations. *Geophys. Res. Lett.* **41**, 1375–1381.

Usanova, M. E., et al. (2010), Conjugate ground and multisatellite observations of compression-related EMIC Pc1 waves and associated proton precipitation, *J. Geophys. Res.*, **115**, A07208, doi:10.1029/2009JA014935.

Van Allen, J. A. and L. A. Frank (1959), Radiation around the Earth to a radial distance of 107,400 km. *Nature*, **183**, 430–434, doi:10.1038/183430a0.

Waters, C. L. and F. W. Menk, Monitoring magnetospheric waves from the ground [Chapter 7 of this volume].

1

Wave and Particle Measurements in Earth's Neighborhood: A Historical Mission Overview

D. N. Baker

■ Abstract

The first scientific measurements of the Space Age were made more than 50 years ago. Those observations established that Earth is enshrouded in powerful high-energy radiation belts. Over the ensuing decades, there has been an impressive number and variety of spacecraft missions that have characterized the particles and waves that define the terrestrial space environment. This chapter provides a brief chronological description of many of the space missions that have shaped our scientific understanding of the wave-particle domain of the Earth. From the earliest Sputnik and Explorer detections of the trapped radiation zones to the latest sophisticated constellations boasting comprehensive payload suites, researchers have systematically improved the physical understanding of acceleration, transport, and loss of particles as mediated by plasma and electromagnetic waves. This record has provided tremendous insight into processes of prime astrophysical significance.

Waves, Particles, and Storms in Geospace. First Edition. Georgios Balasis, Ioannis A. Daglis, Ian R. Mann.

1.1 Introduction

Understanding of the energetic particle environment of the Earth and determining the role of wave phenomena in shaping this particle environment has been a truly international undertaking. Since this type of research dates back to the earliest beginnings of space exploration, one might presume that our physical understanding is essentially complete. However, new data and new theoretical insights have assured a steady—or even quickening—pace of discovery and understanding in very recent times.

The goal of this chapter is to provide a broad historical overview of the programs and missions that have informed our understanding of the geospace environment. While no chapter can even begin to treat in depth this more than half a century of exploration and discovery, it is hoped that the present narrative can provide a guide to key missions and accomplishments during the space era. The present chapter is organized in decadal segments progressing to the latest launches and near-term prospects. With this historical overview as a backdrop, the subsequent chapters dealing with specific topics and themes should bring readers to a more complete and up-to-date understanding of geospace wave-particle phenomena.

1.2 1950s: A new dawn—The "Space Age"

The idea of placing artificial satellites in Earth orbit dates back to well before the first missiles capable of launching such spacecraft. There were science fiction notions of humans in space quite early on and there were ideas of applied communication systems placed in the geostationary Earth orbit (GEO). But one of the first articles about placing an unmanned scientific satellite into orbit dates back to the early 1950s (Singer, 1952). The adoption of the idea of using artificial satellites for geophysical studies can be traced to the International Geophysical Year (IGY) of 1957–58. It became the stated policy of the United States (U.S.) and the Union of Soviet Socialist Republics (U.S.S.R.) to use their technological prowess to launch satellites for peaceful purposes during the IGY period (e.g., Van Allen, 1983).

Work dating back to the post-World War II days had been done using captured German V-2 rockets to study upper atmospheric, cosmic ray, and auroral physics questions (e.g., Van Allen, 1983). As the U.S. and the U.S.S.R. prepared for the IGY era, instruments were conceived that could extend cosmic ray measurements up to much higher altitudes than had been possible with balloon investigations or even with the "rockoons" of Van Allen and coworkers (Van Allen, 1957). These were rockets sent to great heights (>80 km) by launching them from balloons that had first been lofted to initial altitudes of 20 km or more.

In a stunning move, the U.S.S.R. successfully launched the first satellites into orbit, Sputnik 1 being launched on October 4, 1957, and Sputnik 2 being launched on November 7, 1957. The Sputnik 2 spacecraft had instrumentation on board from Moscow

State University (e.g., Vernov, 1958) that was quite capable of detecting the radiation belts around the Earth. However, the Soviets were not able to receive real-time radio telemetry from high enough altitudes to discern the intense trapped radiation of the belts surrounding Earth. Scientists in Australia recorded telemetry from Sputnik 2 and requested the information to decode the data. The Soviets refused. The Soviets demanded the recorded telemetry from the Australians. The Australians refused. Thus, the possibility of the discovery of the "Vernov Radiation Belts" slipped into history (e.g., Dessler, 1984).

On January 31, 1958, the U.S. launched the Explorer 1 spacecraft into orbit. This mission, and the subsequently successful Explorer 3 mission, showed quite conclusively that Earth was enshrouded in toroids of intense, high-energy radiation (Van Allen et al., 1958, 1959). Subsequent work with instruments on board the Pioneer series of very high altitude spacecraft (Van Allen and Frank, 1959) showed the physical extent and configuration of the inner and outer "Van Allen" radiation belts. This remarkable work became the first great discovery of the Space Age. Figure 1.1 shows the iconic image of Wernher von Braun, James Van Allen, and William Pickering holding aloft the engineering model of the Explorer 1 spacecraft. The inset shows the Sputnik 1 spacecraft picture as well.

Figure 1.1 *(Main figure) William Pickering (JPL), James Van Allen, and Wernher von Braun (left to right) holding aloft the engineering model of the Explorer 1 spacecraft. (Inset) The Sputnik 1 spacecraft model.*

1.3 1960s: Our place in space—morphology and plasma domains

Surprisingly quickly after the first successful space missions, a whole host of satellite programs were undertaken by the U.S., the U.S.S.R., and other countries (that generally relied on the U.S. for rides into space). The latter included the Ariel 1 spacecraft of the United Kingdom (U.K.) that was launched aboard a Thor-Delta vehicle from Cape Canaveral, Florida on April 26, 1962. This spacecraft had six scientific instruments onboard, all designed by British scientists. This success made the U.K. the world's third space-faring nation. There were five subsequent Ariel spacecraft with the last (Ariel 6) being launched from Wallops Island, Virginia, onboard a Scout vehicle in 1979.

Another highly successful collaborative effort in those early days was the U.S.–Canadian joint program called Alouette (Jelly, 1996). The principal goal of Alouette was to understand and improve radio communications. It was understood that ionospheric irregularities and disruptions interfered with radio links and that measurements of waves and particles in near-Earth space could help improve the situation. The Alouette program was carried out between the National Research Council of Canada and counterparts from NASA in the U.S. A topside sounder was developed and the payload also carried energetic particle detection systems. Alouette 1 was launched on September 29, 1962, from Vandenberg Air Force Base, California, aboard a Thor-Agena B rocket. This made Canada the fourth space-faring nation after the U.S., the U.S.S.R., and the U.K.

Interestingly, the "applications" side of space research also took root very quickly after the dawn of the Space Age. Almost immediately after the Sputnik and Explorer program successes, private industry began to plan and develop spacecraft to accomplish practical goals in space. A key example was the Telstar spacecraft that was designed and developed by the AT&T Bell Laboratories (see Dickieson, 1963). The operational goal of Telstar 1 was to provide real-time trans-oceanic communication. However, it was also richly instrumented to study and understand the space environment, which—it was argued—could disrupt or confound the spacecraft operations (see Brown, 1963). Figure 1.2 shows a photo of the Telstar spacecraft demonstrating the roughly spherical spacecraft design. Telstar 1 was launched aboard a Thor-Delta spacecraft from Cape Canaveral on July 10, 1962.

Both Ariel 1 and Telstar 1 fell victim to another aspect of the early space program. On July 9, 1962, one day before the Telstar launch, the U.S. detonated a hugely powerful nuclear weapon within the Earth's radiation belts. This so-called Starfish high-altitude nuclear test introduced massive fluxes of highly relativistic electrons into the Earth's natural radiation belt environment. The fission-generated high-energy electrons became trapped on geomagnetic field lines and actually persisted for a period of many years as a clearly discernable population of particles well above the (presumed) natural background (e.g., Walt, 1997). The Starfish electrons were of such intensity that they clearly

Figure 1.2 *The Telstar spacecraft, an early communication satellite that was also well instrumented for space environmental measurements. Launched in July 1962.*

damaged such things as solar panels and spacecraft electronics for satellites operating in space at the time. Thus, both Ariel 1 and Telstar 1 were degraded in performance by the Starfish experiment. The Explorer 4 spacecraft was built in less than three months and launched on July 26, 1958, to characterize the artificial radiation belts resulting from these so-called Argus nuclear tests (Walt, 1997).

The Cold War and the concerns about nuclear weapons in space also led to missions by other agencies of the U.S. government besides NASA. The Vela satellite series was a group of spacecraft that was developed to monitor compliance with the 1963 Partial Test Ban Treaty between the U.S. and the U.S.S.R. Vela is short for *velador* which means "watchman" in Spanish. It started as a small-budget research program in 1959 and ended 26 years later as a highly effective collaboration between the U.S. military and the U.S. Department of Energy (DOE and its predecessor organization). In total, 12 Vela satellites were built and deployed. They were placed in quite high-altitude orbits (circular at 118,000 km altitude). From these orbits—well beyond the Van Allen belts—the Vela spacecraft could stare back at the Earth to look for atmospheric nuclear bursts and could also monitor space around the Earth for "exoatmospheric" nuclear detonations. The spacecraft had a complete set of nuclear detection instruments from the Los Alamos and Sandia Laboratories, but also had remarkable space environment sensors (e.g., Coon, 1966). The plasma analyzers on the Vela spacecraft were among the first (and arguably the best) to characterize the plasma (Bame et al., 1967) and energetic particle (Singer et al., 1964) properties at relatively large geocentric distances.

The mid-1960s also saw application-oriented efforts by NASA to improve communications, weather observations, and understanding of the space environment. The Applications Technology Satellite (ATS) series of spacecraft played an important role in improving sensors and subsystems that led to the highly effective Geostationary Operational Environmental Satellites (GOES) program of today that provides weather images of the Earth and provides space weather information on a regular basis. The ATS-1 spacecraft was launched on December 7, 1966, and was still functioning over 20 years later. It was finally deactivated on December 1, 1978. The ATS-2 mission failed on launch (April 6, 1967) but ATS-3 launched successfully on November 5, 1967. Again, ATS-4 and ATS-5 failed on launch, but the ATS-6 mission (launched from Cape Canaveral on May 30, 1974) was highly successful. It had extremely capable particle and fields instruments (see Fritz and Neeley, 1982) that successfully characterized the geostationary orbit environment.

1.4 1970s: System dynamics—storms and substorms

Important "families" of spacecraft got their start in the 1960s, but reached their full flower in the 1970s. While the earlier decades of the space era were devoted to exploration and discovery—largely establishing a magnetospheric roadmap—the 1970s became a decade of detailed and quantitative understanding. Much of this arose (and remains with us today) because researchers and managers understood that one needed comprehensive particles and fields payloads in carefully tailored orbits.

One of the key mission lines for solar-terrestrial research was the Orbiting Geophysical Observatory (OGO) series (Hartt, 1960). The odd-numbered OGO platforms (OGO 1, OGO 3, and OGO 5) were in near-equatorial orbits. The even-numbered missions (OGO 2, OGO 4, and OGO 6) were in highly inclined, low polar orbits. OGO 1 was launched on September 5, 1964, from Cape Canaveral. It is still in orbit today, some 50 years later. OGO 3 was launched on June 7, 1966, and OGO 5 was launched on March 4, 1968, and re-entered Earth's atmosphere on July 2, 2011.

The low-Earth orbit (LEO) OGO missions made many key cosmic ray and solar energetic particle observations in their relatively brief lifetimes. However, it was the long-lived equatorial OGO missions that really advanced our understanding of the dynamics of the Earth's magnetosphere. These data—technically obtained in the late

1960s, but published in the early 1970s—provided fundamental insights into how geomagnetic storms really work. Perhaps the most important insight came from the OGO 5 data in space being closely coupled and correlated with ground-based observations (McPherron et al., 1973).

As shown here in Figure 1.3, McPherron and coworkers used the OGO 5 plasma, magnetic field, and wave data to understand how large-scale currents flow between the magnetosphere and ionosphere. They were able to show how "substorm" phenomena in the middle magnetosphere and magnetotail relate to the "current wedge" and auroral phenomena that had long been seen from ground vantage points. The work of McPherron and coworkers gave rise to a phenomelogical model of substorm processes that focused centrally upon the "near-Earth neutral line" (NENL) as a magnetic reconnection driver of magnetospheric dynamical behavior. While there have been refinements and additions to the picture, the NENL model remains the best and most successful framework for interpreting solar wind-magnetosphere-ionosphere coupling processes (see Baker et al., 1996).

Similar in some respects to the OGO series, the Interplanetary Monitoring Platform (IMP) series of spacecraft began in the 1960s, but reached remarkable fruition with the extremely long-lived IMP-8 spacecraft that was launched on October 25, 1973 (King, 1982). Last commands were sent to IMP-8 on October 28, 2001. In the prime operational years, IMP-8 played a huge role in monitoring the solar wind energy input to the magnetosphere and in detecting energetic particle and plasma processes in Earth's middle magnetotail (e.g., Sarris and Axford, 1979; Baker and Stone, 1976).

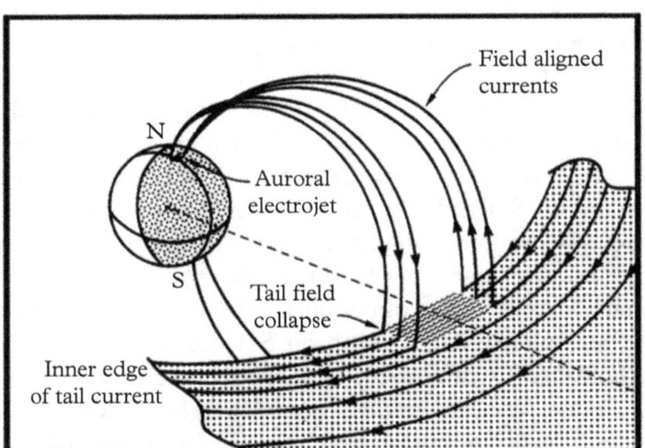

Figure 1.3 *A diagram relating magnetotail currents, field-aligned currents, and auroral electrojet properties as were inferred to occur during magnetospheric substorms using OGO 5 spacecraft observations.*
(McPherron et al., 1973.)

In the same time frame as the IMP spacecraft, and as a follow-on to the Vela nuclear detection satellites, the U.S. Air Force and Los Alamos National Laboratory provided an entire series of spacecraft at geostationary orbit. These LANL measurements initially were focused on energetic particle measurements (Baker et al., 1982) and they provided key knowledge about particle acceleration, transport, and loss. The detailed, high-quality LANL data became a perfect monitoring tool for studying substorm onset location and timing. Moreover, the data provided extremely strong corroborating evidence for the neutral line model of substorms proposed by McPherron et al. (1973). Interestingly, the LANL data sets at higher energies also greatly invigorated radiation belt studies (Baker et al., 1982) that went on to frame questions now being addressed by the Radiation Belt Storm Probes (or Van Allen Probes) mission (see Mauk et al., 2013).

The European Space Agency (ESA) complemented quite substantially the U.S. operational satellite fleet by launching the two Geostationary Earth Orbit Satellite (GEOS) missions. GEOS-1 was launched on April 20, 1977, and had a comprehensive package of plasma, fields, and energetic particle detectors. Somewhat fortuitously for science, it failed to reach geostationary orbit and GEOS-1 instead remained in a highly elliptical transfer orbit. This gave remarkable radial looks at magnetospheric processes (see Knott, 1982). GEOS-2 was identical to GEOS-1 in configuration and payload. It was successfully launched into a perfect geostationary orbit on June 15, 1978. The combination of GEOS-1, GEOS-2, LANL spacecraft, and the ATS-GOES missions at geostationary orbit has proven to be immensely important for magnetospheric understanding.

During the era of the 1970s, basic science was well served by operational spacecraft launched and operated by the U.S. military. This included the P78-2 spacecraft (also known as SCATHA), which was launched in February 1979. This mission had very good plasma, energetic particle, and field measurement capabilities and returned data from $r \sim 5.3\ R_E$ geocentric distance to $r \sim 7.8\ R_E$ geocentric distances (see Stevens and Vampola, 1978). Another key and highly successful space mission was the S3-3 program (meaning Small Scientific Satellite -3) that was launched in July 1976 (see Cattell, 1982). The S3-3 mission played a key role in understanding magnetosphere-ionosphere coupling and was particularly notable for providing some of the first plasma ion composition measurement ever obtained in space. Finally, it should be noted that during the 1970s the Defense Meteorological Satellite Program (DMSP) of the U.S. Air Force and the National Oceanic and Atmospheric Administration TIROS program (subsequently the Polar-Orbiting Environmental Satellite, POES, program) provided what have been essential continuing observations of the near-Earth space environment (see Kroehl, 1982).

Rounding out the incredibly active 1970s were the International Sun-Earth Explorer (ISEE) spacecraft missions. The ISEE-1 and ISEE-2 satellites were launched together on October 22, 1977, into closely matched, highly elliptical Earth orbit with apogees near 23 Earth radii (R_E). The U.S. spacecraft (ISEE-1) was well instrumented for both particle and field measurements (see Ogilvie et al., 1978). The ESA spacecraft (ISEE-2)

(Formisano, 1982) was similarly (but not identically) instrumented. By flying in close formation, the ISEE-1, -2 pair made huge contributions to collisionless shock physics, particle acceleration studies, and substorm understanding by separating spatial from temporal phenomena.

1.5 1980s: Universal processes—reconnection and acceleration

As noted above, the ISEE-1, -2 pair of spacecraft were launched in late 1977. The purposes of the mission were: (1) to investigate solar-terrestrial relationships at the outermost boundaries of the Earth's magnetosphere; (2) to examine in detail the structure of the solar wind near the Earth and the shock wave that forms the interface between the solar wind and Earth's magnetosphere; (3) to investigate motions of and mechanisms operating in the plasma sheet; and (4) to continue the investigation of cosmic rays and solar flare emissions in the interplanetary region near 1 AU. The spacecraft carried a number of complementary instruments for making measurements of plasmas, energetic particles, waves, and fields. The mission thus extended the investigations of previously mentioned IMP spacecraft. The launch of the coordinated spacecraft in this mission permitted the separation of spatial and temporal effects and continued well into the 1980s.

The third spacecraft in the ISEE family was quite different from the other two. This heliocentric spacecraft had a spin axis normal to the ecliptic plane and a spin rate of about 20 rpm. It was initially placed into an elliptical halo orbit about the Lagrangian libration point (L1) 235 Earth radii on the sunward side of the Earth, where it continuously monitored changes in the near-Earth interplanetary medium (Farquhar et al., 1985). In conjunction with the ISEE-1, -2 spacecraft, which had eccentric geocentric orbits, ISEE-3 explored the coupling and energy transfer processes between the incident solar wind and the Earth's magnetosphere. In addition, the heliocentric ISEE 3 spacecraft provided a near-Earth baseline for making cosmic ray and other planetary measurements for comparison with corresponding measurements from deep-space probes. ISEE-3 was the first spacecraft to use the halo orbit.

In 1982 ISEE-3 began the magnetotail and comet encounter phases of its mission. A maneuver was conducted on June 10, 1982, to remove the spacecraft from the halo orbit around the L1 point and place it in a transfer orbit involving a series of passages between Earth and the L2 (magnetotail) Lagrangian libration point. After several passes through the Earth's magnetotail, with gravity assists from lunar flybys in March, April, September, and October of 1983, a final close lunar flyby (119.4 km above the Moon's surface) on December 22, 1983, ejected the spacecraft out of the

Earth-Moon system and into a heliocentric orbit ahead of the Earth, on a trajectory intercepting that of Comet Giacobini–Zinner. At this time, the spacecraft was renamed International Cometary Explorer (ICE). A total of fifteen propulsive maneuvers (four of which were planned) and five lunar flybys were needed to carry out the transfer from the halo orbit to an escape trajectory from the Earth-Moon system into a heliocentric orbit.

The phase of the ISEE-3 orbit from late 1982 to late 1983 was a spectacular one for magnetospheric science. As shown in Figure 1.4, ISEE-3 sensors detected with remarkable regularity the expulsion of large plasma bubbles called "plasmoids" associated with magnetospheric substorms (Baker et al., 1987). ISEE-3 was also able to study deep-magnetotail particle acceleration and transport with exceptional effectiveness (see Scholer et al., 1984). Overall, ISEE-3 paved the way for the highly successful Geotail mission from Japan (see Nishida, 1994).

In the 1980s, the operational needs for space situational awareness once again demanded that the DOE national laboratories step up to provide particle information. Satellites of the U.S. Air Force Global Positioning System (GPS), which were placed into highly inclined (55°–63°) circular orbits at an altitude of 11,000 nautical miles, pass twice per 12-h orbit near the peak-intensity region of magnetically trapped energetic particles. Radiation damage to the satellite components would have been detrimental to the system's planned longevity. Available experimental or theoretical knowledge of particle fluxes at GPS orbits was inadequate for predicting the extent of radiation damage, largely because the resident trapped particle populations vary and, in addition, depend on the level of solar activity.

A secondary mission of the GPS program, beginning with the launch of the satellite FSV-6 in late 1980, was to carry instrumentation in support of the Integrated Operational Nuclear-Test Detection System program. This instrumentation includes an x-ray burst detector (BDX) and a logic-control and data-storage unit (BDP). It was decided that the simplest and most effective way to obtain the required particle-flux data was to equip some satellites with an appropriate dosimeter rather than a BDX. Because this decision was made when the BDP and the BDX were already in the hardware stage, restrictions were placed on the dosimeter (BDD) design to be as simple as possible. Such requirements were met with no change in the BDP hardware and with the addition of only a single hard wire to the connecting cable. The level on this wire determined the subset of BDP firmware appropriate to processing data from a BDX or a BDD (Argo et al., 1980).

For the purpose at that time, the most useful form of dosimeter data was energy spectra with absolute intensity as a function of time and position in orbit. From such data one could calculate the radiation dose received by a satellite component and so predict the expected degradation of its performance with time. Within the design restrictions mentioned previously, the LANL team devised a four-channel spectrometer that provided electron-flux data of the energy range from 0.3 to 10 MeV and proton flux from 6 to 100 MeV (Argo et al., 1980).

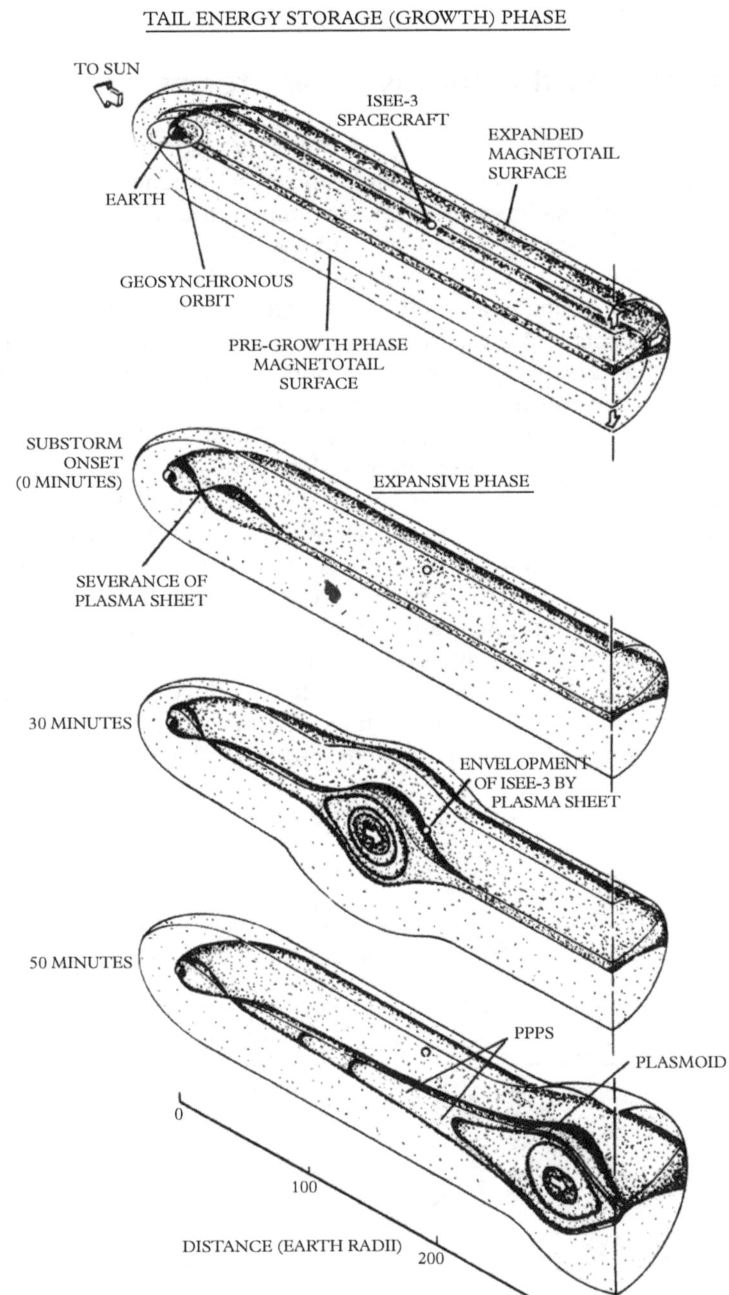

Figure 1.4 *Growth and expansive phase of substorms as inferred from ISEE-3 deep magnetotail measurements (from Baker et al., 1987). After plasma and energy have built up in the magnetotail during the growth phase, a portion of the plasma sheet is severed by reconnection close to the Earth causing substorm onset. The plasmoid thus formed moves rapidly down the tail enveloping ISEE-3 some 200 R_E downtail 20–40 min after substorm onset.*

1.6 1990s: Follow the energy—solar-terrestrial coupling

With the successes of the ISEE-1, -2 missions studying magnetospheric processes concurrently with ISEE-3 monitoring solar wind driving, the science community moved in the 1990s toward a systematic study of mass, momentum, and energy flow from the Sun/solar wind through the magnetosphere, and down into the various layers of Earth's atmosphere. This program became known as the International Solar-Terrestrial Physics, or ISTP, Program (see Acuña et al., 1995). The international qualities of ISTP sprang from having deep magnetotail measurements made by the Japanese Geotail spacecraft (see Nishida, 1994) and from having strong ground-based involvement from many nations outside the U.S. The European Solar and Heliospheric Observatory (SOHO) and the European–U.S. Cluster missions were also central to the ISTP goals.

Figure 1.5 (from Acuña et al., 1995) very well captures the spirit of the ISTP science program. The goal was to observe solar energy generation with the SOHO spacecraft, document actual solar wind drivers using the WIND, SOHO, and IMP-8 spacecraft, and then observe geospace consequences using myriad spacecraft and ground-based facilities. Key parts of the geospace system were the inner magnetosphere, the distant magnetotail, and the ionosphere-atmosphere system. Overall—and eventually—the ISTP program proved to be highly successful in its science goals, but delays in implementing various pieces of the program definitely strained the coordination aspects. SOHO, Polar, Wind, and Geotail (along with ground-based and modeling aspects of the program) were working well by the late 1990s and the vision in Figure 1.5 was largely underway by that time.

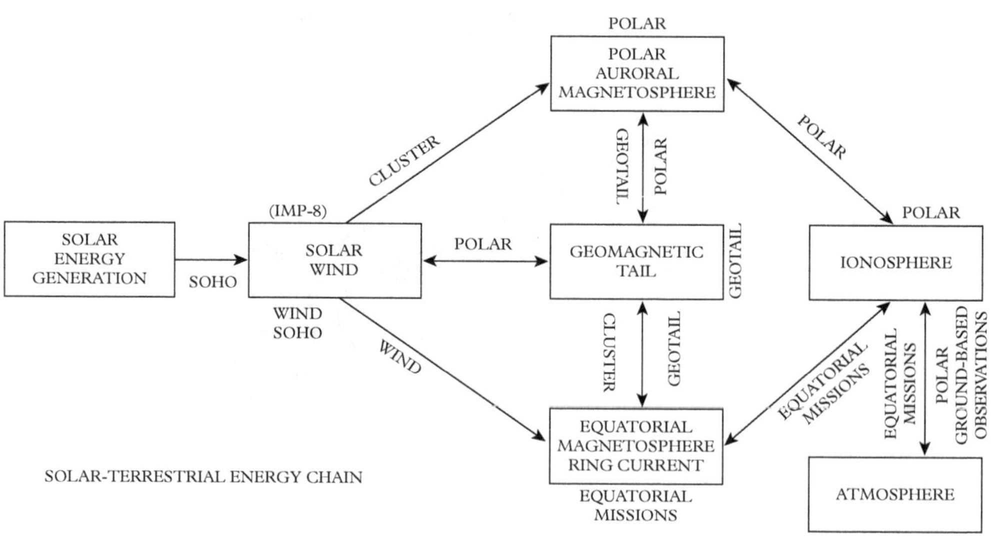

Figure 1.5 *The energy flow and redistribution studied during the International Solar-Terrestrial Physics (ISTP) program.*
(Acuña et al., 1995.)

Another fascinating aspect of the 1990s was a resurgence of interest in radiation belt science. The joint NASA–Air Force mission called the Combined Release and Radiation Effects Satellite (CRRES) was launched on July 25, 1990 (see Johnson and Kierein, 1992). CRRES was very well instrumented to measure plasmas, waves, fields, and energetic particles throughout the inner and outer Van Allen zones. Unfortunately, due to mismanagement of the spacecraft batteries and power system, CRRES failed after only 14 months on orbit. Nonetheless, it showed the rich possibilities for radiation belt research.

An example is shown here in Figure 1.6 (taken from Li et al. (1993)). At 3:41 UT on March 24, 1991, an almost instantaneous formation of a new electron radiation belt was observed by CRRES which was situated at a distance of 2.55 R_E near the

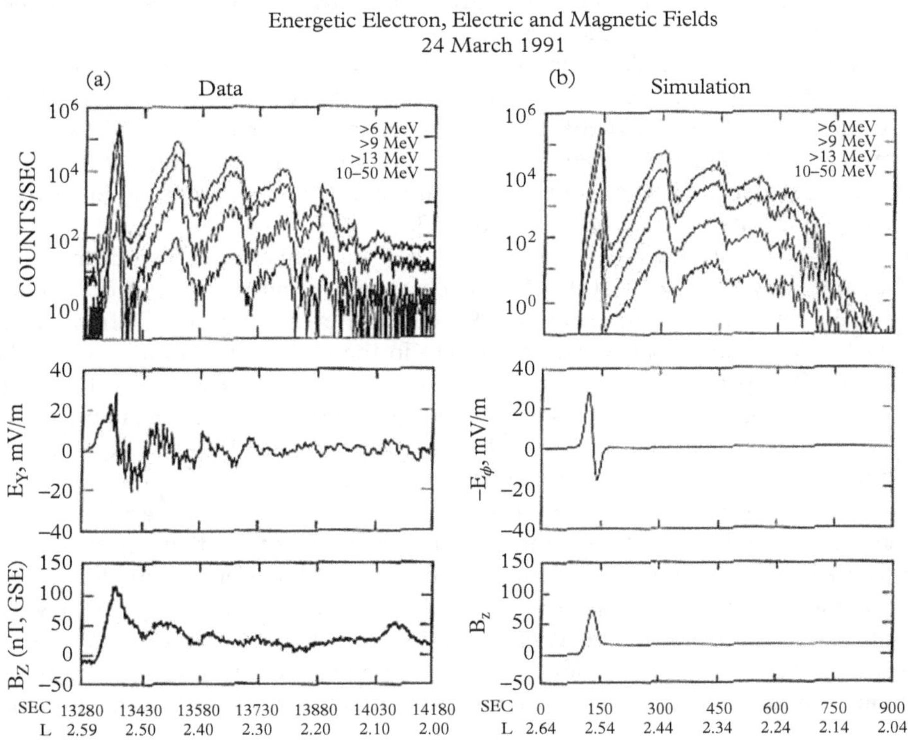

Figure 1.6 *(a) Data from the CRRES satellite at the time of the March 24, 1991, SSC. Top panel shows count rates as a function of time from four energetic electron channels measuring integral counts above a threshold energy indicated. Middle and bottom panels show the measured electric field E_y in a co-rotational frame and the B_z magnetic field component with a model magnetic field subtracted, in GSE coordinates over the same time interval; (b) simulated results in the same format as (a) measured at a spatial location corresponding to the trajectory of the CRRES satellite, including the Aerospace detector responses in the top panel. (Li et al., 1993.)*

equatorial plane at a local time of 03:00 MLT (Li et al., 1993). The creation of the radiation belt was accompanied by an electron and ion drift-echo event from which it was deduced that the injected electrons were at energies predominantly above 15 MeV. This radiation belt lasted beyond the end of the CRRES mission six months later. The event was initiated by an interplanetary shock that compressed the magnetosphere inside geostationary orbit. This event—observed by CRRES and remarkably well modeled by Li et al.—showed how dramatically (and rapidly) the Earth's radiation belts can be altered.

At about the same time that the CRRES spacecraft was under development, the Institute of Space and Astronautical Science (ISAS) in Japan was developing the low-altitude EXOS-D (or "Akebono") spacecraft. Akebono had very capable energetic particle and field instruments (Oya and Tsuruda, 1990) and was launched successfully on February 22, 1989. Thus, this mission overlapped entirely with the CRRES mission and has continued to be productive to the present time.

Finally, the first of NASA's Small Explorer (SMEX) missions also got underway in the early 1990s. This mission was called SAMPEX (Solar, Anomalous, and Magnetospheric Particle Explorer). SAMPEX was launched into a nearly polar, ~500 km altitude orbit on July 3, 1992 (see Baker et al., 1993). The mission revolutionized our views of radiation belt dynamics and continued to monitor the radiation belts for 20 years until its re-entry to the atmosphere on November 13, 2012.

One of the largest and most studied radiation belt enhancements occurred in late October and early November 2003. This Halloween Storm event (or set of events) was similar in its revelations to the March 1991 CRRES event. Figure 1.7 shows a record of SAMPEX data (and solar wind speed data in the top plot) from a paper by Baker et al. (2004). The period of time covered is from 2003 to early 2004. Figure 1.7 (bottom) shows quite obviously the sudden appearance (or at least, powerful reemergence) of a "new" belt of electrons associated with the Halloween Storm (Baker et al., 2004). Over at least the next two years, the inner zone belt decayed away.

The original powerful acceleration event in late October 2003 was analyzed in detail by Baker et al. (2004). As shown in the interpretive sketches from that paper, the Earth's outer Van Allen zone was virtually annihilated by the Halloween Storm. The complete disappearance of multi-MeV electrons at L~4 was quickly followed by the new generation of MeV electrons in the heart of the slot region (which usually is devoid of such electrons). Baker et al. (2004) showed that the plasmasphere was scoured away in this event, and this allowed a complete (if temporary) reconfiguration of the radiation belts.

1.7 2000s: Multiscale phenomena—the telescope-microscope duality

The four-spacecraft Cluster mission was originally proposed in 1982 and it became an official ESA mission (joint with the U.S.) targeted for launch in 1996. As a cost-saving measure, the identical set of satellites were to be launched free of charge on the first test flight of the Ariane-5 launch vehicle. Tragically, the Ariane performance on launch

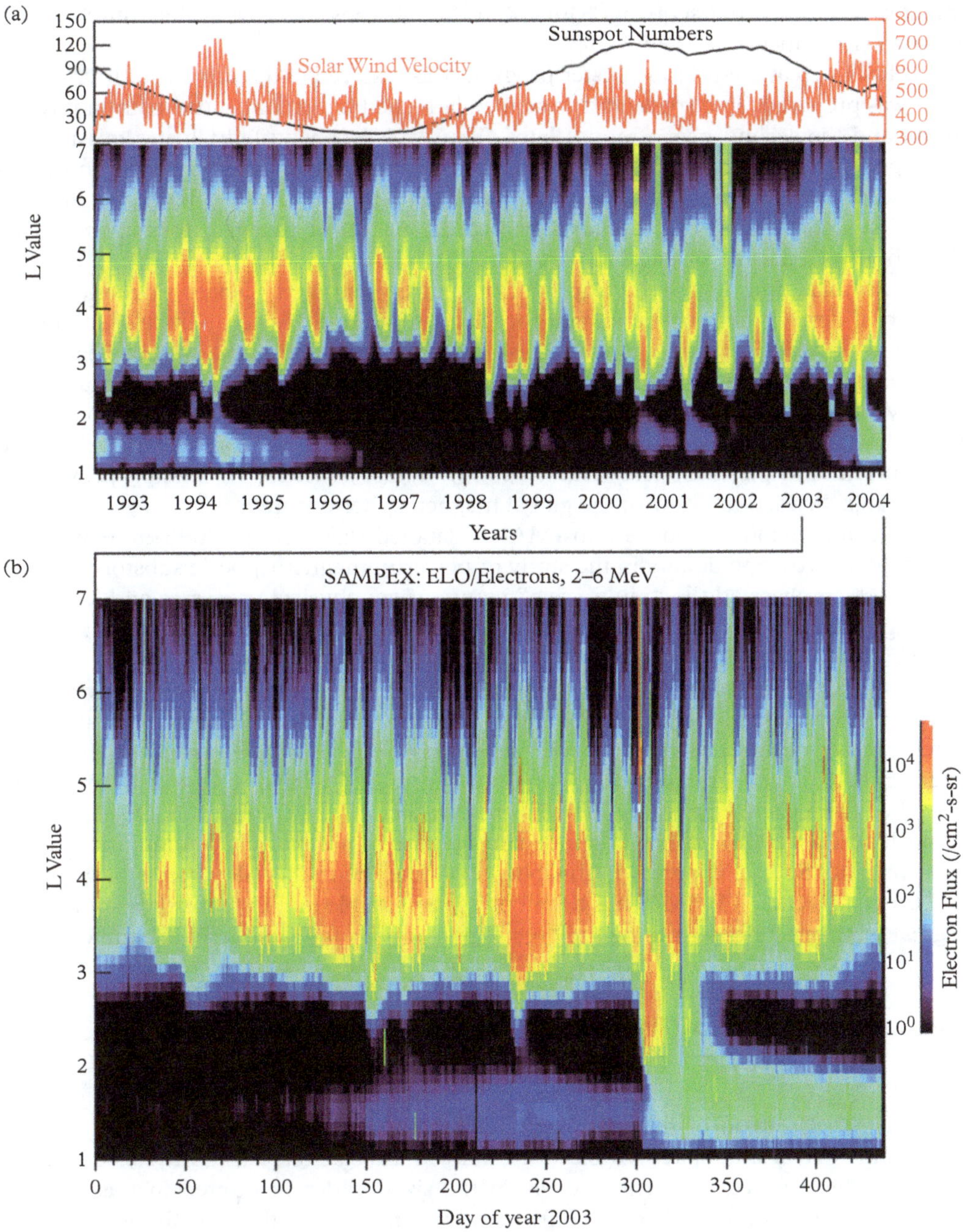

Figure 1.7 *A long run of SAMPEX data (panel a) and associated solar wind speed and sunspot number data (top). A blowup (panel b) showing details of the Halloween Storm period.*
(Baker et al., 2004.)

initiated a self-destruct sequence (June 4, 1996) thereby destroying the entire Cluster payload (Escoubet et al., 1997).

To ESA's great credit, the Cluster program was restored and a new set of four spacecraft was prepared. The spacecraft were then launched in two pairs, one pair on July 16, 2000, and the second pair a month later. Thus, the reconstituted Cluster mission has been providing microphysical views of magnetospheric cusp, plasma sheet, and magnetosheath processes since 2000 from the vantage point of a 4 $R_E \times 19$ R_E highly inclined orbit. This constellation of four closely aligned spacecraft flying in formation can be thought of as a plasma physical "microscope."

At a similar time to the development of the reconstituted Cluster mission, NASA supported development of the Imager for Magnetopause-to-Aurora Global Exploration (IMAGE) spacecraft (Burch, 2000). This spacecraft had a comprehensive payload of ultraviolet imagers as well as neutral atom imaging capability. The spacecraft was launched from Vandenberg AFB on March 25, 2000, into an elliptical polar orbit. From this orbit, IMAGE could provide marvelous images of the entire auroral oval as well as the ring current and plasmasphere (Burch, 2000). Thus, the IMAGE mission provided a remarkable "telescope" view of the global magnetospheric system.

The combination of Cluster and IMAGE data led Baker et al. (2002) to exploit the telescope-microscope duality in the study of powerful magnetospheric substorm events. This work emphasized the "multi-scale" nature of geophysical processes and led to the striking image (Figure 1.8) that graced the cover of *Geophysical Research Letters* for the issue containing the Baker et al. (2002) study.

This focus on both the global aspects of substorms and upon the microphysical aspects of substorms has been continued with the Time History of Events and Macroscale Interactions during Substorms (THEMIS) mission. THEMIS consisted of five small spacecraft deployed into separate, carefully considered orbits. The constellation was launched from Cape Canaveral, Florida, on February 17, 2007, aboard a Delta II rocket (Angelopoulos, 2008). It has continued to operate in an evolved configuration to the present time. Several chapters in this book examine the multiscale aspects of geophysical events. Overall, THEMIS has been both a remarkable tool for substorm studies and a great platform for determining the "seed" populations for radiation belt studies.

1.8 2010s: The Sun-Earth system—space weather and beyond

As a last brief section of this overview chapter, it seems appropriate to note that the discipline of solar and space physics (also known as solar-terrestrial physics or "heliophysics") has progressed tremendously in the six decades recounted in this chapter. The discipline has moved from rudimentary exploration and the most basic of discoveries to being a field with mature understanding of the underlying physical processes and principles. This has given the discipline the chance to make predictive forecasts of major "space weather" events and join other geoscience disciplines in understanding the hazards that confront our technological society.

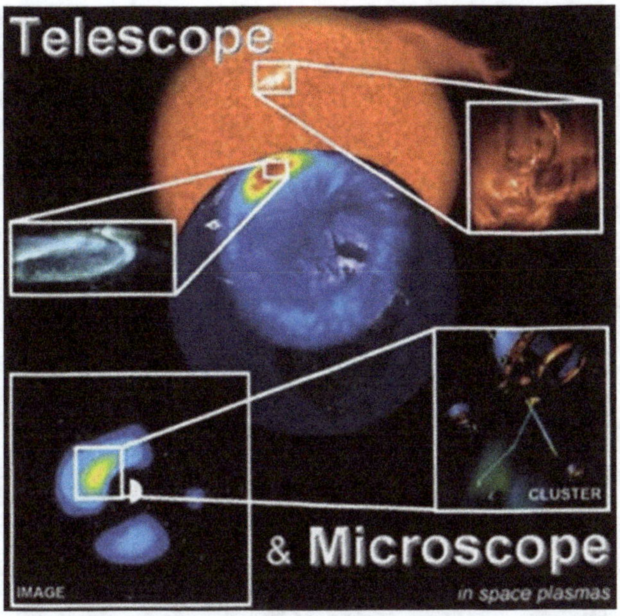

Figure 1.8 *The telescope-microscope concept as described by Baker et al. (2002) and as portrayed on the cover of the* Geophysical Research Letters *issue containing that paper. (Baker et al., 2002).*

The newest major mission underway in magnetospheric science right now is the Radiation Belt Storm Probes (renamed the Van Allen Probes) mission of NASA. As described by Mauk et al. (2013), the Van Allen Probes program represents an elegant closure with the topic that opened our discussion in this chapter: Radiation belt physics. The goal of the Van Allen Probes mission is to glean much deeper understanding of the wave-particle interactions that shape the Van Allen belts. Many subsequent chapters in this book address exactly those issues.

The high-energy electron Van Allen Probes instruments worked well from the moment they were turned on (September 1, 2012). They made observations of freshly accelerated particles trapped in the belts, recording their high energies as the belts increased in size. Then something happened that no one had reported before: The particles settled into a new configuration, showing a third belt extending out into space (see Figure 1.9). Within mere days of launch, the spacecraft showed something that would require rewriting textbooks about how the Van Allen belts can be configured (Baker et al., 2013a): Instead of just an inner and outer Van Allen zone, the data clearly showed three distinct belts. Although hinted at before in low-altitude data such as those from SAMPEX, mentioned earlier, the temporal, spatial, and energy resolution of the new measurements by the Van Allen Probes has proven crucially important.

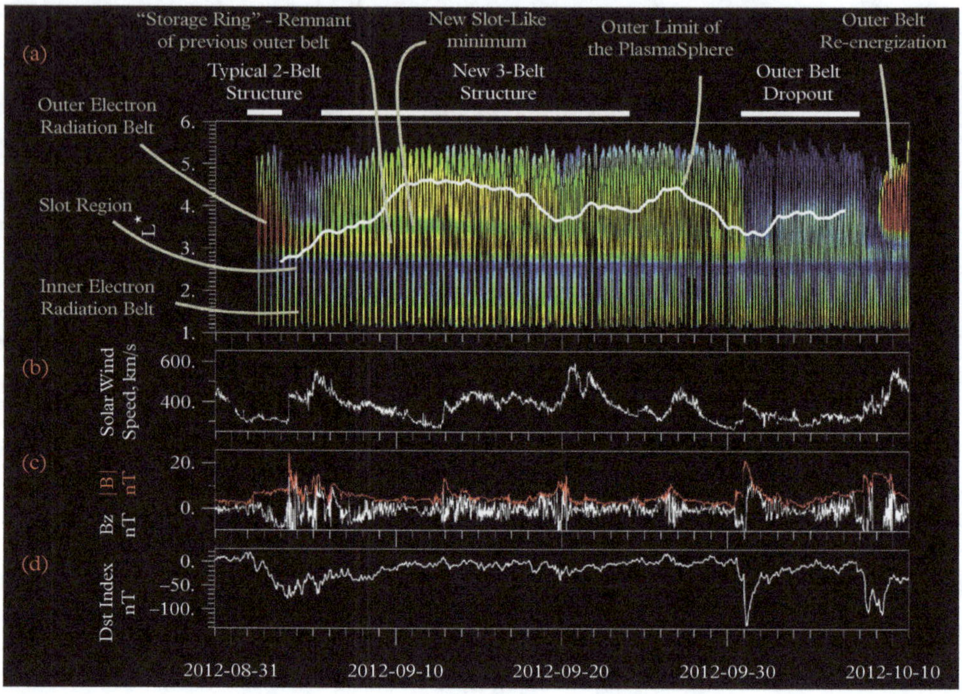

Figure 1.9 *Data from the Van Allen Probes spacecraft showing development of the relativistic electron storage ring. (a) Panel showing the plasmapause, the outer boundary of the plasmasphere for the period September 1 to October 7, 2012. The white curve overplotted on the color-coded electron particle flux data in Panel (a) shows the modeled, 3-day averaged plasmapause radial location that is in agreement with spacecraft plasma wave data. (b) Concurrently measured solar wind speed upstream of Earth's magnetosphere. (c) Interplanetary magnetic field for the interval under study. |B| is the magnitude of the interplanetary magnetic field in nanotesla [nT]; Bz, the value of the north-south component. (d) Geomagnetic activity index [Dst] for the period under study.* (*Baker et al., 2013a.*)

There is every indication that RBSP/Van Allen Probes will meet or exceed every expectation that was laid out for it. What a wonderful monument for J.A. Van Allen that his namesake mission is providing striking new insights to radiation belt science five and a half decades after the pioneering discoveries of Explorer 1 (Baker et al., 2013b).

REFERENCES

Acuña, M.H., et al. (1995), The Global Geospace Science Program and its Investigations, *Space Sci. Rev.*, **71**, 5–21.

Angelopoulos, V. (2008), The THEMIS Mission, *Space Sci. Rev.*, **141**, 5–34.

Argo, H.V., et al. (1980), The BDD: A Dosimeter for the Global Positioning System, LA-8421-MS, Informal Report, Los Alamos National Laboratory, Los Alamos, NM.

Baker, D.N. and E.C. Stone (1976), Energetic electron anisotropies in the magnetotail: identification of open and closed field lines, *Geophys. Res. Lett.*, **3**, 557–560.

Baker, D.N., et al. (1982), The Los Alamos synchronous orbit data set, in *The IMS Source Book* (eds. C.T. Russell and D.J. Southwood), American Geophysical Union, Washington, D.C.

Baker, D.N., R.C. Anderson, R.D. Zwickl, and J.A. Slavin (1987), Average plasma and magnetic field variations in the distant magnetotail associated with near-earth substorm effects, *J. Geophys. Res.*, **92**, 71–81.

Baker, D.N., G.M. Mason, O. Figueroa, G. Colon, J. Watzin, and R. Aleman (1993), An overview of the SAMPEX mission, *IEEE Trans. Geosci. Elec.*, **31**, 541.

Baker, D.N., T.I. Pulkkinen, V. Angelopoulos, W. Baumjohann, and R.L. McPherron (1996), The neutral line model of substorms: past results and present view, *J. Geophys. Res.*, **101**, 12995.

Baker, D.N., et al. (2002), A telescopic and microscopic view of a magnetospheric substorm on 31 March 2001, *Geophys. Res. Lett.*, **29**, 91–94.

Baker, D.N., S.G. Kanekal, X. Li, S.P. Monk, J. Goldstein, and J.L. Burch (2004), An extreme distortion of the Van Allen belt arising from the 'Hallowe'en' solar storm in 2003, *Nature*, **432**, 878, doi:10.1038/nature03116.

Baker, D.N., et al. (2013a), A long-lived relativistic electron storage ring embedded within the Earth's outer Van Allen Radiation Zone, *Science*, **340**, 6129, 186–190.

Baker, D.N., et al. (2013b), James Van Allen and His Namesake NASA Mission, *Eos*, **94**, 49, 469–484.

Bame, S.J., et al. (1967), Characteristics of the plasma sheet in the Earth's magnetotail, *J. Geophys. Res.*, 72, 113–129.

Brown, W.L. (1963), The spacecraft radiation experiments, *Bell Syst. Tech. J.*, **42**, 899–941.

Burch, J.L. (2000), The IMAGE Mission, edited by J.L. Burch, reprinted from *Space Science Reviews*, **91**, 1–2.

Cattell, C.A. (1982), S3-3 satellite instrumentation and data, in *The IMS Source Book* (eds. C.T. Russell and D.J. Southwood), American Geophysical Union, Washington, D.C.

Coon, J.H. (1966), Vela Satellite Measurements of Particles in the Solar Wind and the Distant Geomagnetosphere, in *Radiation Trapped in the Earth's Magnetic Field* (ed. B.M. McCormac), 231, Reidel Publishing Co., Dordrecht.

Dessler, A.J. (1984), The Vernov Radiation Belt (Almost), *Science*, **226**, 915.

Dickieson, A.C. (1963), The Telstar experiment, *Bell Syst. Tech. J.*, **42**, 739–746.

Escoubet, C.P., R. Schmidt, and M.L. Goldstein (1997), Cluster—science and mission overview, *Space Sci. Rev.*, **79**, 11–32.

Farquhar, R., et al. (1985), Trajectories and orbital maneuvers for the ISEE-3/ICE Comet Mission, *J. Astron. Sci.*, **33**, 235–254.

Formisano, V. (1982), The International Sun Earth Explorer Mission – ISEE-2, in *The IMS Source Book* (eds. C.T. Russell and D.J. Southwood), American Geophysical Union, Washington, D.C.

Fritz, T.A. and C.A. Neeley (1982), Geostationary Satellites ATS-6 and SMS/GOES: Description, Position and Data Availability during IMS, in *The IMS Source Book* (eds. C.T. Russell and D.J. Southwood), American Geophysical Union, Washington, D.C.

Hartt, J. (1960), First 'Streetcar' Satellites to be built for NASA, *Milwaukee Sentinel*, 26 December.

Jelly, D. (1996), The Alouette Satellite Program; Placing Canada on the Celestial Map, *Friends of CRC*.

Johnson, M.H. and J. Kierein (1992), Combined release and radiation effects satellite (CRRES): spacecraft and mission, *J. Spacecraft Rockets*, **29**, 556–563.

King, J.H. (1982), Availability of IMP-7 and IMP-8 data for the IMS Period, in *The IMS Source Book* (eds. C.T. Russell and D.J. Southwood), American Geophysical Union, Washington, D.C.

Knott, K. (1982), The availability of GEOS data for IMS research, in *The IMS Source Book* (eds. C.T. Russell and D.J. Southwood), American Geophysical Union, Washington, D.C.

Kroehl, H.W. (1982), Space environment monitoring by low-altitude operational satellites, in *The IMS Source Book* (eds. C.T. Russell and D.J. Southwood), American Geophysical Union, Washington, D.C.

Li, X., I. Roth, M. Temerin, J.R. Wygant, M.K. Hudson, and J.B. Blake (1993), Simulation of the prompt energization and transport of radiation belt particles during the March 24, 1991 SSC, *Geophys. Res. Lett.*, **20**, 22, 2423–2426.

McPherron, R.L., et al. (1973), Satellite studies of magnetospheric substorms on August 15, 1968:9. Phenomenological model for substorms, *J. Geophys. Res.*, **78** (16), 3131–3149.

Mauk, B.H., et al. (2013), Science objectives and rationale for the radiation belt storm probes mission, *Space Sci. Rev.*, **179** (104), 3–27.

Nishida, A. (1994), The GEOTAIL mission, *Geophys. Res. Lett.*, **21**, 2871–2873.

Ogilvie, K.W., et al. (1978), Descriptions of experimental investigations and instruments for the ISEE spacecraft, *IEEE Trans. Geosci. Electron.*, **GE_16**, 151–153.

Oya, H. and K. Tsuruda (1990), Introduction to the Akebono (EXOS-D) satellite observations, *J. Geomag. Geoelectr.*, **42**, 367–370.

Sarris, E.T. and W.I. Axford (1979), Energetic protons near the plasma sheet boundary, *Nature*, 277, 460–462.

Scholer, M., et al. (1984), Simultaneous observation of the plasma sheet in the near Earth and distant magnetotail: ISEE-1 and ISEE-3, *Geophys. Res. Lett.*, **11**, 1034–1037.

Singer, S.F. (1952), A minimum orbiting unmanned satellite of the Earth (MOUSE), *J. Brit. Interplan. Soc.*, **11**, 61.

Singer, S.F., et al. (1964), *Proc. Intern. Space Sci. Symp.*, 5, Florence.

Stevens, J.R. and A.L. Vampola (1978), Description of the space test program P78-2 spacecraft and payloads, Air Force Space and Missile Systems Organization report SAMSO TR-78-24.

Van Allen, J.A. (1957), Direct Detection of Auroral Radiation with Rocket Equipment, *Prod. Natl. Acad. Sci., U.S.A.*, **43** (1), 57–62.

Van Allen, J.A., et al. (1958), Observation of high intensity radiation by satellites 1958 Alpha and Gamma, *Jet Propuls.*, **28**, 588–592.

Van Allen, J.A. (1959), The geomagnetically trapped corpuscular radiation, *J. Geophys. Res.*, **64**, 1683–1689.

Van Allen, J.A. and L.A. Frank (1959), Radiation around the Earth to a radial distance of 107, 400 km, *Nature*, **183**, 430–434.

Van Allen, J.A. (1983), *Origins of Magnetospheric Physics*, Smithson. Ins. Press, Washington, DC.

Vernov, S.N. (1958), Artificial satellite measurements of cosmic radiation, *Doklady Akad. Nauk. SSSR*, **120**, 1231–1233 (in Russian).

Walt, M. (1997), From nuclear physics in space physics by way of high altitude nuclear tests, in *Discovery of the Magnetosphere* (eds. C. Steward Gillmor and J.R. Spreiter), History of Geophysics 7, 235–264, American Geophysical Union, Washington, D.C.

2

Introduction to Wave-Particle Interactions and their Impact on Energetic Particles in Geospace

Kazue Takahashi and Yoshizumi Miyoshi

■ Abstract

The magnetosphere is a natural plasma laboratory that contains a wide variety of waves excited by both disturbances in the solar wind and instabilities within the magnetosphere. The waves interact with particles in the outer zone radiation belt and ring current, thereby playing a fundamental role in redistributing the particles in spatial coordinates and energy as geomagnetic activity changes. We illustrate how observations from spacecraft and ground-based experiments can be used to advance our knowledge on the behavior of the particles. Examples include previous spacecraft observations of particle interactions with ultra-low-frequency waves, previous ground-based observations of electron precipitation caused by electromagnetic ion cyclotron (EMIC) waves, and planned satellite observations of nonlinear EMIC and whistler wave-particle interactions.

Waves, Particles, and Storms in Geospace. First Edition. Georgios Balasis, Ioannis A. Daglis, Ian R. Mann.

2.1 Introduction

The dynamic evolution of the ring current and radiation belts during geomagnetic storms is an important phenomenon not only as a target for scientific research but also as a practical matter of concern (space weather effects) for human activity both on the ground and in space. Essential to understanding and hopefully to predicting how the ring current and the radiation belts behave is observational determination of local wave-particle interactions and their global consequences. There are other processes that can change the particle distribution, for example, global reconfiguration of the magnetic field by large-scale structures in the solar wind. However, there is no question that wave-particle interactions make a critical contribution to particle acceleration and transport at times of high geomagnetic activity.

The geospace is a natural laboratory in which we can conduct in situ measurements of wave-particle interactions covering a wide range of wave frequencies, wavelengths, and particle energies. However, unlike in laboratories on the ground, we cannot control the boundary conditions or measurement locations in the magnetosphere to determine the causal relationship between various phenomena. Fortunately, modern spacecraft missions, some consisting of multiple probes, allow us to gain new information on waves and particles in the inner magnetosphere that could not be obtained a decade ago. For example, the Van Allen Probes mission consists of two probes that can separate spatial and temporal variations of the phase space density of energetic electrons. This, combined with ground-based remote sensing techniques and numerical simulations, provides us with opportunities to advance our knowledge about wave-particle interactions.

The objective of this chapter is to briefly review some of the wave-particle interactions that are believed to impact the outer radiation belt ($3 < L < 7$) during storms and how these processes can be studied using observational tools that are available now. For comprehensive overviews of the radiation belts, geomagnetic storms, and wave-particle interactions, readers are referred to recent articles (e.g., Hudson et al. 2008, Shprits et al. 2008a, 2008b, Thorne 2010, Ukhorskiy and Sitnov 2012, Miyoshi et al. 2012) as well as to other chapters in this book.

2.2 Impact of wave-particle interaction on electron acceleration and loss

Figure 2.1 is a summary diagram showing processes discussed much in the literature that lead to acceleration and transport of electrons in the inner magnetosphere. In this figure, the horizontal and vertical axes indicate distance from the Earth and electron kinetic energy, respectively. The effects of plasma waves on electrons are as follows. Interaction of electrons with Pc5 ultra-low-frequency (ULF) waves (1–10 mHz) excited by external source mechanisms conserves the first adiabatic invariant (magnetic moment) of the electrons but violates the third adiabatic invariant. As a consequence, plasmasheet electrons ($L > 7$, ~keV energy) are transported through the radial diffusion process (Fälthammar 1965, Schulz and Lanzerotti 1974) to the radiation belt ($L < 7$), where the energy can reach ~MeV. Here the energy exchange between electrons and waves occurs through the drift resonance, $\omega - m\omega_d = 0$ (if an axisymmetric magnet field is assumed), where ω is the wave frequency, m is the azimuthal wave number, and ω_d is the bounce averaged drift frequency (e.g., Elkington et al. 1999). Magnetosonic waves, excited between f_{cH^+} and f_{LH} (typically in the 1–10 Hz band) (Russell et al. 1970) possibly by proton ring distributions (Meredith et al. 2008), can locally accelerate electrons through the Landau resonance $\omega - k_{\|}v_{\|} = 0$ (Horne et al. 2007) or transit-time scattering (Bortnik and Thorne 2010), where f_{cH^+} is the

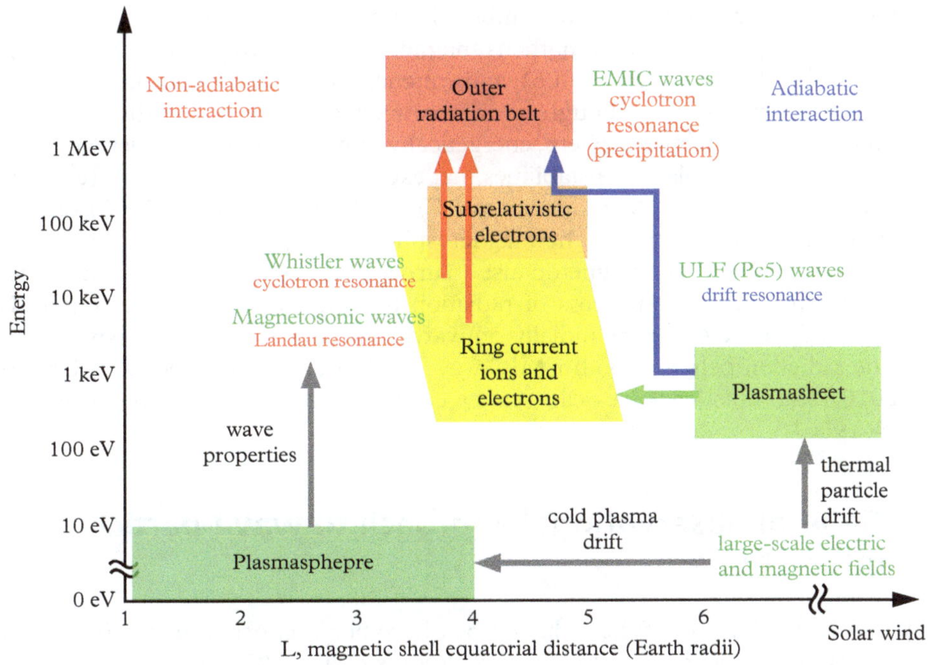

Figure 2.1 *Diagram illustrating the relationship between magnetospheric regions, near-dc and wave fields, and acceleration and transport of electrons (slightly modified from Figure 2 of Miyoshi et al. (2012)). Domains in L and particle energy are shown by boxes, electromagnetic fields including waves are shown in green fonts, the processes that regulate the plasma domain and wave properties are shown by gray arrows, and electron acceleration/transport is shown by blue (adiabatic process) and red (non-adiabatic process) arrows.*

proton gyro frequency, f_{LH} is the lower hybrid resonance frequency, and $k_{\|}$ and $v_{\|}$ are the wavenumber and particle velocity parallel to the magnetic field, respectively. Whistler mode chorus waves (Tsurutani and Smith, 1974), excited in two bands, $\sim$0.1–0.5 f_{ce} and $\sim$0.5–0.7 f_{ce}, by hot (10–100 keV) ring-current energy electrons, can locally accelerate the electrons to $\sim$MeV via the electron cyclotron resonance $\omega - k_{\|}v_{\|} = 2\pi f_{ce}/\gamma$ (e.g., Summers et al. 2002), where f_{ce} is the electron cyclotron frequency and γ is the relativistic factor. Electromagnetic ion cyclotron (EMIC) waves, excited below f_{cH^+}, can also be in cyclotron resonance with electrons, which leads to pitch angle scattering of $\sim$MeV electrons into the atmospheric loss cone (Thorne and Kennel, 1971).

The main point of Figure 2.1 is that the wave phenomena are interconnected with the solar wind driving and internal kinetic and fluid processes. For example, high solar wind velocity enhances the Kelvin-Helmholtz instability on the magnetopause and generates ULF waves in the Pc5 band (e.g., Claudepierre et al. 2008). At the same time, the high solar wind velocity, combined with the southward component of the interplanetary magnetic filed, strengthens magnetospheric convection, causes the plasmasphere to shrink (e.g., Li et al. 2006), and enhances transport of plasmasheet particles to the inner magnetosphere though substorm injections (e.g., Miyoshi et al. 2013). Meanwhile, the inner magnetosphere, energetically pumped up by this solar wind action, will experience various plasma instabilities that excite waves spanning ULF (e.g., Ozeke and Mann 2001) to very-low-frequency (VLF) (e.g., Meredith et al. 2002) bands. Finally, configurational changes of the magnetosphere alter particle drift orbits and lead to particle loss through the magnetopause (Turner et al. 2013). Wave-particle interaction is also involved in the process of radiation belt electron loss. For example, ULF waves can diffuse the electrons radially outward if there is a peak phase space density within the radiation belt, and both EMIC waves (Thorne and Kennel 1971, Ukhorskiy et al. 2010) and whistler waves can scatter electrons into the loss cone (Kennel and Petschek 1966).

2.3 General observational approach to wave-particle interactions

Vastly different behaviors of the radiation belt have been reported for storms with similar histories of ring current evolution (e.g., Reeves et al. 2003). This finding could be understood only if we could keep track of each of the processes such as those illustrated in Figure 2.1 and then find their relative importance.

To tackle the above radiation belt mystery, the following approaches can be taken. (a) Determine the wave properties (frequency, amplitude, wave vector, and polarization) with in situ measurements. (b) Determine the local response (both linear and nonlinear) of particles to the waves with in situ measurements. (c) Evaluate the global consequence of the spatiotemporal variation of wave fields and particle fluxes with multipoint observations of selected events or with statistical analyses of data from a small number of experimental resources. (d) Determine the connection of the solar wind input to the radiation belt dynamics to be able to predict the radiation belt behavior.

2.4 Experimental resources

Orbital properties of spacecraft dictate what we can achieve by in situ observations. Spacecraft with low-inclination orbits are best suited for exploring the radiation belt and ring current, because particles with all mirror latitudes pass the magnetic equator and the equator is the site where many wave modes are excited. Figure 2.2 shows orbits of

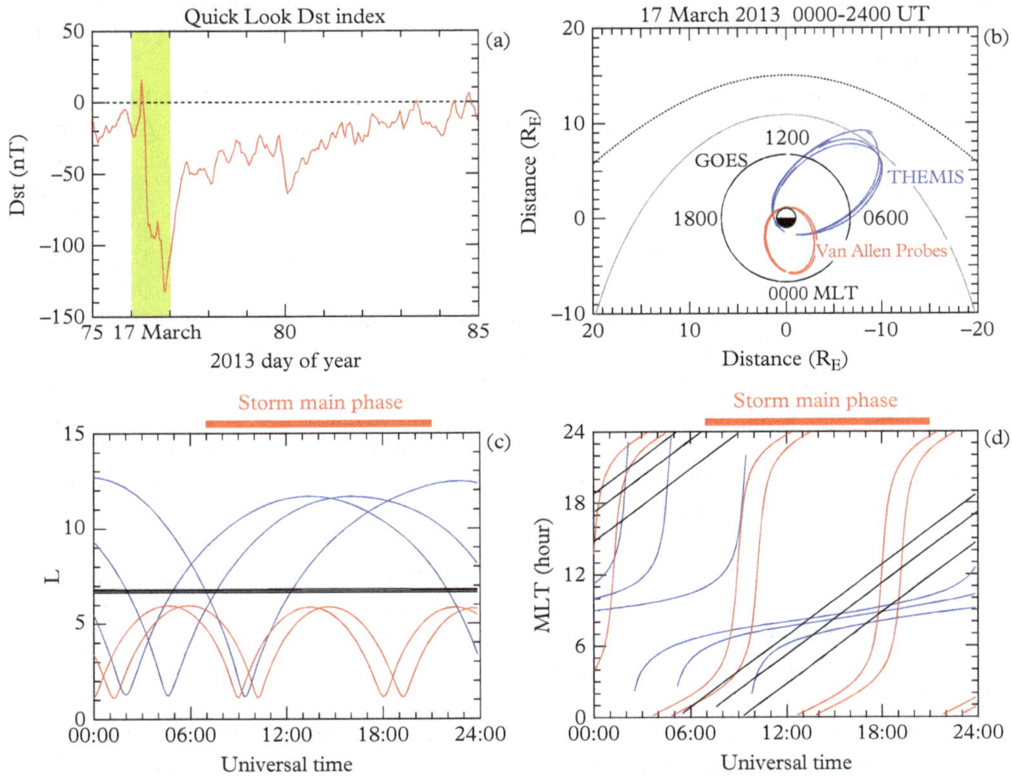

Figure 2.2 *Position of Van Allen Probe-A and -B, THEMIS-A, -D, and -E, and GOES-13, -14, and -15 for a 24-hour period during a geomagnetic storm in March 2013. (a) Quick look Dst index (source: World Data Center for Geomagnetism, Kyoto). The shading indicates the period for which the satellite orbits are plotted. (b) L (radius) versus magnetic local time (MLT, azimuth) plots of the spacecraft orbits. The magnetic coordinates are defined using a dipole magnetic field. (c) Time series plots of L of the spacecraft. (d) Time series plots of MLT of the spacecraft.*

such spacecraft including the two Van Allen Probes (orbital period ~9 hours, inclination ~10°) (Mauk et al. 2013), the three near-Earth THEMIS probes (orbital period ~24 hours, inclination 5–12°) (Angelopoulos 2008, Frey et al. 2008), and three GOES geostationary spacecraft (Singer et al. 1996). Orbits are shown for a 24-hour period encompassing the main phase of a geomagnetic storm (Figure 2.2a). The equatorial projection of the orbits (Figure 2.2b) indicates that the Van Allen Probes were in good position to observe substorm activity near midnight, while the THEMIS probes were positioned to detect direct influence of the solar wind as well as chorus emission expected on the dayside. Most importantly, the short orbital periods of the Van Allen Probes allow us to obtain several radial cuts of the radiation belts during the storm main phase (Figure 2.2c). The local time separation also varies (Figure 2.2d), so it is possible to study the azimuthal propagation velocity or wavelength of waves when two spacecraft are on the same L shell. The spacecraft separation can be much smaller than 1 R_E, making it possible to determine wavelengths comparable to the Larmor radius (~100 km) of ring current ions.

Comprehensive instrumentation is required on spacecraft to capture the different wave-particle interactions illustrated in Figure 2.1. If we take the Van Allen Probes as an example (Figure 2.3), the experiments on each spacecraft include a fluxgate magnetometer (to detect magnetohydynamic (MHD) and EMIC waves), a search coil magnetometer (whistler waves), an electric field detector (MHD, EMIC, and whistler waves), a plasma wave analyzer (plasma waves up to the upper hybrid band), a plasma analyzer (plasmasphere and plasmasheet), and energetic particle detectors (ring current and radiation belts) (Wygant et al. 2013, Kletzing et al. 2013, Funsten et al. 2013, Baker et al. 2013, Blake et al. 2013, Mitchell et al. 2013, Mazur et al. 2013). The major ions, H^+, He^+, and O^+, are distinguished because they behave differently during geomagnetic storms (Hamilton et al. 1988). Electron number density is routinely determined from the spacecraft potential or from the spectral features in plasma wave spectra, usually the cutoff frequency of the continuum radiation, the plasma frequency, and the upper hybrid resonance frequency. Burst mode data sampling (up to 35 k samples/s) is included for time domain analysis of whistler and other waves. The Japanese Exploration of energization and Radiation in Geospace (ERG) spacecraft, planned for launch in 2015, will have a similar instrumentation with the important addition of the Software-Wave Particle Interaction Analyzer (S-WPIA), which determines $\mathbf{v} \cdot \mathbf{E}$ (= rate of particle energization) at the time scale of the whistler wave period (Fukuhara et al. 2009, Miyoshi et al. 2012).

It is necessary to combine other experimental resources to obtain a more global view of the inner magnetosphere and address ionospheric effects on magnetospheric phenomena. Experiments commonly used in conjunction with equatorial satellites include the following: low earth orbit (LEO) satellites (e.g., Kohnert et al. 2011); balloons (e.g., Millan et al. 2013); radars (e.g., Baker et al. 2010); magnetometers (e.g., Ozeke et al. 2012); LF-radio standard networks (Clilverd et al. 2009); and GPS receivers (e.g., Foster and Erickson 2013).

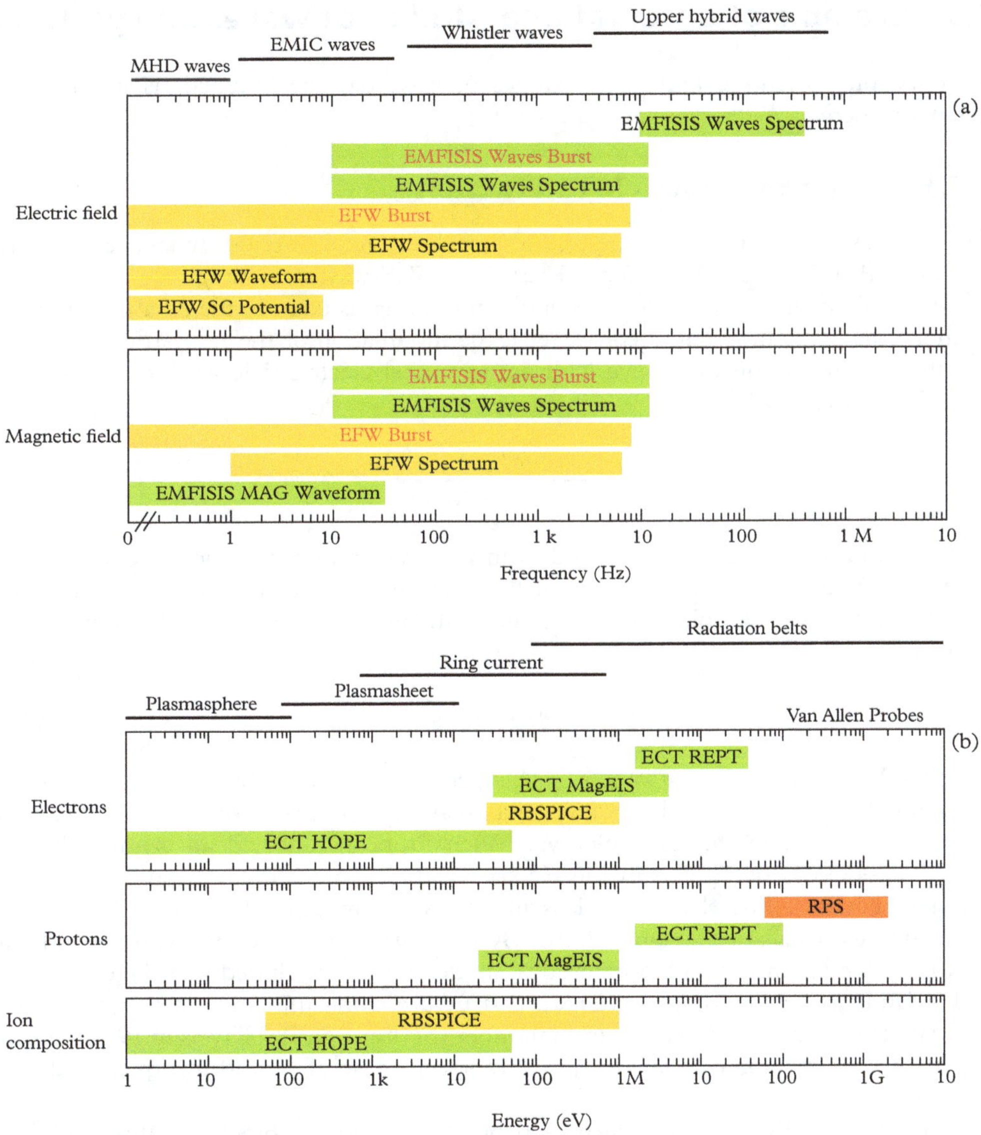

Figure 2.3 *Van Allen Probes experiment capabilities and their relations to waves and particles. (a) Field experiments. Typical frequency bands occupied by different wave modes are shown at the top. (b) Particle experiments. Typical energy ranges of particles in different magnetospheric regions are shown at the top.*

2.5 Example of observational studies of waves and particles

This section presents examples of studies of wave-particle interactions in the inner magnetosphere.

2.5.1 ULF wave-particle interaction

The duty cycle (one energy sweep) of a particle detector is typically shorter than the period of Pc4-5 waves (45–600 s). Therefore, ULF waves are unique in that we can detect the signatures of particle resonances in time series data. In the adiabatic approximation (magnetic moment is conserved), the modulation of the flux, δF, of particles executing drift motion in a wave with an azimuthal electric field δE_φ is expressed as (e.g., Southwood 1976)

$$\delta F \propto \frac{i\delta E_\varphi}{(\omega - m\omega_\mathrm{d} - N\omega_\mathrm{b})} e^{-i(\omega t - m\varphi)} \tag{2.1}$$

where ω and m are the frequency and azimuthal wavenumber of the wave, ω_d and ω_b are the drift and bounce frequencies of the particles, and N is an integer. This equation predicts that the amplitude of δF peaks at the particle energy corresponding to the resonance

$$\omega - m\omega_\mathrm{d} - N\omega_\mathrm{b} = 0 \tag{2.2}$$

and that the phase of δF relative to δE_φ changes by 180° across the resonance energy.

Figure 2.4 shows in-situ observation of a ULF wave-particle resonance phenomenon (Dai et al. 2013). In this example, Van Allen Probe-A detected an azimuthally polarized 84-s (Pc4) oscillation in the electric field (Figure 2.4a) not too far from the magnetic equator. The electric field oscillation is accompanied by a radially polarized magnetic field oscillation (Figure 2.4b). By comparing the observed wave period and the δE_φ-δB_r relationship with a theoretical model, it was concluded that the oscillations had a fundamental standing Alfvén wave structure along the ambient magnetic field. This structure requires N to be an even number (Southwood and Kivelson 1982), meaning that the drift resonance $\omega - m\omega_\mathrm{d} = 0$ (a special case of $N = 0$ in Equation 2.2) is possible.

The theoretically expected particle behavior is evident. First, the fluxes of ions (assumed to be protons) measured at 7 energy steps between 60 keV and 170 keV all oscillate with the same period as the wave (Figure 2.4c). Second, the amplitude of the flux oscillation is energy dependent, with the maximum amplitude occurring at ~90 keV. Finally, there is a shift of the oscillation phase with energy. The cross phase relative to δE_φ increases with energy, with the zero crossing occurring at ~90 keV.

Drift resonance with protons requires the wave to be propagating in the same direction as the proton guiding center drift. This immediately tells us that the wave was

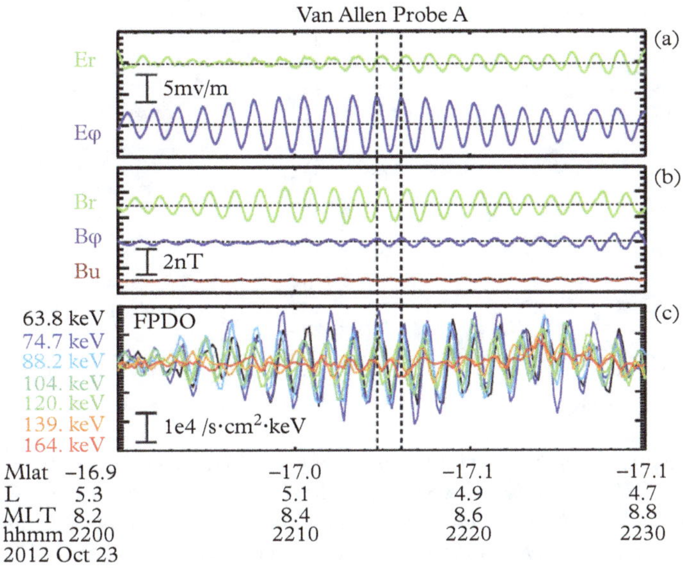

Figure 2.4 *Observation of ion drift resonance with Van Allen Probe-A (Dai et al. 2013). (a) The radial* (r) *and azimuthal* (φ) *components of the electric field during a monochromatic 84-s (Pc4) ULF wave event. Two vertical dashed lines are drawn through the peaks of the* E_φ *component. (b) The three components of the magnetic field. The* B_μ *component is along the average magnetic field. (c) Omnidirectional ion (assumed to be protons) flux measured in 7 energy channels.*

propagating westward. Furthermore, with the measured wave frequency and the known dependence of ω_d on particle energy, the azimuthal wave number is found to be –70. The large magnitude of m is consistent with the wave being a poloidal mode (radial oscillations of field lines) and points to an internal instability. In fact, the examination of the ion phase space density indicated an inward gradient that can destabilize a fundamental poloidal wave through drift resonance (Southwood 1976, Chen and Hasegawa 1988).

2.5.2 Two-spacecraft measurements of wave propagation

We need to know both the frequency and wave vector to determine how waves interact with particles. With a single spacecraft we can measure the frequency but not the wavenumber unless there is a clear particle resonance signature as seen in Figure 2.4. Observational determination of wavelengths or wavenumbers is a major challenge, for example, in evaluating the effectiveness of electron radial diffusion by ULF waves (e.g., Tu et al. 2012). When we have multiple spacecraft, it is possible to determine the wavenumber using phase delay analysis.

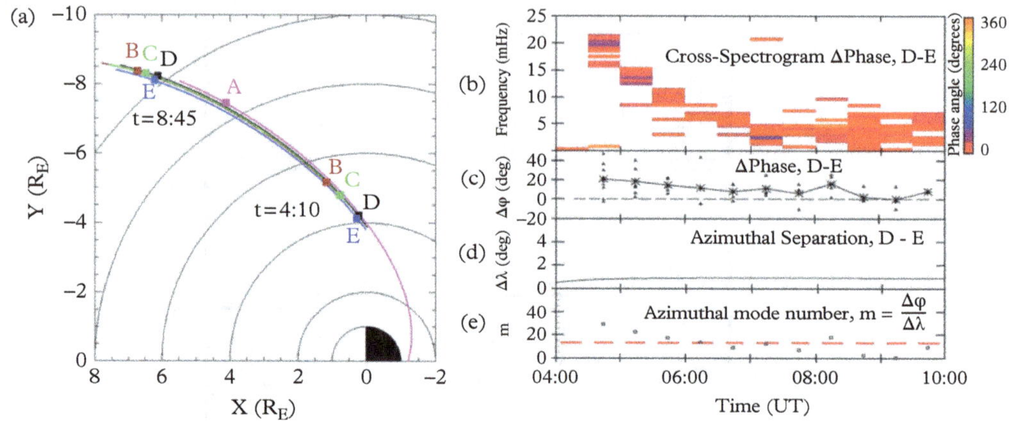

Figure 2.5 *Wavenumber analysis for ULF waves using THEMIS spacecraft (Sarris et al. 2009). (a) Location of the THEMIS probes on 4 September 2007 in the X – Y plane of the geocentric solar ecliptic coordinate system. The dots indicate the locations at 0410 UT and 0845 UT. (b) Cross phase between the radial components of the electric field measured by THEMIS-D and THEMIS-E, plotted as a function of time and frequency. A color key is used to indicate the cross phase value. (c) Phase difference between the two probes. (d) Azimuthal separation of the two probes. (e) Azimuthal wave number.*

Figure 2.5 illustrates an example of phase delay analysis applied to ULF waves observed from THEMIS-D and -E (Sarris et al. 2009). The frequency of the waves decreased with L (Figure 2.5b) as expected for standing Alfvén waves excited on local field lines. From comparison of the radial electric field and the azimuthal magnetic field components, it was concluded that the oscillations were fundamental toroidal waves (azimuthal oscillations of field lines). The spacecraft maintained a small azimuthal separation (Figure 2.5d) as they moved from $L = 4$ (0400 UT) to $L = 12$ (1000 UT). The small separation eliminates the $2n\pi$ ambiguity in cross phase. The toroidal waves observed by the two spacecraft had high coherence, and it was possible to determine the phase delay continuously (Figure 2.5c). The m number derived from the phase delay is mostly near 15, which is substantially smaller than that of the fundamental poloidal wave shown in Figure 2.4. Low m values are expected for waves driven by the solar wind.

2.5.3 Ground-satellite observation of EMIC waves, aurora, and particle precipitation

With the above two examples, we described how spacecraft measurements in the equatorial magnetosphere are used to study wave properties and wave-particle interactions. However, the consequence of the local wave-particle interaction on global particle population is difficult to address unless we make additional measurements. The question is whether waves cause a sufficient amount of particle transport to produce large changes in the particle flux in the radiation belt and ring current. For example, are a large number of particles lost through wave-particle interactions? For loss mechanisms involving particle precipitation into the atmosphere, ground-based observation and low-altitude satellite or balloon observations provide valuable information.

Figure 2.6 shows the result of a study (Miyoshi et al. 2008) taking the above approach. In this example, a search coil magnetometer on the ground detected EMIC waves while an all-sky camera took 2d images of emissions caused by precipitating ions (tens of keV). The EMIC waves occurred in the helium band, and the all-sky camera detected localized proton aurora, which were taken as evidence that the EMIC waves

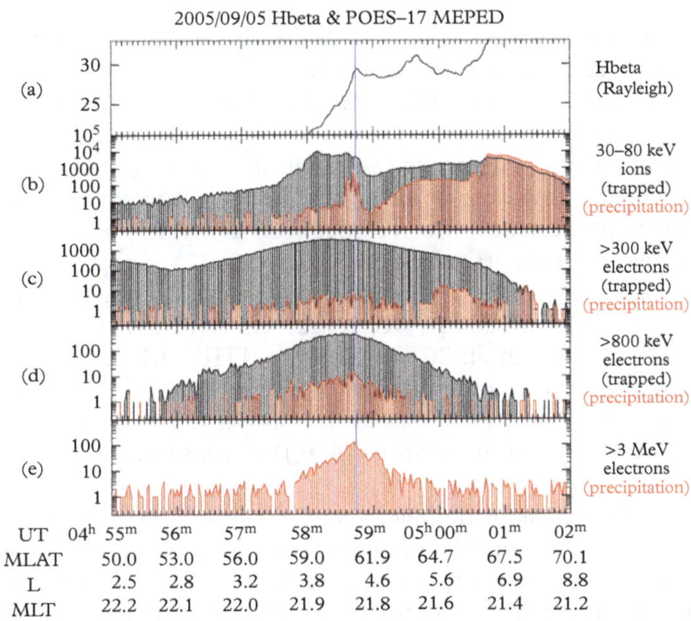

Figure 2.6 *Observation of particle precipitation associated with EMIC waves observed on the ground (Miyoshi et al. 2008).*
(a) Emission at the Hydrogen Balmer line captured by an all-sky camera and averaged over 0430–0500 UT on September 5, 2005. The field of view of the camera covered the magnetic field footpoint of the POES-17 satellite for 0455–0502 UT, and the plot shows the time-averaged emission intensity evaluated along the footpoint over this time interval. The location of the satellite is shown at the bottom of the figure using magnetic latitude (MLAT), magnetic shell L value, and magnetic local time (MLT). The vertical blue line indicates the time when the satellite footprint crossed the stable isolated proton aurora. (b–e) Intensity of the flux (detector count rate) of electrons and ions measured by the POES-17 satellite, plotted as a function of universal time. Trapped and precipitating particles are shown in black and light grey, respectively.

pitch angle scattered the protons into the loss cone. Making this event even more unique, the low-altitude satellite POES-17 traversed the field of view of the all-sky camera. The particle instruments on this satellite detected simultaneously occurring peaks in the flux of precipitating ions (30–80 keV, Figure 2.6b) and precipitating relativistic electrons (>3 MeV, Figure 2.6e), which spatially matched an intensity peak of the proton aurora (Figure 2.6a). These observations led Miyoshi et al. to conclude that the EMIC waves caused precipitation of both ions and electrons. This type of wave-particle interaction was predicted theoretically (Cornwall et al. 1970; Thorne and Kennel 1971).

2.5.4 Nonlinear wave-particle interactions

Until now, linear and quasilinear theories have mainly been employed to describe wave-particle interactions in the inner magnetosphere. However, nonlinear processes could play a major role, and future experimental studies need to address them. To emphasize this point, we consider recent theoretical studies and computer simulation of nonlinear wave-particle interactions associated with whistler (chorus) and EMIC waves. First, the so-called nonlinear trapping process produces electromagnetic electron and ion holes in the velocity phase space, which respectively generate whistler and EMIC waves through the resonant current (Omura et al. 2010, 2012). Second, computer simulations produced triggered whistler and EMIC emissions with rising tones (e.g., Hikishima et al. 2009, Shoji and Omura 2013), which are very similar to real observations (Santolík et al. 2003, Sakaguchi et al. 2013). Third, a THEMIS study confirmed that the observed variations of chorus amplitude are consistent with the nonlinear theory (Kurita et al. 2012). Finally, nonlinear processes lead to strong electron accelerations by chorus waves and rapid precipitation of electrons by EMIC waves (Katoh et al. 2008, Omura and Zhao 2013).

To understand the details of nonlinear wave-particle interactions, it is necessary to conduct time domain measurements of both wave fields and particle fluxes. Specifically, the measurements need to be made with a time resolution higher than the period of relevant waves. The fields experiments on the Van Allen Probes are capable of making the required fields measurements in the EMIC and whistler bands, but particle fluxes are not measured in high enough time resolution. Therefore, it is expected that the S-WPIA experiment on the ERG spacecraft (see Section 2.4) will shed a new light on the nonlinear processes involving whistler waves.

2.6 Summary

In summary, we have presented examples of observational studies of wave-particle interactions in the magnetosphere. The purpose was to provide information on the experimental resources available to us and how these resources can be used. Important factors in experimental studies are the orbits of spacecraft and instrumentation. We reviewed studies of wave-particle interactions involving ULF, EMIC, and VLF waves. The focus was on how to take advantage of multi-instrument observation and simultaneous observations from multiple spacecraft. We also emphasized the importance of additional experimental resources to address the global consequence of wave-particle interactions.

Acknowledgments

Work at JHU/APL was supported by NASA grants NNX10AK93G and NNX14AB97G and by the Van Allen Probes mission. Work at the Solar-Terrestrial Environment Laboratory at Nagoya University was supported by Grants-in-Aid for Scientific Research (23340146, 23224011) of the Japan Society for the Promotion of Science.

REFERENCES

Angelopoulos, V. (2008). The THEMIS Mission. *Space Sci. Rev.*, **141**, 5–34, doi:10.1007/s11214-008-9336-1.

Baker, D. N., et al. (2013). The Relativistic Electron-Proton Telescope (REPT) instrument on board the Radiation Belt Storm Probes (RBSP) spacecraft: Characterization of Earths radiation belt high-energy particle populations. *Space Sci. Rev.*, **179**, 337–381, doi:10.1007/s11214-012-9950-9.

Baker, J. B. H., et al. (2010). Monitoring ionospheric space weather with the Super Dual Auroral Radar Network (SuperDARN). *Radar Conference*, **2010 IEEE**, 1414–1417, doi:10.1109/RADAR.2010.5494396.

Blake, J. B., et al. (2013). The Magnetic Electron Ion Spectrometer (MagEIS) instruments aboard the Radiation Belt Storm Probes (RBSP) spacecraft. *Space Sci. Rev.*, **179**, 383–421, doi:10.1007/s11214-013-9991-8.

Bortnik, J., and R. M. Thorne (2010). Transit time scattering of energetic electrons due to equatorially confined magnetosonic waves. *J. Geophys. Res.*, **101**(A1), 479–493, doi:10.1029/95JA02985.

Chen, L., and A. Hasegawa (1988), On magnetospheric hydromagnetic waves excited by energetic ring-current particles. *J. Geophys. Res.*, **93**(A8), 8763–8767, doi:10.1029/JA093iA08p08763.

Claudepierre, S. G., S. R. Elkington, and M. Wiltberger (2008). Solar wind driving of magnetospheric ULF waves: Pulsations driven by velocity shear at the magnetopause. *J. Geophys. Res.*, **113**, A05218, doi:10.1029/2007JA012890.

Clilverd, M. A., et al. (2009), Remote sensing space weather events: Antarctic-Arctic Radiation-belt (Dynamic) Deposition-VLF Atmospheric Research Konsortium network. *Space Weather*, 7, S04001, doi:10.1029/2008SW000412.

Cornwall, J. M., F. V. Coroniti, and R. M. Thorne (1970), Turbulent loss of ring current protons. *J. Geophys. Res.*, **75**(25), 4699–4709.

Dai, L., et al. (2013). Excitation of poloidal standing Alfvén waves through drift resonance wave-particle interaction. *Geophys. Res. Lett.*, **40**, 4127–4132, doi:10.1002/grl.50800.

Elkington, S. R., M. K. Hudson, and A. A. Chan (1999). Acceleration of relativistic electrons via drift-resonant interaction with toroidal-mode Pc-5 ULF oscillations. *Geophys. Res. Lett.*, **26**(21), 3273–3276.

Fälthammar, C. G. (1965). Effects of time-dependent electric fields on geomagnetically trapped radiation. *J. Geophys. Res.*, **70**(11), 2503–2516.

Foster, J. C., and P. J. Erickson (2013). Ionospheric superstorms: Polarization terminatoreffects in the Atlantic sector. *J. Atmos. Solar-Terr. Phys.*, **103**, 147–156, doi:10.1016/j.jastp.2013.04.001.

Frey, S., V. Angelopoulos, M. Bester, J. Bonnell, T. Phan, and D. Rummel (2008). Orbit Design for the THEMIS Mission. *Space Sci. Rev.*, **141**, 61–89, doi:10.1007/s11214-008-9441-1.

Fukuhara, H., H. Kojima, Y. Ueda, Y. Omura, Y. Katoh, and H. Yamakawa (2009). A new instrument for the study of wave-particle interactions in space: One-chip Wave-Particle Interaction Analyzer. *Earth Planets Space*, **61**, 765–778.

Funsten, H. O., et al. (2013). Helium, Oxygen, Proton, and Electron (HOPE) Mass Spectrometer for the Radiation Belt Storm Probes Mission. *Space Sci. Rev.*, **179**, 423–484, doi:10.1007/s11214-013-9968-7.

Hamilton, D. C., G. Gloeckler, and F. M. Ipavich (1988). Ring current development during the great geomagnetic storm of February 1986. *J. Geophys. Res.*, **93**(A12), 14343–14355.

Hikishima, M., S. Yagitani, Y. Omura, and I. Nagano (2009). Full particle simulation of whistler-mode rising chorus emissions in the magnetosphere. *J. Geophys. Res.*, **114**, A01203, doi:10.1029/2008JA013625.

Horne, R. B., R. M. Thorne, S. A. Glauert, N. P. Meredith, D. Pokhotelov, and O. Santolík (2007), Electron acceleration in the Van Allen radiation belts by fast magnetosonic waves. *Goephys. Res. Lett.*, **34**, L17107, doi:10.1029/2007GL030267.

Hudson, M. K., B. T. Kress, H.-R. Mueller, J. A. Zastrow, and J. B. Blake (2008). Relationship of the Van Allen radiation belts to solar wind drivers. *J. Atmos. Solar-Terr. Phys.*, **70**(5), 708–729.

Katoh, Y., Y. Omura, and D. Summers (2008). Rapid energization of radiation belt electrons by nonlinear wave trapping. *Ann. Geophys.*, **26**, 3451–3456, doi:10.5194/angeo-26-3451-2008.

Kennel, C. F., and H. E. Petschek (1966). Limit on stably trapped particle fluxes. *J. Geophys. Res.*, **71**, 1.

Kletzing, C. A., et al. (2013). The Electric and Magnetic Field Instrument Suite and Integrated Science (EMFISIS) on RBSP. *Space Sci. Rev.*, doi:10.1007/s11214-013-9993-6.

Kohnert, R., S. Palo, and X. Li (2011). Small space weather research mission designed fully by students. *Space Weather*, **9**(4), doi:10.1029/2011SW000668.

Kurita, S., Y. Katoh, Y. Omura, V. Angelopoulos, C. M. Cully, O. Le Contel, and H. Misawa (2012). THEMIS observation of chorus elements without a gap at half the gyrofrequency. *J. Geophys. Res.*, **117**, A11223, doi:10.1029/2012JA018076.

Li, X., D. N. Baker, T. P. O'Brien, L. Xie, and Q. G. Zong (2006). Correlation between the inner edge of outer radiation belt electrons and the innermost plasmapause location. *Geophys. Res. Lett.*, **33**, L14107, doi:10.1029/2006GL026294.

Mauk, B. H., N. J. Fox, S. G. Kanekal, R. L. Kessel, D. G. Sibeck, and A. Ukhorskiy (2013). Science Objectives and rationale for the Radiation Belt Storm Probes Mission. *Space Sci. Rev.*, **179**, 3–27, doi:10.1007/s11214-012-9908-y.

Mazur, J., et al. (2013). The Relativistic Proton Spectrometer (RPS) for the Radiation Belt Storm Probes Mission. *Space Sci. Rev.*, **179**, 221–261, doi:10.1007/s11214-012-9926-9.

Meredith, N. P., R. B. Horne, and R. R. Anderson (2008). Survey of magnetosonic waves and proton ring distributions in the Earth's inner magnetosphere. *J. Geophys. Res.*, **113**, A06213, doi:10.1029/2007JA012975.

Meredith, N. P., R. B. Horne, R. H. A. Iles, R. M. Thorne, D. Heynderickx, and R. R. Anderson (2002). Outer zone relativistic electron acceleration associated with substorm-enhanced whistler mode chorus. *J. Geophys. Res.*, **107**(A7), 1144, doi:10.1029/2001JA900146.

Millan, R. M., et al. (2013). The Balloon Array for RBSP Relativistic Electron Losses (BARREL). *Space Sci. Rev.*, **179**, 503–530, doi:10.1007/s11214-013-9971-z.

Mitchell, D. G., et al. (2013). Radiation Belt Storm Probes Ion Composition Experiment (RBSPICE). *Space Sci. Rev.*, **179**, 263–308, doi:10.1007/s11214-013-9965-x.

Miyoshi, Y., et al. (2012). The Energization and Radiation in Geospace (ERG) project, in *Dynamics of the Earth's Radiation Belts and Inner Magnetosphere*, vol. 119, edited by D. Summers et al., AGU, Washington, D. C.

Miyoshi, Y., K. Sakaguchi, K. Shiokawa, D. Evans, J. Albert, M. Connors, and V. Jordanova (2008). Precipitation of radiation belt electrons by EMIC waves, observed from ground and space. *Geophys. Res. Lett.*, **35**, L23101, doi:10.1029/2008GL035727.

Miyoshi, Y., R. Kataoka, Y. Kasahara, A. Kumamoto, T. Nagai, and M. F. Thomsen (2013). High-speed solar wind with southward interplanetary magnetic field causes relativistic electron flux enhancement of the outer radiation belt via enhanced condition of whistler waves. *Geophys. Res. Lett.*, **40**, 4520–4525.

Omura, Y., D. Nunn, and D. Summers (2012). Generation processes of whistler mode chorus emissions: Current status of nonlinear wave growth theory, in *Dynamics of the Earth's Radiation Belts and Inner Magnetosphere, Geophys. Monogr. Ser.*, vol. 199, edited by D. Summers et al., pp. 243–254, AGU, Washington, D. C., doi:10.1029/2012GM001347.

Omura, Y., J. Pickett, B. Grison, O. Santolík, I. Dandouras, M. Engebretson, P. M. E. Dcrau, and A. Masson (2010). Theory and observation of electromagnetic ion cyclotron triggered emissions in the magnetosphere. *J. Geophys. Res.*, **115**, A07234, doi:10.1029/2010JA015300.

Omura, Y., and Q. Zhao (2013). Relativistic electron microbursts due to nonlinear pitch angle scattering by EMIC triggered emissions. *J. Geophys. Res.*, **118**, 5008–5020, doi:10.1002/jgra.50477.

Ozeke, L. G., and I. R. Mann (2001). Modeling the properties of high-*m* Alfvén waves driven by the drift-bounce resonance mechanism. *J. Geophys. Res.*, **106**, 15583–15597.

Ozeke, L. G., I. R. Mann, K. R. Murphy, I. J. Rae, D. K. Milling, S. R. Elkington, A. A. Chan, and H. J. Singer (2012). ULF wave derived radiation belt radial diffusion coefficients. *J. Geophys. Res.*, **117**, A04222, doi:10.1029/2011JA017463.

Reeves, G. D., K. L. McAdams, R. H. W. Friedel, and T. P. O'Brien (2003). Acceleration and loss of relativistic electrons during geomagnetic storms. *Geophys. Res. Lett.*, **30**(10), 1529, doi:10.1029/2002GL016513.

Russell, C. T., R. E. Holzer, and E. J. Smith (1970). OGO 3 observations of ELF noise in the magnetosphere: 2. The nature of the equatorial noise. *J. Geophys. Res.*, **75**(4), 755–768, doi:10.1029/JA075i004p00755.

Sakaguchi, K., Y. Kasahara, M. Shoji, Y. Omura, Y. Miyoshi, T. Nagatsuma, A. Kumamoto, and A. Matsuoka (2013). Akebono observations of EMIC waves in the slot region of the radiation belts. *Geophys. Res. Lett.*, **40**, 1–5, doi:10.1002/2013GL058258.

Santolík, O., D. A. Gurnett, J. S. Pickett, M. Parrot, and N. Cornilleau-Wehrlin (2003). Spatio-temporal structure of storm-time chorus. *J. Geophys. Res.*, **108**, 1278, doi:10.1029/2002JA009791.

Sarris, T. E., W. Liu, K. Kabin, X. Li, S. R. Elkington, R. Ergun, R. Rankin, V. Angelopoulos, J. Bonnell, K. H. Glassmeier, and U. Auster (2009), Characterization of ULF pulsations by THEMIS. *Geophys. Res. Lett.*, **36**, L04104, doi:10.1029/2008GL036732, 2009.

Schulz, M., and L. Lanzerotti (1974). *Particle Diffusion in the Radiation Belts*. Springer, Berlin.

Shoji, M., and Y. Omura (2013). Triggering process of electromagnetic ion cyclotron rising tone emissions in the inner magnetosphere. *J. Geophys. Res.*, **118**, 5553–5561, doi:10.1002/jgra.50523.

Shprits, Y. Y., D. A. Subbotin, N. P. Meredith, and S. R. Elkington (2008b). Review of modeling of losses and sources of relativistic electrons in the outer radiation belt II: Localacceleration and loss. *J. Atmos. Solar-Terr. Phys.*, **70**, 1694–1713.

Shprits, Y. Y., S. R. Elkington, N. P. Meredith, and D. A. Subbotin (2008a). Review of modeling of losses ands ources of relativistic electrons in the outer radiation belt I: Radial transport. *J. Atmos. Solar-Terr. Phys.*, **70**, 1679–1693.

Singer, H. J., L. Matheson, R. Grubb, A. Newman, and S. D. Bouwer (1996). Monitoring space weather with the GOES magnetometers. *Proc. SPIE Int. Soc. Opt. Eng.* **2812**, 299–308.

Southwood, D. J. (1976). A general approach to low-frequency instability in the ring current plasma. *J. Geophys. Res.*, **81**(19), 3340–3348.

Southwood, D. J., and M. G. Kivelson (1982). Charged particle behavior in low-frequency geomagnetic pulsations, 2, Graphical approach. *J. Geophys. Res.*, **87**(A3), 1707–1710.

Summers, D., C. Ma, N. P. Meredith, R. B. Horne, R. M. Thorne, D. Heynderickx, and R. R. Anderson (2002). Model of the energization of outer-zone electrons by whistler-mode chorus during the October 9, 1990 geomagnetic storm. *Geophys. Res. Lett.*, **29**(24), 2174, doi:10.1029/2002GL016039.

Thorne, R. M. (2010). Radiation belt dynamics: The importance of wave-particle interactions. *Geophys. Res. Lett.*, **37**, L22107, doi:10.1029/2010GL044990.

Thorne, R. M., and C. F. Kennel (1971). Relativistic electron precipitation during magnetic storm main phase. *J. Geophys. Res.*, **76**(19), 4446–4453.

Tsurutani, B. T., and E. J. Smith (1974). Postmidnight chorus: A substorm phenomenon. *J. Geophys. Res.*, **79**(1), 118–127.

Tu, W., S. R. Elkington, X. Li, W. Liu, and J. Bonnell (2012). Quantifying radial diffusion coefficients of radiation belt electrons based on global MHD simulation and spacecraft measurements. *J. Geophys. Res.*, **117**, A10210.

Turner, D. L., V. Angelopoulos, W. Li, M. D. Hartinger, M. Usanova, I. R. Mann, J. Bortnik, and Y. Shprits (2013). On the storm-time evolution of relativistic electron phase space density in Earth's outer radiation belt. *J. Geophyts. Res.*, **118**(5), 2196–2212, doi:10.1002/jgra.50151.

Ukhorskiy, A. Y., and M. I. Sitnov (2012). Dynamics of Radiation Belt Particles. *Space Sci. Rev.*, doi:10.1007/s11214-012-9938-5.

Ukhorskiy, A. Y., Y. Y. Shprits, B. J. Anderson, K. Takahashi, and R. M. Thorne (2010). Rapid scattering of radiation belt electrons by storm-time EMIC waves. *Geophys. Res. Lett.*, **37**, L09101, doi:10.1029/2010GL042906.

Wygant, J. R., et al. (2013). The Electric Field and Waves Instruments on the Radiation Belt Storm Probes Mission. *Space Sci. Rev.*, **179**, 183–220, doi:10.1007/s11214-013-0013-7.

3

Geospace Magnetic Storms and the Van Allen Radiation Belts

Geoffrey D. Reeves and Ioannis A. Daglis

■ Abstract

The dynamics of geospace magnetic storms and changes in the Van Allen radiation belts are intimately intertwined. At the simplest level, the build-up and decay of the storm-time ring current and of the radiation belts are fundamentally similar: electrons and ions are injected into the inner magnetosphere, where they are further accelerated to form the ring current and radiation belts. Both populations decay gradually, after elevated magnetospheric activity has ceased. The ring current decays primarily by charge exchange and the radiation belts decay primarily by atmospheric precipitation. Digging even slightly deeper than this simplistic description, however, uncovers many unanswered questions, which are addressed in this chapter.

Waves, Particles, and Storms in Geospace. First Edition. Georgios Balasis, Ioannis A. Daglis, Ian R. Mann.

3.1 Introduction

The two major charged particle populations that are trapped by the geomagnetic field in geospace are the Van Allen radiation belts and the ring current. They are distinguished by their high energies (~10 to 500 keV for the ring current, ~100 keV to 20 MeV / 200 MeV for the radiation belt electrons/protons), when compared to much lower energies of the surrounding plasmas, and by their intricate relationship during the most complex phenomenon of solar-terrestrial coupling—the geospace magnetic storms. The lower energy end of these populations—the ring current—is the main cause of the defining feature of magnetic storms: the worldwide reduction of the geomagnetic field horizontal component, which is evident in the storm-time Dst index profile. The higher energy end of these populations—the radiation belts—was the first component of magnetospheric plasma, and space plasma in general, to be discovered at the very beginning of the space era.

The milestone discovery of the radiation belts was made through the measurements of Geiger–Müeller tubes of the group of Professor James Van Allen on board the Explorer 1 and 3 satellites in 1958 [Van Allen et al., 1958]. Van Allen correctly interpreted those measurements as a result of intense corpuscular radiation [Van Allen, 1959]. The first source proposed for the origin of the radiation belts was decay of neutrons produced by the interaction of cosmic rays with the Earth's atmosphere [Singer, 1958; Kellogg, 1959; Vernov et al., 1959]. Balloon and rocket measurements, however, showed that the loss rates of radiation belt electrons through precipitation into the atmosphere [e.g., Winckler et al., 1962] would have demanded improbably high rates of replenishment if cosmic rays had been the source. The time dependence of outer belt electron fluxes was even more problematic [see, e.g., Walt, 1996], but, until spectral measurements became available, variations in detector count rates could potentially be explained by more subtle changes in spectral slope—an ambiguity that would resurface, for different reasons, in the observations from the 1990s and 2000s. In contrast, the proposal by Kellogg [1959] that neutron albedo might be responsible for protons in the radiation belts is now widely accepted [e.g., Selesnick et al., 2007]. To this date, we generally think of the Earth's radiation belts as consisting of two belts of energetic particles—a very stable inner belt in the region $L <\sim 2$ composed of both energetic electrons and ions and a very dynamic outer belt in the region $L >\sim 3$ that is composed almost entirely of electrons. However, recent results have shown that multiple belt structures exist [Baker et al., 2013] and that the nature of the slot and the inner belt–even their existence–is highly energy-dependent [Fennell et al., 2015; Reeves, 2015]. The region of relatively lower radiation levels in the region $L \approx 2$–3 is referred to as the slot region. In this chapter we focus primarily on the dynamic outer belt electron population.

The other major energetic particle population in geospace is the ring current—the electric current responsible for the global depression of the magnetic field's horizontal component on Earth's surface. Although the ring current is carried by both electrons and ions, most of its energy density is due to medium-energy (tens of keV) ions [e.g., Williams, 1981; Daglis et al., 1993]. The existence of a giant diamagnetic ring current encircling Earth had been predicted long before its actual observation in space [e.g., Chapman and Ferraro, 1930, 1931]. Early OGO-3 satellite measurements in the 1960s analysed by Frank [1967] confirmed the existence of the ring current and showed that the ring current is dominated by ions with energies around 50 keV. The detailed composition and energy of the ring current were not clarified until the Active Magnetospheric Particle Tracer Explorer (AMPTE) mission of the late 1980s [e.g., Hamilton et al., 1988; Daglis et al., 1993]. A few years later, the Combined Release and Radiation Effects Satellite (CRRES) confirmed and expanded the conclusions of the AMPTE mission, especially with regard to intense magnetic storms [e.g., Daglis, 1997].

The dynamics of geospace magnetic storms and of the ring current and changes in the Earth's radiation belts are intimately intertwined. At the most simple level the build-up and decay of both the ring current and radiation belts are fundamentally similar: electrons and ions are injected into the inner magnetosphere where they are further accelerated to form the ring current and radiation belts. Both gradually decay after activity has stopped. The ring current decays primarily by charge exchange and the radiation belts decay primarily by atmospheric precipitation. Digging even slightly deeper than this simplistic description, however, uncovers many unanswered questions. How do injected electrons and ions get trapped in the inner magnetosphere? What processes accelerate these 'seed' populations? What controls the structure and the evolution of the ring current and radiation belts? Where are they most intense and why does that vary from event to event? How are the overall intensities of the ring current and radiation belts related? How do ring current processes affect the radiation belts? Those questions, and their implications for space weather, were among the primary motivations for the NASA Van Allen Probes mission, launched in August 2012 [Mauk et al., 2012].

In this chapter we will attempt to provide a few representative examples of the developments that have led to our current understanding as well as examples that point to some of the holes in our understanding [following the treatment in Reeves, 2015]. Although the radiation belts may play some role in controlling ring current dynamics, the dominant coupling is in the opposite direction. Therefore we will primarily focus on aspects of ring current dynamics that have important effects on the radiation belts. The most prominent effects are those of various ring current driven electromagnetic waves on radiation belt electrons [e.g., Bortnik et al., 2006; Ozeke and Mann, 2008]. We also emphasize that this paper does not provide a review of recent results from the ongoing Van Allen Probes mission but direct the interested reader to http://rbspgway.jhuapl.edu/biblio for a current bibliography.

3.2 Electron motion in Earth's radiation belts

Very quickly after the discovery of the radiation belts, Northrop and Teller [1960] had worked out the equations of motion that govern adiabatic motion in the geomagnetic field and derived the invariants associated with electron gyromotion (μ), bounce between mirror points (J or K), and drift azimuthally around the Earth (Φ or $L^\star$). These invariants are approximately conserved under slow changes to the magnetic field, fully describe relativistic electron motion in an arbitrary geomagnetic field configuration, and are fundamental to understanding the dynamics of the radiation belts and their coupling to the ring current.

The first invariant, μ, is associated with the electron gyro-motion around the magnetic field and is given by

$$\mu = \frac{p_\perp^2}{2m_0 B}$$

where m_0 is the rest mass, and $p_\perp$ is the relativistic momentum in the direction perpendicular to the magnetic field, B.

The second invariant is associated with the bounce motion between the Northern and Southern mirror points:

$$\mathcal{J} = \oint p_\parallel ds = 2\int_{s_n}^{s_s} p_\parallel ds.$$

When μ is conserved, the second invariant can also be expressed in a convenient form that is independent of energy

$$K = \frac{\mathcal{J}}{2\sqrt{2m_0\mu}} = \int_{s_s}^{s_n} \sqrt{B_m - B(s)}\, ds$$

where B_m is the magnetic field strength at the mirror point and s is the path length along the field line. The third invariant is an integral around the drift path and between the mirror points:

$$\Phi = \oint \vec{B} \bullet d\vec{A}$$

which conserves the total magnetic flux enclosed within a drift shell. In radiation belt physics it is more common to use the $L^\star$ parameter [Roederer, 1970]

$$L^* = \frac{2\pi M}{\Phi R_E}$$

where M is the Earth's magnetic moment and R_E is the radius of the Earth (6,371 km). In a dipole magnetic field $L^\star = L$ (sometimes called 'McIlwain L'). However, when the field is not a pure dipole $L^\star$ is not equivalent to L. In a realistic field, electrons

Figure 3.1 *A visualization of radiation belt drift shells based on numerical particle tracing in a realistic geomagnetic field [Reeves et al., 2012]. Mirror points for different equatorial pitch angles (in 10° steps) are shown spaced along the field. As electrons drift around the Earth, conservation of μ and K requires that the electrons follow different drift paths as indicated by the colour-coded field lines. Drift shell splitting is particularly apparent in the upper right of the figure. Electrons measured at the satellite, with 30° pitch angles, move along the copper-coloured surface which is the drift shell for those electrons. On that drift shell, μ, K, and L* are constant but pitch angle, energy, and radial distance are not.*

with different pitch angles follow different drift paths—an effect known as drift shell splitting (Figure 3.1). Therefore, electrons with different pitch angles, measured at a given point in space (such as a specific satellite location), have different third invariants, Φ, and therefore different $L^\star$. That means that, while a given point in space can have a single value of L for all electrons on the field line, it cannot be assigned a single value of $L^\star$.

3.3 Effects of geospace magnetic storms on the radiation belts

One of the primary drivers of change and spatial asymmetry in the Earth's magnetic field (particularly deep in the inner magnetosphere) is the storm-time ring current [Daglis et al., 1999; Daglis, 2006]. The ring current dramatically modifies the geomagnetic field and thus 'causes' geospace magnetic storms. Strong geomagnetic field variations influence, in turn, dramatically the properties of radiation belt electrons, because they affect all three periodic motions of electrons.

Conservation of the first adiabatic invariant implies that changes in the local magnetic field magnitude must also produce changes in the perpendicular momentum, which changes the electrons' energy and pitch angle and therefore their magnetic mirror points.

Magnetic storms typically involve intense stretching and subsequent dipolarization of the night-side magnetic field. Since the second invariant is an integral along the field between the mirror points, conservation of J will also redistribute radiation belt electrons in a strongly pitch-angle dependent way. Stretching of the night-side field increases drift shell splitting, moving equatorially mirroring electrons closer to Earth on the night-side and field-aligned electrons away from the Earth. Combined with the typical radial gradients, the stretching of the field produces a change in the local pitch angles toward a 'butterfly' distribution with a minimum at 90°. At the same time a satellite can find itself on field lines that now map much further down the tail and toward the outer boundary of the radiation belts.

The third invariant Φ or $L^\star$ is an integral of the magnetic flux enclosed in a drift path. An increase in the storm-time ring current decreases the magnetic field strength in the inner magnetosphere and electrons must move outward in order for their drift orbits to enclose the same flux – an effect known as the 'Dst effect' [Kim and Chan, 1997] (Figure 3.2). This outward motion moves electrons to even lower local field magnitudes, further reducing their energy. Measured at a fixed energy, this appears as a reduction (or dropout) of fluxes.

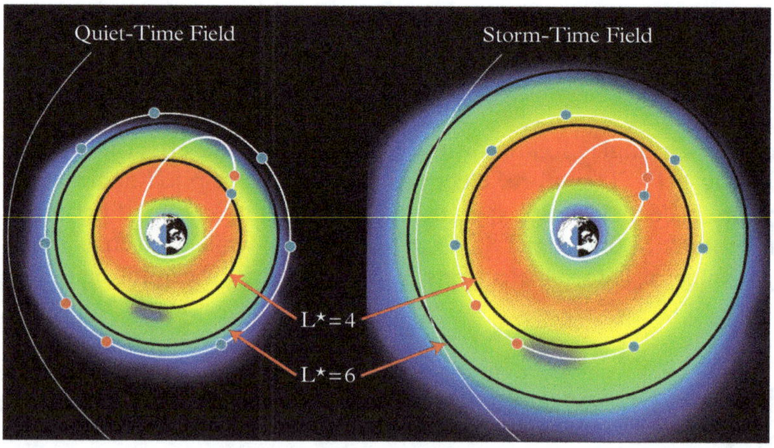

Figure 3.2 *A schematic visualization of the Dst effect [Reeves et al., 2012]. As the ring current intensifies and the magnetic field strength in the inner magnetosphere decreases, radiation belt electrons move outward to conserve the third adiabatic invariant. Fluxes (in colour) are arbitrary and L-shells are not shown quantitatively.*

On storm time scales (hours to days), changes in the geomagnetic field change motion of radiation belt electrons adiabatically. In other words there is no net, permanent effect on the electrons due to changes in the field alone and when the field returns to its pre-storm state the radiation belt electrons will too. Therefore any other changes in the radiation belts have to be understood in the context of those adiabatic changes. To understand radiation belt processes, one must disentangle both adiabatic and non-adiabatic storm-time processes that, typically, occur simultaneously.

For example, as shown in Figure 3.2, some drift paths (constant L*) moving adiabatically outward encounter the magnetopause and electrons on those drift shells are permanently lost from the radiation belts. When the field will relax to its pre-storm configuration those drift shells will exhibit a net (non-adiabatic) loss of electrons, while other drift shells will exhibit only a temporary (adiabatic) decrease in fluxes with no net loss. Other non-adiabatic storm-time processes that affect the radiation belt include intensification of ULF waves that enhance radial diffusion of radiation belt electrons, injection of intensified population of substorm-accelerated electrons that can form a seed population for further acceleration to relativistic energies, formation of unstable plasma distributions producing VLF waves that resonantly interact with radiation belt electrons, and changes in ion composition that can also affect plasma instabilities and field-line resonant frequencies.

All of the processes affecting radiation belt electrons (both adiabatically and non-adiabatically) are typically enhanced during magnetic storms. Therefore it was natural to assume that intense magnetic storms would produce big radiation belt enhancement events. Reeves [1998] attempted to demonstrate this relationship, but found instead that, while most radiation belt enhancements (at geosynchronous orbit) occur during storms, the intensity of radiation belt enhancements is not correlated with the intensity of the storm-time ring current (Figure 3.3).

Shortly thereafter Onsager et al. [2002] showed that magnetic storms do not just enhance the radiation belts—they can also produce non-adiabatic, permanent loss (or dropout) of electrons from the radiation belts. Additionally, they demonstrated that the dropouts typically showed an admixture of adiabatic and non-adiabatic processes that produced different responses at different times.

In a follow-on study Reeves et al. [2003] surveyed 276 moderate and intense storms (min Dst < –50 nT) from 1989 through 2000, showing examples of storms that intensified the radiation belts, storms that depleted the radiation belts, and storms that produced no net change in the radiation belts (Figure 3.4). They also compared statistically the maximum fluxes at geosynchronous orbit 1–3 days before the storm with the maximum flux 1–5 days after the storm and found that 53% of storms increased, 19% decreased, and 28% had no net effect on the intensity of the radiation belt fluxes.

Many studies have focused on the radiation belt response during storms but few studies have looked systematically at the ring current conditions during radiation belt events. This is in part due to the fact that it is easy to categorize storms by minimum Dst but there is no widely accepted way to quantitatively define a radiation belt event. On the other hand, some recent studies showed that radiation belt enhancements could be observed even in the absence of magnetic storms [Schiller et al., 2014; Su et al., 2014b].

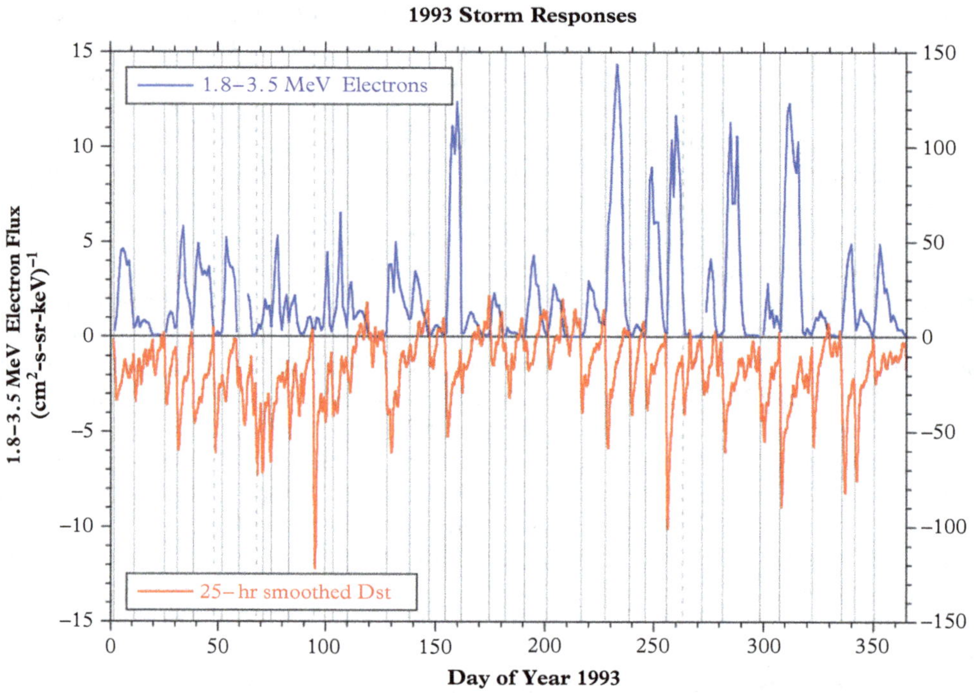

Figure 3.3 *Relativistic electron fluxes measured at geosynchronous orbit (blue) and 25-hour smoothed Dst (red) in 1993—a year dominated by recurrent coronal hole high-speed streams. During all phases of the solar cycle, radiation belt electron enhancements typically occur in association with storm-time decreases in Dst but the intensities are not correlated. Strong storms do not necessarily produce strong radiation belt enhancements.*

In contrast to the studies discussed earlier, Tverskaya et al. [2003] found a rather consistent and repeatable radiation belt response to storms. They examined the location of the peak of the radiation belts as a function of Dst. Using data from the Polar, HEO, and SAMPEX satellites they confirmed and extended the results of an earlier study [Tverskaya, 1986] showing that the location of peak ~2 MeV electron flux could be described by the relationship $|Dst|_{max} = 2.75 \times 10^4 / L^4_{max}$, independently of the actual flux of radiation belt electrons (Figure 3.5). The reason why this remarkable relationship should exist has not yet been fully explained. Similarly the results of Tverskaya et al. [2003] have not been reconciled with the results of Reeves et al. [2003], who reported that the percentage of storms that increased (or decreased) the intensity of the belts was independent of L-shell (see also Figure 3.5).

Other evidence of the relationship between magnetic storms and the radiation belts came from studies of the seasonal dependence of relativistic electron fluxes [Baker et al., 1999]. Averages performed over three-month intervals, centred on the spring and fall equinoxes and on the winter and summer solstices, showed that equinoctial fluxes of

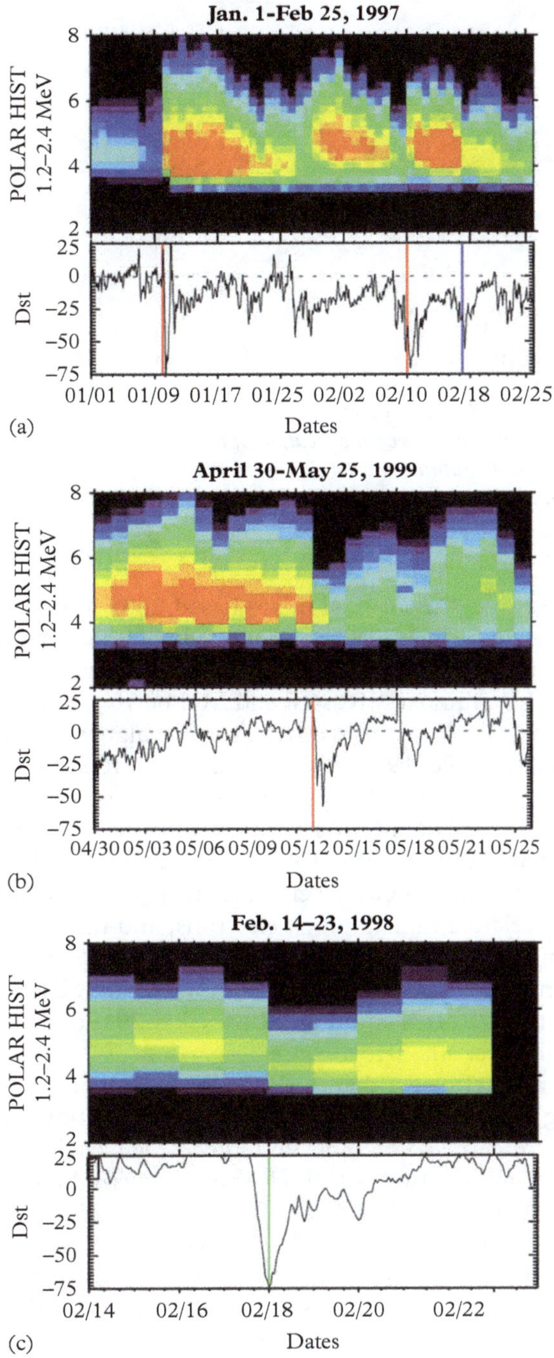

Figure 3.4 *Examples of storms that (A) increased radiation belt fluxes, (B) decreased fluxes, and (C) left the belts relatively unchanged (less than a factor of 2 net change in flux) [from Reeves et al., 2003].* *©John Wiley and Sons 2003.*

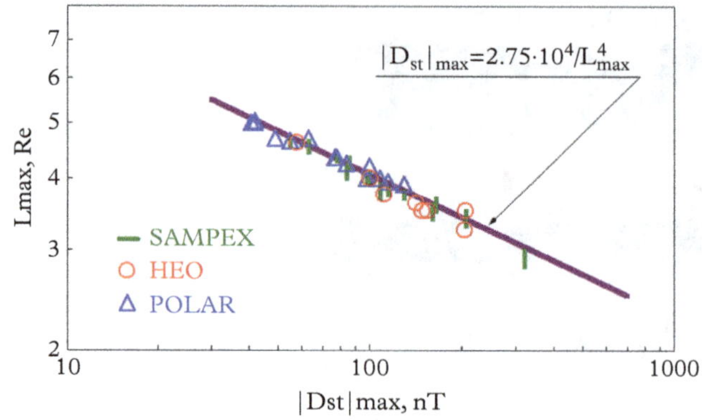

Figure 3.5 *Tverskaya et al. [2003] showed a remarkably consistent and repeatable relationship between the location of the peak flux of 2 MeV electrons in the radiation belts (L_{max}) with maximum magnetic storm intensity ($|Dst|_{max}$).*
Tverskaya et al. [2003]

electrons are nearly a factor of three larger than the solstitial fluxes based upon a superposed epoch analysis for 1992–1999. They concluded that both the radiation belts and the ring current respond similarly to more efficient merging of the interplanetary and the magnetospheric magnetic fields near the equinoxes [Russell and McPherron, 1973].

It has long been known that radiation belt electron fluxes tend to be higher when solar wind velocities are higher [Paulikas and Blake, 1979]. Solar wind speeds tend to be highest in the declining phase of the solar cycle when driving can be dominated by co-rotating interaction regions (CIRs) produced by high-speed streams (HSS) from solar coronal holes [Blake et al., 1997; Baker et al., 1998; Reeves, 1998]. This relationship led researchers to consider whether specific structures in CIRs were particularly effective in driving magnetospheric processes that accelerate radiation belt electrons, and therefore whether increases and decreases in radiation belt fluxes during storms might be related to the type of solar wind drivers [Miyoshi and Kataoka, 2005; Borovsky and Denton, 2006; Clilverd et al., 2010; Borovsky and Denton, 2011; Denton and Borovsky, 2012; Xiao-Fang et al., 2013; Hietala et al., 2014; Kilpua et al., 2015].

In general, CME-driven storms are more effective than CIR-driven storms in producing a strong enhancement of the ring current—even when the integrated southward IMF is approximately the same—but there can also be other differences. Miyoshi and Kataoka [2005] selected storms with -130 nT < Dst < -100 nT and categorized them as CME-driven or CIR-driven (Figure 3.6). They found that 'CIRs are significantly more effective for the evolution of the outer belt than CMEs' and speculated that this was due to 'the existence of a series of particle injections driven by Alfvén waves within the fast coronal hole stream following the CIR'. Miyoshi and Kataoka also selected a third category—of CME-driven great storms with Dst < -130 nT, which produced no stronger radiation belt response than CME-driven intense storms (-130 nT < Dst

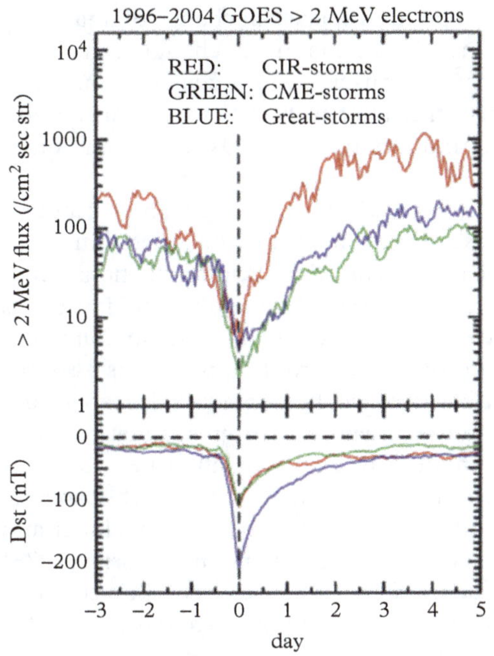

Figure 3.6 *Miyoshi and Kataoka [2005] showed that when CIR- and CME-driven storms are selected based on the same minimum Dst, CIR storms 'produce' on average a greater intensification of radiation belt electrons. CME-driven great storms with Dst < −130 nT did not produce any stronger radiation belt response than CME-driven storms with −130 nT < Dst < −100 nT.*

< −100 nT). On the other hand, it is important to note that only CME-driven great storms (Dst < −130 nT) were effective in enhancing the flux of RB electrons in the inner portion at L < 4 where the slot region exists before the storm onset. This is consistent with previous conclusions that the peak position of the outer belt strongly depends on the Dst index (i.e., the storm intensity) and on the plasmapause position [O'Brien et al., 2003; Miyoshi et al., 2004].

The findings of Miyoshi and Kataoka indicated that storms of different interplanetary causes influence radiation belts differently even when their intensity is comparable. This implied that different storm drivers favour specific storm-associated processes that have distinct impacts on belt dynamics. It became clear that a complete understanding of the relationship between magnetic storms and radiation belt electrons would come from a comprehensive understanding of the various storm-associated processes that have synergetic or competing effects on relativistic electron acceleration and losses [Friedel et al., 2002 and Millan and Thorne, 2007 respectively].

3.4 Local acceleration and radial diffusion

Radiation belt particles are typically measured as particle number flux—a function of spatial position, particle energy, and pitch angle, j = j(**R**, E, alpha)—which is not an adiabatically conserved quantity. As discussed previously, in a time-varying geomagnetic field the flux of radiation belt electrons at a fixed point in space can change dramatically, as electrons adjust their motion to conserve their invariants in the changing magnetic field. Unlike flux, phase space density, $f = j/p^2$, is conserved in a changing magnetic field. Therefore when $j = j(\mathbf{R}, E, \text{alpha})$ is transformed to $f(\mu, K, L^\star)$ we can distinguish non-adiabatic electron acceleration, transport, and loss from the 'Dst effect' and other purely adiabatic effects.

Adiabatic motion, by definition, results in no net changes to the electron distribution but Fälthammar [1965] and, later on, other researchers showed that acceleration of radiation belt electrons could occur through radial diffusion when field fluctuations occur on the time scale of electron drifts. He concluded: 'when the first and second invariants are conserved, any radial motion of the particles is associated with a corresponding energy change. Some particles move outward and others inward; but, if there is a source in the outer magnetosphere and a sink farther in, there will be net inward transport and associated net energy gain'. Dungey [1963], on the other hand, proposed that resonant interaction between electromagnetic whistler-mode waves and radiation belt electrons could form a sink for electrons by pitch angle scattering electrons into the atmospheric loss cone. He concluded that 'the effect is found to be important between the inner and outer belts'—the region that is called the slot region near $L \approx 2$. Lyons and Thorne [1973] showed, quantitatively, that the quiet-time structure of the outer electron belt could be 'explained on the basis of a balance between pitch angle scattering loss and inward radial diffusion from an average outer zone source'. These studies and others, including the comprehensive treatment by Schulz and Lanzerotti [1974], established the fundamental paradigm for radiation belt dynamics: acceleration by inward radial diffusion and loss by pitch angle scattering into the atmosphere.

However, electrons with energies >5 MeV are often enhanced after a storm—even at geosynchronous orbit where the distance to the trapping boundary and, therefore, the amount of energy that can be gained by inward transport from the plasma sheet are small. Energetic electrons in the near-Earth plasma sheet quickly gradient-curvature drift out of the magnetotail so there is never a 'reservoir' of multi-MeV electrons outside the trapping boundary. How then, can we account for the geosynchronous observations? Baker et al. [1989] suggested that the 'recirculation model' developed to explain electron observations at Jupiter [Nishida, 1976] might also apply in the Earth's radiation belts and this modification of the standard radial diffusion paradigm was widely adopted by radiation belt scientists. In the recirculation model (Figure 3.7) electrons radially diffuse inward gaining energy while conserving the first two adiabatic invariants (step 1). In the slot region the pitch angle scattering rates exceed the radial diffusion rates and the particles are scattered toward the atmosphere (step 2). At low altitude the electrons are scattered across field lines without substantial loss of energy (step 3). They are then free to repeat the cycle. While the details were problematic (particularly step 3) the recirculation model had several advantages. It naturally accounted for the 2–3 day build-up of electron fluxes following the commencement of a storm and the very high energies seen at geosynchronous orbit. Recirculation could also explain acceleration up to arbitrarily high energies even at geosynchronous orbit.

Challenges to the recirculation model began to emerge as simultaneous multi-satellite measurements started to become widely available. Energetic particle measurements from

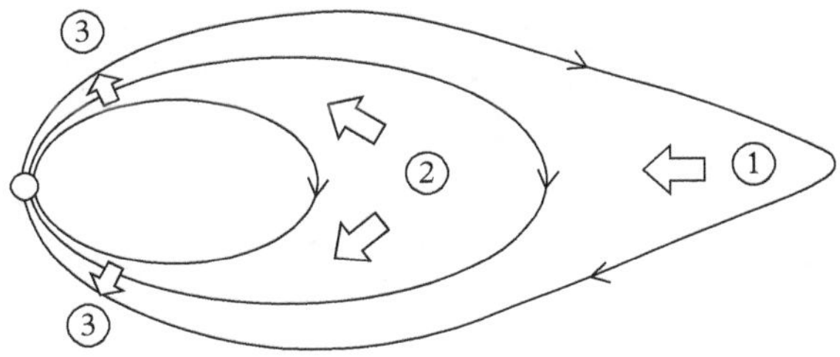

Figure 3.7 *A schematic of the 'recirculation' model for radiation belt acceleration [from Nishida, 1976].*

GOES and Los Alamos National Laboratory (LANL) instruments at geosynchronous orbit provided near-global coverage, albeit in a narrow range of L-shells. Combining geosynchronous measurements with data from other orbits (SAMPEX, GPS, HEO, and others) started to fill in a three-dimensional picture of the radiation belts [Friedel et al., 1999]. Perhaps even more importantly, multi-satellite studies could overcome the inherent limitation of satellite *orbital periods (typically 12 to >24 hours)* and provide measurements at multiple L-shells on time scales less than an hour. Multi-satellite analysis made new, detailed studies of space weather events possible. One of the first such studies was a radiation belt enhancement event in November 1993 [Li et al., 1997].

Another such event, in January 1997, was exceptionally well studied [Baker et al., 1998; Goodrich et al., 1998; Jordanova et al., 1998; Kamide et al., 1998; Reeves et al., 1998a; Reeves et al., 1998b; Jordanova et al., 1999; Hudson et al., 2000; Lyons et al., 2000; Hudson et al., 2001]. Following a procedure similar to that used by Li et al. [1997], Reeves et al. [1998a] binned data from 11 satellites in different orbits according to L-shell. Near geosynchronous orbit, L≈6.6, the 2 MeV radiation belt electrons showed their 'typical' behaviour—a gradual build-up of fluxes over ~4 days (with a less typical shock compression early in the event). The behaviour was startlingly different in the heart of the outer radiation belt at L≈4.6—where the GPS satellites cross the equatorial plane (Figure 3.8a). There, the 2 MeV electrons shot up by over two orders of magnitude in less than 12 hours. Following the initial acceleration, fluxes remained remarkably constant and showed no evidence of the gradual build-up seen at geosynchronous orbit. In fact, the acceleration was so rapid that the radial profiles seen by GPS satellites moving away from the equator (i.e., outward in L) measured significantly different fluxes than on the inbound part of the orbit even though inbound and outbound measurements at the same L-shell were separated by less than an hour (Figure 3.8b, top). The gradual build-up of fluxes seen near geosynchronous orbit appeared to be the result of outward radial diffusion from the newly accelerated population near L≈4.6 (figure 3.8b, top). It is interesting to contrast these high-time resolution measurements with observations from a single satellite. The Polar satellite, with an ~18-hour orbit, captured the state of the belts before and after the acceleration (Figure 3.8b, bottom) but could not resolve the acceleration event itself.

Reeves et al. [1998a] drew two conclusions from the January 1997 event. Firstly, the observations were not consistent with the recirculation model. One of the strengths of the recirculation model was that it naturally explained the gradual build up to a delayed peak that is characteristic of radiation belt observations at geosynchronous orbit. For January

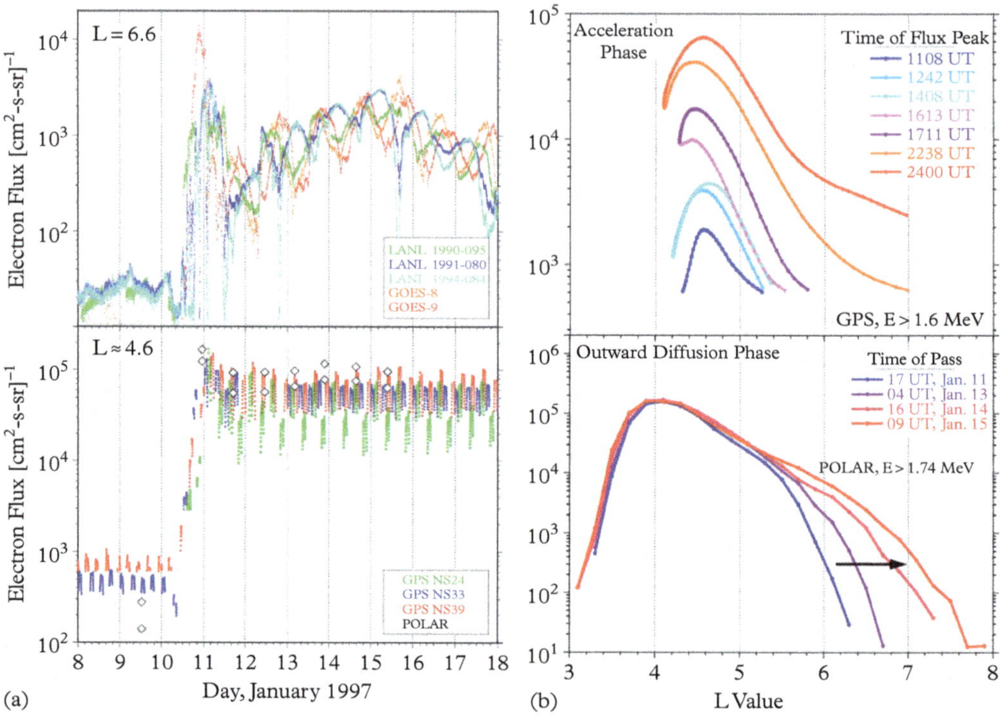

Figure 3.8 *Multi-satellite electron observations during the January 1997 radiation belt event [from Reeves et al., 1998a]. Rapid acceleration in the heart of the radiation belts ($L \approx 4.6$) and geosynchronous orbit ($L \approx 6.6$) suggested local acceleration and subsequent outward radial diffusion. But, without phase space density observations, the results were not definitive.*

1997, the geosynchronous observations still showed that delay. In the recirculation model that delay should be observed at all L-shells as electrons complete the energization loop sketched in Figure 3.7. Instead at L≈4.6 the energization was not only much faster but also short-lived, with almost no change in flux after January 12. Secondly, if the electrons were accelerated inside geosynchronous orbit and diffused outward then the acceleration region would have to be confined to a remarkably narrow range of L-shells, well inside geosynchronous orbit, in the heart of the outer belt. If true then the source of the MeV radiation belt electrons would not be the magnetotail, but rather the source would be lower-energy electrons in that same spatial region—somewhere between the plasmasphere and the geosynchronous orbit.

At the same time that new missions and multi-satellite studies were providing a new window on radiation belt dynamics, new theories of relativistic electron acceleration were being developed and included localized acceleration processes deep in the radiation belts. The new studies built on the pioneering work of Dungey [1963] and

Kennel and Petschek [1966], who showed that the VLF whistler-mode waves which pitch-angle scattered electrons into the atmosphere could be produced by plasma distributions with temperature anisotropies unstable to the growth of whistler-mode waves. The (suprathermal) electron temperature anisotropy excites whistler-mode waves, which cause pitch-angle scattering of electrons toward the loss cone. In the late 1990s Summers et al. [1998] proposed that cyclotron resonant interactions at much higher (relativistic) energies have the opposite effect, diffusing electrons toward 90° while simultaneously increasing their energy. At MeV energies, gradients in the energy dimension implied that the flow of energy is from the waves to the particles providing a mechanism for accelerating electrons in situ from the local population of lower-energy electrons.

This 'local acceleration' scenario suggested a more complex and interdependent set of processes might be responsible for acceleration of radiation belt electrons to MeV energies. First electrons must be transported from the magnetotail to the inner magnetosphere by processes such as substorm injections and enhanced large-scale convection—the same processes that produce the storm-time ring current. In the process a fraction of the magnetotail electrons are energized to form a seed population of electrons with energies from a few keV to a few 100s keV. Temperature anisotropies at lower energies (<30 keV) provide a source of free energy for the generation of whistler-mode waves while higher-energy electrons (≈100–500 keV) provide a 'seed' population that can be accelerated to MeV energies by those same waves.

These theories of local acceleration were to undergo further subsequent development [e.g., Horne and Thorne, 1998; Horne et al., 2000; Summers and Ma, 2000; Horne, 2002; Summers et al., 2002; Horne and Thorne, 2003; Summers et al., 2004; Albert and Young, 2005; Horne et al., 2005a; Horne et al., 2005b; Shprits et al., 2006; Omura et al., 2007; Summers et al., 2007; Omura et al., 2008; Albert et al., 2009; Thorne, 2010]. One characteristic has remained invariant though; the resonance between VLF waves and relativistic electrons takes place locally, on a given flux tube, without transporting the electrons radially. This local acceleration is therefore fundamentally different from radial-diffusive acceleration in both the energization process and the source population.

However, prior to the launch of the Van Allen Probes satellites, the measurements needed to clearly distinguish between the two types of acceleration process. A radially localized peak in electron fluxes (at a given energy) can be produced by either local or radial-diffusive acceleration. It was widely recognized in the radiation belt community that it would be necessary to convert measurements of electron flux, j, to phase space density, f. The two quantities are related by the simple formula $f = j/p^2$. The challenge comes in converting flux from its natural coordinate system (position, energy, and pitch angle) to phase space density as a function of the adiabatic invariants (μ, K, L*) that define the canonical coordinate system for relativistic electron motion [Schulz, 1996; see also GEM Inner Magnetosphere and Storms Campaign Working Group report, 1999 (http://www-ssc.igpp.ucla.edu/gem/work_groups/ims_wg2.html)]. In any realistic magnetic field the conversion from $j(\vec{r}, E, \alpha)$ to $f(\mu, K, L^*)$ is both unique and time dependent for any point along a satellite's orbit. Fortunately, new global magnetic field models, combined with improvements in computational power, have made this conversion significantly more efficient and robust. Advanced magnetic field models, in particular, are especially important, as accurate phase space density calculations are highly dependent on the accuracy of the field models [e.g., Turner and Li, 2008; Turner et al., 2012].

3.5 Phase space density gradients

Radial diffusion occurs whenever various processes produce field fluctuations on the time scale of electron drifts. At some level, these fluctuations always exist so radial diffusion is always operating in the Earth's radiation belts—either at rapid rates or slower rates. Since diffusion always moves a quantity from a region of higher density to lower, the radial gradients of phase space density (i.e., $f(L^*)$ at fixed μ and K) will be positive (increasing with radial distance) if electrons are diffusing inward from a magnetotail source population, whereas they will be negative (decreasing with radial distance) in the outer part of the belt if the electrons are accelerated locally and the magnetotail/magnetopause acts as a sink. Another advantage of analysing radial gradients of phase space density is that the adiabatic changes are 'automatically' accounted for. When phase space density is plotted as a function of L⋆ (and L⋆ is calculated in a realistic field model), phase space density on that drift shell does not change when the magnetic field changes. In the absence of acceleration or loss, L⋆ (for a given μ, K) defines a surface of constant phase space density, $f(\mu, K, L^*) = const.$ Therefore the only way to unambiguously identify radial diffusion, acceleration, or loss is to look for changes in phase space density, e.g., $f(\mu, K, L^*, t)$, which are the signatures of violation of one or more adiabatic invariants.

The next couple of figures illustrate, schematically, the changes in phase space density profiles expected from radial diffusion (D_{LL}, Figure 3.9) and local acceleration (D_{EE}, Figure 3.10). Phase space density profiles are represented for three different times with the lower, middle, and upper lines respectively. If the phase space density at the source and sink do not vary with time, then increasing radial diffusion will move electrons inward, flattening the gradient at high L⋆ where diffusion is strongest (Figure 3.9). In the inner regions (L⋆ ≈ 4–5) the phase space density increases locally but the increase is due to the radial transport of electrons. Stated another way, the phase space density is not conserved, $f(\mu, K, L^*, t) \neq const$, because 'new' electrons are added from the source region. Note, however, that regardless of how strong the radial diffusion is, the radial gradient will always be positive and monotonic from the sink (slot region) to the source (magnetotail).

VLF waves energize electrons locally through energy diffusion and produce a distinctively different change in phase space density profiles—a radially localized peak that grows from the background distribution to produce negative radial gradients at high L⋆ (Figure 3.10). Here again the phase space density increases in the inner regions (L⋆ ≈ 4–5) but the increase is due to the local acceleration of electrons. In this case, $f(\mu, K, L^*, t) \neq const$ because acceleration (energy diffusion) does not conserve μ.

There are, however, complications to what otherwise would be a fairly straightforward measurement as illustrated in Figures 3.11 and 3.12. Firstly, the fluxes of electrons and

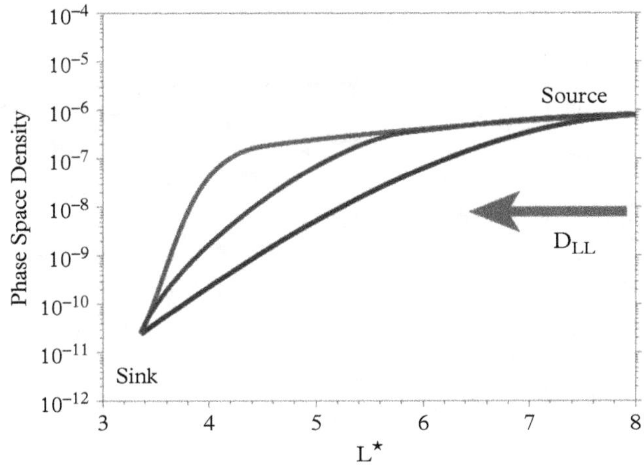

Figure 3.9 *A schematic of radial diffusion from a source region (the magnetotail) to a sink region (the slot). An increase in radial diffusion will increase the transport rate and increase the phase space densities in the heart of the outer radiation belt ($L^\star \approx$ 4–5) [from Reeves, 2015].*

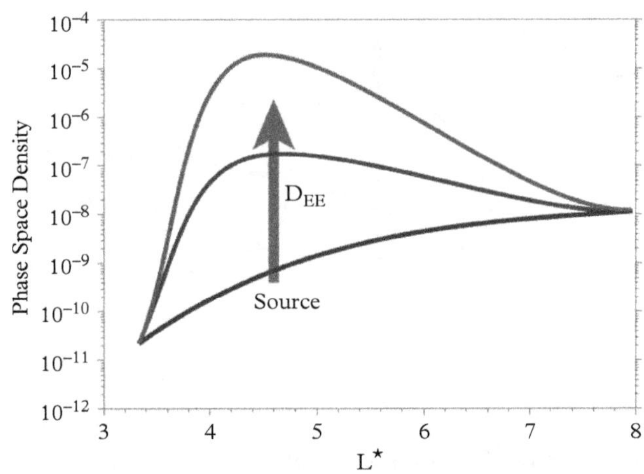

Figure 3.10 *A schematic of local acceleration. Wave-particle interactions produce energy diffusion that accelerates electrons. In this case the source of the relativistic electrons is the local lower-energy electron population [from Reeves, 2015].*

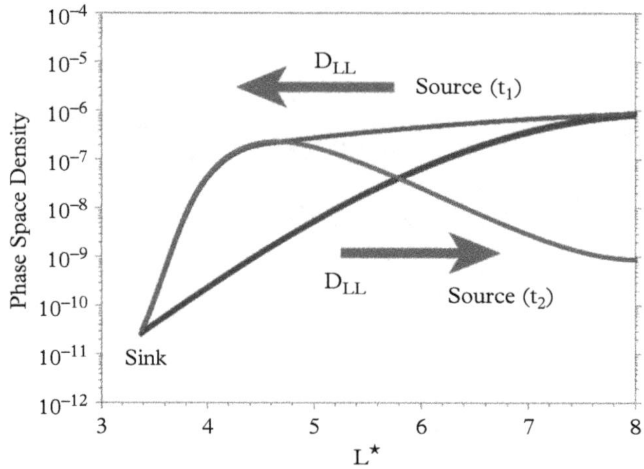

Figure 3.11 *A schematic showing how peaks in phase space density can be formed by a combination of radial diffusion and a variable source at high L* [from Reeves, 2015].*

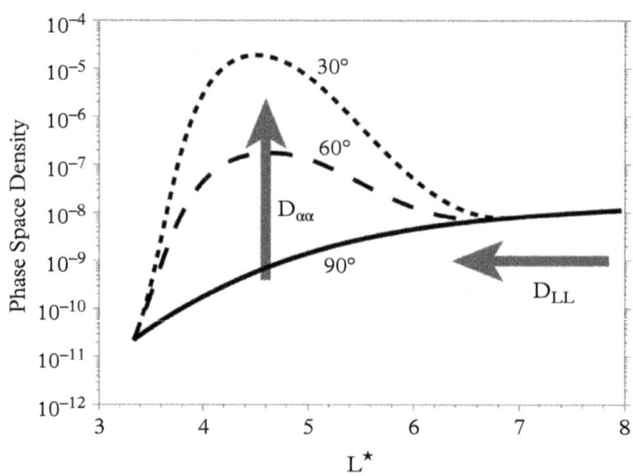

Figure 3.12 *A schematic of the different radial profiles of phase space density that can be produced for different equatorial pitch angles (K values) through a combination of radial diffusion and localized pitch angle diffusion [from Reeves, 2015].*

ions at the inner edge of the plasma sheet are not constant. Therefore the 'source' in Figure 3.9 can be time-variable. As illustrated in Figure 3.11, if the source at time t1 is high then radial diffusion will transport electrons earthward. If at time t2 the source is lower then diffusion will transport electrons outward. Since diffusion is faster at higher L than at lower L a phase space density peak can be produced without local acceleration. Phase space density peaks can also be produced by a combination of radial diffusion and pitch angle diffusion, without energy diffusion (Figure 3.12). This possibility leaves ambiguities in studies using off-equatorial measurements.

3.6 Studies with the Van Allen Probes: insights into the effects of wave-particle interactions and the ring current influence

NASA's Van Allen Probes mission (formerly Radiation Belt Storm Probes—RBSP) was designed, in part, to overcome the observational limitations of previous missions and differentiate clearly between radial-diffusive acceleration and local wave-particle acceleration [Kessel et al., 2012; Mauk et al., 2012]. The Van Allen Probes are a pair of satellites with a comprehensive set of particle and fields instruments in an equatorial, elliptical orbit with apogee at $\approx$5.8 R_E. Their orbital period is $\approx$9 hours but is slightly different for the two satellites. This allows for the phasing in the orbits to vary with time. At times the satellites are perfectly out of phase with apogees separated by only 4.5 hours, passing through the heart of the belts every 2–3 hours. This unprecedented temporal resolution, near-equatorial orbit, and paired satellite configuration is providing views of radiation belt dynamics that were not previously possible.

In order to resolve whether or not local acceleration by VLF waves was a viable mechanism for accelerating radiation belt electrons, Reeves et al. [2013] analysed phase space density profiles for a storm that occurred on October 8–9, 2012, early in the Van Allen Probes mission. The October 2012 storm produced a strong and rapid enhancement of the fluxes in the outer electron belt and the first opportunity to test whether or not local wave-particle interactions are a viable mechanism for relativistic electron acceleration. The 2 MeV electron fluxes at L $\approx$ 4.2 increased by nearly three orders of magnitude in approximately 12 hours. The unique configuration of the Van Allen Probes was able to capture the rapid transformation of the belts in a way that definitively ruled out the other possibilities for producing radial peaks in phase space density during this particular storm.

Figure 3.13 shows profiles of phase space density as a function of L* for fixed values of μ and K. Reeves et al. [2013] used the TS04 [Tsyganenko and Sitnov, 2005] magnetic field model to calculate the adiabatic invariants. In Figure 3.13 each radial pass is labelled by the time at which the satellite crosses L* = 4.2. Here phase space densities are for μ = 3433 MeV/G and K = 0.11 $G^{1/2}$ R_E (E $\approx$ 2–6 MeV and 50° $< \alpha_e <$ 90° depending on the point on the orbit). Between 23:17 UT on October 8 and 03:12 UT on October 9,

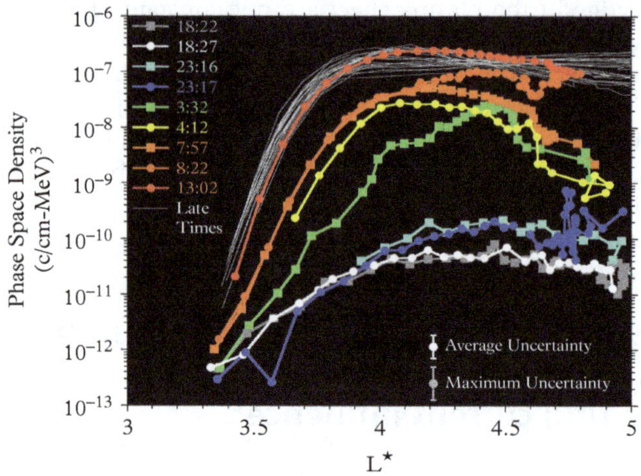

Figure 3.13 *Radial profiles of phase space density measured by the two NASA Van Allen Probes satellites on October 8–9, 2012. A rapid increase in phase space density in a region centred near L$^\star \approx$ 4.2, a slower increase in the outer portion of the belt (L*>4.7), and the resulting radial peaks in phase space density are the signatures expected for local wave-particle acceleration [from Reeves et al., 2013].*

phase space densities increased by approximately two orders of magnitude at L* = 4.2 and continued increasing until about 13:02 UT when the IMF turned northward.

On October 8–9, 2012, the Van Allen Probes were nearly perfectly out of phase in their orbits. This provided the fastest possible time resolution but also provided simultaneous measurements at high and low L*. During this time, RBSP-B measured a positive radial gradient at L*<4.2. At the same time, RBSP-A measured a negative radial gradient at L*>4.2, showing that the radial peak in phase space density was not an artefact of spatial-temporal aliasing. At 08:12 UT, by coincidence, the orbital dynamics were such that both spacecraft reached L$^\star \approx$ 4.2 at the same time. Although the satellites were at quite different local times, the satellites should measure the same phase space density when they are on the same L* (a 'drift shell conjunction'). Any mismatches provide limits on uncertainties due to instrument calibration and also to uncertainties in the calculation of the invariant integrals [Morley et al., 2013]. Reeves et al. used this (and other techniques) to determine that the average and worst-case uncertainties were a factor of 1.4 and 2.0 respectively. They were also able to rule out time dependence of the plasma sheet source population (Figure 3.11) using simultaneous, multi-satellite geosynchronous measurements. These measurements pointed to local wave-particle acceleration as the only viable process and strongly suggested that acceleration by VLF chorus was the proximate cause.

Thorne et al. [2013], studying the same event, used measured VLF chorus wave amplitudes and frequency distributions to calculate the energy (momentum) and pitch angle diffusion coefficients in the heart of the acceleration region at $L^{\star} \approx 4.2$. They applied a 2D diffusion solver to the initial pitch angle and energy distribution measured by the REPT instrument [Baker et al., 2012] and found that resonant interaction with chorus waves could fully explain the evolution of the MeV electron distribution.

Tu et al. [2014] were able to investigate the October 2012 event in even more detail using the DREAM-3D diffusion code and a remarkably complete set of observations to constrain the boundary conditions and free parameters. They were able to quantitatively illustrate the complex interplay of chorus acceleration, electron losses, and radial diffusion. A particularly important component of the simulation turned out to be the low-energy 'seed population'—a topic further investigated by Boyd et al. [2014]. Studies of relativistic electron acceleration by VLF chorus continue to produce a more-and-more detailed picture of this mechanism [Foster et al., 2014; Li et al., 2014c; Su et al., 2014a; Liu et al., 2015].

Within the plasmasphere, VLF waves are primarily observed in the form of broadband, incoherent hiss. While chorus and hiss are both whistler-mode waves, hiss is relatively inefficient at producing acceleration and primarily produces losses through pitch angle scattering. Common understanding would suggest that ring current electrons only penetrate the plasmasphere in the plume regions but Van Allen observations have shown that ring current electrons can sometimes (but rarely) be directly injected into the plasmasphere where they can provide a source of free energy for the generation of hiss [Chen et al., 2014; Li et al., 2014a; Li et al., 2014b; Ni et al., 2014]. Other studies have shown that plasmaspheric hiss can also result from whistler-mode chorus waves that are generated at higher L-shells, outside the plasmasphere, and subsequently refracted at high latitudes into the plasmasphere [e.g., Bortnik et al., 2007; 2009].

As the Van Allen Probes are equipped with a complete complement of fields and waves instruments [Kletzing et al., 2013; Wygant et al., 2014], more comprehensive investigations of ring current associated fields and waves that are important for radiation belt dynamics have become possible. One example are storm-time mechanisms for the acceleration of the initial seed population of 100s keV electrons. Mozer et al. [2014] and Malaspina et al. [2014] found evidence of local acceleration from non-linear electric field structures associated with substorm injections and the transport of electrons and ions into the ring current.

Much attention has been focused on the production of whistler waves by unstable plasma distributions in storm-time substorm injections and the electron component of the ring current. Furthermore, there are other important radiation belt processes associated with the ion component of the ring current. The Van Allen Probes measurements have been able to confirm and add considerable detail to the importance of magnetopause losses caused, not only by earthward motion of the magnetopause, but also by the outward, adiabatic transport of radiation belt electrons [Shprits et al., 2013; Hudson et al., 2014; Turner et al., 2014a; Turner et al., 2014b]. Unstable ring current ion distributions also produce radiation belt-effective waves and are being studied in unprecedented detail. Electromagnetic ion cyclotron (EMIC) waves, excited by the injection of ring current ions, can cause rapid precipitation loss of relativistic electrons on timescales of a few hours [e.g., Bortnik et al., 2006], but may only act on the most energetic part of the radiation belts (although the precise energy range is hotly debated). New observations from the Van Allen Probes shed light not only on the effect of EMIC waves but also on how they are generated and what their characteristics are [Allen et al., 2013; Shprits et al., 2013; Usanova et al., 2014; Zhang et al., 2014].

Magnetosonic waves [Ma et al., 2014; Zhao et al., 2014; Zhou et al., 2014] are another candidate for radiation belt electron acceleration—although it is not clear whether they can efficiently accelerate relativistic electrons. Last but not least, radial diffusion operates to redistribute and accelerate (or decelerate) electrons [Baker et al., 2014; Li et al., 2014d] producing the remarkable spatial coherence observed in the radiation belts. The Van Allen Probes have also provided the most detailed observational evidence to date for drift resonant interactions between radiation belt electrons and ULF waves [Claudepierre et al., 2013; Dai et al., 2013; Mann et al., 2013], which may be driven by low-frequency mirror and drift instabilities of ring current ion populations [Ozeke and Mann, 2008; Ge et al., 2011].

3.7 Summary

We have presented an overview of the complex influence of geospace magnetic storms on the dynamics of the Van Allen radiation belts—in particular the outer (electron) belt. Although the basic build-up and decay of the storm-time ring current and of the radiation belts are fundamentally similar, the acceleration of energetic 'seed' electrons to relativistic energies and their loss through atmospheric precipitation are rather distinct. The comprehensive suite of particles and fields instruments on-board the Van Allen Probes, in combination with the advances in computational power and magnetic field modelling, allow the efficient conversion of fluxes to phase space densities and the extraction of wave characteristics. Thus, the impacts of storm-associated fields and waves on the acceleration and loss of radiation belt electrons can be identified and assessed, making now-casting and forecasting of radiation belt content and intensity a realistic goal for the not-too-distant future.

Acknowledgments

The work leading to this paper was supported by RBSP-ECT funding under NASA's Prime contract no. NAS5- 01072, by the Dynamic Radiation Environment Assimilation Model (DREAM) project at Los Alamos National Laboratory, and by the European Union's Seventh Framework Programme (FP7-SPACE-2011-1) grant agreement no. 284520 for the 'Monitoring, Analyzing and Assessing Radiation Belt Energization and Loss' (MAARBLE) collaborative research project.

REFERENCES

Albert, J. M., and S. L. Young (2005), Multidimensional quasi-linear diffusion of radiation belt electrons, *Geophys. Res. Lett.*, **32**, L14110, doi:10.1029/2005GL023191.

Albert, J. M., et al. (2009), Three-dimensional diffusion simulation of outer radiation belt electrons during the 9 October 1990 magnetic storm, *J. Geophys. Res.*, **114**, A09214, doi:10.1029/2009JA014336.

Allen, R. C., et al. (2013), Multiple bidirectional EMIC waves observed by Cluster at middle magnetic latitudes in the dayside magnetosphere, *J. Geophys. Res.: Space Phys.*, **118**, 10, doi:10.1002/Jgra.50600.

Baker, D. N., et al. (1989), Relativistic electrons near geostationary orbit: Evidence for internal magnetospheric acceleration, *Geophys. Res. Lett.*, **16**, 6, doi:10.1029/GL016i006p00559.

Baker, D. N., et al. (1998), Coronal mass ejections, magnetic clouds, and relativistic magnetospheric electron events: ISTP, *J. Geophys. Res.: Space Phys.*, **103**, A8, doi: 10.1029/97ja03329.

Baker, D. N., et al. (1999), Equinoctial and solstitial averages of magnetospheric relativistic electrons: A strong semiannual modulation, *Geophys. Res. Lett.*, **26**, 20, doi:10.1029/1999gl003638.

Baker, D. N., et al. (2012), The relativistic electron-proton telescope (REPT) instrument on board the Radiation Belt Storm Probes (RBSP) spacecraft: Characterization of Earth's radiation belt high-energy particle populations, *Space Sci. Rev.*, doi:10.1007/s11214-012-9950-9.

Baker, D. N., et al. (2013), A long-lived relativistic electron storage ring embedded in Earth's outer Van Allen belt, *Science*, **340**, 6129, doi:10.1126/science.1233518.

Baker, D. N., et al. (2014), Gradual diffusion and punctuated phase space density enhancements of highly relativistic electrons: Van Allen Probes observations, *Geophys. Res. Lett.*, **41**, 5, doi:10.1002/2013gl058942.

Blake, J. B., et al. (1997), Correlation of changes in the outer-zone relativistic electron population with upstream solar wind and magnetic field measurements, *Geophys. Res. Lett.*, **24**, 8, doi:10.1029/97GL00859.

Borovsky, J. E., and M. H. Denton (2006), Differences between CME-driven storms and CIR-driven storms, *J. Geophys. Res.*, **111**, A07S08, doi:10.1029/2005JA011447.

Borovsky, J. E., and M. H. Denton (2011), A survey of the anisotropy of the outer electron radiation belt during high-speed-stream-driven storms, *J. Geophys. Res.-Space*, **116**, doi:10.1029/2010ja016151.

Bortnik, J., R. M. Thorne, T. P. O'Brien, J. C. Green, R. J. Strangeway, Y. Y. Shprits, and D. N. Baker (2006), Observation of two distinct, rapid loss mechanisms during the 20 November 2003 radiation belt dropout event, *J. Geophys. Res.*, **111**, A12216, doi:10.1029/2006JA01 1802.

Bortnik, J., R. M. Thorne, N. P. Meredith, and O. Santolik (2007), Ray tracing of penetrating chorus and its implications for the radiation belts, *Geophys. Res. Lett.*, **34**, L15109, doi:10.1029/2007GL030040.

Bortnik, J., W. Li, R. M. Thorne, V. Angelopoulos, C. M. Cully, J. W. Bonnell, O. LeContel, and A. Roux (2009), An observation linking the origin of plasmaspheric hiss to discrete chorus emissions, *Science*, **324**, 775–778.

Boyd, A. J., et al. (2014), Quantifying the radiation belt seed population in the 17 March 2013 electron acceleration event, *Geophys. Res. Lett.*, **41**, 7, doi: 10.1002/2014gl059626.

Chapman, S., and V. C. A. Ferraro (1930), A new theory of magnetic storms, *Nature*, **126**, 129–130.

Chapman, S., and V. C. A. Ferraro (1931), A new theory of magnetic storms, I. The initial phase, *Terrest. Magn. Atmosph. Elec.*, **36**, 77–97, 171–186.

Chen, L., et al. (2014), Generation of unusually low frequency plasmaspheric hiss, *Geophys. Res. Lett.*, **41**, 16, doi:10.1002/2014gl060628.

Claudepierre, S. G., et al. (2013), Van Allen Probes observation of localized drift resonance between poloidal mode ultra-low frequency waves and 60 kev electrons, *Geophys. Res. Lett.*, **40**, 17, doi:10.1002/Grl.50901.

Clilverd, M. A., et al. (2010), Energetic outer radiation belt electron precipitation during recurrent solar activity, *J. Geophys. Res.*, **115**, A8, doi:10.1029/2009ja015204.

Daglis, I. A. (1997), The role of magnetosphere-ionosphere coupling in magnetic storm dynamics, in *Magnetic Storms, Geophys. Monogr. Ser., vol. 98*, edited by B. T. Tsurutani et al., pp. 107–116, AGU, Washington, D. C.

Daglis, I. A., E. T. Sarris, and B. Wilken (1993), AMPTE/CCE CHEM observations of the ion population at geosynchronous altitudes, *Ann. Geophys.*, **11**, 685–696.

Daglis, I. A., R. M. Thorne, W. Baumjohann, and S. Orsini (1999), The terrestrial ring current: Origin, formation, evolution and decay, *Rev. Geophys.*, **37**(4), 407–438, doi:10.1029/1999RG900009.

Daglis, I.A. (2006), Ring current dynamics, *Space Sci. Rev.*, **124**, 183–202, doi: 10.1007/s11214-006-9104-z.

Dai, L., et al. (2013), Excitation of poloidal standing alfvén waves through drift resonance wave-particle interaction, *Geophys. Res. Lett.*, **40**, 16, doi:10.1002/grl.50800.

Denton, M. H., and J. E. Borovsky (2012), Magnetosphere response to high-speed solar wind streams: A comparison of weak and strong driving and the importance of extended periods of fast solar wind, *J. Geophys. Res.*, **117**, doi:10.1029/2011ja017124.

Dungey, J. W. (1963), Loss of van allen electrons due to whistlers, *Planet. Space Sci.*, **11**, 6, doi:10.1016/0032-0633(63)90166-1.

Fälthammar, C. G. (1965), Effects of time-dependent electric fields on geomagnetically trapped radiation, *J. Geophys. Res.*, **70**, 11, doi:10.1029/Jz070i011p02503.

Fennell, J. F., et al. (2015), Van Allen Probes show that the inner radiation zone contains no MeV electrons: ECT/MagEIS data, *Geophys. Res. Lett.*, **42**, 5, doi:10.1002/2014gl062874.

Foster, J. C., et al. (2014), Prompt energization of relativistic and highly relativistic electrons during a substorm interval: Van Allen Probes observations, *Geophys. Res. Lett.*, **41**, 1, doi:10.1002/2013gl058438.

Frank, L. A. (1967), On the extraterrestrial ring current during geomagnetic storms, *J. Geophys. Res.*, 72, 3753–3767.

Friedel, R. H. W., et al. (1999), A multi-spacecraft synthesis of relativistic electrons in the inner magnetosphere using LANL, GOES, GPS, SAMPEX, HEO and Polar, *Radiat. Meas.*, **30**, 5, doi:10.1016/S1350-4487(99)00230-9.

Friedel, R.H.W., et al. (2002), Relativistic electron dynamics in the inner magnetosphere - a review, *J. Atmos. Sol.-Terr. Phy.*, **64**, 2, doi:Pii S1364-6826(01)00088-8.

Ge, Y.S., J.P. McFadden, J. Raeder, V. Angelopoulos, D. Larson, and O.D. Constantinescu (2011), Case studies of mirror-mode structures observed by THEMIS in the near-Earth tail during substorms, *J. Geophys. Res.*, **116**, A01209, doi:10.1029/2010JA015546.

Goodrich, C. C., et al. (1998), An overview of the impact of the January 10–11 1997 magnetic cloud on the magnetosphere via global MHD simulation, *J. Geophys. Res.*, **25**, 14, doi:10.1029/98GL01159.

Hamilton, D. C., G. Gloeckler, F. M. Ipavich, W. Stüdemann, B. Wilken, and G. Kremser (1988), Ring current development during the great geomagnetic storm of February 1986, *J. Geophys. Res.*, **93**, 14,343–14,355.

Hietala, H., et al. (2014), Depleting effects of ICME-driven sheath regions on the outer electron radiation belt, *Geophys. Res. Lett.*, **41**, 7, doi:10.1002/2014gl059551.

Horne, R. B., and R. M. Thorne (1998), Potential waves for relativistic electron scattering and stochastic acceleration during magnetic storms, *Geophys. Res. Lett.*, **25**, 15, doi:10.1029/98GL01002.

Horne, R. B., et al. (2000), Proton and electron heating by radially propagating fast magnetosonic waves, *J. Geophys. Res.*, **105**, 27597–27610.

Horne, R. B. (2002), The contribution of wave particle interactions to electron loss and acceleration in the earth's radiation belts during geomagnetic storms, in *Review of Radio Science 1999–2002*, edited by W. R. Stone, chapter 33, pp. 801–828, Wiley, New York City.

Horne, R. B., and R. M. Thorne (2003), Relativistic electron acceleration and precipitation during resonant interactions with whistler-mode chorus, *Geophys. Res. Lett.*, **30**, 10, doi:10.1029/2003GL016973.

Horne, R. B., et al. (2005a), Timescale for radiation belt electron acceleration by whistler mode chorus waves, *J. Geophys. Res.: Space.*, **110**, A3, doi:10.1029/2004ja010811.

Horne, R. B., et al. (2005b), Wave acceleration of electrons in the van allen radiation belts, *Nature*, **437**, 8, doi:doi:10.1038/nature03930.

Hudson, M. K., et al. (2000), Increase in relativistic electron flux in the inner magnetosphere: Ulf wave mode structure, *Adv. Sp. Res.*, **25**, 12, doi:10.1016/S0273-1177(99)00518-9.

Hudson, M. K., et al. (2001), Radiation belt electron acceleration by ULF wave drift resonance: Simulation of 1997 and 1998 storms in *Space Weather*, Geophys. Monogr. Ser., vol. 125, edited by P. Song et al., pp. 289–296, AGU, Washington, D. C.

Hudson, M. K., et al. (2014), Simulated magnetopause losses and Van Allen Probe flux dropouts, *Geophys. Res. Lett.*, doi:10.1002/2014gl059222.

Jordanova, V. K., et al. (1998), Effect of wave-particle interactions on ring current evolution for January 10 - 11, 1997: Initial results, *Geophys. Res. Lett.*, **25**(15), 2971–2974, 10.1029/98GL00649.

Jordanova, V. K., et al. (1999), Ring current activity during the early Bz<0 phase of the January 1997 magnetic cloud, *J. Geophys. Res.*, **104**(A11), 24895–24914, doi: 10.1029/1999JA900339.

Kamide, Y., J.-H. Shue, X. Li, G. Lu, M. Brittnacher, G. Parks, and G. D. Reeves (1998), Internally and externally triggered substorms: A case study of the January 10, 1997 events, in *Proceedings of the 4th International Conference on Substorms*, edited by S. Kokubun and Y. Kamide, pp. 305–309, Kluwer Academic Publishers.

Kellogg, P. J. (1959), Possible explanation of the radiation observed by Van Allen at high altitude in satellites, *Il Nuovo Cimento*, **11**, pp. 48–66.

Kennel, C. F., and H. E. Petschek (1966), Limit on stably trapped particle fluxes, *J. Geophys. Res.*, **71**, 1, doi:10.1029/JZ071i001p00001.

Kessel, R. L., et al. (2012), The Radiation Belt Storm Probes (RBSP) and space weather, *Space Sci. Rev.*, doi:10.1007/s11214-012-9953-6.

Kilpua, E. K. J., H. Hietala, D. L. Turner, H. E. J. Koskinen, T. I. Pulkkinen, J. V. Rodriguez, G. D. Reeves, S. G. Claudepierre, and H. E. Spence (2015), Unraveling the drivers of the storm time radiation belt response, *Geophys. Res. Lett.*, **42**, doi:10.1002/2015GL063542.

Kim, H. J., and A. A. Chan (1997), Fully adiabatic changes in storm-time relativistic electron fluxes, *J. Geophys. Res.: Space*, **102**, A10, doi:10.1029/97ja01814.

Kletzing, C. A., et al. (2013), The electric and magnetic field instrument suite and integrated science (EMFISIS) on RBSP, *Space Sci. Rev.*, doi:10.1007/s11214-013-9993-6.

Li, W., et al. (2014a), Evidence of stronger pitch angle scattering loss caused by oblique whistler-mode waves as compared with quasi-parallel waves, *Geophys. Res. Lett.*, doi:10.1002/2014gl061260.

Li, W., et al. (2014b), Quantifying hiss-driven energetic electron precipitation: A detailed conjunction event analysis, *Geophys. Res. Lett.*, **41**, 4, doi:10.1002/2013gl059132.

Li, W., et al. (2014c), Radiation belt electron acceleration by chorus waves during the 17 March 2013 storm, *J. Geophys. Res.*, **119**, 6, doi: 10.1002/2014ja019945.

Li, X. L., et al. (1997), Multisatellite observations of the outer zone electron variation during the November 3–4, 1993, magnetic storm, *J. Geophys. Res.*, **102**, A7, doi: 10.1029/97ja01101.

Li, Z., et al. (2014d), Radial diffusion comparing a THEMIS statistical model with geosynchronous measurements as input, *J. Geophys. Res.: Space Phys.*, doi:10.1002/2013ja019320.

Liu, S., et al. (2015), Van Allen Probes observations linking radiation belt electrons to chorus waves during 2014 multiple storms, *J. Geophys. Res.: Space Phys.*, *120*, 938–948, doi:10.1002/2014JA020781.

Lyons, L. R., and R. M. Thorne (1973), Equilibrium structure of radiation belt electrons, *J. Geophys. Res.*, 78, 13, doi:10.1029/Ja078i013p02142.

Lyons, L. R., et al. (2000), Auroral disturbances during the January 10, 1997 magnetic storm, *Geophys. Res. Lett.*, **27**, 20, doi:10.1029/1999gl000014.

Ma, Q., et al. (2014), The trapping of equatorial magnetosonic waves in the earth's outer plasmasphere, *Geophys. Res. Lett.*, **41**, 18, doi:10.1002/2014gl061414.

Malaspina, D. M., et al. (2014), Nonlinear electric field structures in the inner magnetosphere, *Geophys. Res. Lett.*, **41**, 16, doi:10.1002/2014gl061109.

Mann, I. R., et al. (2013), Discovery of the action of a geophysical synchrotron in the earth's van allen radiation belts, *Nat. Commun.*, **4**, Doi:10.1038/Ncomms3795.

Mauk, B. H., et al. (2012), Science objectives and rationale for the Radiation Belt Storm Probes mission, *Space Sci. Rev.*, doi:10.1007/s11214-012-9908-y.

Millan, R. M., and R. M. Thorne (2007), Review of radiation belt relativistic electron loss, *J. Atmos. Sol.-Terr. Phys.*, **69**.

Miyoshi, Y. S., et al. (2004), Solar cycle variations of the electron radiation belts: Observations and radial diffusion simulation, *Space Weather*, **2**, S10S02, doi:10.1029/2004SW000070.

Miyoshi, Y., and R. Kataoka (2005), Ring current ions and radiation belt electrons during geomagnetic storms driven by coronal mass ejections and corotating interaction regions, *Geophys. Res. Lett.*, **32**, 21, doi:10.1029/2005gl024590.

Morley, S. K., et al. (2013), Phase space density matching of relativistic electrons using the Van Allen Probes: REPT results, *Geophys. Res. Lett.*, **40**, 18, doi:10.1002/Grl.50909.

Mozer, F. S., et al. (2014), Direct observation of radiation-belt electron acceleration from electron-volt energies to megavolts by nonlinear whistlers, *Phys. Rev. Lett.*, **113**, 3, doi:10.1103/PhysRevLett.113.035001.

Ni, B. B., et al. (2014), Resonant scattering of energetic electrons by unusual low- frequency hiss, *Geophys. Res. Lett.*, **41**, 6, doi:10.1002/2014gl059389.

Nishida, A. (1976), Outward diffusion of energetic particles from the Jovian radiation belt, *J. Geophys. Res.*, **81**, 10, doi:10.1029/JA081i010p01771.

Northrop, T., and E. Teller (1960), Stability of the adiabatic motion of charged particles in the earth's field, *Phys. Rev.*, **117**, 1, doi:10.1103/PhysRev.117.215.

O'Brien, T. P., et al. (2003), Energization of relativistic electrons in the presence of ULF power and MeV microbursts: Evidence for dual ULF and VLF acceleration, *J. Geophys. Res.*, **108** (A8), 1329, doi:10.1029/2002JA009784.

Omura, Y., et al. (2007), Relativistic turning acceleration of resonant electrons by coherent whistler mode waves in a dipole magnetic field, *J. Geophys. Res.*, **112**, A06236, doi:10.1029/2006JA012243.

Omura, Y., et al. (2008), Theory and simulation of the generation of whistler-mode chorus, *J. Geophys. Res.*, **113**, A04223, doi:10.1029/2007JA012622.

Onsager, T. G., et al. (2002), Radiation belt electron flux dropouts: Local time, radial, and particle-energy dependence, *J. Geophys. Res.*, **107**, A11, doi: 10.1029/2001ja000187.

Ozeke, L. G., and I. R. Mann (2008), Energization of radiation belt electrons by ring current ion driven ULF waves, *J. Geophys. Res.*, **113**, A02201, doi:10.1029/2007JA012468.

Paulikas, G. A., and J. B. Blake (1979), Effects of the solar wind on magnetospheric dynamics: Energetic electrons at the synchronous orbit, in *Quantitative Modeling of Magnetospheric Processes*, edited by W.P. Olson, pp. 180–202.

Reeves, G. D. (1998), Relativistic electrons and magnetic storms: 1992–1995, *Geophys. Res. Lett.*, **25**, 11, doi:10.1029/98gl01398.

Reeves, G. D., et al. (1998a), The global response of relativistic radiation belt electrons to the January 1997 magnetic cloud, *Geophys. Res. Lett.*, **25**, 17, doi:10.1029/98gl02509.

Reeves, G. D., et al. (1998b), The relativistic electron response at geosynchronous orbit during the January 1997 magnetic storm, *J. Geophys. Res.*, **103**, A8, doi:10.1029/97JA03236.

Reeves, G. D., et al. (2003), Acceleration and loss of relativistic electrons during geomagnetic storms, *Geophys. Res. Lett.*, **30**, 10, doi:10.1029/2002gl016513.

Reeves, G. D., et al. (2012), Dynamic radiation environment assimilation model: DREAM, *Space Weather*, **10**, doi:10.1029/2011sw000729.

Reeves, G. D., et al. (2013), Electron acceleration in the heart of the van allen radiation belts, *Science*, **341**, 6149, doi:10.1126/science.1237743.

Reeves, G. D. (2015), Radiation belt electron acceleration and role of magnetotail, in *Magnetotails in the Solar System*, edited by A. Keiling, C. M. Jackman, and P. A. Delamere, John Wiley & Sons, Inc., doi: 10.1002/9781118842324.

Reeves, G. D., et al. (2016), Energy-dependent dynamics of keV to MeV electrons in the inner zone, outer zone, and slot regions, *J .Geophys. Res.: Space Phys.*, **121**, 1, doi:10.1002/2015ja021569.

Roederer, J. G. (1970), *Dynamics of Geomagnetically Trapped Radiation*, Springer-Verlag, New York.

Russell, C. T., and R. L. McPherron (1973), Semiannual variation of geomagnetic activity, *J. Geophys. Res.*, 78, 1, doi:10.1029/JA078i001p00092.

Schiller, Q., et al. (2014), A nonstorm-time enhancement of relativistic electrons in the outer radiation belt, *Geophys. Res. Lett.*, **41**, 1, doi:10.1002/2013gl058485.

Schulz, M., and L. J. Lanzerotti (1974), *Particle Diffusion in the Radiation Belts*, Springer-Verlag, New York.

Schulz, M. (1996), Canonical coordinates for radiation belt modeling, in *Radiation Belt Models and Standards*, edited by J. F. Lemaire et al., pp. 153–161, American Geophysical Union, Washington DC.

Selesnick, R. S., et al. (2007), A theoretical model of the inner proton radiation belt, *Space Weather*, 5, S04003, doi:10.1029/2006SW000275.

Shprits, Y. Y., et al. (2006), Bounce-averaged diffusion coefficients for field-aligned chorus waves, *J. Geophys. Res.*, **111**, A10225, doi:10.1029/2006JA011725.

Shprits, Y. Y., et al. (2013), Unusual stable trapping of the ultrarelativistic electrons in the van allen radiation belts, *Nat. Phys.*, **9**, 11, doi:10.1038/Nphys2760.

Singer, S. F. (1958), Trapped albedo theory of the radiation belt, *Phys. Rev. Lett.*, **1**, 5, doi:10.1103/Physrevlett.1.181.

Su, Z., et al. (2014a), Quantifying the relative contributions of substorm injection and chorus waves to the rapid outward extension of electron radiation belt, *J. Geophys. Res.*, *submitted*, doi:10.1002/2014JA020709.

Su, Z. P., et al. (2014b), Nonstorm-time dynamics of electron radiation belts observed by the Van Allen Probes, *Geophys. Res. Lett.*, **41**, 2, doi:10.1002/2013gl058912.

Summers, D., et al. (1998), Relativistic theory of wave-particle resonant diffusion with application to electron acceleration in the magnetosphere, *J. Geophys. Res.*, **103**, A9, doi:10.1029/98JA01740.

Summers, D., and C. Ma (2000), A model for generating relativistic electrons in the earth's inner magnetosphere based on gyroresonant wave-particle interactions, *J. Geophys. Res.* **105**.

Summers, D., et al. (2002), Model of the energization of outer-zone electrons by whistler-mode chorus during the October 9, 1990, geomagnetic storm, *Geophys. Res. Lett.*, **29**, 24, doi:10.1029/2002gl016039.

Summers, D., et al. (2004), Competition between acceleration and loss mechanisms of relativistic electrons during geomagnetic storms, *J. Geophys. Res.*, **109**, A04221, doi:10.1029/2004JA010437.

Summers, D., et al. (2007), Timescales for radiation belt electron acceleration and loss due to resonant wave-particle interactions: 1. Theory, *J. Geophys. Res.*, **112**, A04206, doi:10.1029/2006JA011801.

Thorne, R. M. (2010), Radiation belt dynamics: The importance of wave-particle interactions, *Geophys. Res. Lett.*, **37**, doi:10.1029/2010gl044990.

Thorne, R. M., et al. (2013), Rapid local acceleration of relativistic radiation belt electrons by magnetospheric chorus, *Nature*, **504**, 7480, doi:10.1038/nature12889.

Tsyganenko, N. A., and M. I. Sitnov (2005), Modeling the dynamics of the inner magnetosphere during strong geomagnetic storms, *J. Geophys. Res.*, **110**, doi:10.1029/2004JA010798.

Tu, W. C., et al. (2014), Event-specific chorus wave and electron seed population models in DREAM3D using the Van Allen Probes, *Geophys. Res. Lett.*, **41**, 5, doi:10.1002/2013gl058819.

Turner, D.L., and X. Li (2008), Radial gradients of phase space density of the outer radiation belt electrons prior to sudden solar wind pressure enhancements, *Geophys. Res. Lett.*, **35**, L18101, doi:10.1029/2008GL034866.

Turner, D. L., V. Angelopoulos, Y. Shprits, A. Kellerman, P. Cruce, and D. Larson (2012), Radial distributions of equatorial phase space density for outer radiation belt electrons, *Geophys. Res. Lett.*, **39**, L09101, doi:10.1029/2012GL051722.

Turner, D. L., et al. (2014a), Competing source and loss mechanisms due to wave-particle interactions in Earth's outer radiation belt during the 30 September to 3 October 2012 geomagnetic storm, *J. Geophys. Res.: Space Phys.*, **119**, 3, doi:10.1002/2014ja019770.

Turner, D. L., et al. (2014b), On the cause and extent of outer radiation belt losses during the 30 September 2012 dropout event, *J. Geophys. Res.: Space Phys.*, 119, 3, doi:10.1002/2013ja019446.

Tverskaya, L. V. (1986), On the boundary of electron injection into the magnetosphere, *Geomagn. Aeron.*, **26**, 5.

Tverskaya, L. V., et al. (2003), Predicting the L-position of the storm-injected relativistic electron belt, *Adv. Sp. Res.*, **31**(4).

Usanova, M. E., et al. (2014), Effect of EMIC waves on relativistic and ultrarelativistic electron populations: Ground-based and Van Allen Probes observations, *Geophys. Res. Lett.*, **41**, 5, doi:10.1002/2013gl059024.

Van Allen, J. A., et al. (1958), Observation of high intensity radiation by satellites 1958 Alpha and Gamma, *Jet Propulsion*, **28**, 588–592.

Van Allen, J. A. (1959), The geomagnetically trapped corpuscular radiation, *J. Geophys. Res.*, **64**, 1683–1689.

Vernov, S. N., et al. (1959), Possible mechanism of production of terrestrial corpuscular radiation under the action of cosmic rays, *Sov. Phys., Dokl.,* **4**, 154.

Walt, M. (1996), Source and loss processes for radiation belt particles, in *Radiation Belt Models and Standards,* edited by J. F. Lemaire et al., pp. 1–15, American Geophysical Union, Washington DC.

Williams, D. J. (1981), Ring current composition and sources: An update, *Planet. Space Sci.,* **29**, 1195-1203.

Winckler, J. R., et al. (1962), A study of the precipitation of energetic electrons from the geomagnetic field during magnetic storms, *J. Geophys. Res.,* **67**, 10, doi:10.1029/JZ067i010p03717.

Wygant, J. R., et al. (2014), The electric field and waves instruments on the Radiation Belt Storm Probes mission, in *The Van Allen Probes Mission,* edited by N. Fox and J. Burch, pp. 183–220, Springer US.

Xiao-Fang, Z., et al. (2013), Effect of interplanetary disturbances and geomagnetic activities on relativistic electrons at geosynchronous orbit, *Chin. J. Geophys.,* **56**, 5, doi:10.1002/cjg2.20051.

Zhang, J. C., et al. (2014), Excitation of EMIC waves detected by the Van Allen Probes on 28 April 2013, *Geophys. Res. Lett.,* **41**, 12, doi:10.1002/2014gl060621.

Zhao, H., et al. (2014), Peculiar pitch angle distribution of relativistic electrons in the inner radiation belt and slot region, *Geophys. Res. Lett.,* **41**, 7, doi:10.1002/2014gl059725.

Zhou, Q., et al. (2014), Excitation of nightside magnetosonic waves observed by Van Allen Probes, *J. Geophys. Res. Space Phys.,* **119**, 9125–9133, doi:10.1002/2014JA020481.

4

The Role of Pc-5 ULF Waves in the Radiation Belts: Current Understanding and Open Questions

Scot R. Elkington and Theodore E. Sarris

■ Abstract

Pc-5 frequency magnetospheric ULF waves have been closely tied to dynamical variations in the energetic electron fluxes observed in the Van Allen radiation belt. The mode of interaction between these global waves and the trapped energetic particles comprising the belts is through a radial transport process that results from breaking of the drift invariant of trapped particle motion, while conserving the bounce and gyro invariants. Most often this is described as a stochastic process, whereby particles being diffusively transported to regions of stronger magnetic field strength increase their energy, while those transported outward to regions of lower magnetic field strength lose energy. In this chapter we describe the properties of ULF waves required to quantitatively determine their effect on radiation belt particles.

4.1 The role of Pc-5 waves in the radiation belts

Variations in the strength and direction of magnetic and electric fields in near-Earth space occur in a wide range of frequencies and plasma wave modes. In particular,

Waves, Particles, and Storms in Geospace. First Edition. Georgios Balasis, Ioannis A. Daglis, Ian R. Mann.

a category of magnetohydrodynamic (MHD) waves known as Pc-5 ULF pulsations (Jacobs et al., 1964) occur at mHz frequencies, and have been established as being strongly correlated with variations in the trapped energetic particle populations that comprise the Van Allen radiation belts (e.g. Baker et al., 1998; Rostoker et al., 1998; Mathie and Mann, 2000; Mann et al., 2004).

In ideal MHD theory a cold plasma will support two basic wave modes: the magnetosonic or compressional wave, and the shear Alfvén wave (e.g. Sturrock, 1994). In their simplest description, the Alfvén mode consists of transverse oscillations propagating parallel to the local magnetic field, while the magnetosonic mode is an isotropically-propagating transverse oscillation. In a cylindrically-symmetric geometry the shear wave is characterized by purely azimuthal magnetic field perturbations, B_ϕ, and a radial electric field, E_r, and is commonly referred to as a toroidal-mode Alfvén wave. The compressional wave in this geometry is characterized by oscillations in the total magnetic field strength B_z and an azimuthal electric field E_ϕ, and is referred to as a poloidal-mode wave.

ULF waves in the Pc-5 category may be driven either by internal (magnetospheric) plasma processes, or may originate in the dynamic interactions of the Earth's magnetosphere with the solar wind. Internally-generated ULF waves may result from mirror (Hasegawa, 1969) or drift-bounce (Southwood et al., 1969; Chen and Hasegawa, 1991) instabilities, or as a result of anisotropies in the perpendicular ring current (Takahashi et al., 1985). Externally-driven ULF waves may result from shear flow instabilities along the flanks of the magnetopause (Kivelson and Pu, 1984; Cahill and Winckler, 1992; Mann et al., 1999; Claudepierre et al., 2008), as a result of variations in solar wind pressure (Kivelson and Southwood, 1988; Kepko and Spence, 2003; Claudepierre et al., 2009; Sarris et al., 2010), or as a result of variations in the large-scale convective motion of the magnetosphere driven by changes in the global magnetospheric reconnection rate and the flow of solar wind around the magnetopause (Ridley et al., 1998; Ruohoniemi and Greenwald, 1998).

There are a number of reviews in the literature that discuss proposed modes of interaction between radiation belt particles and Pc-5 ULF waves (e.g. Friedel et al., 2002; Elkington, 2006; Shprits et al., 2008). Large-scale field variations at frequencies commensurate with the particle drift frequency break the drift invariant while maintaining the gyro and bounce invariants associated with trapped particle motion (Roederer, 1970), resulting in a radial drift across field lines and concurrent change in energy. Radial diffusion is a particular case that describes the stochastic motion of an ensemble of particles under the effect of a continuum of ULF waves near the particle drift frequency. Depending on the global particle distribution with respect to the radial coordinate, radial diffusion may act to either increase flux levels in the inner magnetosphere as particles diffuse inward from regions of higher phase space density near the trapping boundary (e.g. Hudson et al., 2012), or, if there is a paucity of electrons at the trapping boundary (e.g. Green et al., 2004; Shprits et al., 2006; Loto'aniu et al., 2010; Turner et al., 2012), act as a loss process as the aggregate particles drift outward and are lost to the magnetopause.

In this diffusive case, the effect of ULF waves on trapped energetic particles may be quantitatively determined by solving the Fokker-Planck equation, including appropriate diffusion coefficients governing the radial transport of particles that results from interaction with global ULF waves. For field variations that break the third adiabatic invariant only (Schulz, 1996), the Fokker-Planck equation reduces to the well-known radial diffusion form,

$$\frac{\partial f}{\partial t} = L^2 \frac{\partial}{\partial L}\left[\frac{D_{LL}}{L^2}\frac{\partial f}{\partial L}\right] - \frac{f}{\tau_L}, \tag{4.1}$$

where f is the drift-averaged distribution function or reduced phase space density and τ_L represents the particle lifetime.

Early analytic formulations of the relevant transport coefficients D_{LL} have been given by Fälthammar (1965, 1966a, b, 1968), Cornwall (1968), and Schulz and Lanzerotti (1974), among others. Assuming a time-varying potential (i.e. curl-free) electric field, such as may result from the uniform global convection associated with the Dungey cycle (Dungey, 1961), the form of the radial diffusion coefficient is

$$D_{LL}^E = \frac{1}{8B_E^2 R_E^2} L^6 \sum_m P_m^E (m\omega_d). \tag{4.2}$$

The transport coefficients that result from variations in the magnetic field is given as

$$D_{LL}^B = \frac{M^2}{8q^2\gamma^2 B_E^2 R_E^4} L^4 \sum_m m^2 P_m^B (m\omega_d), \tag{4.3}$$

(Fei et al., 2006). In these expressions, m is the azimuthal mode number of the ULF waves driving the diffusion, M is the first adiabatic invariant, γ is the Lorentz factor, q is the particle charge, $B_E = 0.31$ Gauss, and R_E is an Earth radius. $\mathcal{P}_m^B$ and $\mathcal{P}_m^E$ are the power spectral densities of the compressional magnetic field and the potential electric field, both evaluated at the resonant wave frequency

$$\omega = m\omega_d, \tag{4.4}$$

where ω_d is the particle drift frequency (Elkington et al., 1999, 2003). It should be noted that this drift-resonant frequency condition specifies the population of particles that will be affected by a given frequency band of ULF waves, since the drift frequency is energy-dependent.

Note that the form (4.3) is relativistically correct and general in terms of the power spectral density, in contrast to the form offered by Fälthammar (1966a, 1968), which was for nonrelativistic particles and assumed a power spectrum that decreased as ω^2. Under the assumptions used in Fälthammar, an explicit L^{10} dependence appears in the diffusion coefficient, in contrast to the L^4 dependence exhibited in Equation (4.3).

4.2 Questions: Quantitative determination of transport rates

There have been numerous and ongoing efforts to quantify the radial transport rates suggested in Equations (4.2) and (4.3) based on observations of electric and magnetic field variations. For example, Brautigam et al. (2005) used electric field measurements from the CRRES spacecraft to calculate radial diffusion coefficients based on a functional fit of $\mathcal{P}_m^E$ as a function of radial position, L, and the magnetospheric activity parameter K_p. They found that the diffusion coefficient, D_{LL}, could vary by 1–2 orders of magnitude between low activity ($K_p = 1$) and high activity ($K_p = 6$), depending on L and the first (gyro) invariant. However, the nature of the single-point electric field measurements forced them to consider only the diffusive effects of the curl-free part of the electric field variations, to assume only the $m = 1$ azimuthal resonance, and that the power was uniformly distributed in local time. Rae et al. (2012) used ground magnetometers to examine the local time distribution of ULF wave power in the electric field as a function of solar wind speed, and Ozeke et al. (2012) used a combination of ground- and space-based observations to extend these analyses to magnetic field fluctuations. In this latter work they obtained coefficients using assumed values of m, held constant at all radial distances and frequencies. These results also contrasted with earlier empirical efforts to quantify radial diffusion by Brautigam and Albert (2000), in that the effects of electric field activity at ULF frequencies were determined to be more important than the magnetic field variations in driving particle dynamics.

Key to quantitatively describing the effect of radial transport on radiation belt particles is understanding the global spectral and spatiotemporal characteristics of magnetospheric ULF waves. In the remainder of this chapter we outline the fundamental questions that must be addressed to fully describe the relevant radial diffusion transport coefficients, Equations (4.2) and (4.3), and discuss recent progress in understanding the characteristics of ULF waves and their resulting effects on radiation belt dynamics.

4.2.1 What is the power spectrum as a function of frequency?

There are several explicit and implicit functional dependencies in Equations (4.2) and (4.3). The power spectral density terms $\mathcal{P}_m^{E,B}$ prominently suggest the need for a quantitative measure of the ULF wave activity. In particular, this activity should be quantified at the drift resonant frequency given by Equation (4.4) for the particle population in question, rather than a frequency-integrated measure of the wave spectrum. Appropriate measures of the power spectral density for use in Equations (4.2) and (4.3) can be obtained in a relatively straightforward fashion using the periodogram spectral density methods described in Press et al. (1992).

The means of measuring the relevant electric and magnetic field fluctuations presents a main challenge in calculating relevant diffusion coefficients. To understand the transport problem throughout the radiation belts, one would ideally have access to the time-dependent field fluctuations at multiple simultaneous locations throughout space. In practice, measurements are limited to what one can infer about ULF fluctuations in space from ground magnetometers (e.g. Mann et al., 2008), and from in situ observations located at only a few points in space. Geosynchronous orbit provides an appropriate platform for simultaneous measurements of magnetic ULF power at a number of local times along a nominal particle drift orbit (e.g. Perraut et al., 1978; Singer et al., 1996; Huang et al., 2010; Clausen et al., 2011; Ozeke et al., 2012). For example, Bloom and Singer (1995) used geosynchronous measurements to examine trends in the power spectral density as a function of frequency, finding generally increasing power with decreasing frequency in the Pc-2 (~0.1 Hz) through Pc-5 (mHz) bands, as indicated in Figure 4.1.

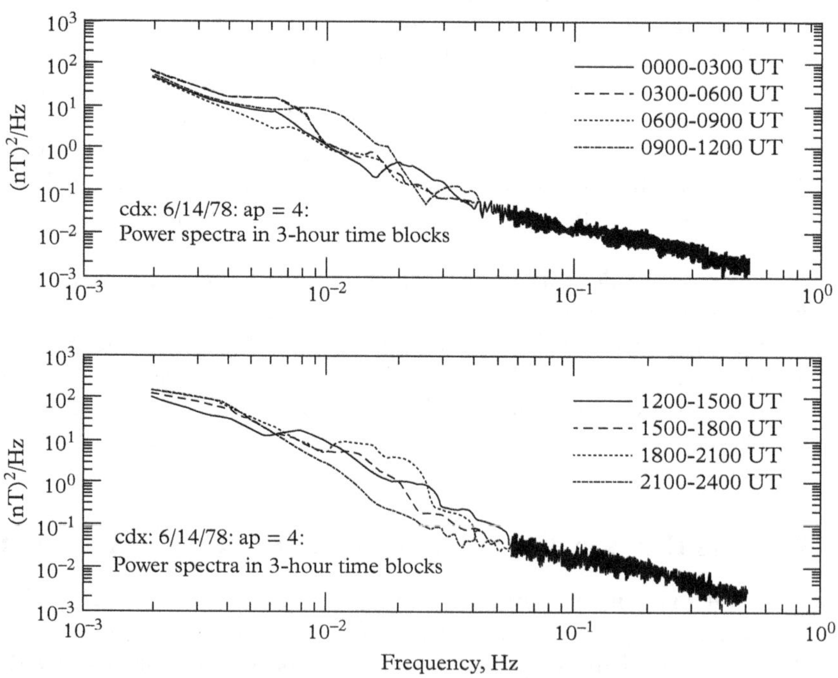

Figure 4.1 *Magnetic ULF wave power at geosynchronous orbit as a function of frequency. After Bloom and Singer (1995).*

We note that the power spectral density as a function of wave frequency alone is insufficient to determine the resonant frequency at a particular point in space given by Equation (4.4) for a given particle population; one also needs some estimate of the distribution of energy across azimuthal modes, discussed below. However, in low-mode number situations such as are afforded in "breathing mode" magnetic fluctuations driven by solar wind pressure variations, an $m = 1$ assumption may be sufficient to quantitatively tie the wave frequency to the particle drift frequency (Tu et al., 2012).

4.2.2 What is the radial profile of the ULF activity?

In addition to the explicit radial dependence appearing in each term in Equations (4.2) and (4.3) (i.e. L^4 in the magnetic term, L^6 in the electric term), the power spectral density terms $\mathcal{P}_m^{E,B}$ have an implicit radial dependence. This may be particularly pronounced in regions near Field Line Resonances (FLRs), quasisinusoidal signals with relatively high wave power at confined radial distances. The absence or presence of ULF wave power at radial locations relative to any peaks in the phase space density profile (e.g. Shprits et al., 2006; Turner et al., 2012) of trapped radiation belt electrons determine whether Pc-5 ULF waves will act as an acceleration or loss process in the radiation belts. Thus, it is necessary to understand the radial structure of ULF wave power in the magnetosphere to determine the effect of transport in the radiation belts.

The CRRES (Wygant et al., 1992; Singer et al., 1992), Polar (Russell, 1995), Cluster (Escoubet et al., 1997), and THEMIS (Sibeck and Angelopoulos, 2008) missions are among those that have provided insight into electric and magnetic field power at different L values along their respective orbital tracks (Brautigam et al., 2005; Sarris et al., 2009a; Liu et al., 2009a, 2010; Loto'aniu et al., 2010; Ozeke et al., 2012), providing a global picture of ULF wave power spectral density in local time and L over the course of their respective precession times. An example from Liu et al. (2009b) is indicated in Figure 4.2, which shows magnetic and electric field amplitudes and power spectral densities along a single radial profile traversed by the spacecraft THEMIS-D from the nominal plasmapause position at $\sim$4.5 R_E to the magnetopause at $\sim$9.5 R_E. Although there are particular peaks in the power spectral density corresponding to the coherent FLR activity evident in the field amplitude panels, the general trend is for higher power at lower frequency, consistent with the geosynchronous observations suggested in Figure 4.1. Another noticeable feature is for the peak power spectral density to trend to lower frequencies as the spacecraft traverses to higher L values, reflecting that the resonant frequency for a given field line (at constant density) decreases as the Alfvén speed decreases with decreasing B, and also decreases as the total field line length increases (Kivelson and Southwood, 1988). Finally, the power increases somewhat with increasing radial distance, an effect most clearly pronounced in the radial and compressional magnetic field components at L values approaching the magnetopause.

Ionospheric conductivity effects will tend to modify the polarization and amplitude of ULF wave activity observed on the ground (Hughes, 1974), but not the frequency spectrum of the fluctuations. This allows a relative measure of space-based magnetic

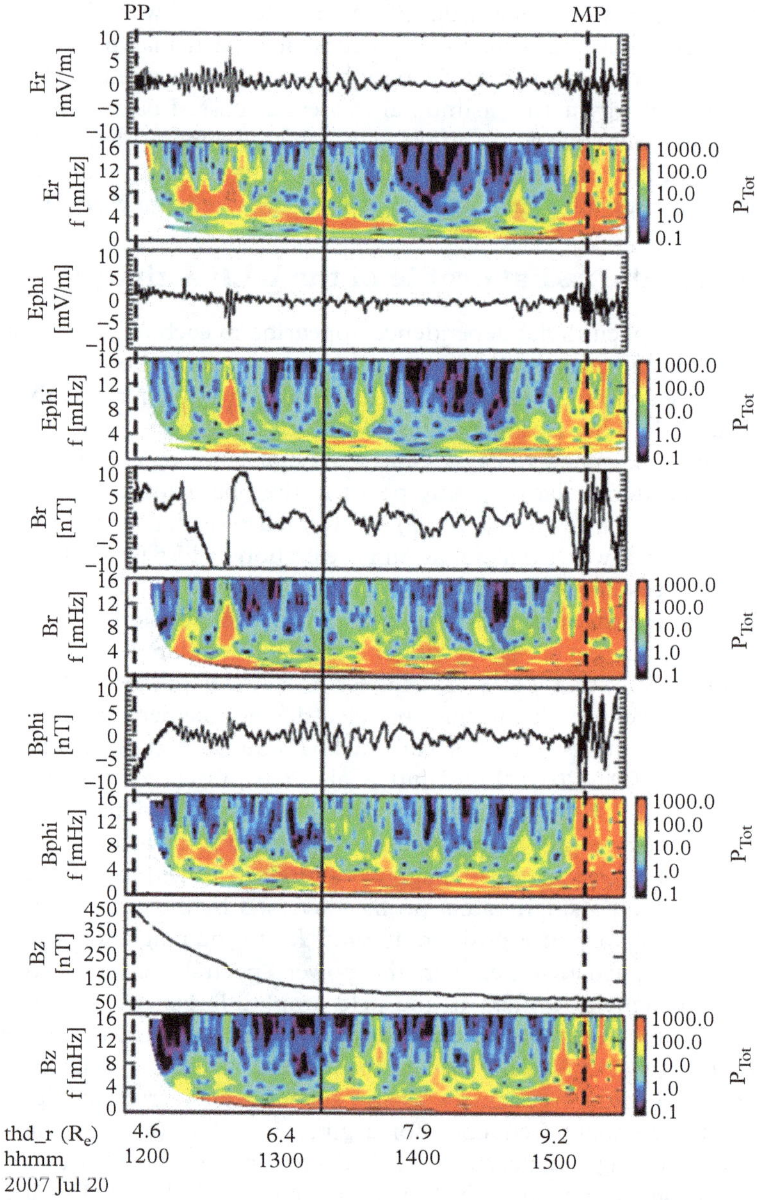

Figure 4.2 *Magnetic and electric ULF wave power along a radial cut from the THEMIS-D spacecraft (Liu et al., 2009b).*

field variations over a wide range of latitudes and local times, but does not provide absolute amplitude of the associated compressional magnetic field components in space required in Equation (4.3). However, the work of Ozeke et al. (2009) provides a means of mapping magnetometer fluctuations corresponding to guided poloidal and toroidal mode Alfvénic electric fluctuations (Kivelson and Southwood, 1988). This technique allows simultaneous observation of the equatorial electric field fluctuations over a distribution of local times and radial distances in space. Rae et al. (2012) and Ozeke et al. (2012) have used this technique to investigate the global dependence of ULF electric wave power on local time, radial distance, and geomagnetic activity based on magnetometer observations from the CARISMA magnetometer array. An example is indicated in Figure 4.3, which for moderate driving conditions again indicates the features discussed above; namely, decreasing power with increasing frequency and increasing power with increasing radial distance.

4.2.3 What is the azimuthal mode structure?

Explicit in the radial transport rates given by Equations (4.2) and (4.3) is the dependence on the azimuthal mode structure of a given population of waves, characterized by the mode number m. This dependence dictates which particles will be effectively transported in the interaction, based on the variation of drift velocity with particle energy and the drift resonance condition $\omega - m\omega_d = 0$. Figure 4.4 provides a simple, idealized sketch showing how the mode structure may be interpreted for the $m = 1$ and $m = 2$ global ULF modes, where the phase here provides symmetry about the Earth–Sun line. Here the shaded regions represent either a positive or negative field perturbation, while the unshaded regions show field perturbations of the opposite sign. The line of transition between the shaded and unshaded region represent nodes between the perturbed regions. The azimuthal mode number, m, is a measure of the wavelength, and is calculated based on the number of azimuthal wavelengths that would fit within a particle drift orbit.

In the simplest application, unambiguously determining the azimuthal mode structure to a particular mode number m requires $2m$ uniform azimuthal measurements. Because an appropriate azimuthal distribution of in situ measurements is rare in the magnetosphere, it is common to make an explicit or implicit assumption about the mode distribution of the ULF wave power. For example, it is often implicitly or explicitly assumed that all power is in the $m = 1$ mode of the waves being measured (e.g. Shprits and Thorne, 2004; Shprits et al., 2009; Brautigam et al., 2005). By contrast, Hudson et al. (1999, 2001) and Loto'aniu et al. (2006) explicitly assumed an $m = 2$ mode spectrum.

While most studies by necessity assume a single mode number, notable exceptions may be found in the work by Fei et al. (2006), Tu et al. (2012) and Elkington et al. (2012), who used mode spectrum results obtained from a Lyon-Fedder-Mobarry (LFM) MHD simulation (Lyon et al., 2004) to calculate the ULF wave azimuthal mode numbers. They found that, even though the magnetic power calculated at $m = 1$ in the MHD simulations was dominant for some parts of the events they studied, at times,

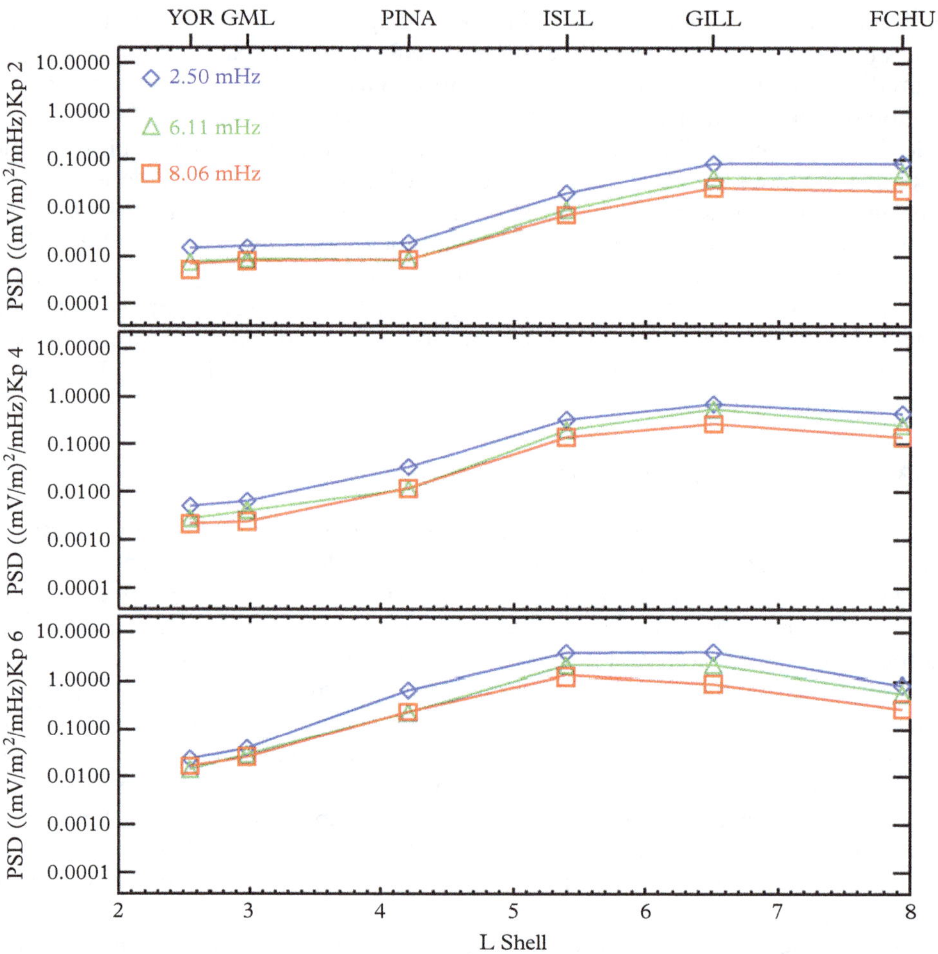

Figure 4.3 *Electric ULF wave power in the equatorial plane, derived from magnetic fluctuations observed at a number of latitudinally-distributed ground magnetometers (Ozeke et al., 2012).*

assuming that all power comes from $m = 1$ may may lead to overestimation of the power capable of drift-resonantly interacting with a given particle population via the $\omega = \omega_d$ resonance, and underestimating the transport of higher-energy particles that may interact with $m > 1$ resonances. This was particularly true during the main phase of a geomagnetic storm. These studies also found that, overall, assuming that the total B_z power from the $m = 1$ mode (related to the solar wind dynamic pressure variations) is a reasonable approximation, while the electric field had substantial $m > 1$ power, which is mostly related to night-side geomagnetic activity associated with substorm-induced flow channels and global magnetospheric convective patterns.

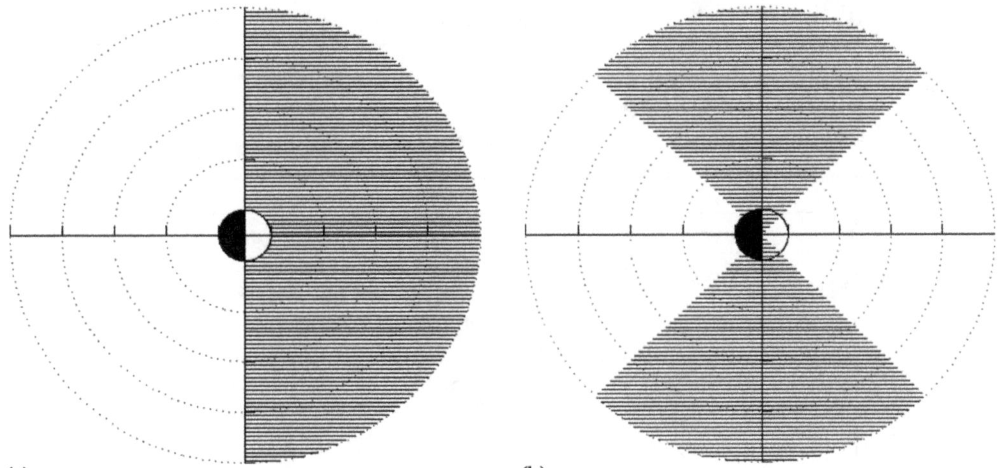

(a) (b)

Figure 4.4 *Simple interpretation of the nodal structure of the m = 1 and m = 2 modes of global ULF wave activity at an instant of time. In this figure the shaded regions represent regions of positive or negative field perturbation, while the unshaded regions represent local times with oppositely-signed perturbation. The line of transition between the shaded and unshaded region represent nodes between the perturbed regions.*

Figure 4.5 shows the fraction of the total power spectral density in each mode, for the equatorial azimuthal electric field and magnetic field for the September 24–26, 1998 geomagnetic storm studied by Elkington et al. (2012). The integrated power represents the power spectral density summed over all mode numbers representable on the azimuthal MHD grid, $\sum_{m=0}^{24} P_m$, integrated in frequency up to the temporal Nyquist frequency. The fractional power in each mode for the $m = 1, 2$, and 3 modes are indicated in the lower of each pair of plots by the solid, dotted, and dashed lines, respectively, and the fractional power of the sum of the first five modes plotted using "+" symbols.

We see in Figure 4.5 that the $m = 1$ mode is generally dominant in both the electric and magnetic fields during the recovery phase, with much smaller contributions from the $m = 2$ and 3 modes. However, during the main phase the electric field power exhibits relatively large contributions from the $m = 2$ and 3 modes; for a period of time the power in these modes is actually larger than the power in the $m = 1$ mode. In this event we see much more structure in the electric field than the magnetic field during the main phase, with the power in the first three mode numbers accounting for only ~60% of the total power, in contrast to closer to ~80% in the case of the magnetic power. During the recovery phase the power is concentrated in the lower mode numbers for both the electric and magnetic fields, with the lowest three mode structures accounting for ~80% of the power in both cases. These results are consistent in the simulations both at geosynchronous altitudes and at equatorial radial distances of 4.2 R_E, corresponding to altitudes of the GPS constellation of spacecraft.

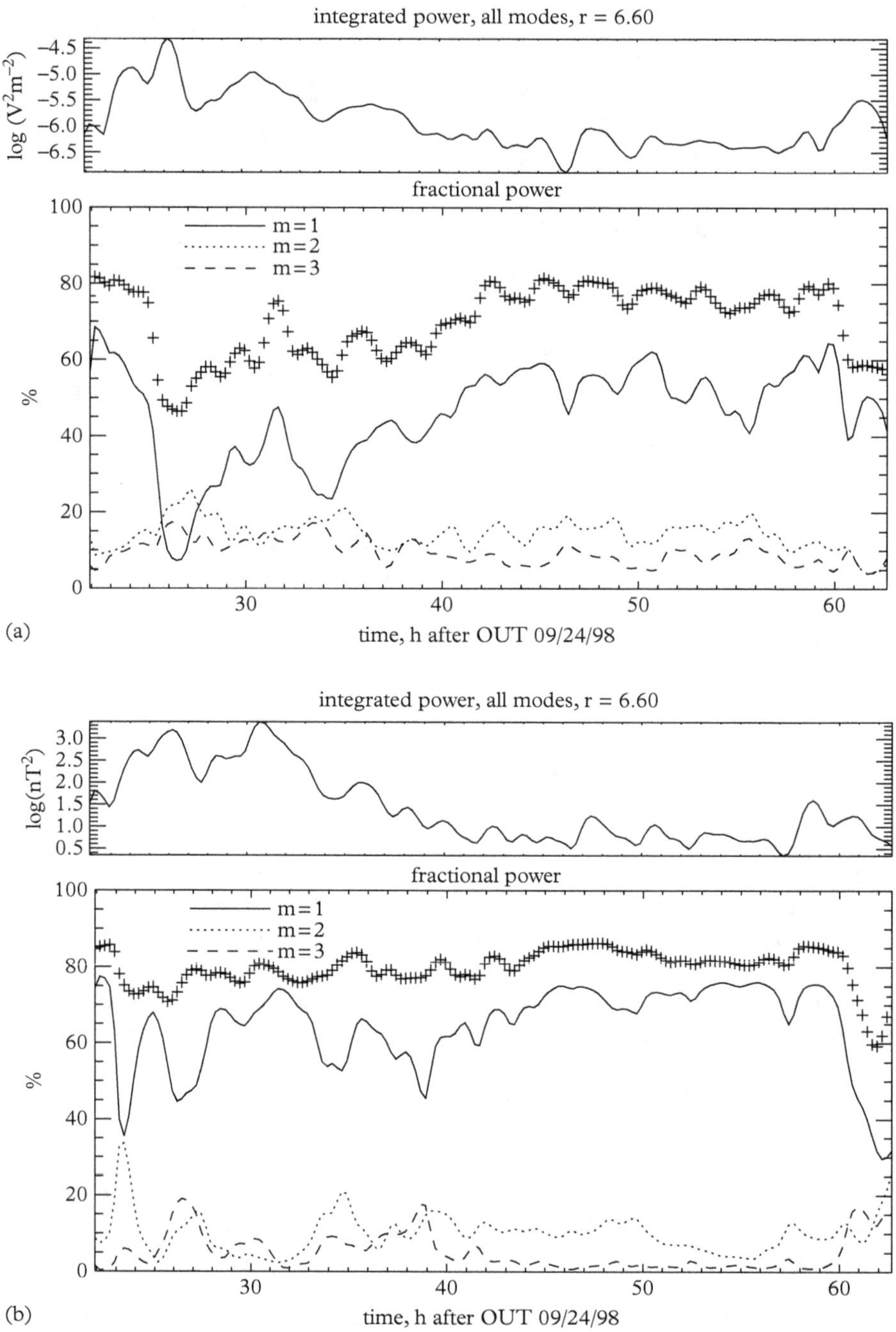

Figure 4.5 *Fraction of the power in the lowest three mode numbers for the electric (a) and magnetic (b) ULF wave fields. In the top of each pair of plots we indicate the total power spectral density over all azimuthal modes as a function of time. The '+' symbols in the bottom of each pair of plots indicates the fraction of power in the sum of the first 5 modes as a fraction of the total power; the remaining lines indicate the power in each of the first three modes. (Elkington et al., 2012).*

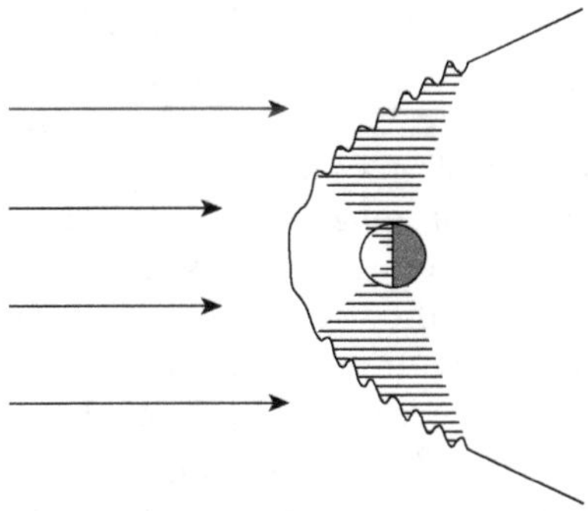

Figure 4.6 *Sketch of Kelvin-Helmholz waves generated by sheer interactions along the flank magnetopause, illustrating the difference between azimuthal extent (indicated by the shaded regions) and azimuthal wave number (indicated by the recurring wavelengths along the magnetopause).*

4.2.4 What is the azimuthal extent?

Distinct from the azimuthal mode structure of the waves is the azimuthal extent of the wave power. This distinction is illustrated in Figure 4.6, which shows a schematic of shear-driven ULF waves on the flanks. The azimuthal extent of the waves is indicated by the shaded region and extends over several hours of local time on the dawn and dusk flanks. The mode number, by contrast, is indicated by the wavelength, λ, of the waves, with several cycles occurring over the region where the shear waves are induced. Quantitatively, $m \equiv 2\pi/\lambda$. Simulations of magnetopause Kelvin-Helmholz waves undertaken by Claudepierre et al. (2008) suggest that typical mode numbers of these waves lie in the range m = 12-15, corresponding to 1–2 h of local time, while the azimuthal region where the waves occur typically subtends several hours of local time.

Geosynchronous observations of magnetic Pc-5 fluctuations via the NOAA GOES spacecraft have provided continual, long-term measurements of magnetic ULF wave power, covering all local times once per day. Huang et al. (2010) conducted a statistical study of ULF wave power in all three components of the magnetic vector perturbations as a function of frequency, local time, solar wind driving, and geomagnetic activity. They found a preponderance of power in the compressional magnetic field on the dayside resulting from pressure fluctuations that increased with increasing solar wind velocity, and additional regions of power in the midnight sector probably associated with substorm activity. Tu et al. (2012) studied a particular magnetic storm associated with high speed solar wind regions, and similarly found peaks in the compressional component of magnetic ULF wave activity in the noon sector extending through dusk. Clausen et al. (2011) studied 17 storms associated with variable radiation belt electron populations, and also found that most power in the frequency ranges associated with particle drift frequencies occurred in the compressional mode in the noon sector.

Liu et al. (2009b) examined THEMIS magnetic field data to derive a statistical model of Pc-5 ULF wave power in FLRs at all local times for a range of L shells. In contrast to the earlier work of Anderson et al. (1990), who used AMPTE/CCE data to find that shear Pc-5 waves were more frequently seen on the dawnside magnetosphere, Liu et al. suggested that compressional field line resonances were more prevalent on the duskside but that shear-mode Pc-5 waves were evenly distributed in azimuth.

The azimuthal distribution of Alfvénic electric field variations were obtained from ground magnetometers by Rae et al. (2012) using the mapping techniques described in Ozeke et al. (2009). Figure 4.7 indicates the results obtained for the GILL magnetometer, magnetically mapping to a point near geosynchronous orbit. Consistent with the in situ geosynchronous observations of magnetic field fluctuations, they found increasing power at all local times with increasing solar wind speed. Consistent with the Anderson et al. (1990) results, they found significantly more power on the dawnside in the shear mode (associated with radial electric fields) and in ground measurements associated with azimuthal electric fields. Notable in Figure 4.7 are the peaks in power spectral density in the 3–5 mHz range in the noon, dawn, and dusk sectors, likely associated with driven FLR activity. Aside from the peak in power associated with FLRs seen at noon, differences in the power spectral densities seen at noon and midnight were roughly of the same order of magnitude.

4.2.5 What is the propagation direction?

The wave frequency ω in Equation (4.4) may take on either positive or negative values, signifying the direction of phase propagation of the ULF waves. Understanding the propagation direction is key in assigning wave power appropriately to particle populations, as the drift-resonant nature of the interaction underlying transport expressions (4.2) and (4.3) requires that a particle drift with a phase speed commensurate with the phase propagation speed of the wave. That is, the particle should see a field with a relatively constant sign associated with the wave as the particle and wavefront move together in azimuth. For trapped electrons, drifting in a right-hand sense about the Earth, the wave frequency (4.4) should be positive; for protons, with their oppositely-directed drift sense, the frequency should be negative.

Mann et al. (1999) discussed the excitation of magnetospheric waveguide modes, whereby wave energy was converted from plasma flows in the magnetosheath to enhanced ULF wave energy propagating tailward through the magnetosphere. For protons drifting in a counter-clockwise sense about the Earth, most of the ULF-related acceleration under this scenario would occur in the dawn sector. Conversely, electrons would experience most of their acceleration from this tailward propagation of energy in the dusk sector.

Properly assigning a propagation direction to a given wave observation may be difficult. For waves originating from a known source, such as for shear waves propagating tailward along the flanks of the magnetopause, assigning propagation direction may be intuitively clear. This assumption is obviously complicated in situations where E and B fluctuations may be arising from multiple sources, such as in the dawn-midnight or dusk-midnight sectors, where one may reasonably expect to see tailward-propagating waves resulting from solar wind pressure perturbations at the magnetopause or Kelvin-Helmholz waves generated at the flanks, mixed with sunward-propagating waves resulting from substorm activity or global convection. Descriptions of the Poynting vector hold promise as a means of examining wave propagation direction, but can

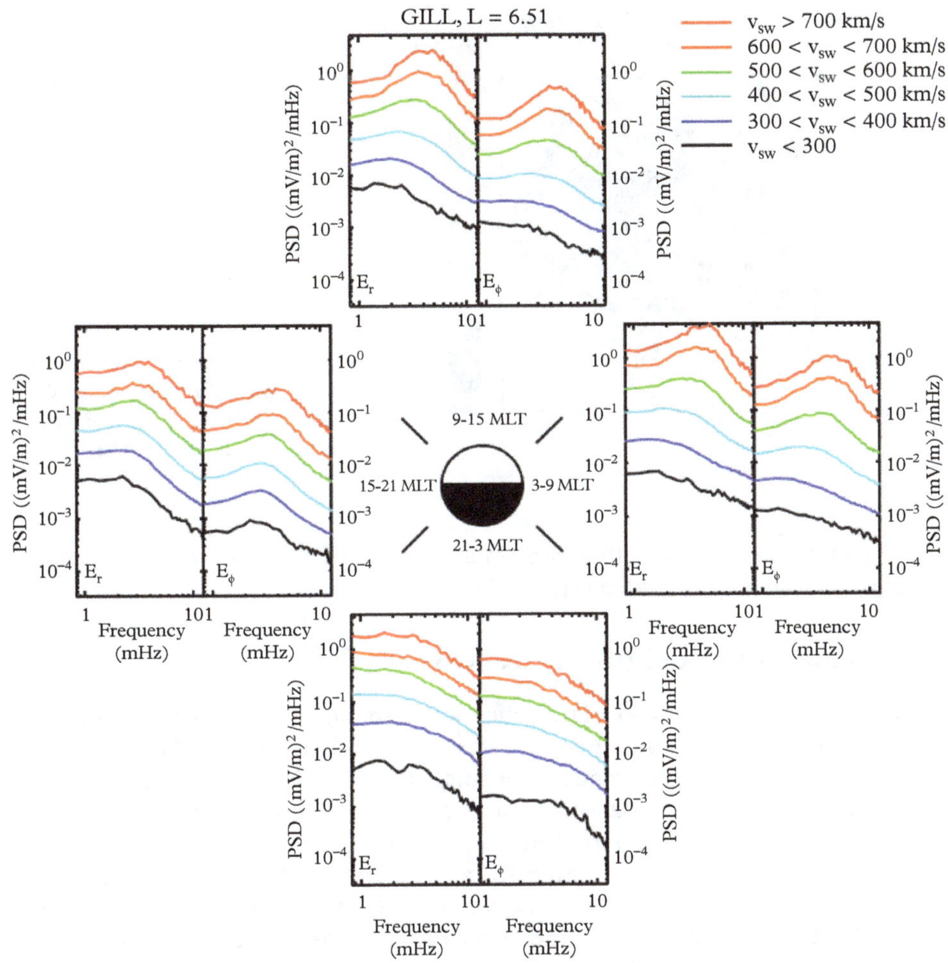

Figure 4.7 *Electric ULF wave power near geosynchronous as a function of solar wind velocity and local time, for toroidal shear (E_ϕ) and poloidal (E_r) ULF wave modes (Rae et al., 2012).*

be similarly complicated at points in space where a superposition of wave activity arising from multiple sources is present.

Figure 4.8 shows initial work to quantify ULF energy propagation in the inner magnetosphere via calculation of a mean Poynting vector at different locations in the magnetosphere (Hartinger et al., 2013). This study found significant scatter in Poynting vector direction at all locations and frequencies studied. However, for Pc-5 waves propagating with an azimuthal component likely to transport energetic particles trapped in the inner magnetosphere, there was a preference for duskward and earthward energy transfer on the dayside, and a preference for sunward Poynting vectors in the predawn

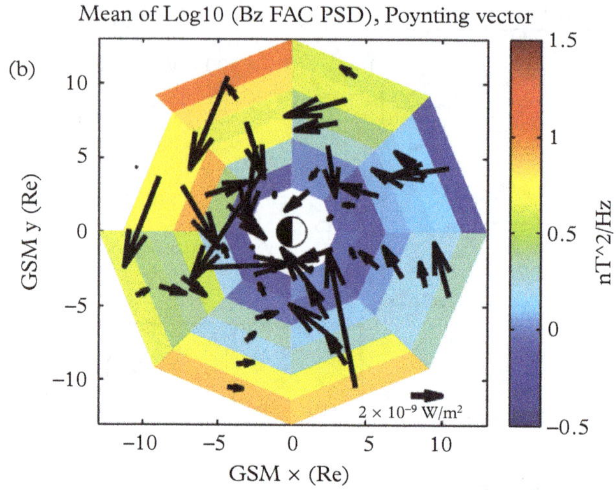

Figure 4.8 *6mHz parallel magnetic power spectral density on the color scale, along with of the GSM projection of the mean Poynting vector. Via Hartinger et al., (2013)*

sector potentially associated with substorm activity. Both of these situations would tend to transport/accelerate trapped energetic electrons.

4.2.6 What is the origin of the waves?

Finally, one needs to assign an electrostatic or electromagnetic character to the waves in order to properly quantify the transport rates via Equations (4.2) or (4.3), respectively. This is difficult under circumstances of a single point in-situ measurement, as electromagnetic waves will have non-zero curl in the electric field and will require additional assumptions regarding spatial structure to properly resolve the global field character. For example, potential waves resulting from variations in the convection electric field were envisioned in the formulation of Equation (4.2) in the original derivation by Fälthammar (1965). However, in the same paper Fälthammar includes the effect of electric fields induced by magnetic field variations in the equation analogous to Equation (4.3). In this formulation, magnetic field fluctuations were responsible for 7/15 of the resulting radial diffusion, while the electric field fluctuations induced by the magnetic field were responsible for the remainder of the transport term.

A single-point measurement of E and B cannot disambiguate how much of the electric field power is appropriate to assign to Equation (4.2) and how much of the electric field power is implicit in Equation (4.3), leading to the potential for "overcounting" the effect of measured electric fields. Some work, for example Fei et al. (2006), assumed that the electric and magnetic field fluctuations were completely decoupled, and assigned all electric field power to Equation (4.2) assuming magnetic field fluctuations were the only term contributing to Equation (4.3). However, this is an imperfect approach due to the clear phase relationships between the driven magnetic variations and the corresponding induced electric field variations (Perry et al., 2005).

According to the Helmholz theorem, a vector may be completely specified in terms of its divergence, curl, and a normal component described over a boundary (e.g. Arfken, 1985). Assuming charge neutrality, the problem in space physics is reduced to that of measuring a curl and specifying appropriate boundary conditions. Constellations of at least four spacecraft, such as are available via the Cluster mission (Escoubet et al., 1997), are capable of uniquely determining the curl of the electric field on spatial scales appropriate to the spacecraft separation. However, determination of an appropriate boundary condition among the dynamic E and B fields in the near-Earth space environment, allowing specification of the curl of the electric field at all spatial scales, remains a significant challenge in sorting out this problem.

4.3 Conclusions and future possibilities

Complete specification of the effects of Pc-5 ULF wave-driven transport requires global knowledge of the character of ULF waves, quantified by the questions outlined in this chapter. Previous in situ observations, both at geosynchronous and across L shells, have provided statistical characterizations of the spatial (i.e. radial and azimuthal) and spectral characteristics of wave activity under different driving conditions. These quantities have been incorporated into empirical models of radial diffusion coefficients, for example, provided by Brautigam and Albert (2000) or Ozeke et al. (2012). The quality of these empirical diffusion coefficient models can be traced back to the comprehensiveness of the data that was used to build them, and are not always consistent with each other (for example, Brautigam and Albert (2000) suggests a dominant role for magnetic diffusion, Equation (4.3), while Ozeke et al. (2012) suggests a larger role for electric diffusion, Equation (4.2)). However, these empirical specifications have been used with some success in driving dynamic models of the radiation belts (e.g. Shprits et al., 2005; Lam et al., 2007; Tu et al., 2013). The recent launch of the Van Allen Probe mission (Mauk et al., 2013), consisting of a pair of identically-instrumented spacecraft in a "string of pearls" geotransfer orbit, promises to do much to increase our understanding of the occurrence and characteristics of the low frequency waves that lead to transport through Equation (4.1).

Quantifying dynamic changes in the radiation belts via application of the diffusion Equation, (4.1), requires not only careful quantification of the diffusion coefficients, but also of the spatial characteristics of the distribution function, f. This is evidenced by the dependence of Equation (4.1) on the radial gradient of the phase space density on the right-hand side of Equation (4.1). Here again the Van Allen Probes hold promise to improve our understanding of the spatial evolution of the distribution function, including quiescent configurations that may be used as initial conditions in dynamic models of the radiation belts.

Remaining questions necessary to quantify the ULF-driven transport of radiation belts in terms of the transport coefficients (4.2) and (4.3) include better understanding of the mode structure, propagation direction of ULF waves, and the electromagnetic character of waves arising from different sources. As mentioned in Section 4.3, unambiguously determining the mode structure to mode number m requires $2m$ measurements. As it is unlikely that the number of in-situ measurements required to evaluate any more than the lowest-order mode structure will be achieved in the near future, alternative methods are being developed. For example, cross-wavelet analysis between time series is used increasingly in geophysical data analysis in order to examine if regions in time and frequency with large common power also have a consistent phase relationship (e.g. Grinsted et al., 2004; Sarris et al., 2009b, 2012). These methods may be used to infer or constrain the global mode structure based on a few azimuthally-separated measurements. In the case of magnetospheric electric variations, these techniques will be augmented through the mapping techniques suggested by Ozeke et al. (2009). Additional information about the mode structure may be obtained by combining global MHD models of the magnetospheric system (Tu et al., 2012; Elkington et al., 2012) with these new measurements.

Determining the propagation direction and electromagnetic character of measured waves remain significant challenges. Poynting vector calculations (Hartinger et al., 2013) hold some promise in evaluating propagation direction, but can be confounded in regions where waves from multiple sources may be propagating in different directions through the magnetosphere. Presumably better estimates of propagation direction will be obtained as we improve understanding of the various global drivers and dynamics of the magnetospheric system. Determining the electromagnetic character of the waves also presents a challenge. Again, this may be relatively easy to do in cases where clear phase relations between measured E and B variations are available, such as in the case of strong FLR activity. In the more general case, separating the curl-free component of the electric field from the total electric field requires knowledge of the spatial distribution of the fields around the point in question. Constellation missions such as provided by Cluster, perhaps combined with global physical simulations of the geospace environment, may hold promise in disentangling the appropriate partitioning of measured electric fields between the potential and induced electric fields explicit in Equation (4.2) and implicit in Equation (4.3).

Acknowledgments

Work has been supported by NSF grant ATM0842388 and NASA grants NNX09AI05G, NNX10AQ50G, NNX13AE39G, NNX10AQ48G and NNX14-AC04G at the University of Colorado/LASP. Additional support provided by the European Union (European Social Fund—ESF) "Investing in Knowledge" and Greek national funds through the Operational Program "Education and Lifelong Learning" of the National Strategic Reference Framwork (NSRF) Research Funding Program: Thales. The authors thank I.R. Mann, D.K. Milling and the rest of the CARISMA team for data. CARISMA is operated by the University of Alberta, funded by the Canadian Space Agency.

REFERENCES

Anderson, B. J., M. J. Engebretson, S. P. Rounds, L. J. Zanetti, and T. A. Potemra (1990), A statistical study of pulsations observed by the AMPTE/CCE magnetic fields experiment—1. Occurrence distributions, *J. Geophys. Res.*, **95**, 10,495.

Arfken, G. (1985), *Mathematical Methods for Physicists*, 985 pp., Academic Press, Boston.

Baker, D. N., et al. (1998), A strong CME-related magnetic cloud interaction with the Earth's magnetosphere: ISTP observations of rapid relativistic electron acceleration on May 15, 1997, *Geophys. Res. Lett.*, **25**, 2975.

Bloom, R. M., and H. J. Singer (1995), Diurnal trends in geomagnetic noise power in the Pc2 through Pc5 bands at low geomagnetic latitude, *J. Geophys. Res.*, **100**, 14,943.

Brautigam, D. H., and J. M. Albert (2000), Radial diffusion analysis of outer radiation belt electrons during the October 9, 1990, magnetic storm, *J. Geophys. Res.*, **105**(A1), 291.

Brautigam, D. H., G. P. Ginet, J. M. Albert, J. R. Wygant, D. E. Rowland, A. Ling, and J. Bass (2005), CRRES electric field power spectra and radial diffusion coefficients, *J. Geophys. Res.*, **110**(A2), A02,214, doi:10.1029/2004JA010612.

Cahill, L. J., Jr., and J. R. Winckler (1992), Periodic magnetopause oscillations observed with the GOES satellites on March 24, 1991, *J. Geophys. Res.*, **97**(A6), 8239.

Chen, L., and A. Hasegawa (1991), Kinetic theory of geomagnetic pulsations 1: Internal excitations by energetic particles, *J. Geophys. Res.*, **96**(A2), 1503–1512.

Claudepierre, S. G., S. R. Elkington, and M. Wiltberger (2008), Solar wind driving of magnetospheric ULF waves: Pulsations driven by velocity shear at the magnetopause, *J. Geophys. Res.*, **113**, A05218, doi:10.1029/2007JA012890.

Claudepierre, S. G., M. Wiltberger, S. R. Elkington, and M. K. Hudson (2009), Magnetospheric cavity modes driven by solar wind dynamic pressure fluctuations, *Geophys. Res. Lett.*, **36**, L13101, doi:10.1029/2009GL039045.

Clausen, L. B. N., J. B. H. Baker, J. M. Ruohoniemi, and H. J. Singer (2011), ULF wave characteristics at geosynchronous orbit during the recovery phase of geomagnetic storms associated with strong electron acceleration, *J. Geophys. Res.*, **116**(A9), A09203, doi:10.1029/2011JA016666.

Cornwall, J. M. (1968), Diffusion processes influenced by conjugate-point wave phenomena, *Radio Sci.*, **3**(7), 740.

Dungey, J. W. (1961), Interplanetary magnetic field and the auroral zones, *Phys. Rev. Lett.*, **6**, 47.

Elkington, S. R. (2006), A review of ULF interactions with radiation belt electrons, in *Magnetospheric ULF waves: Synthesis and new directions*, vol. 169, edited by K. Takahashi, P. J. Chi, R. E. Denton, and R. L. Lysak, p. 177, AGU, Washington, D.C.

Elkington, S. R., M. K. Hudson, and A. A. Chan (1999), Acceleration of relativistic electrons via drift-resonant interaction with toroidal-mode Pc-5 ULF oscillations, *Geophys. Res. Lett.*, **26**(21), 3273.

Elkington, S. R., M. K. Hudson, and A. A. Chan (2003), Resonant acceleration and diffusion of outer zone electrons in an asymmetric geomagnetic field, *J. Geophys. Res.*, **108**(A3), 1116, doi:10.1029/2001JA009202.

Elkington, S. R., A. A. Chan, and M. Wiltberger (2012), Global structure of ULF waves during the 2426 September 1998 geomagnetic storm, in *Dynamics of the Earths radiation belts and inner magnetosphere*, vol. 199, edited by D. Summers, I. R. Mann, D. N. Baker, and M. Schulz, p. 127, AGU, Washington, D.C.

Escoubet, C. P., C. T. Russell, and R. Schmidt (Eds.) (1997), *The Cluster and Pheonix Missions*, Kluwer Academic, Boston, (reprinted from *Space Sci. Rev.*, **79**, nos 1–2, 1997).

Fälthammar, C.-G. (1965), Effects of time-dependent electric fields on geomagnetically trapped radiation, *J. Geophys. Res.*, **70**(11), 2503.

Fälthammar, C.-G. (1966a), On the transport of trapped particles in the outer magnetosphere, *J. Geophys. Res.*, **71**(5), 1487.

Fälthammar, C.-G. (1966b), Coefficients of diffusion in the outer radiation belts, in *Radiation trapped in the Earth's magnetic field*, edited by B. M. McCormac, p. 398, NATO Advanced Study Institute, Reinhold Book Corp., New York.

Fälthammar, C.-G. (1968), Radial diffusion by violation of the third adiabatic invariant, in *Earth's particles and fields*, edited by B. M. McCormac, p. 157, NATO Advanced Study Institute, Reinhold Book Corp., New York.

Fei, Y., A. A. Chan, S. R. Elkington, and M. J. Wiltberger (2006), Radial diffusion simulation of relativistic electron transport by ULF waves in the September 1998 storm, *J. Geophys. Res.*, **111**, A12,209, doi:10.1029/2005JA011211.

Friedel, R. H. W., G. D. Reeves, and T. Obara (2002), Relativistic electron dynamics in the inner magnetosphere – a review, *J. Atmos. Solar Terr. Phys.*, **64**(2), 265.

Green, J. C., T. G. Onsager, T. P. O'Brien, and D. N. Baker (2004), Testing loss mechanisms capable of rapidly depleting relativistic electron flux in the Earth's outer radiation belt, *J. Geophys. Res.*, **109**(A12), A12211, doi:10.1029/2004JA010579.

Grinsted, A., J. C. Moore, and S. Jevrejeva (2004), Application of the cross wavelet transform and wavelet coherence to geophysical time series, *Nonlin. Process Geophys.*, **11**, 51.

Hartinger, M. D., M. B. Moldwin, K. Takahashi, J. W. Bonnell, and V. Angelopoulos (2013), Survey of the UFL wave Poynting vector near the Earth's magnetic equatorial plane, *J. Geophys. Res.*, **118**, 6212, doi:10.1002/jgra.50591.

Hasegawa, A. (1969), Drift mirror instability in the magnetosphere, *Phys. Fluids*, **12**(12), 2642.

Huang, C.-L., H. E. Spence, H. J. Singer, and W. J. Hughes (2010), Modeling radiation belt radial diffusion in ULF wave fields: 1. Quantifying ULF wave power at geosynchronous orbit in observations and in global MHD model, *J. Geophys. Res.*, **115**, A06215, doi:10.1029/2009JA014917.

Hudson, M. K., S. R. Elkington, J. G. Lyon, C. C. Goodrich, and T. J. Rosenberg (1999), Simulation of radiation belt dynamics driven by solar wind variations, in *Sun-Earth Plasma Connections*, vol. 109, edited by J. L. Burch, R. L. Carovillano, and S. K. Antiochos, p. 171, AGU, Washington, D.C.

Hudson, M. K., S. R. Elkington, J. G. Lyon, M. J. Wiltberger, and M. Lessard (2001), Radiation belt electron acceleration by ULF wave drift resonance: Simulation of 1997 and 1998 storms, in *Space Weather*, vol. 125, edited by P. Song, H. J. Singer, and G. L. Siscoe, p. 289, AGU, Washington, D. C.

Hudson, M. K., T. Brito, S. R. Elkington, B. T. Kress, Z. Li, and M. Wiltberger (2012), Radiation belt 2D and 3D simulations for CIR-driven storms during Carrington Rotation 2068, *J. Atmos. Solar Terr. Phys.*, **83**, 51, doi:10.1016/j.jastp.2012.03.017.

Hughes, W. J. (1974), The effect of the atmosphere and ionosphere on long-period magnetospheric micropulsations, *Planet. Space Sci.*, **22**, 1157.

Jacobs, J. A., Y. Kato, S. Matsushita, and V. A. Troitskaya (1964), Classification of geomagnetic micropulsations, *J. Geophys. Res.*, **69**, 180.

Kepko, L., and H. E. Spence (2003), Observations of discrete, global magnetospheric oscillations directly driven by solar wind density variations, *J. Geophys. Res.*, **108**(A6), 21–1, doi:10.1029/2002JA009676.

Kivelson, M. G., and Z.-Y. Pu (1984), The Kelvin-Helmholtz instability on the magnetopause, *Planet. Space Sci.*, **32**(11), 1335.

Kivelson, M. G., and D. J. Southwood (1988), Hydromagnetic waves in the ionosphere, *Geophys. Res. Lett.*, **15**, 1271.

Lam, M. M., R. B. Horne, N. P. Meridith, and S. A. Glauert (2007), Modeling the effects of radial diffusion and plasmaspheric hiss on outer radiatio belt electrons, *Geophys. Res. Lett.*, **34**(20), L20112, doi:10.1029/2007GL031598.

Liu, W., T. E. Sarris, X. Li, S. Elkington, R. Ergun, V. Angelopoulos, J. Bonnell, and K. H. Glassmeier (2009a), Electric and magnetic field observations of Pc4 and Pc5 pulsations in the inner magnetosphere: A statistical study, *J. Geophys. Res.*, **114**, A12206, doi:10.1029/2009JA014243.

Liu, W., et al. (2009b), Observation and modeling of the injection observed by THEMIS and LANL satellites during the 23 march 2007 substorm event, *J. Geophys. Res.*, **114**(A00C18), doi:10.1029/2008JA013498.

Liu, W., T. E. Sarris, X. Li, R. Ergun, V. Angelopoulos, J. Bonnell, and K. H. Glassmeier (2010), Solar wind influence on Pc4 and Pc5 ULF wave activity in the inner magnetosphere, *J. Geophys. Res.*, **115**, A12201, doi:10.1029/2010JA015299.

Loto'aniu, T. M., I. R. Mann, L. G. Ozeke, A. A. Chan, Z. C. Dent, and D. K. Milling (2006), Radial diffusion of relativistic electrons into the radiation belt slot region during the 2003 Halloween geomagnetic storms, *J. Geophys. Res.*, **111**(A4), A04218, doi:10.1029/2005JA011355.

Loto'aniu, T. P., H. J. Singer, C. L. Waters, V. Angelopoulos, I. R. Mann, S. R. Elkington, and J. W. Bonnell (2010), Relativistic electron loss due to ultralow frequency waves and enhanced outward radial diffusion, *J. Geophys. Res.*, **115**, A112245, doi:10.1029/2010JA015755.

Lyon, J. G., J. A. Fedder, and C. M. Mobarry (2004), The Lyon–Fedder–Mobarry (LFM) global MHD magnetospheric simulation code, *J. Atmos. Solar Terr. Phys.*, **66**(15), 1333, doi:10.1016/j.jastp.2004.03.020.

Mann, I. R., A. N. Wright, K. J. Mills, and V. M. Nakariakov (1999), Excitation of magnetospheric waveguide modes by magnetosheath flows, *J. Geophys. Res.*, **104**(A1), 333.

Mann, I. R., T. P. O'Brien, and D. K. Milling (2004), Correlations between ULF wave power, solar wind speed, and relativistic electron flux in the magnetosphere: solar cycle dependence, *J. Atmos. Solar Terr. Phys.*, **66**, 187, doi:10.1016/j.jastp.2003.10.002.

Mann, I. R., et al. (2008), The upgraded CARISMA magnetometer array in the THEMIS era, *Space Sci. Rev.*, **141**(1-4), 413, doi:10.1007/s11214-008-9457-6.

Mathie, R. A., and I. R. Mann (2000), A correlation between extended intervals of ULF wave power and storm-time geosynchronous relativistic electron flux enhancements, *Geophys. Res. Lett.*, **27**(20), 3261.

Mauk, B. H., N. J. Fox, S. G. Kanekal, R. L. Kessel, D. G. Sibeck, and A. Ukhorskiy (2013), Science objectives and rationale for the Radiation Belt Storm Probes mission, *Space Sci. Rev.*, **179**(1-4), 3, doi:10.1007/s11214-012-9908-y.

Ozeke, L. G., I. R. Mann, and I. J. Rae (2009), Mapping guided Alfv'en waves magnetic field amplitudes observed on the ground to equatorial electric field amplitudes in space, *J. Geophys. Res.*, **114**, A01214, doi:10.1029/2008JA013041.

Ozeke, L. G., I. R. Mann, K. R. Murphy, I. J. Rae, D. K. Milling, S. R. Elkington, A. A. Chan, and H. J. Singer (2012), ULF wave derived radiation belt radial diffusion coefficients, *J. Geophys. Res.*, **117**(A4), A04221, doi:10.1029/2011JA017335.

Perraut, S., R. Gendrin, P. Robert, A. Roux, C. de Villedary, and D. Jones (1978), ULF waves observed with magnetic and electric sensors on GEOS-1, *Space Sci. Rev.*, **22**, 347, doi:10.1007/BF00210873.

Perry, K. L., M. K. Hudson, and S. R. Elkington (2005), Incorporating spectral characteristics of Pc5 waves into three-dimensional radiation belt modeling and the diffusion of relativistic electrons, *J. Geophys. Res.*, **110**(A3), A03,215, doi:10.1029/2004JA010760.

Press, W. H., S. A. Teukolsky, W. T. Vetterling, and B. P. Flannery (1992), *Numerical recipes in Fortran 77: The art of scientific computing, Fortran Numerical Recipes*, vol. 1, 2 ed., 922 pp., Cambridge Univ. Press, New York.

Rae, I. J., I. R. Mann, K. R. Murphy, L. G. Ozeke, D. K. Milling, A. A. Chan, S. R. Elkington, and F. Honary (2012), Ground-based magnetometer determination of in situ Pc4–5 ULF electric field wave spectra as a function of solar wind speed, *J. Geophys. Res.*, **117**(A4), A04221, doi:10.1029/2011JA017335.

Ridley, A. J., G. Lu, C. R. Clauer, and V. O. Papitashvili (1998), A statistical study of the ionospheric convection response to changing interplanetary magnetic field conditions using the assimilative mapping of ionospheric electrodynamics technique, *J. Geophys. Res.*, **103**, 4023.

Roederer, J. G. (1970), *Dynamics of Geomagnetically Trapped Radiation*, 166 pp., Springer-Verlag, New York.

Rostoker, G., S. Skone, and D. N. Baker (1998), On the origin of relativistic electrons in the magnetosphere associated with some geomagnetic storms, *Geophys. Res. Lett.*, **25**, 3701.

Ruohoniemi, J. M., and R. A. Greenwald (1998), The response of high-latitude convection to a sudden southward IMF turning, *Geophys. Res. Lett.*, **25**, 2913.

Russell, C. T. (Ed.) (1995), *The Global Geospace Mission*, Kluwer Academic, Boston, (reprinted from *Space Sci. Rev.*, **71**, nos 1–4, 1995).

Sarris, T. E., et al. (2009a), Characterization of ULF pulsations by THEMIS, *Geophys. Res. Lett.*, **36**(4), L04104, doi:10.1029/2008GL036732.

Sarris, T. E., X. Li, and H. J. Singer (2009b), A long-duration narrowband Pc5 pulsation, *J. Geophys. Res.*, **114**, A01213, doi:10.1029/2007JA012660.

Sarris, T. E., W. Liu, X. Li, K. Kabin, E. R. Talaat, R. Rankin, V. Angelopoulos, J. Bonnell, and K.-H. Glassmeier (2010), THEMIS observations of the spatial extent and pressure-pulse excitation of field line resonances, *Geophys. Res. Lett.*, **37**, L15104, doi:10.1029/2010GL044125.

Sarris, T. E., X. Li, W. Liu, E. Argyriadis, and R. Ergun (2012), Mode number calculations of ULF field line resonances using ground magnetometers and THEMIS measurements, *J. Geophys. Res. Space Physics*, **118**, 6986–6997, doi:10.1002/2012JA018307.

Schulz, M. (1996), Canonical coordinates for radiation-belt modeling, in *Radiation Belts: Models and Standards*, vol. 97, edited by J. F. Lemaire, D. Heynderickx, and D. N. Baker, p. 153, AGU, Washington, D.C.

Schulz, M., and L. J. Lanzerotti (1974), *Particle Diffusion in the Radiation Belts, Physics and Chemistry in Space*, 7, 215 pp., Springer-Verlag, New York.

Shprits, Y. Y., and R. M. Thorne (2004), Time dependent radial diffusion modeling of relativistic electrons with realistic loss rates, *Geophys. Res. Lett.*, **31**, L08805, doi:10.1029/2004GL019591.

Shprits, Y. Y., R. M. Thorne, G. D. Reeves, and R. Friedel (2005), Radial diffusion modeling with empirical lifetimes: comparison with CRRES observations, *Ann. Geophys.*, **23**(4), 1467.

Shprits, Y. Y., R. M. Thorne, R. Friedel, G. D. Reeves, J. Fennell, D. N. Baker, and S. G. Kanekal (2006), Outward radial diffusion driven by losses at magnetopause, *J. Geophys. Res.*, **111**, A11,214, doi:10.1029/2006JA011657.

Shprits, Y. Y., S. R. Elkington, N. P. Meredith, and D. A. Subbotin (2008), Review of modeling of losses and sources of relativistic electrons in the outer radiation belt I: Radial transport, *J. Atmos. Solar Terr. Phys.*, **70**(14), 1679, doi:10.1016/j.jastp.2008.06.008.

Shprits, Y. Y., D. Subbotin, and B. Ni (2009), Evolution of electron fluxes in the outer radiation belt computed with the VERB code, *J. Geophys. Res.*, **114**(A11), A11209, doi:10.1029/2008JA013784.

Sibeck, D. G., and V. Angelopoulos (2008), THEMIS science objectives and mission phases, *Space Sci. Rev.*, **141**, 35, doi:10.1007/s11214-008-9393-5.

Singer, H., L. Matheson, R. Grubb, A. Newman, and D. Bouwer (1996), Monitoring space weather with the GOES magnetometers, *Proc. SPIE*, **2812**, 299, doi:10.1117/12.254077.

Singer, H. J., W. P. Sullivan, P. Anderson, F. Mozer, P. Harvey, and J. Wygant (1992), Fluxgate magnetometer instrument on CRRES, *J. Spacecr. Rockets*, **29**, 599.

Southwood, D. J., J. W. Dungey, and R. J. Etherington (1969), Bounce resonant interaction between pulsations and trapped particles, *Planet. Space Sci.*, **17**, 349.

Sturrock, P. A. (1994), *Plasma Physics*, 335 pp., Cambridge Univ. Press, Cambridge.

Takahashi, K., C. T. Russell, and R. R. Anderson (1985), ISEE 1 and 2 observations of the spatial structure of a compressional Pc 5 wave, *Geophys. Res. Lett.*, **12**, 613.

Tu, W., S. R. Elkington, X. Li, W. Liu, and J. Bonnell (2012), Quantifying radial diffusion coefficients of radiation belt electrons based on global MHD simulation and spacecraft measurements, *J. Geophys. Res.*, **117**, A10210, doi:10.1029/2012JA017901.

Tu, W., G. S. Cunningham, Y. Chen, M. G. Henderson, E. Camporeale, and G. D. Reeves (2013), Modeling radiation belt electron dynamics during GEM challenge intervals with the DREAM3D difusion model, *J. Geophys. Res.*, **118**(10), 6197, doi:10.1002/jgra.50560.

Turner, D. L., V. Angelopoulos, Y. Shprits, A. Kellerman, P. Cruce, and D. Larson (2012), Radial distributions of equatorial phase space density for outer radiation belt electrons, *Geophys. Res. Lett.*, **39**(9), L09101, doi:10.1029/2012GL051722.

Wygant, J. R., P. R. Harvey, D. Pankow, F. S. Mozer, N. Maynard, H. Singer, M. Smiddy, W. Sullivan, and P. Anderson (1992), CRRES electric field/Langmuir probe instrument, *J. Spacecr. Rockets*, **29**, 601.

5

Modeling the Energetic Particles of the Inner Magnetosphere

S. Bourdarie, V. K. Jordanova, M. Liemohn, and T. P. O'Brien

■ Abstract

To model the energetic charged particle of the inner magnetosphere, several possibilities are available depending on particle properties (trapped or untrapped). The global modeling of particle transport is very difficult to perform, first because it is a six-dimensional system, second the initial populations are often not known in the entire magnetosphere, third because global electric and magnetic fields are needed everywhere as a function of time, and fourth the populations are not measured everywhere in the magnetosphere during the events. Of course in recent years the global knowledge of the inner magnetosphere has greatly increased, especially with the recent launch of the Van Allen Probes (Mauk et al., 2014). The transport of low-energy particles by changes in magnetic and electric fields leads to the rise of the ring current population, which is composed of particles in the keV-energy range. Then the ring current, together with the magnetic and electric perturbations and VLF waves, modifies the high-energy populations (MeV range). So to perform this entire modeling, keV particle transport must be determined, as well as high-energy particle transport. Further, the cold (eV) plasma mediates wave-particle interactions at higher energies by modifying the propagation and dispersion relations for waves.

Waves, Particles, and Storms in Geospace. First Edition. Georgios Balasis, Ioannis A. Daglis, Ian R. Mann.

Trapped particle transport theory will be discussed in Section 5.1 followed by more detailed descriptions of the ring current modeling and radiation belts in Sections 5.2 and 5.3.

5.1 Background

In the Earth's magnetosphere, the magnetic field is the sum of two terms: one of internal (main component) and the other of external origin. Realistic models of the internal field exist, the old models such as Jensen and Cain 1962 (Jensen and Cain, 1962), GSFC 12/66, and the modern International Geophysical Reference Field model (IGRF) (Cain et al., 1967). Up to a distance of some Earth radii, 5 to 6, the magnetic field is close to a dipolar field and the magnetosphere is more or less in revolution with the Earth. Beyond that distance, the external fields become less and less negligible and contribute to the deformation of the dipolar internal field. These external fields are the sum of several components: magnetopause currents, tail currents, field-aligned currents, and the ring current. Several empirical models of the external field exist: we can mention for example the models of Tsyganenko (1987, 1989, 1996, 2003, 2005), Olson and Pfitzer (1977), Pfitzer et al. (1988), and Alexeev et al. (2000).

All charged particles immersed in an electromagnetic field will be subject to the Lorentz force:

$$\vec{F} = q(\vec{v} \times \vec{B} + \vec{E}) \tag{5.1}$$

where q is the particle's charge, $\vec{v}$ its speed, $\vec{B}$ the magnetic field, and $\vec{E}$ the electric field.

If the magnetic field is very strong and the energy of the particles is great (and therefore their speed too) then the effect of the electric field can be ignored in eqn 5.1 and the Lorentz force is reduced to

$$\vec{F} \approx q(\vec{v} \times \vec{B}) \tag{5.2}$$

Under these conditions, the movement of the high-energy particles can be generally broken down into three basic periodic movements (Northrop, 1963): the gyration around the field line, the bounce between two mirror points, and the drift around the dipole axis. In this case, the periods associated with the three motions are quite different and a variable splitting is possible to solve the equation of motion, eqn 5.2. The global motion is quasi-periodic and the particle is trapped by the magnetic field. Such particles compose the radiation belts or Van Allen belts.

If the particle energy is low (in the keV range) then electric field effects can no longer be ignored and eqn 5.1 must be solved. Such particles are denoted as ring current particles in the following. The motion of these particles is governed by the magnetospheric convection and corotation electric fields, as well as by drifts related to the gradient and curvature of the geomagnetic field. The outer boundary of the ring current is located where the magnetic field is no longer able to maintain closed particle orbits around the Earth (stable trapping), while the inner boundary is determined by the dense atmosphere. Thus for typical ring current energies 1–300 keV the region under consideration extends between 2 and 8 Earth radii.

5.1.1 Trapped particle transport theory

In order to understand and reproduce the dynamics of the charged particles present in the magnetosphere, it is common to define magnetic coordinates (Fig. 5.1). r is the distance from the center of the dipole to the point under consideration, λ its latitude (and θ its colatitude: $\theta = \pi/2 - \lambda$), and ϕ its magnetic longitude. A field line (or force line) is defined by the McIlwain parameter, L (McIlwain, 1961) roughly equal (only true with a dipolar field) to the distance (expressed in planet radii) from the center of the planet to the intersection point of that force line with the magnetic equatorial plane. A point on a force line is defined by the B parameter, modulus of the magnetic field at the point under consideration (B being minimum at magnetic equator). B and L then represent a coordinate system linked to the model of the magnetic field under consideration.

The trapped particle motion evolution is described by eqn 5.2 which can be re-written as:

$$\vec{F} = \frac{d\vec{p}}{dt} = q(\vec{v} \times \vec{B}) \tag{5.3}$$

where $\vec{p}$ is the particle momentum and t the time.

The solution of eqn 5.3 defines the particle motion evolution in a magnetic field. In the case of energetic trapped (the $\vec{E}$ field contribution is negligible) charged particles,

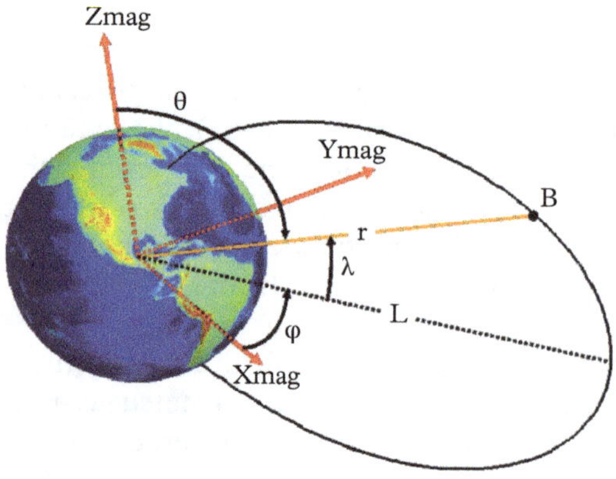

Figure 5.1 *Magnetic coordinates. Zmag is the dipole axis.*

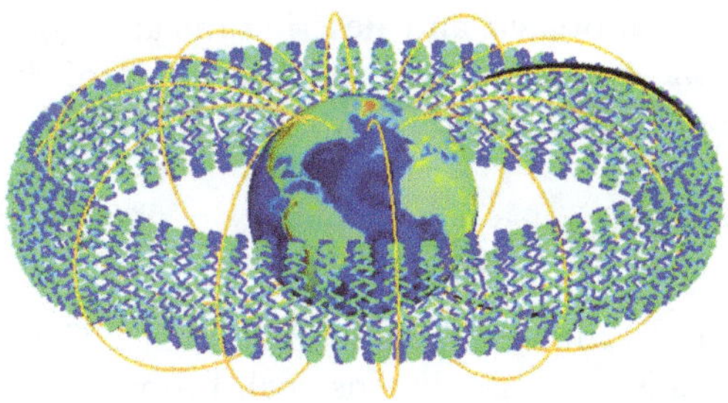

Figure 5.2 *Trapped charged particle motion in the Earth's magnetic field.*

its motion may be decomposed into three elementary periodic motions: the gyration around the field line, the bounce between two mirror points, their location being related to the particle pitch angle, and the drift around the dipole axis (Fig. 5.2). Note that in this case, the Larmor (gyration) radius of the particle must be smaller than the field curvature radius, otherwise the particle escapes from the terrestrial environment. A charged particle submitted to these three basic (Northrop, 1963) and periodic movements then moves through torus shaped surfaces around the Earth, which are commonly called drift shells. The periods associated with each of these basic movements for a 3 MeV electron at $L = 3$ are respectively 2.14×10^{-4} s, 0.19 s, and 504 s. The disparity between the periods is very great; a factor of the order of 1000 should be noted between each of them going from the gyration movement to the drift movement.

To solve eqn 5.3 a time step less than 0.1 ms is required and a calculation duration of at least 500 s must be performed to make sure that the particle is trapped. This is only feasible with a few particles. To describe an ensemble of trapped particles, further hypotheses are necessary. It is convenient to perform this calculation under the guiding center approximation (GCA): the center of the Larmor radius of the particle is taken as a reference. A particle's trajectory then follows a drift shell. In this approximation, eqn 5.3 becomes

$$\frac{d\vec{p}}{dt} = q\vec{v} \times \vec{B} - \mu\vec{\nabla}B \tag{5.4}$$

where $\vec{v}$ is the velocity of the guiding center, $\vec{B}$ the magnetic field at particle location, and μ the relativistic magnetic particle momentum (which is constant along its trajectory). One must note that under the GCA approximation, the gyration phase of the particle is unknown (i.e., this information is lost). The fundamental relationship dynamics (eqn 5.4) cannot be solved for a number of particles large enough so that it perfectly samples the radiation belt population, and therefore a macroparticle weighting must be used to approximate the full distribution.

Then, it is convenient to adopt a statistical approach and describe radiation belt particle transport using the Vlasov formulation. In this case, the time evolution of the phase space distribution function, $f(\vec{r}, \vec{p}, t)$, is governed by the following equation:

$$\frac{\partial f}{\partial t} + \vec{v}\vec{\nabla}_{\vec{r}} f + q(\vec{v} \times \vec{B})\vec{\nabla}_{\vec{p}} f = 0 \tag{5.5}$$

where the distribution function is defined in the position-momentum phase space $(\vec{r}, \vec{p})$. However, eqn 5.5 is still difficult to solve because it is six dimensional. Properties of trapped particle motion with three periodic motions are then very helpful to simplify the problem. For each basic motion (gyration, bounce, and drift) an adiabatic invariant can be defined as (Landau and Lifchitz, 1968):

$$\mathcal{J}_i = \frac{1}{2\pi} \oint \vec{\Pi}_i d\vec{r}_i \tag{5.6}$$

where $\vec{\Pi}$ is the generalized momentum conjugated with $\vec{r}$ and i denotes the elementary periodic motion (i=1 is the gyration, i=2 is the bounce motion, and i=3 is the drift). The integration is performed over one full period of the periodic motion considered. $\mathcal{J}_i$ is constant (is an invariant) as long as all problem parameters are varying on times scales longer than the corresponding motion period. Note that the first adiabatic invariant is proportional to the relativistic magnetic moment, M, of the particle and the third one is proportional to the magnetic flux, Φ, contained in the drift shell. L^* (Roederer, 1970) is defined as:

$$L^* = 2\pi \frac{B_o}{\Phi} \tag{5.7}$$

where B_o is the field magnitude on the equator at the Earth's surface.

A canonical transformation is then easily made from $(\vec{r}, \vec{p})$ phase space to $(\vec{\mathcal{J}}, \vec{\varphi})$ phase space (Landau and Lifchitz, 1968). The $\mathcal{J}_i (i = 1, 2, 3)$ variables are then called the action variables, and the conjugated quantities are the angle variables (or phase) φ_i. In this new phase space, the Vlasov eqn 5.5 is written as:

$$\frac{\partial f}{\partial t} + \frac{d\vec{\mathcal{J}}}{dt}\frac{\partial f}{\partial \vec{\mathcal{J}}} + \frac{d\vec{\varphi}}{dt}\frac{\partial f}{\partial \vec{\varphi}} = 0 \tag{5.8}$$

where $f(\vec{\mathcal{J}}, \vec{\varphi}, t)$ is the distribution function. Note that for a stably trapped particle then the action variables are exactly the adiabatic invariants and $\frac{d\mathcal{J}_i}{dt} = 0 (i = 1 : 3)$. Because energetic charged particles trapped in the inner magnetosphere are affected by macroscopic (like magnetic field perturbations) and microscopic (presence of waves, presence of other particles, atoms and molecules) perturbations, their trajectory is then modified (e.g., Roederer, 1970; Boscher et al., 1996). Finally, particles can collide with other particles, atoms or molecules, leading to processes like charge exchange or nuclear interactions.

Most of these phenomena may be well described by diffusion processes. For example, when a wave propagates along a field line, a gyrating particle will interact with the wave field if the wave frequency (corrected by the Doppler shift) is a multiple of the particle gyrofrequency. This interaction is called a resonant interaction. Such a microscopic perturbation is faster than the particle gyration period (consequently faster than bounce and drift periods), so the first invariant is violated and consequently the second and the third: that leads to pitch angle and energy diffusion.

From the Vlasov point of view, the microscopic processes are taken into account by modifying the distribution function. Equation 5.8 becomes

$$\frac{\partial f}{\partial t} + \frac{d\vec{\mathcal{J}}}{dt}\frac{\partial f}{\partial \vec{\mathcal{J}}} + \frac{d\vec{\varphi}}{dt}\frac{\partial f}{\partial \vec{\varphi}} = \left(\frac{df}{dt}\right)_{interaction} \tag{5.9}$$

The new right-hand side term is in general calculated by a Fokker–Planck operator for simplicity. The resulting Fokker–Planck equation (or diffusion equation) with all microscopic processes accounted for is

$$\frac{\partial f}{\partial t} + \sum_{i=1}^{3}\frac{d\mathcal{J}_i}{dt}\frac{\partial f}{\partial \mathcal{J}_i} + \sum_{i=1}^{3}\frac{d\varphi_i}{dt}\frac{\partial f}{\partial \varphi_i} = -\sum_{i=1}^{3}\frac{\partial (F_i f)}{\partial \mathcal{J}_i} + \sum_{i=1}^{3}\sum_{j=1}^{3}\frac{\partial}{\partial \mathcal{J}_i}\left(D_{ij}\frac{\partial f}{\partial \mathcal{J}_i}\right) + Sources - Losses \tag{5.10}$$

where the processes are taken into account by diffusion coefficients D_{ij}, friction terms F_i, and internal losses and sources. For trapped particles, frictional processes are due to the deceleration of particles which interact by Coulomb collisions with free and bounded geocoronal electrons. Diffusion terms are essentially the pitch angle diffusion by Coulomb collisions and wave-particle interactions, the energy diffusion is due to wave-particle interactions, and the radial diffusion is due to field perturbations.

Equation 5.10 still describes a 6D problem. But from there, it is very easy to reduce the dimensionality of the problem following appropriate hypotheses. For example, one can typically neglect gyrophase structure, which averages out in a few gyration periods, so eqn 5.10 can be averaged over the gyration phase. The problem is then reduced to a 5D problem and the distribution function, f, depends only on $\mathcal{J}_1, \mathcal{J}_2, \mathcal{J}_3, \varphi_2, \varphi_3$ and time. Following this strategy, the dynamics of trapped particles is not accessible on time scales shorter than the particle gyration period. It is possible to pursue the same strategy and average eqn 5.10 over the gyration and bounce phase to get a 4D description of trapped particle dynamics (see for example Bourdarie et al., 1997; Jordanova et al., 1994). It is then equivalent to 2D in space and 2D in velocity. A common formulation for radiation belt particles is a 3D description (equivalent to 2D in space and 1D in velocity) where eqn 5.10 is averaged over the three phases (see for example: Schulz and Lanzerotti, 1974; Beutier and Boscher, 1995; Beutier et al., 1995; Bourdarie et al., 1996; Selesnick et al., 2007; Albert et al., 2009; Subbotin et al., 2010; Xiao et al., 2010). In this case, particle distributions are assumed to be phase-invariant and the dynamics of trapped particles is not accessible on time scales shorter than the particle drift period (i.e., a couple of hours to tens of minutes for energies in the range 100 keV to 1 MeV).

5.2 Modeling ring current particles

The major sources of magnetospheric plasma are the solar wind and the ionosphere. Particles injected far downstream in the magnetic tail are transported sunward into regions of stronger magnetic field and are energized. This process, called magnetospheric convection, is governed by the $\vec{E} \times \vec{B}$ particle drift due to the large-scale electric field that extends from dawn to dusk across the nightside magnetosphere. On the other hand, the Earth's magnetic dipole field corotating with the Earth induces an electrostatic field known as corotation electric field, whose equipotential surfaces are closed. Cold (eV) plasma, following the closed equipotential surfaces and corotating with the Earth, forms the plasmasphere. The last closed equipotential surface can be associated with the plasmapause. Inside the plasmasphere ionospheric plasma builds up and densities of the order of 10^3–$10^4 cm^{-3}$ are attained.

Superimposed on the drifts of charged particles in the electric fields are drifts resulting from the gradient of the magnetic field strength and the curvature of the magnetic field lines. In contrast with the electric drifts, magnetic drifts are energy dependent; therefore they are negligible for low energies $\sim$ 1 eV (thermal plasma) and they are dominant for high energies $\sim$ 1 MeV (radiation belt particles). For intermediate energy particles $\sim$ 1–200 keV, both electric and magnetic drifts have comparable importance in the determination of the particle trajectories. In addition, since the gradient-curvature drift is charge dependent, ions drift westward around the Earth, while electrons drift eastward and a current is formed around the Earth called the ring current. This current produces a magnetic field depression observable at the surface of the Earth that is expressed through the *Dst* index.

The simplest modeling approach to study ring current particle dynamics is the simulation of the single particle motion, which solves the equation of motion (eqn 5.1) for an individual particle under the influence of external electric and magnetic fields. Ring current particles follow either open or closed drift paths depending on their energy, pitch angle, and location in space. Their trajectories have been studied during different magnetospheric conditions using adiabatic drift theory by several authors (e.g., Chen, 1970; Takahashi and Iyemori, 1989; Kistler and Larson, 2000). Such investigations are very useful in order to explain not only the process of transport, access, and formation of the stormtime ring current, but also some features in the measured energy spectra of the ring current ion species like dips in different local time sectors (McIlwain, 1972) and nose events (Smith and Hoffman, 1974; Ejiri, 1978). Figure 5.3 shows the bounce-averaged ion drift paths obtained by Kistler et al. (1999) for the geomagnetic storm of February 18, 1998, using a dipole magnetic field combined either with the Kp-dependent Volland (1973)–Stern (1975) electric field or with the IMF-dependent Weimer (1996) electric field. Over the plotted energy range from $\sim$ 3 to 35 keV, the ion drift changes from the eastward and sunward $\vec{E} \times \vec{B}$ drift at lower energies, to the westward gradient and

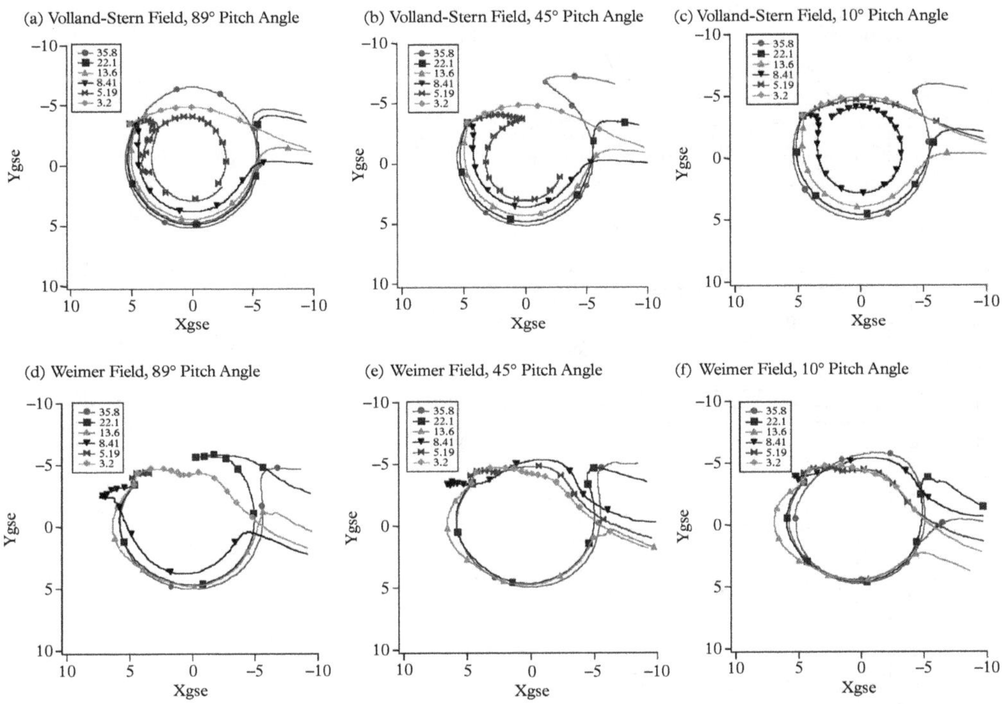

Figure 5.3 *Bounce-averaged drift paths for ions which arrive at L = 5.75, ~ 10MLT with 6 different energies (36., 22., 14., 8., 5., and 3. keV). The paths were calculated using a dipole magnetic field and either a Volland-Stern electric field (a–c) or a Weimer electric field (d–f) for 89., 45., and 10. degree pitch angle, respectively. The legend gives the ion energy in keV.*
(Kistler et al., 1999).

curvature drifts at higher energies. Note that the transition region between the eastward and westward ion drift paths is between 3 and 5 keV for the Volland-Stern electric field, while for the Weimer electric field it is at a little higher energy ~ 8 to 13 keV, particularly at smaller pitch angles. Other differences are that the eastward drift in the Weimer field is significantly slower than in the Volland-Stern field and that the westward Volland-Stern drift paths tend to go closer to the Earth than the Weimer paths. Characteristic features called stagnation dipsİ appear in measured ion spectra at energies where particles have little or no azimuthal velocity due to the compensation of the eastward and westward ion drifts (e.g., Lennartsson et al., 1979). Comparisons with observations from the Equator-S ion composition instrument (ESIC) indicated that simulations using the Weimer electric field reproduced much better the observed spectral features; neither electric field model, however, accurately predicted the energies of the observed minima (Kistler et al., 1999).

Another example is shown in Figure 4, comparing the total energy content of the keV-energy-range protons against the observed Dst index time series for a magnetic storm

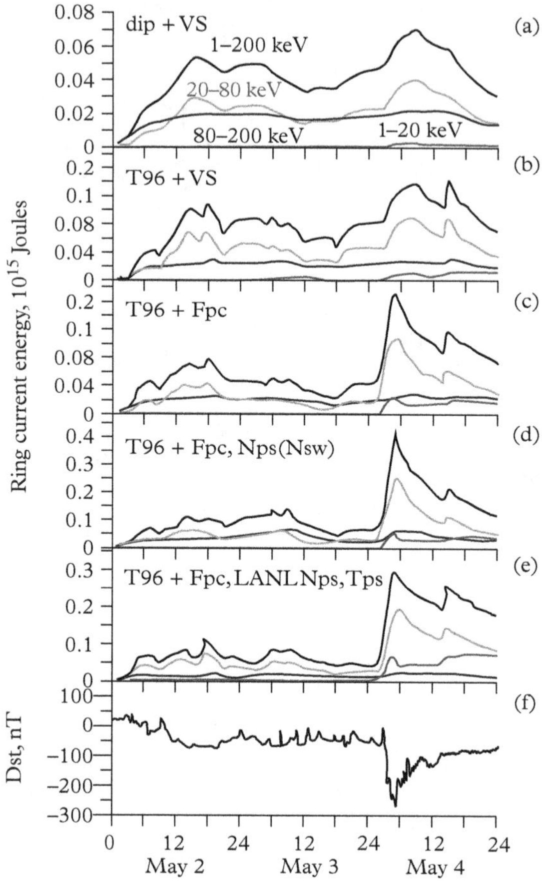

Figure 5.4 *Simulation results from a particle tracking model, integrating the results throughout the simulation domain for comparison with Dst (bottom panel). (From Ganushkina et al., 2005.)*

(Ganushkina et al., 2005). Each panel in this figure shows model results with a different configuration of the magnetic field (dipole or T96 model), electric field (with intensity defined by Kp or solar wind), and plasma outer boundary condition (constant, solar wind driven or as observed by LANL geosynchronous satellites).The results illustrate the usefulness of this approach to quantify the realism of various outer boundary conditions at reproducing the near-Earth magnetic perturbation, and the relative contribution of specific energy ranges to this distortion.

5.2.1 Kinetic ring current models

As discussed in Section 5.1, a different approach to study plasma dynamics is adopted by the kinetic theory. Instead of solving the equation of motion for each individual particle, it uses a statistical approach and looks at the development of the distribution function for the system of particles. The average values of the various physical quantities of interest, such as number density, flow velocity, particle flux, and so on, can be deduced by taking the moments of the distribution function, and used to describe the macroscopic behavior of plasma. The effects of collisions are incorporated into the kinetic equation through a collision term, usually the Fokker–Planck diffusion term.

Most of the present ring current models are 4D (gyro- and bounce-averaged), and use radial distance from Earth and magnetic local time (MLT) as spatial variables, but either energy and pitch angle (e.g., Jordanova et al., 1994) or the first and second adiabatic invariants (e.g., Fok et al., 1995; Chen et al., 1998) as the other two independent variables for the phase space distribution function f. Using the brackets <> to denote bounce-averaging, the kinetic eqn 5.10 becomes

$$\frac{\partial f}{\partial t} + \frac{1}{R_o^2}\frac{\partial}{\partial R_o}\left(R_o^2\left\langle\frac{dR_o}{dt}\right\rangle f\right) + \frac{\partial}{\partial \varphi}\left(\left\langle\frac{d\varphi}{dt}\right\rangle f\right) + \frac{1}{\gamma p}\frac{\partial}{\partial E}\left(\gamma p\left\langle\frac{dE}{dt}\right\rangle f\right)$$
$$+ \frac{1}{h\mu_o}\frac{\partial}{\partial \mu_o}\left(h\mu_o\left\langle\frac{d\mu_o}{dt}\right\rangle f\right) = \left\langle\left(\frac{df}{dt}\right)_{ce}\right\rangle + \left\langle\left(\frac{df}{dt}\right)_{cc}\right\rangle + \left\langle\left(\frac{df}{dt}\right)_{wp}\right\rangle + \left\langle\left(\frac{df}{dt}\right)_{atm}\right\rangle \quad (5.11)$$

Here R_o is the radial distance in the magnetic equatorial plane, φ is the geomagnetic east longitude, E is the kinetic energy p is the relativistic momentum of the particle, γ is the relativistic factor, and $\mu_o = cos(\alpha_o)$, where α_o is the equatorial pitch angle and $h(R_o, \varphi, \mu_o)$ is proportional to the half-bounce path length (e.g., Jordanova et al., 1996). The left-hand side of eqn 5.11 describes the transport and acceleration of charged particles along adiabatic drift paths in time-dependent electric ($\vec{E}$) and magnetic ($\vec{B}$) fields. The bounce-averaged $\vec{E} \times \vec{B}$ and magnetic gradient-curvature drift velocities (e.g., Roederer, 1970; Jordanova et al., 2010) for an arbitrary magnetic field are

$$\left\langle \vec{V}_{ED} \right\rangle = \frac{\vec{E}_o \times \vec{B}_o}{B_o^2} \quad (5.12)$$

$$\left\langle \vec{V}_{GC} \right\rangle = \frac{2p}{q\tau_B B_o}\nabla_o(R_o I) \times \vec{b}_o \quad (5.13)$$

where τ_B is the bounce period and $\vec{b}_o$ is the unit vector in the direction of $\vec{B}$. The purely field-geometric integral $I(R_o, \varphi, \mu_o)$ in the above expression is proportional to the second adiabatic invariant.

In addition to ring current changes due to time-dependent convection, several loss mechanisms like charge exchange, Coulomb collisions, wave-particle interactions, and collisions at low altitude with the dense atmosphere contribute to ring current decay. The most important among these loss processes is charge exchange, where ring current ions are neutralized on collision with thermal exospheric hydrogen atoms (e.g., Kistler et al., 1989). The resulting energetic neutral atoms (ENA) are no longer trapped by the Earth's magnetic field and are either lost in space or precipitate down to the atmosphere. ENA were used in the past NASA/IMAGE mission to obtain global images of the ring current (e.g., Burch, 2000). Coulomb collisions between energetic ring current ions and thermal plasma result in energy transfer from the fast moving particles to the thermal particles and in angular deflection of the particles (Wentworth et al., 1959). The Coulomb decay lifetimes are comparable to or smaller than charge exchange lifetimes at low (< 5 keV) energies (Fok et al., 1991). Scattering of ring current particles into the loss cone due to resonant interactions with electromagnetic ion cyclotron (EMIC) waves occurs on short timescales and could contribute significantly to ion loss (Cornwall et al., 1970; Lyons and Thorne, 1972), especially during the main phase of the storm when ring current energy loss timescales may be as short as few hours (Gonzalez et al., 1989). All these mechanisms are included in the right-hand side of eqn 5.11. Losses through the dayside magnetopause are included by allowing free outflow of particles from the dayside boundary. Wave-particle interactions are incorporated according to quasi-linear theory, first calculating the equatorial growth rate of EMIC waves, integrating the local growth rates along wave paths to estimate the wave amplitudes, and then calculating the corresponding diffusion coefficients (e.g., Jordanova et al., 2001).

To investigate the role of various collisional loss processes in ring current dynamics, Jordanova et al. (1999) simulated off-equatorial ring current proton spectra measured by Polar during the moderate storm on March 20–22, 1996. They used a kinetic drift-loss model, which solved the bounce-averaged eqn 5.11 in a dipole magnetic field configuration. The model was driven by a Kp-dependent Volland-Stern electric field. The particle injection from the magnetotail was specified after nightside data measurements from instrument on LANL geosynchronous satellites. The calculated H^+ distribution function was compared with HYDRA, TIMAS, and CEPPAD data for different magnetic latitudes, MLT, and L shells. In order to do such comparisons, the modeled distribution was mapped along magnetic field lines to the position of the spacecraft, conserving the first and second invariant. Several tests were made varying the offset parameter in the Volland-Stern electric field model, and the best results were achieved when the symmetry line of the plasma convection paths was offset by 3 hours eastward. Collisional losses have larger effect on the postnoon spectra since particles experience losses for a longer time until they reach the dayside sector. Charge exchange and Coulomb collision losses reduced significantly the proton distribution at low L shells due to the higher geocoronal and plasmaspheric densities. At large L shells the loss from collisions was noticeable mostly for the slowly drifting ions within the stagnation dip. The agreement at large ($L = 5.5$) shells was satisfactory over a wide energy range except near the dip centered at $\sim$ 4 keV, where the model overestimated the losses. These features were strongly emphasized at low ($L = 3.5$) shells. Comparisons of the postmidnight spectra indicated that the agreement was very good at $L = 5.5$, while at lower ($L = 4.5$) shells, the model overestimated the data between 20 and 60 keV and underestimated the data at higher energies (> 100 keV). Loss processes were negligible in the postmidnight local time sector since the low-energy (< 50 keV) ions were on open drift trajectories and were continuously injected from the plasmasheet. The underestimation of the high-energy population at $L = 4.5$ could be due to the need of a more realistic electric field model or better initial conditions.

The relative importance of various mechanisms in the decay rate of the ring current was investigated by many authors but open questions remain. An initial study of

Hamilton et al. (1988), based on AMPTE/CCE observations, stated that the faster, early stage recovery from the *Dst* minimum is dominated by the O^+ charge-exchange loss rate and the slower, later stage recovery is governed by the H^+ charge exchange loss rate. Daglis (1997) also found observational evidence for this mechanism in CRRES measurements, noticing that the fraction of ring current energy from O^+ drops simultaneously with *Dst*. Recent simulation studies have shown, however, the importance of the convection electric field in controlling ring current evolution not only through the energization of ring current particles but also by affecting the dayside loss of ring current ions through the magnetopause (e.g., Takahashi and Iyemori, 1989; Jordanova et al., 1999; Liemohn et al., 1999). During the main phase of a geomagnetic storm, charged particles that are injected from the magnetotail are transported earthward and accelerated by the enhanced cross-tail convection electric field, thus increasing the total energy content of the preexisting ring current population; particles remaining on open drift paths are lost as they drift through the dayside magnetopause. Liemohn et al. (1999) simulated ring current evolution during both the solar maximum storm of June 4–7, 1991, and the solar minimum storm of September 24–27, 1998, and thus found that convective drift loss out the dayside magnetopause is the dominant process in removing ring current particles. This was especially true for the 1998 storm, where dayside losses became an order of magnitude larger than the next most significant loss process, charge exchange.

The effects of plasma sheet variations on the fast initial ring current decay were also investigated. Simulating the March 31, 2001, storm Jordanova et al. (2003) demonstrated that the drop of plasma sheet density during the early recovery phase is important for the fast ring current decay. Comparisons of the magnetic field disturbance of the ring current on ground, obtained from three simulation runs performed by Jordanova et al., with the measured Dst index and magnetopause currents corrected SYM-H index are shown in Fig. 5.5. In the first simulation (solid) the boundary conditions were updated as the storm evolved according to LANL measurements (Fig. 5.5, top). In the second (dashed-dotted), the boundary conditions were updated until hour 32 of the simulation and were kept at this enhanced level after that. In the third (dotted), the prestorm plasma sheet measurements at the start of the simulation were used as boundary conditions throughout the modeled period. The convection potential was kept the same in all three numerical simulations. The first (solid) simulation using LANL data showed very good agreement with observations. The second (dashed-dotted) simulation demonstrated that an additional ~ 50 nT enhancement in Dst magnitude occurs if the plasma sheet density drop is not taken into account. The third (dotted) simulation demonstrated that there is no preconditioning of the ring current during this storm period; i.e., the second enhancement in the convection electric field was not more effective in building the ring current than the first in the case of the same magnetospheric plasma inflow on the nightside. This study showed that the fast initial ring current decay is controlled not only by the decreased convection electric field, dayside outflow through the magnetopause, and internal loss processes, but also by the time-varying nightside inflow of plasma from the magnetotail. All factors need to be considered for accurate predictions of ring current decay timescales and Dst index.

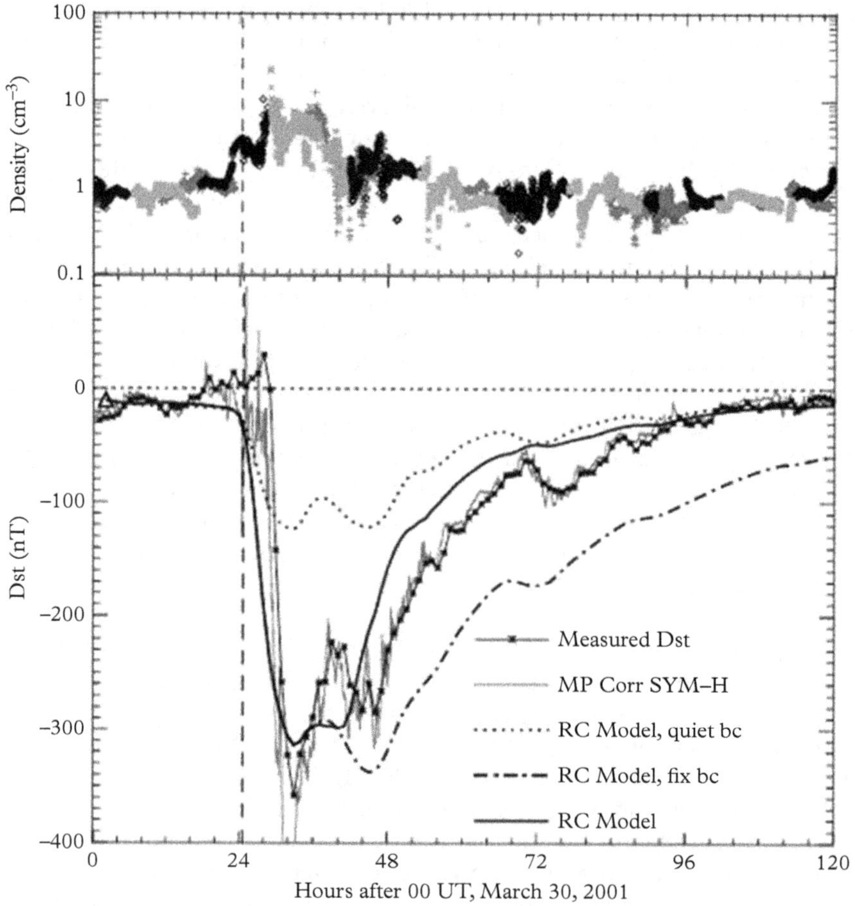

Figure 5.5 *(top) MPA plasma sheet ion density, when the LANL satellites were on the nightside. (bottom) Comparison of three model simulations with Dst and SYM-H measurements.*
(after Jordanova et al., 2003).

Finally, the possible role of precipitation losses in eroding the storm time ring current is still under investigation. Low-altitude satellite measurements indicate that precipitating ion fluxes appear with highly variable magnitude and location (e.g., Cornwall et al., 1971; Hultqvist et al., 1976; Sóraas et al., 1977). Thus, in addition to the main precipitation zone at higher latitudes, characterized by roughly isotropic proton pitch angle distributions, an equatorward region of proton fluxes with lower intensities is observed. During lower geomagnetic activity, the peak of the fluxes usually reduces and shifts outward it can be as high as $L = 6$. These low-altitude regions of energetic ion precipitation presumably map to the ring current/inner plasma sheet region and have been attributed to scattering of ions in stretched magnetic field configurations (Sergeev et al., 1993) or to resonances with ion-cyclotron waves (Williams and Lyons, 1974a, Williams and Lyons, 1974b). Using a combination of observations and model results Kozyra et al. (1998) investigated the ring current decay during the February 6–10, 1986, great magnetic storm when intense ion precipitation was observed during the rapid early recovery phase. The estimated precipitation losses from NOAA-6 and DMSP observations were thus found to be of equal magnitude to the O^+ charge exchange losses during the early recovery phase. The longer decay timescales during the late recovery phase resulted from a combination of O^+ and H^+ charge exchange losses. Jordanova (2005) simulated ring current evolution during the large storm of October 21–25, 2001, and calculated the H^+ energy losses due to charge exchange, Coulomb collisions, and ion precipitation. It was found that charge exchange losses were the largest, while Coulomb collisions losses remained about two orders of magnitude smaller. Resonant interactions with He^+ band EMIC waves scattered high-energy particles into the loss cone, which resulted in significant proton precipitation and more than 10% reduction of the total ring current energy. Ion precipitation losses became comparable to charge exchange losses near Dst minima and accounted for a substantial fraction of the losses during the early recovery phase of the storm. The effects of plasma wave scattering on the heavy ion components have not been investigated yet, nor the scattering by H^+ or O^+ band EMIC waves. All these topics need to be addressed in future studies. Plasma wave measurements from the recently launched August 2012 Van Allen Probes mission (Mauk et al., 2014) will help determining the spatial extent of EMIC waves during storm time and evaluating the global impact of ion precipitation losses to ring current decay.

5.2.2 Self-consistent models

A better physics description of the coupled inner magnetosphere processes is provided in numerical models that calculate self-consistently the electric and magnetic fields in which the particles drift. Recent modeling efforts focusing on the development of self-consistent calculations of the magnetospheric electric and magnetic fields are briefly described below.

5.2.2.1 Self-consistent electric field models

Because the keV-energy ions dominate the plasma pressure of the inner magnetosphere, this population controls the magnetic and electric field distortions of near-Earth space (e.g., Parker and Stewart, 1967; Wolf, 1974). This is included by calculating the currents associated with the hot-ion pressures within the simulation domain; Parker (1957) provided the formulas for relating particle pressure to the perpendicularly flowing electric currents, both from the gradient-curvature drift effect:

$$\vec{J}_{\perp,GC} = \frac{\vec{B}}{B^2} \times \left[P_\perp \frac{\nabla B}{B} + P_\parallel \frac{(\vec{B} \cdot \nabla)\vec{B}}{B^2} \right] \tag{5.14}$$

and from the magnetization current:

$$\vec{\jmath}_{\perp,M} = \frac{\vec{B}}{B^2} \times \left[\nabla P_\perp - P_\perp \left(\frac{\nabla B}{B} + \frac{(\vec{B} \cdot \nabla)\vec{B}}{B^2} \right) \right] \tag{5.15}$$

The addition of eqns 5.14 and 5.15 yields

$$\vec{\jmath}_\perp = \frac{\vec{B}}{B^2} \times \left[\nabla P_\perp + (P_\parallel - P_\perp) \frac{\nabla B}{B} \right] \tag{5.16}$$

By enforcing current closure, the divergence of $\mathcal{J}_\perp$ from eqn 5.16 can then be accumulated along each magnetic field line to find the field-aligned current flowing into or out of the ionosphere as a function of magnetic latitude and local time (cf., Liemohn et al., 2001). Some studies bypass this two-step calculation by assuming isotropy and using the Vasyliunas equation, which directly relates the equatorial plasma conditions to field-aligned current intensity (Vasyliunas, 1970; Birmingham, 1992),

$$\vec{\jmath}_{\parallel,iono} = \frac{B_{iono}}{2B_{eq}^2} \left(\nabla P_{eq} \times \vec{B}_{eq} \right) \cdot \nabla_{eq} \left(\int \frac{dl}{B} \right) \tag{5.17}$$

where the subscripts "iono" and "eq" refer to quantities at the ionospheric and equatorial plane locations along the magnetic field line. These field-aligned currents can then be used as source and sink terms in a Poisson equation solution of the ionospheric electric potential Φ,

$$\vec{\jmath}_R = [\nabla_\perp \cdot (\Sigma \cdot \nabla\Phi)_\perp] \tag{5.18}$$

where $\mathcal{J}_R$ is the radial current density and Σ is the height-integrated ionospheric conductance. The specification of conductance plays a critical role in the solution of eqn 5.18 and the location and evolution of the nightside ionospheric conductance can significantly influence the resulting potential pattern (e.g., Wolf, 1970; Ebihara et al., 2004; Liemohn et al., 2005). Solution of eqn 5.18 yields a distribution of electric potentials throughout the latitude range of the simulation, which can then be mapped back out to the magnetosphere (assuming equipotential field lines) for use in the drift physics model that produced the original hot-ion pressures used in eqns 5.16 or 5.17. That is, the electric field within the ring current model can be generated self-consistently with the solution of the particle phase space density, allowing for the nonlinearity of this feedback loop to influence the evolution of the numerical result.

Several simulation approaches have used this technique (e.g., Jaggi and Wolf, 1973; Fok et al., 2001; Liemohn et al., 2004), and they have been used extensively to investigate a number of storm-time phenomena. Self-consistent electric fields successfully describe the pattern of undershielding of high-latitude electric fields during sudden interplanetary magnetic field (IMF) southward turnings (e.g., Harel et al., 1981; Spiro et al., 1988), the formation of a deep potential well near local midnight and the resulting Harang reversal

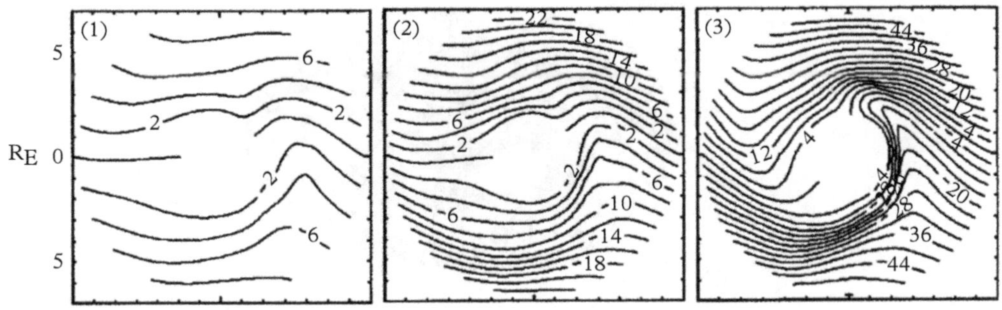

Figure 5.6 *Electric potential distributions in the inner magnetosphere from a self-consistent electric field drift physics simulation of a magnetic storm. The lines are equipotentials, spaced every 4 kV, shown at three moments during the storm main phase. The view is over the north pole, with the Sun to the left and distances given in Earth radii.*
From Fok et al. (2001).

in the convection pattern (e.g., Wolf et al., 1982; Ridley and Liemohn, 2002; Fok et al., 2003), the formation of subauroral polarization streams in the dusk/evening sector (e.g., Fok et al., 2001; Liemohn et al., 2006), overshielding within the inner magnetosphere during northward IMF turnings (e.g., Kelley et al., 1979; Goldstein et al., 2002), interchange instabilities in the inner magnetosphere (e.g., Sazykin et al., 2002; Zheng et al., 2006), and the presence of small-scale structure in the plasmasphere and the hot-ion pressure distribution (Gallagher et al., 2005; Liemohn et al., 2005). Figure 5.6 shows an example of the midnight potential well feature from a self-consistent electric field calculation of a magnetic storm event (Fok et al., 2001). The equipotential lines become highly distorted in the near-Earth nightside, diverting around midnight and forming a region of intense convection in the afternoon-evening sector.

5.2.2.2 Self-consistent magnetic field models

Improved inner magnetosphere models in which the magnetic field is self-consistently computed in force balance with the particle populations were developed as well. Initial efforts toward the development of a magnetically self-consistent model considered the computation of the equatorial magnetic intensity by solving a force balance equation in the equatorial plane, and coupling it to a kinetic drift loss model (Chen et al., 2006; Liu et al., 2006). Simulations with this model of a hypothetical storm showed that the calculated disturbed magnetic field was as large as −150 nT near an equatorial radial distance of 3 R_E. Tracing equatorially mirroring protons and electrons with this model Chen et al. (2006) found that the ring current contribution to the Dst index during the October, 19, 1998 storm was reduced by about 25% in contrast to the dipolar case by making the simulation magnetically self-consistent.

Another modeling effort is the ring current-atmosphere interactions model with self-consistent magnetic field (RAM-SCB) that couples two numerical models: (a) the RAM code which solves the bounce-averaged kinetic equation for the major ring current species (Jordanova et al., 1996; Jordanova et al., 2010), and (b) the 3D

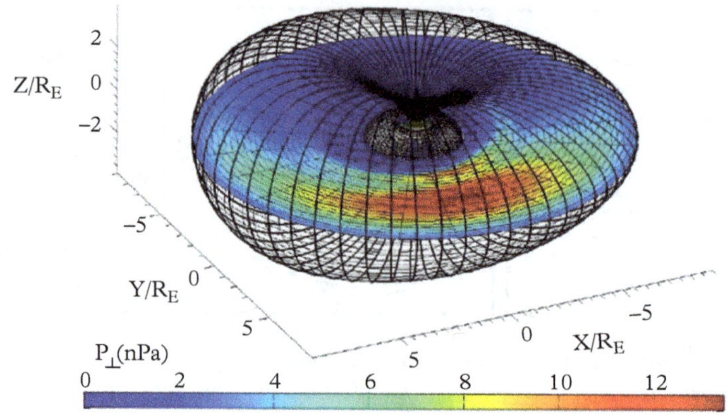

Figure 5.7 *RAM-SCB: coupling of RAM with the 3D plasma equilibrium code. The thick black lines are magnetic field lines on the outer Euler potential surface. RAM pressure is shown on the equatorial plane.*
From Zaharia et al. (2010).

Euler-potential-based plasma equilibrium code (Zaharia et al., 2004; Zaharia et al., 2010). In RAM-SCB, the anisotropic plasma pressure produced by the ring current particles is used to calculate the resulting force balanced 3D magnetic field, which is then used to propagate the phase space distribution function (Fig. 5.7). Preserving the storm time pressure anisotropy in the inner magnetosphere is a distinctive feature of the RAM-SCB model. Such anisotropy naturally develops during enhanced plasma sheet injection and is essential for determining the instability of plasma waves that affect the dynamics of both ring current ions and radiation belt electrons.

Using the RAM-SCB model, the effects of non-dipolar magnetic field configuration and the feedback of a self-consistently computed magnetic field on ring current dynamics were investigated by Jordanova et al. (2010) during the double-dip storm of November 20–21, 2002. Important differences were found when more realistic non-dipolar magnetic field geometry was used. As the fluxes intensified after the injection and earthward transport of plasma sheet ions during storm time, notable day-night asymmetry developed and the high-energy fluxes were displaced closer to Earth on the nightside than on the dayside due to drift-shell splitting. A depression of the ring current fluxes and a confinement of the ring current close to Earth were thus observed on the nightside as the geomagnetic activity increased. The total ring current energy and the Dst index calculated with the self-consistent B field were in best agreement with observations, being smaller compared to the dipolar calculations but larger than the empirical B field predictions.

A unique feature of RAM-SCB is the global simulation of the dominant magnetospheric plasma waves driven by the free energy in the ring current distributions. The temporal and spatial development of the ring current and the subsequent generation of plasma waves during the October 23–26, 2002, high-speed stream driven storm

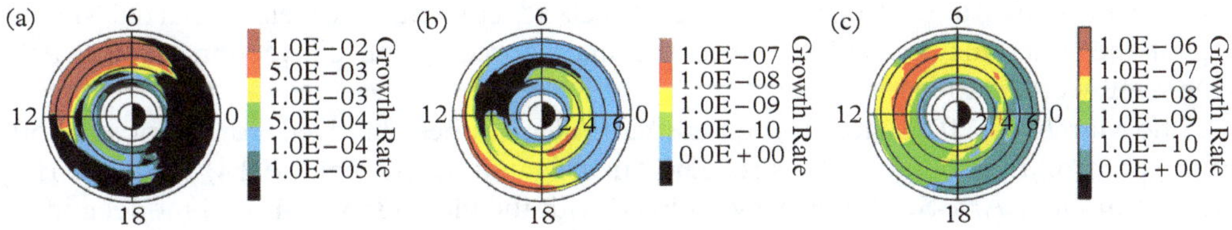

Figure 5.8 *Results from RAM-SCB simulation at 04 UT October 24, 2002: (a) normalized growth rate of whistler mode chorus, (b) convective growth rate (cm^{-1}) of He^+ band EMIC waves, and (c) convective growth rate (cm^{-1}) of magnetosonic waves.*
After Jordanova et al. (2012).

were studied by Jordanova et al. (2012). It was found that as the self-consistent magnetic field was depressed at large ($> 4R_E$) radial distances on the nightside during the storm main phase, the particles' drift velocities increased, and the ion and electron fluxes were reduced. Increased anisotropies developed in a more realistic magnetic field in the ion and electron velocity distributions due to energy-dependent drifts, losses, and dispersed injections. There was thus sufficient free energy to excite whistler mode chorus, electromagnetic ion cyclotron (EMIC), and magnetosonic waves in the equatorial magnetosphere (Fig. 5.8). The linear growth rate of whistler mode chorus intensified in the postmidnight to noon sector, EMIC waves were predominantly excited in the afternoon to midnight sector, and magnetosonic waves were excited over a broad MLT range both inside and outside the plasmasphere. The wave growth rates in a dipolar magnetic field had significantly smaller magnitude and spatial extent than the ones calculated in a self-consistent magnetic field.

5.2.2.3 Fully self-consistent models

Another class of first principle physics-based numerical models, which compute both fields and plasmas fully self-consistently, are the global magnetohydrodynamics (MHD) models (Lyon et al., 1981; Ogino, 1986; Raeder et al., 1995; Gombosi et al., 1998). However, while they are driven by the solar wind and cover virtually the whole magnetosphere, the MHD formalism cannot accurately describe the inner magnetosphere (closer than 10 R_E from Earth) (e.g., De Zeeuw et al., 2004), because it does not include the energy-dependent gradient/curvature drifts and does not have a heat flux (Heinemann and Wolf, 2001). Recent modeling efforts use MHD codes to treat the global magnetosphere in numerical frameworks, and have other models that better describe specific other regions coupled into them; one such effort is the Space Weather Modeling Framework (Tóth et al., 2005). Both two-way (De Zeeuw et al., 2004) and one-way (Toffoletto et al., 2004; Buzulukova et al., 2010; Zaharia et al., 2010; Welling et al., 2011) coupling between global MHD and inner magnetosphere models have been performed. Those couplings, beyond improving the MHD results in the inner magnetosphere, also provide more realistic electric and magnetic fields to the inner magnetosphere models. Simulations using the global MHD BATS-R-US code coupled with the Rice Convection Model (RCM) thus showed significantly higher inner magnetospheric particle pressures and strong region 2 Birkeland currents, as well as partial shielding of the inner magnetosphere from the main force of the convection electric field (De Zeeuw et al., 2004).

The March 17, 2013, storm when the Van Allen Probes observed a dramatic dropout in the electron radiation belts was recently investigated by Yu et al. (2014) using the ring current model RAM-SCB, two-way coupled with the global BATS-R-US model and an ionospheric potential solver. Several substorm dipolarizations and injections were reproduced in the MHD model, in agreement with the timing of shape changes in the AE/AL index. The associated induced electric field transported plasma sheet plasma to geostationary altitudes, providing the boundary plasma source to the ring current model. The model reproduced major dynamic features observed by the Van Allen Probes in the ring current energy spectra, such as the rapid enhancement in the tens to hundreds of keV electron flux. It was found that impulsive plasma sheet injections, together with a large-scale convection electric field, were necessary to develop a strong ring current. After the ring current pressure built up, the magnetic field was depressed on the nightside, as much as 30% compared to a dipolar field even close to L of 4. Such a disturbance implies the outward transport of radiation belt electrons and their subsequent loss through the magnetopause boundary. It was thus demonstrated that with the feedback of ring current pressure, the MHD model improved remarkably in predicting the global magnetic field morphology and the Dst index, as well as the plasma sheet injections during active times.

5.3 Modeling radiation belt particles

5.3.1 3D diffusion model

Diffusion models were used in the past first to know which physical processes are of major importance to shape the radiation belts, second to compare the steady state to mean empirical models, third to understand and predict the dynamics of the belts, especially for electrons in the outer belt. In this section, both proton and electron diffusion models will be described. In the case of 3D models eqn 5.10 is reduced to:

$$\frac{\partial f}{\partial t} = -\sum_{i=1}^{3} \frac{\partial (F_i f)}{\partial \mathcal{J}_i} + \sum_{i=1}^{3} \sum_{j=1}^{3} \frac{\partial}{\partial \mathcal{J}_i} \left(D_{ij} \frac{\partial f}{\partial \mathcal{J}_i} \right) + Sources - Losses \tag{5.19}$$

5.3.1.1 Proton diffusion models

The first important step is to identify the minimum but necessary set of physical processes allowing for the reproduction of the topology of the proton radiation belts. A synoptic view of the proton radiation belt modeling problem is presented in Fig. 5.9. It consists of two levels represented in darker or lighter grey scale and connected by arrows which express the nature of the relation between the various processes, features, or phenomena. The first level (darker shading) corresponds to the physical

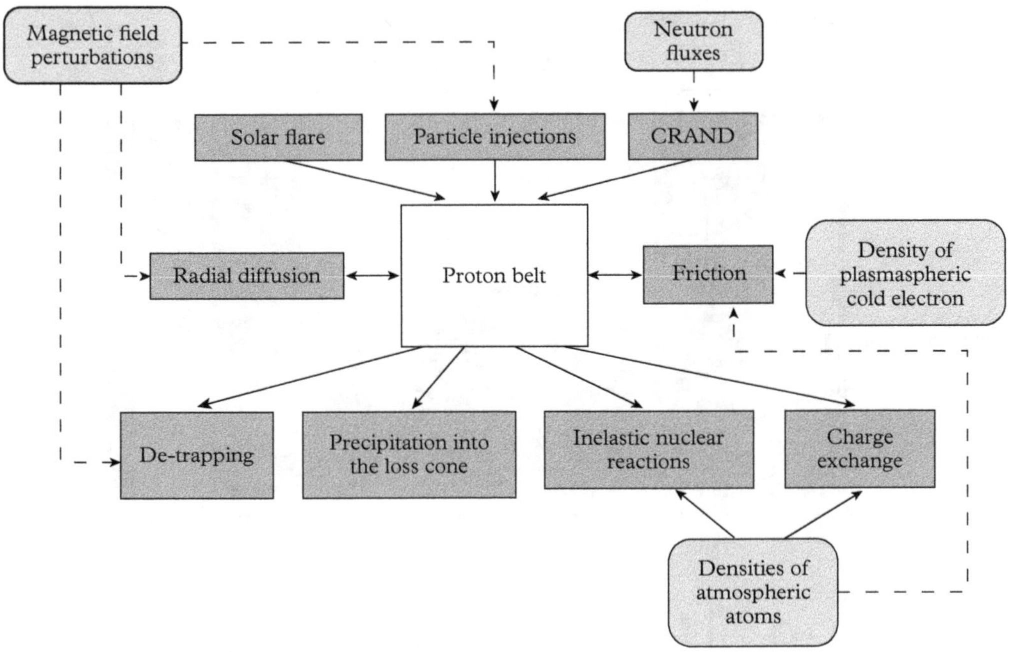

Figure 5.9 *General description of proton radiation belt model.*

processes which occur in establishing the radiation belts. The sources (external and internal) are drawn at the top, the losses at the bottom, the diffusive processes on the left, and the non-diffusive processes on the right. The black solid arrows show the transport of particles. An inward black solid arrow therefore indicates a phenomenon which fills the belt, while an outward arrow indicates a phenomenon which empties it. The second level (lighter shading) shows parameters which influence physical process intensities (dashed black arrows).

The most important internal source process to take into account is the Cosmic Ray Albedo Neutron Decay (CRAND), i.e., the creation of electrons and protons by decay of albedo neutrons resulting from interaction between cosmic rays and atoms of the high atmosphere (Hess et al., 1961; Lenchek and Singer, 1962; Lingenfelter, 1963; Armstrong et al., 1973). This process is driven by the albedo neutron fluxes which are solar cycle dependent: maximum during solar minimum phase while minimum during solar maximum phase. The CRAND process is the major source of protons with energies greater than 40 MeV. Two external source processes are to be considered: the first one being the nightside particle injection during storms and substorms. Freshly injected protons have energies in the keV to few hundreds keV range at the outer trapping boundary of the radiation belts (Reeves et al., 2003b). The second external source comes from the arrival of solar flare particles at Earth at the same time as a strong magnetic storm develops. In this case, solar protons travel to the Earth down to the

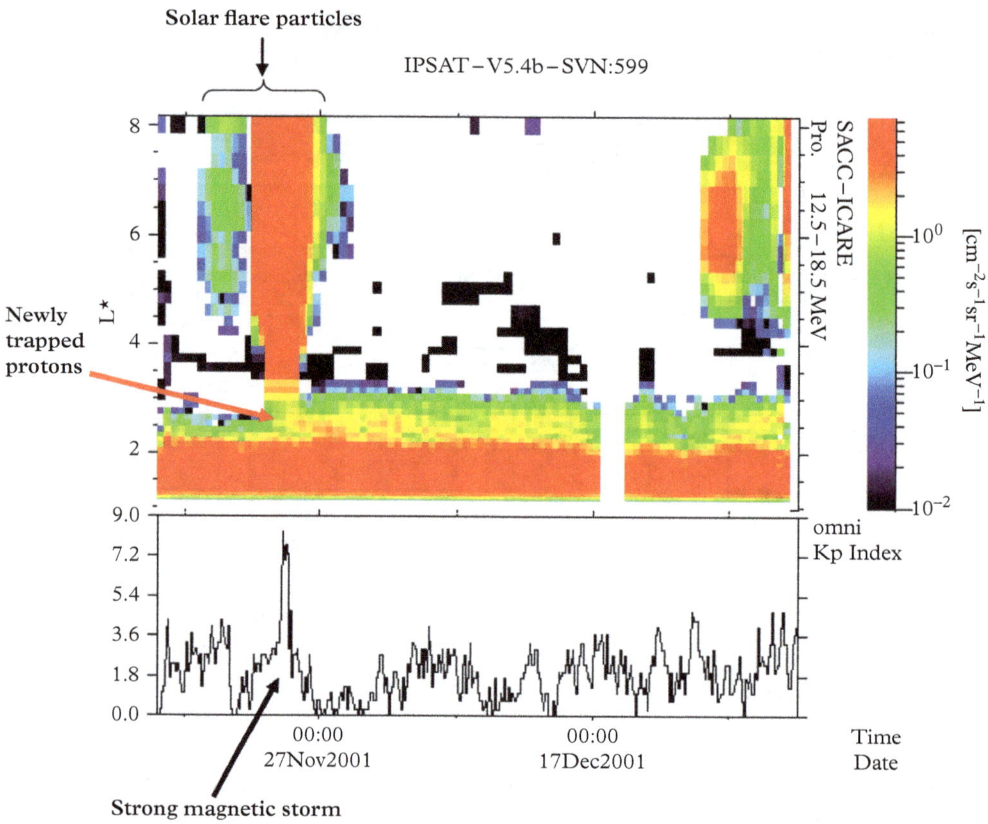

Figure 5.10 *L^* versus time plot of 12.5–18.5 MeV protons measured by SAC-C spacecraft during the November 2001 storm event (top panel) and magnetic activity index (bottom panel).*

magnetospheric shielding boundary. If there is a strong magnetic disturbance at the same time then these protons will be transported inside this limit where they will be trapped (Fig. 5.10).

Several processes contribute to the trapped proton losses. First, the charge exchange loss (Claflin, 1970) comes from a neutral cold hydrogen atom in the exosphere which exchanges its electron with an energetic proton. The resulting proton has a too little energy to be considered part of the radiation belt domain. The charge exchange process is usually expressed as a loss rate term. Second, to correctly reproduce very high energy proton fluxes, inelastic nuclear interactions between trapped protons and atmospheric atoms and molecules have to be added. This process is also commonly expressed as a loss rate term. Third, at the foot of the field lines, trapped protons precipitate into the loss cone; i.e., their mirror point is located in the dense atmosphere so that the particle never bounces back along the field line. And fourth, during very active times, the magnetospheric shielding boundary can be lowered down in L^* so that protons being then

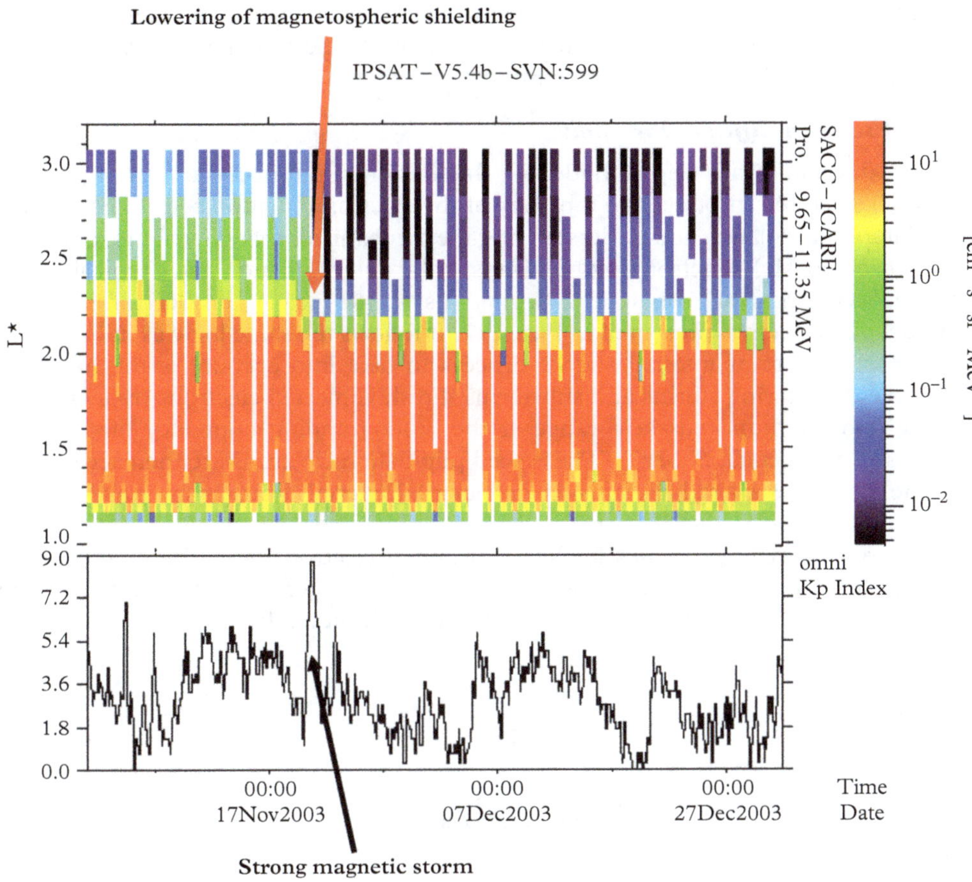

Figure 5.11 *L^* versus time plot of 9.65–11.35 MeV protons measured by SAC-C spacecraft during the November 2003 storm event (top panel) and magnetic activity index (bottom panel).*

outside of this limit are detrapped (Fig. 5.11). They are lost from the radiation belt point of view.

Two more physical processes are very important to account for. First, Coulomb collisions taking place where the cold plasma is dense, i.e., in the plasmasphere and upper atmosphere, induce a continuous non-stochastic friction force. This term is expressed by means of a rate of energy transfer by Coulomb collisions of trapped protons with free and bound electrons (Schulz and Lanzerotti, 1974). Second, radial diffusion transports particles across field lines according to the field disturbance intensities (Fälthammar, 1965; Fälthammar, 1968; Roederer, 1970; Schulz, 1991; Brautigam and Albert, 2000; Ozeke et al., 2012; Lejosne et al., 2012; Lejosne et al., 2013). These two terms rearrange the particle distribution within the proton radiation belt.

Note that in comparison with losses due to charge exchange, pitch angle diffusion effects are insignificant for protons and are neglected; i.e., there are neither wave-particle interactions nor stochastic Coulomb collisions.

5.3.1.2 Electron diffusion models

Again, the first important step is to identify the minimum but necessary set of physical processes needed to reproduce the topology of the electron radiation belts. There is still intense interest in isolating and understanding the mechanisms that contribute to the frequently observed MeV electron flux build-ups in the outer radiation belts, typically during the recovery phase of geomagnetic storms. A synoptic view of the electron radiation belt modeling problem is presented in Fig. 5.12. It follows the same logic as in Fig. 5.13.

The most important internal source process to take into account is the nightside particle injection during storms and substorms. Freshly injected electrons have energies in the keV to few hundreds keV range at the outer trapping boundary of the radiation belts (Reeves et al., 2003a)). These seed population electrons are then accelerated up to MeV energies by radial diffusion or localized energy diffusion by wave-particle interactions.

There are two main processes leading to trapped electron losses. First, at the foot of the field lines trapped electrons precipitate into the loss cone; i.e., their mirror point is located in the dense atmosphere so that the particle never bounces back along the

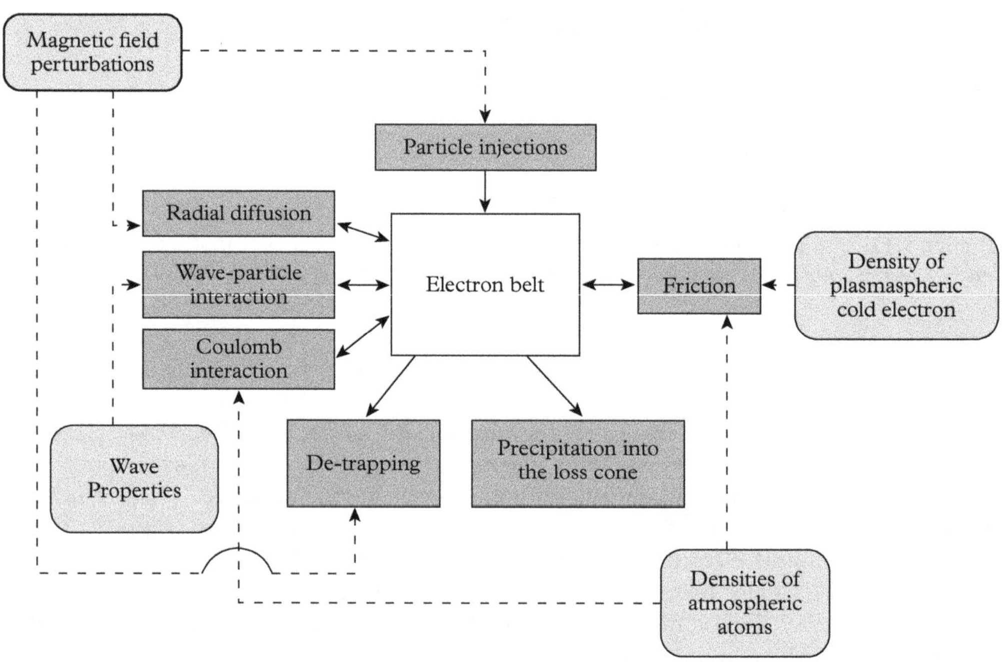

Figure 5.12 *General description of electron radiation belt model.*

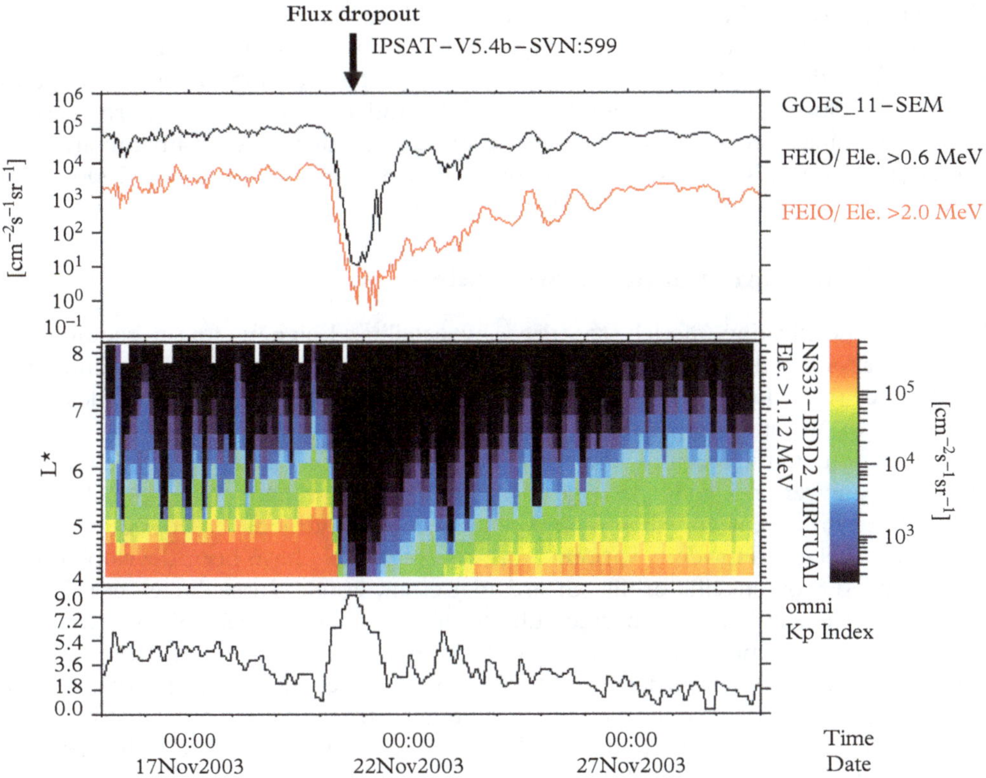

Figure 5.13 *Flux dropout during the November 2003 magnetic storm with measurements at GEO from GOES 11 (top panel), at GPS in a L* versus time map (middle panel) and magnetic activity index (bottom panel).*

field line. And second, during very active times, the last closed drift shell can be lowered down in L* so that electrons being then outside of this limit are detrapped (Fig. 5.13). They are lost from the radiation belt point of view (Reeves et al., 2003b; Bortnik et al., 2006; Turner et al., 2013).

Next friction and diffusion processes are very important to account for. First, the Coulomb collisions taking place where cold plasma is dense, i.e., in the plasmasphere and upper atmosphere, induce (1) a continuous non-stochastic friction force (this term is expressed by means of a rate of energy transfer by Coulomb collisions of trapped electrons with free and bound electrons) (Schulz and Lanzerotti, 1974) and (2) a deviation of the trapped electron path (because the electron is light) according to the differential cross section of the process for an atom or molecule (this term is expressed as a pitch-angle diffusion process and will lead to particle precipitation in the loss cone). Second, wave-particle interactions play a very important role. They induce pitch-angle scattering as well as energy diffusion (Lyons et al., 1972; Abel and Thorne, 1998; Glauert, 2005;

Horne et al., 2005). Various types of wave cohabit in space (e.g. Horne et al., 2014), a popular schematic diagram showing spatial distribution of whistler mode chorus and EMIC waves during magnetic storms in relation to the position of the plasmapause, is given by Summers et al. (1998). Third, the radial diffusion transports particles across field lines according to the field disturbance intensities (Fälthammar, 1965; Fälthammar, 1968; Roederer, 1970; Schulz, 1991; Brautigam and Albert, 2000; Ozeke et al., 2012; Lejosne et al., 2012; Lejosne et al., 2013).

5.3.2 Radiation specification models

Radiation specification models are used by satellite designers to set requirements for total radiation dose (ionizing and non-ionizing), spacecraft charging, and single event effects (see Table 5.1). Such models are usually statistical in nature, describing the mean or worst case particle flux as a function of location, species, and possibly direction. Until recently, the workhorse models have been AE8 (Vette, 1991) and AP8 (Sawyer and Vette, 1976), which provide an estimate of the global trapped radiation environment for solar max or solar min conditions based on data from dozens of satellites. However, various more recent models have been developed to replace AE8/AP8 on specific orbits (Sicard-Piet et al., 2008; Gussenhoven et al., 1997; Bourdarie et al., 2009a), in specific subsets of the inner magnetosphere (Brautigam et al., 1992; Gussenhoven et al., 1993; Huston and Pfitzer, 1998; Huston, 2002; Rodgers et al., 2003; Bourdarie et al., 2009b; Chen et al., 2012; Sandberg et al., 2014), or for worst case conditions (Rodgers et al., 2003; Fennell et al., 2000; O'Brien et al., 2007). Finally, a new model, AE9/AP9/SPM (Ginet et al., 2013), has been developed to fully replace AE8/AP8. This new model is meant to consolidate the various data sources that went into the models developed since AE8/AP8, and will eventually be known as the International Radiation Environment Near Earth (IRENE) as it moves from a primarily U.S. model to an international collaboration and international standard.

There are essentially two broad categories of radiation specification models: those that specify an average environment and those that estimate a worst case. For models of the long-term average environment, the following challenges must be addressed:

Table 5.1 *Space environment hazards, the particles that cause them, and their timescales*

Satellite Hazard	Particle Population	Natural Variation
Surface Charging	0.01–100 keV e^-	Minutes
Surface Dose	0.5–100 keV e^-, H^+, O^+	Minutes
Internal Charging	100 keV–10 MeV e^-	Hours
Ionizing Dose	$>$ 100 keV H^+, e^-	Hours
Single Event Effects	$>$ 10 MeV/amu H^+, Heavy ions	Days
Displacement Damage Dose	$>$ 10 MeV H^+, Secondary n, MeV e^-	Days (Hours e^-)
Nuclear Activation	$>$ 50 MeV H^+, Secondary neutrons	Days (Hours e^-)

- Data cleaning to remove telemetry artifacts, saturation, and severe background
- Inversion to produce flux from count rate
- Intercalbration to remove systematic errors and characterize residual random errors
- Selection of model appropriate coordinates
- Calculation of statistical moments such as mean, standard deviation, and percentiles
- Merging multiple data sources
- Filling in holes in the spatial grid or energy grid
- Representing model uncertainty

For a worst case model and for a model that addresses shorter term averages, it is also necessary to address spatial and temporal dynamics. In the following sections, we will cover each of these challenges in detail.

5.3.2.1 Data cleaning

Most data sets that can be used for specification models have some idiosyncratic problems that must be addressed by removing some data points from further consideration. Examples of such problems include telemetry artifacts, saturation, and overwhelming background. The COSPAR Panel on Radiation Belt Environment Modeling (PRBEM) provides a Data Analysis Procedure which includes data cleaning (Bourdarie et al., 2008) summarized in Friedel et al. (2005). Version 1.2 of the Procedure recommends using a scatter plot of a foreground signal (e.g., electrons) versus a possible background (e.g., protons) to identify when the background overwhelms the foreground. In a scatter plot, such a background usually appears as a diagonal edge to the scatter, as shown in Fig. 5.14. One can then set a threshold in the foreground to background ratio (or any simple function) below which the foreground is considered contaminated and unusable. The PRBEM Procedure also recommends making a histogram of the counts or flux in a channel and examining it for saturation. Saturation appears as a finite upper limit to the flux and may include a pile-up (a peak in the tail of the histogram). In addition to these cleaning methods O'Brien (2013) adds scatter plots of adjacent energy channels and adjacent time points, which can be used to identify telemetry transients or non-physical configurations. In these scatter plots, a region of acceptable data is identified, and all other points are rejected. More involved data cleaning can include time series plots where spacecraft motion has been removed (e.g., by binning in magnetic coordinates), and pass-by-pass plots that could highlight unphysical changes in the radiation belt shape. Bourdarie et al. (2014) add a medianfilter for despiking.

5.3.2.2 Inversion

Inversion is the process by which data are converted from the counts or count rates observed in a detector into the physical quantity of particle flux. In many contexts, inversion consists solely of dividing the count rate by a geometric or energy-geometric factor

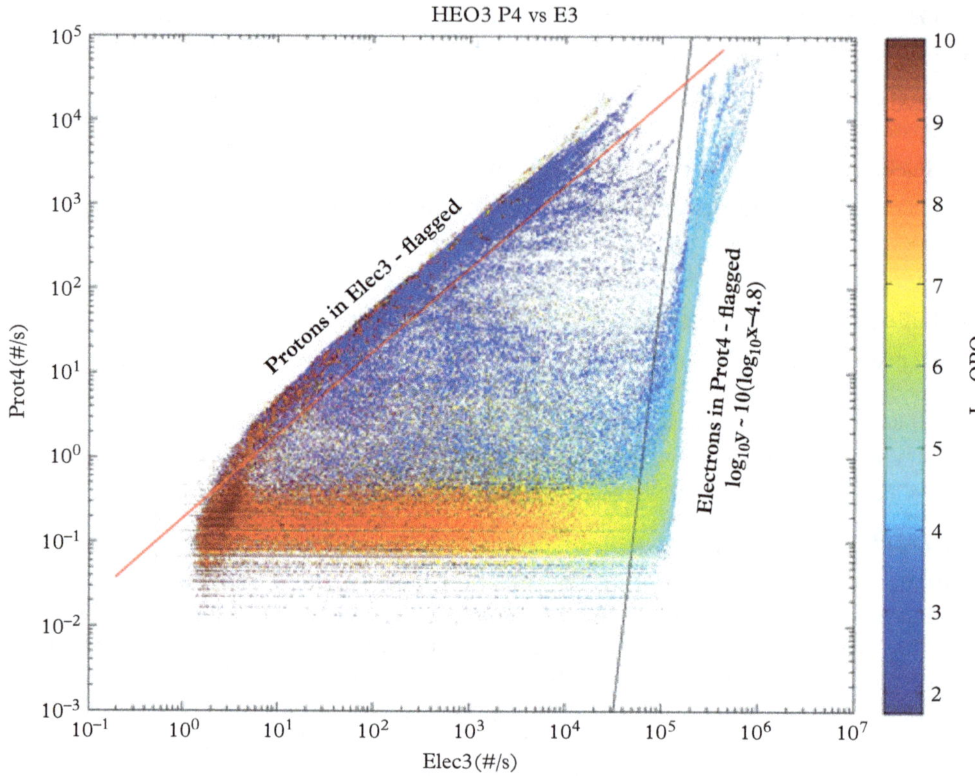

Figure 5.14 *An example data cleaning scatter plot from the HEO3 spacecraft. By plotting an electron channel versus a proton channel, color-coded by L-shell, one can identify regions of background contamination. From O'Brien (2013).*

to obtain particle flux. However, this simple approach is typically inadequate for radiation specifications, which must rely on sensors with complex energy and angle response. Thus some kind of fitting, or inversion, is necessary. Inversion comes in two varieties: energy spectral inversion and angular inversion, which are sometimes combined.

Energy inversion finds the flux $j(E)$ that would produce the observed count rate c_i in several channels:

$$\vec{c} = \frac{1}{\tau}\int_0^\infty j(E)\vec{R}(E)dE + \vec{b} \tag{5.20}$$

where τ is the sensor accumulation time, R_i is the energy response of channel i, and b_i is the background estimate for channel i. The PRBEM Data Analysis Procedure V1.2 describes different methods for identifying and removing background. A common approach is to assume an analytical functional form and then determine the best fit parameters via a least-square error or other optimization technique. Common choices for the analytical function are power law and exponential, with special cases calling for single or double relativistic Maxwellian used at GEO (Cayton et al., 1989), or a power law with an exponential tail (Ginet et al., 2010). More complex approaches rely on a prior model to inform the inversion of the kinds of spectral shapes that are likely. The prior model may simply be a smoothness constraint (Maget et al., 2014) or principal components of variation provided by a precision spectrometer or a simulation (Ginet et al., 2013). The International Radiation Belt Model (IRBEM, irbem.sf.net) library includes an inversion library (invlib) that facilitates several types of spectral inversions.

Angular inversion follows the same format as spectral inversion, with the energy dependence replaced by angular dependence. However, it is typically more complex because the local angular distribution depends more strongly on location (especially distance from the magnetic equator) than does the energy spectrum. Thus, prior information can be hard to obtain. Nonetheless, several analytical models exist that describe the electron angular distribution in the outer zone in terms of sine or Legendre polynomials of the local pitch angle (Vampola, 1997; Gannon et al., 2007; Chen et al., 2014). Now that a directional model like AE9/AP9 exists, it is possible, in principle, to obtain local principal components for either energy or angle inversion. Further, it is possible to use such a model to obtain principal components of 2D energy-angle variation. Two-dimensional inversion follows the same algorithms and numerical schemes as 1D inversion, but the problem is numerically larger and more difficult to set up. However, the extra effort is likely necessary in the case of instruments that rely on passive shielding for collimation (e.g., Violet et al., 1993; Brautigam et al., 2006), because such an approach tends to produce a narrow field of view for the main energies in a given channel, plus a wide-angle background from penetrating higher energy particles.

5.3.2.3 Intercalibration

For many scientific studies temporal and spatial variation is studied, but absolute flux levels are of little consequence. However, for radiation specification models, absolute flux is vital. Further, radiation specification models that combine different data sets can present artificial features in the radiation environment if the data sources are systematically inconsistent with each other. We use intercalibration procedures to minimize the systematic differences between sensors and to gain confidence that we have obtained accurate absolute fluxes.

For energetic protons, the procedure is relatively straightforward (see Bourdarie et al., 2008; Friedel et al., 2005; Guild et al., 2009). Specifically, we assume that the solar particle environment uniformly floods high latitude and high altitude regions of geospace. Thus, vehicles far apart, but immersed in this solar environment, may be assumed to experience the same flux environment. We identify comparable data channels (or create them by interpolation), and then produce a scatter plot of measurements during such events: on one axis, we plot data from the reference sensor, and on the other we plot data from the sensor being calibrated. From the data in the scatter plot, we can extract a median correction factor that brings the less reliable sensor into agreement with

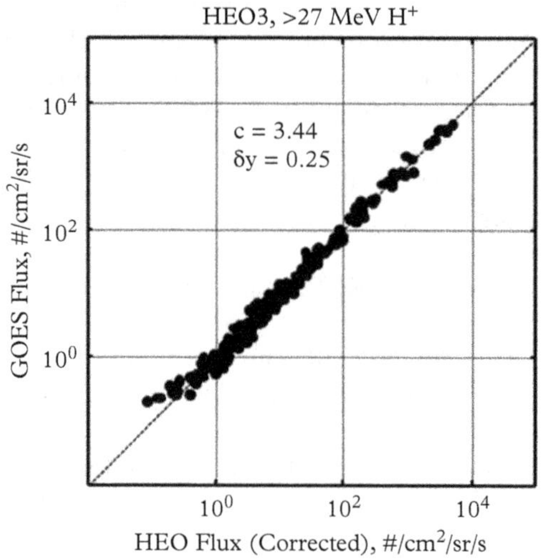

Figure 5.15 *Calibration scatter plot of GOES proton flux and HEO-3 proton flux. The HEO flux was brought into agreement with the GOES flux by applying a factor of 3.44 to the count rate. The residuals are characterized by an r.m.s. error in the natural log flux of 0.25, or approximately a 25% relative error.*

the more reliable sensor, "on average," by which we actually mean that the corrected data have a median ratio of 1 relative to the reference sensor. Figure 5.15 shows this approach for correcting HEO3 to GOES in a set of solar proton events (see Guild et al., 2009). We also compute an error metric that describes the residual scatter: the root mean squared error in the natural log of flux or counts in the comparison. This error metric implicitly assumes the residual random error has a log-normal distribution a reasonable assumption of the relative amount of scatter does not depend strongly on the reference flux (i.e., it has a constant width in a log-log scatter plot). The conjunction approach is robust, except if the trapped spectrum differs dramatically from the typical solar particle spectrum, which happens to some degree with inner zone protons. Further, for lower energy protons, the solar particles do not penetrate very far into the magnetosphere, making it difficult to obtain sufficiently good conjunctions. In such cases, the approach we use for electrons is sometimes a viable alternative.

For energetic electrons, we use the same basic procedure, but we must use a more sophisticated definition of a conjunction. Nominally, we use criteria from Friedel et al. (2005), which are based on L-shell, magnetic latitude, local time, and magnetic activity. Often, the criteria must be adjusted to obtain sufficient sample size.

In the Slot Region Radiation Environment Models (SRREMS, Sandberg et al., 2014), a different approach is used. Namely, data sets are intercalibrated to have the same median in the same bins of model coordinates. If the two data sets cover the same temporal interval, this method should work well. Still, it has not yet been validated by comparison to the conjunction approach. Also, it does not, by itself, produce the residual error metric, although there are doubtless ways to obtain that metric by examining the whole distribution in the coordinate bins (rather than just the median), or by examining multiple bins simultaneously. We should also point out that if the time interval is very short (e.g., a single geomagnetic storm), the storm phase at which the measurements are taken by the two spacecraft may corrupt the procedure.

5.3.2.4 Coordinates

Satellite orbits are usually described in terms of gravitational invariants or near-invariants such as inclination, eccentricity, and semi-major axis (or altitudes of perigee and apogee). The complete set of such invariants would also include analogues to the argument of perigee and longitude of the ascending node. The orbit phase represents the final orbital element, and is not, of course, an invariant. Typically, only the first three invariants are needed, except for special cases like molniya, sun-synchronous, and transfer orbits. In the case of a spiral transfer orbit, the elements are abandoned in favor of a space-time trajectory. Somehow, the radiation specification model must represent the radiation environment in the three- to six-dimensional space of possible orbits, with the additional dimensions of energy and angle of incidence. Models either select a single orbit regime (such as geostationary (Sicard-Piet et al., 2008) or navigation orbits (Bourdarie et al., 2009a)), or they provide the ability to "fly" a proposed satellite mission through the environment. Such "fly in" models must map from the model coordinates to the spacecraft location along its orbit. For regional models, it can be difficult to determine whether a given satellite orbit remains entirely within the model domain, and whether excursions from the domain should be treated as zero flux (outside the radiation belt) or simply unknown (in an unmodeled part of the radiation belt).

Radiation specification models typically do not have enough data density to simply represent the environment as a function of 3D configuration space, energy, and direction (another six-dimensional space). Rather, they typically exploit the repetitive nature of charged particle orbits in a quasi-static magnetic field to represent the particle flux as a function of adiabatic invariants (which are preserved even during slow changes to the global magnetic field), or, at least, drift invariants (which are preserved over a single particle's drift orbit, lasting typically minutes to an hour). Such an approach amounts to implicitly project a single observed particle over its entire 3D trajectory (orbit) through space. As described in Section 5.1, there are three invariants and three associated phase angles for a six-dimensional phase space, e.g., $\left(\vec{\mathcal{J}},\vec{\varphi}\right)$. At typical radiation belt energies, the gyration and bounce phase angles can be neglected, while the drift phase must only be considered under special circumstances: electron flux as a function of longitude from the South Atlantic Anomaly at low altitude. At lower energies, e.g., below 40 keV, the drift phase angle is significant due to the influence of the large-scale electric fields. At still lower energies (or on longer field lines, such as those in the plasma sheet), the bounce phase angle may become important as well. Surprisingly, the gyrophase angle can become important at high energies, or anywhere that the particle's gyroradius is large compared to the local flux gradients with respect to the three invariants. This so-called East-West effect is not actually a non-gyrotropic effect (Lenchek and Singer, 1962). Rather, it is a coordinate transform artifact: the adiabatic coordinates are computed at the center of the particle's gyration, which is offset by a larger gyroradius at higher energy. A steep flux gradient can thus result in an apparent non-gyrotropic distribution.

The adiabatic invariants themselves are rarely used as specification model coordinates. Rather, coordinates derived from the adiabatic invariants, or even just drift invariants, are used instead as a means of simplifying the coordinate transforms or aligning natural gradients with the model coordinates. It is very common to replace the first adiabatic invariant with particle energy (E), which is only a drift invariant. The other two adiabatic invariants are often represented by: equatorial pitch angle α_o and McIlwain L (L_m) (drift invariants in an axisymmetric field), Kafumann's K and Φ (adiabatic invariants), and K and minimum altitude h_{min} (drift invariants). The $\alpha_o - L_m$ system has much appeal to the physical intuition of the physicist, while the $K - \Phi$ system best approximates the physical motion in an arbitrary magnetic field. The $K - h_{min}$ system is very effective at capturing the low-altitude structure caused by the atmosphere.

Whatever coordinate system is used, there is a significant question as to what magnetic field model to use. Modern data analysis exploits complex, dynamic models of the magnetic field due to currents external to the Earth (e.g., Tsyganenko and Sitnov, 2005) and a time-evolving International Geomagnetic Reference Field (IGRF) model for the field due to currents inside the Earth (Finlay et al., 2010). A dynamic model is often out of the question for a specification model for satellite design: one would also need some way to predict the values of the dynamic model inputs (space weather parameters) many years into the future. A quiet time model can be used instead (e.g., Olson and Pfitzer, 1977), which accepts some unnatural variability in the model bins. In the future, playback models, which fly a mission through a past dynamic environment, may use a dynamic external field model because the space weather inputs would usually be known for past intervals.

5.3.2.5 Statistical parameters

It is customary to compute statistical parameters within each bin and energy of the specification model grid. Of these, the most useful statistical parameter is the sample mean. The average flux over an orbit (or a few days) can be scaled to the full mission length to compute a fluence. Fluence can then be fed into an effects code like SHIELDOSE-2 (Seltzer, 1994) to compute a cumulative radiation effect. However, because the radiation environment varies, more recent models have tracked other statistical parameters, such as the sample standard deviation (AE8), specific percentiles (AE9/AP9), or even the entire histogram (Roeder et al., 2005; Ginet et al., 2010; Sandberg et al., 2014). For example, AE8 included a standard deviation parameter over its spatial grid. Unfortunately, the AE8 effort and most recent efforts at describing the variation through statistical parameters often fall short because they fail to account for correlation. (The AE8 case is also interesting because it employs a least-squares fit to log fluxes, suggesting that it represents a median rather than a mean, if the fluxes are distributed approximately log-normal. Based on published values in AE8, treating AE8 as a mean rather than a median represents approximately a factor of 2 underestimate of the electron flux.)

The standard deviation in each model bin can, in principle, be used to estimate error bars on the mean or fluence of an orbit through many bins. However, a problem arises when one attempts to combine variances of multiple correlated variables: flux is correlated at different points in space, and each spatial grid point that the spacecraft traverses is a separate, but correlated, variable, in the average (O'Brien et al., 2007). The variance

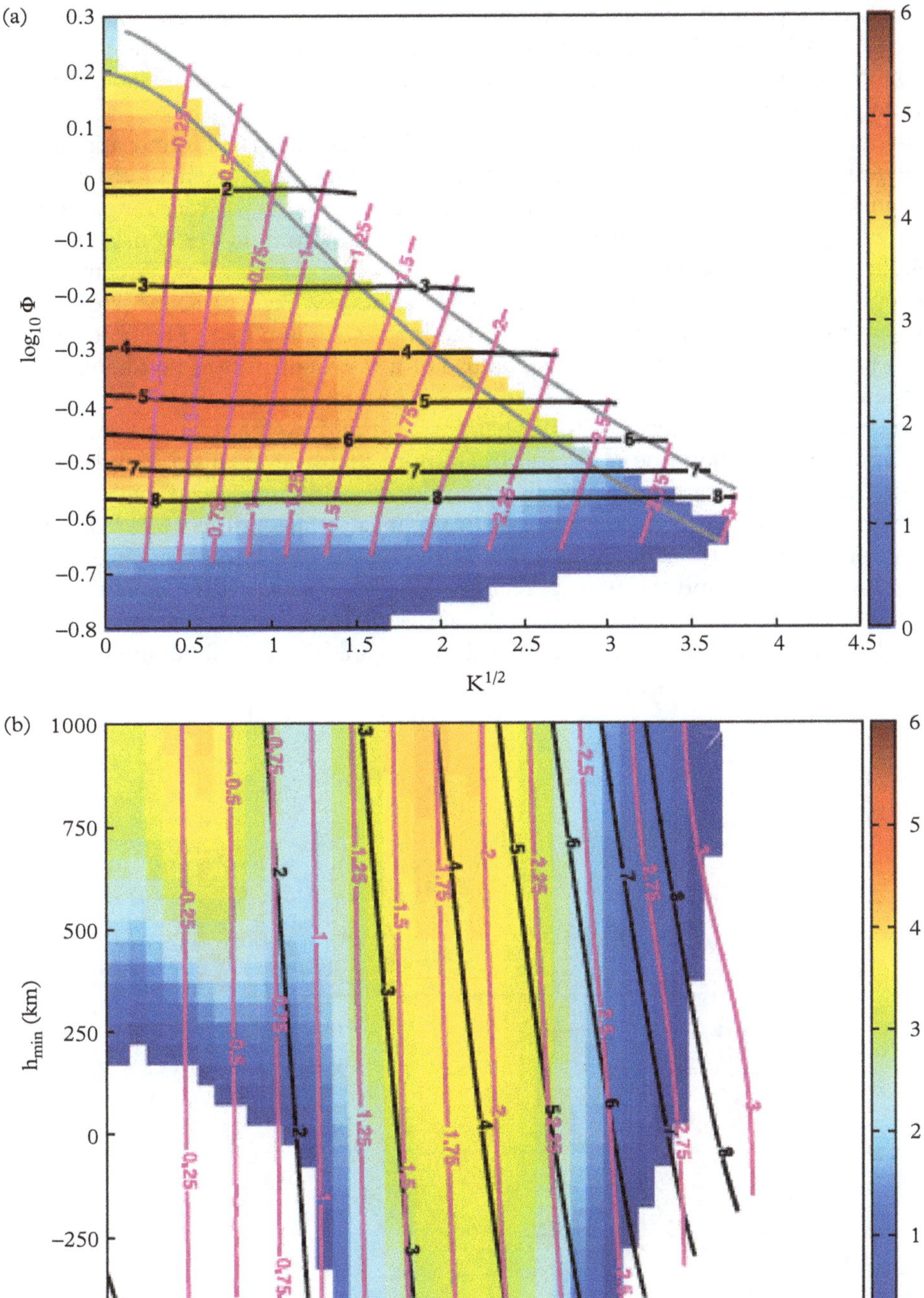

Figure 5.16 *AE9/AP9 fluxes in two coordinate systems. Black and purple traces indicate contours of constant L_m and $\sin^{-2}\alpha_o$, respectively. Also shown in panel (a) are gray contours of h_{min} = 0, 1000 km, the boundaries of panel (b). K has units of $R_E G^{1/2}$, and Φ has units of $R_E^2 G$. From Ginet et al., 2013. © Springer 2013. With permission of Springer.*

of the sum of N perfectly correlated variables is N times the variance of the sum of N uncorrelated variables. There is, thus, a potentially large error in computing the uncertainty on the mean or fluence along an orbit from model bin standard deviations without accounting for correlation.

By themselves, the percentiles (without correlations) are also not directly applicable to satellite design because the environment does not simultaneously reach a fixed percentile at all points on the grid, nor does it remain at a fixed percentile for very long. However, the percentiles can be used to evaluate whether new in situ data are consistent with a model: if some new in situ data correspond to a very high or very low percentile in the corresponding model bin, then either the model or data are likely to be incorrect. Figure 5.17 shows example maps of proton flux in the inner radiation belt. The bottom half is the 95th percentile, which shows the signature of a transient proton belt about 1.5 R_E above the Eath's surface.

If one is to attempt to account for dynamics, and address correlation (see Sections 5.3.2.7 and 5.3.2.8), then one must still track more than the mean in each bin. Computing the entire histogram in the bin is one approach, although it can make it difficult to fill in spatial gaps without creating statistically invalid histograms. AE9/AP9 adopts a somewhat simpler approach. Namely, only two statistics are computed in each bin: a truncated mean, $\overline{x}_{<95}$, and the 95th percentile, x_{95}. The truncated mean is the sample average of all fluxes up to the 95th percentile, and is expected to be more stable to saturation and residual bad points than is a full sample mean. The actual parameters used in

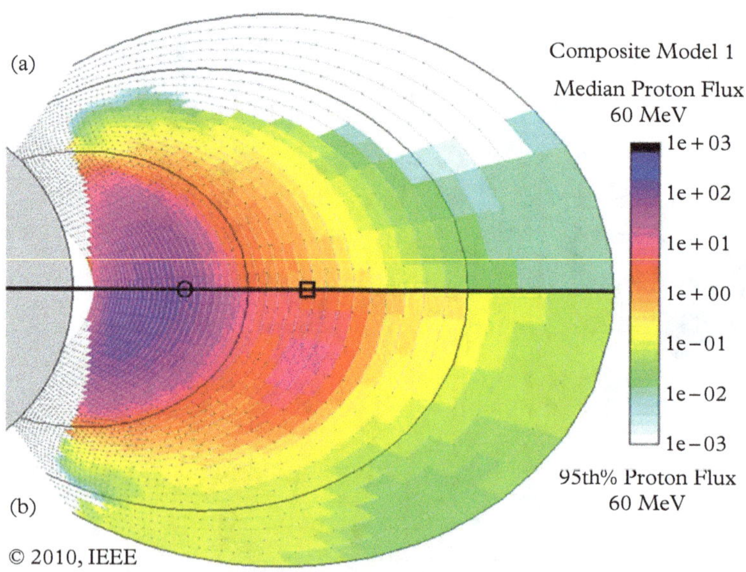

Figure 5.17 *Example median (a, top half) and 95 percentile (b, bottom half) 60 MeV proton fluxes in model coordinate bins. The partial gray circle is the Earth. Thick black arcs indicate 1 RE increments.*
From Ginet et al. (2010).

the AE9/AP9 model are θ_1, the natural log of the median x_{50}, and θ_2, the natural log of the difference between the median and the 95th percentile. This log/difference transform ensures that the 95th percentile will always exceed the median for any two real values (i.e., $-\infty$ to ∞) of the model parameters (θ_1, θ_2):

$$\theta_1 = \ln x_{50} \tag{5.21}$$

$$\theta_2 = \ln(x_{95} - x_{50}) \tag{5.22}$$

From two parameters, e.g., $\overline{x}_{<95}$, and x_{95}, it is possible to compute the coefficients of any two-parameter statistical distribution, and thereby any other percentile (e.g., x_{50}) or statistic (e.g., mean). To determine which statistical distribution to use, one examines the distributions within the model bins to determine what two-parameter distribution is most appropriate. In AE9/AP9 the log-normal and Weibull distributions are used, depending on the species and energy range. Of course, it may be that more than two parameters are needed, and then more parameters of the distribution will need to be calculated from data, and transformed in such a way as to guarantee proper ordering for any set of real values. For example, a future model may use the three-parameter generalized extreme value distribution, which would allow the model to estimate the existence and value of finite upper limits (see, e.g., O'Brien et al., 2007).

5.3.2.6 Filling in holes

Spectral interpolation and extrapolation uses a hypothetical analytical form to specify energies that are not measured in every coordinate bin. Sometimes, this is done as part of the inversion (see Section 5.3.2.2). Spatial filling is much more difficult, as the spatial structure tends to have subtle, multi-variate, and non-monotonic features. For example, the atmospheric loss cone is reasonably an exponential in altitude, but such a shape is not a simple analytical function in most spatial coordinate systems. Similarly, both the inner zone and the outer zone exhibit local maxima in the flux distribution as a function of drift shell. Further, the location and intensity of the maximum depend on whatever coordinates are used in place of the first and second adiabatic invariants. To address the spatial interpolation and extrapolation challenges, the AE9/AP9 team developed the use of *templates*, which are maps of the shape, but not the intensity, of the radiation belt. Each template is one reasonable guess at what the shape is. By using multiple templates in the gap filling procedure, it is possible to estimate the uncertainty associated with gap filling. Templates are typically a combination of in situ observation, analytical fitting, and even physical simulation. They are the means by which experts provide the model derivation calculations information about what reasonable shapes of the belts are.

In AE9/AP9, the quantity that is interpolated is the relative deviation of the candidate flux map with the template. The concept is that if the template says that flux in the hole is 20% higher than a nearby point where there are data, then we should fill the hole with a flux value that is 20% higher than the nearby point. This procedure is done by averaging over multiple nearest neighbors and multiple templates. Defining what constitutes a nearest neighbor is something of an art, given that different dimensions in model coordinates (e.g., E, K, Φ) have fundamentally different scales.

5.3.2.7 *Merging multiple data sources*

Many radiation specification models incorporate data from multiple satellite missions. Ideally, this would occur seamlessly: the various data sets would agree with each other wherever there is overlap, and would seamlessly blend into each other at their edges. Unfortunately, this ideal is often not the case: sensor intercalibration is imperfect, and different sensor data sets cover different intervals of magnetic activity. Where two or more data sets provide information, it can be a fairly simple matter to combine them, perhaps with a weighting based on the assumed or calculated error in each data set. Sometimes one data set is adjusted up or down to force a match, but it is hard to do this in multiple dimensions simultaneously.

Often times, this combination is done with the statistical model parameters, rather than with the raw fluxes. The AE9/AP9 model went a step farther, doing the combination after the data sets are "filled in" (see previous section). This tends to ensure a smooth transition where one data set ends but another continues. That is because the uncertainty associated with the filling-in drops off somewhat slowly as a function of distance away from the edge of any individual data set. Thus, an uncertainty-weighted combination of filled maps will smoothly accommodate the data set edge.

5.3.2.8 *Addressing dynamics*

One of the two biggest challenges in the newer model architectures is addressing dynamics. Earlier sections in this chapter discussed the fact that the radiation and plasma environments vary enormously. Satellite designers must account for this: they need indications of uncertainty in the specification models in order to add appropriate margin for bad luck (e.g., launching into a severe, long-lived, but ultimately transient environment like that after March, 1991). Section 5.3.2.5 described the problems that arise when using standard deviations or percentiles to account for dynamics: without correlations, there can be large mistakes. Thus, addressing dynamics often comes down to addressing spatial and temporal correlations.

Various methods have been tried over the years. AE8/AP8 provides solar min and solar max states to account for solar cycle variation. Scientists have since learned that, especially for AE8, this dichotomy is overly simplistic and wrong in important details. Later models have retained this categorical approach, but have attempted to use better states: CRRESPRO (Gussenhoven et al., 1993) defines a quiet and active state, and IGE (Sicard-Piet et al., 2008) provides three severity levels. Still other models have attempted to produce a parametric climatology: varying the particle environment in response to one or more indices of geomagnetic activity: CRRESELE (Brautigam et al., 1992) used the Ap index, and TPM-1 (Huston, 2002) used the F10.7 index.

The most realistic way to represent dynamics is with a reanalysis: a long-term, dynamic reconstruction of the global radiation environment as a function of space and time using in situ data and a physical model, combined via data assimilation. This approach has been partially implemented with yearly averages in the OZONE electron belt model (Bourdarie et al., 2009b). However, it does not yet represent shorter-term dynamics, which cause internal charging. Further, even a long-term reanalysis does not tend to account for uncertainty in what will happen in the future. Therefore, a supplemental statistical approach has been used in AE9/AP9. It is assumed that the radiation environment state can be represented by a small "principal component" state vector (~ 10 elements) $\vec{q}$. A matrix of principal component shapes $\underline{\underline{Q}}$ projects the state vector onto the global model grid. The matrix $\underline{\underline{Q}}$ is a decomposition of the observed or simulated spatial covariance matrix $\underline{\underline{\Sigma}}$ on the model grid, such that $\underline{\underline{QQ^T}} = \underline{\underline{\Sigma}}$. The temporal variation of the state vector is then generated as an autoregressive moving average of a white noise time series $\vec{\eta}$:

$$\vec{q}_{t+1} = \underline{\underline{C}}\vec{\eta}_{t+1} + \sum_{i=1}^{N_G} \underline{\underline{G}}_i \vec{q}_{t-\tau_i} \tag{5.23}$$

The $\underline{\underline{C}}$ and $\underline{\underline{G}}$ matrices are constructed to reproduce the observed or simulated spatio-temporal correlation at various lags τ_i (O'Brien, 2012). Using this approach, it is possible to generate an arbitrary number of Monte Carlo (surrogate) time series of the principal components and, from those, in turn, the global flux state. Flying a planned mission through multiple time series provides a way to assess the likelihood of mission success. Compared to the other methods outlined above, however, the Monte Carlo approach is very computationally intense: one is essentially simulating the entire mission, in detail, many times, rather than simulating a few orbits in an enhanced radiation environment. However, the Monte Carlo method is the only one available today that can assign probability to specific outcomes.

5.3.2.9 Model uncertainties

Most models treat dynamics as the primary source of uncertainty. However, the model itself has uncertainty associated with the limitations of the data that went into it. AE9/AP9 is the only model that specifically and quantitatively represents this type uncertainty explicitly. In the AE9/AP9 model, uncertainty starts with the original flux measurements. The intercalibration procedure establishes the uncertainty on each flux. As the flux data are processed into statistical parameters, the errors are also propagated forward, usually in a bootstrap fashion, as the statistical manipulations are often nonlinear. Finally, the uncertainties are used in merging the data from different missions into a single flux map on the model grid.

At this stage, an additional bootstrap is performed, which selects (with replacement) an alternate set of missions with which to compute an alternate flux map. Many alternate flux maps are created, thus characterizing the uncertainty in the flux map due to use of different data sets. Ideally, this set of alternatives captures not only the true uncertainty in the flux map, but also bounds the flux map one would derive in the future by adding a new data set. As more data sets are added, the uncertainty drops because the influence of any individual data set diminishes.

With these alternatives in hand, it is possible to compute a matrix $\underline{\underline{S}}_\theta$ that represents the correlated uncertainty in $\vec{\theta}$ across all the model bins. The $\underline{\underline{S}}_\theta$ matrix can then be used to perturb the global flux map for each Monte Carlo scenario, or to produce perturbed

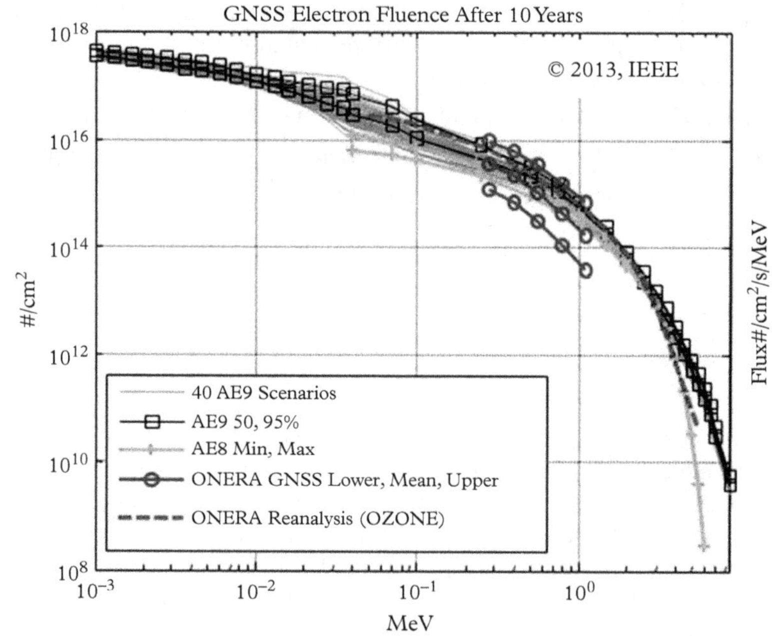

Figure 5.18 *Several models of the electron fluence at a GNSS (navigation) orbit.*
Modified from Xapsos et al. (2013).

mean flux maps. By flying a short interval through just the perturbed mean flux maps, it is possible to approximate very long missions in the Monte Carlo scenarios. In any case, the perturbed flux maps ensure that even after all the dynamics average out, the different scenarios still represent slightly different possible (but realistic) mean or worst case environments.

5.3.2.10 Example: GNSS Orbit

Figure 5.18 shows an example of the fluence spectrum from several different models for the global navigation satellite system (GNSS) orbit. The ONERA GNSS model (Bourdarie et al., 2009a) was based on actual data from that orbit, while the other models synthesize data from multiple missions. AE9 is somewhat unique because it provides many (in this case 40) scenarios of the fluence, from which 50th and 95th percentiles are computed at each energy.

Acknowledgments

S. Bourdarie contribution was supported by ONERA Space Environment department (DESP) internal resources.

Work at Los Alamos was conducted under the auspices of the U.S. Department of Energy with partial support from LDRD, NASA and NSF.
M. Liemohn thanks NASA and NSF for support for this work.
Development of Section 5.3.2 was supported by The Aerospace Corporation's Technical Investment Program.

REFERENCES

Abel, Bob and Thorne, Richard M. (1998). Electron scattering loss in earth's inner magnetosphere: 1. dominant physical processes. *J. Geophys. Res.: Space Physics*, **103**(A2), 2385–2396.

Albert, Jay M., Meredith, Nigel P., and Horne, Richard B. (2009). Three-dimensional diffusion simulation of outer radiation belt electrons during the 9 october 1990 magnetic storm. *J. Geophys. Res.: Space Physics*, **114**(A9), A09214.

Alexeev, Igor I., Belenkaya, Elena S., and Clauer, C. Robert (2000). A model of region 1 field-aligned currents dependent on ionospheric conductivity and solar wind parameters. *J. Geophys. Res.: Space Physics*, **105**(A9), 21119–21127.

Armstrong, T. W., Chandler, K. C., and Barish, J. (1973). Calculations of neutron flux spectra induced in the eart's atmosphere by galactic cosmic rays. *J. Geophys. Res.*, **78**(16), 2715–2726.

Beutier, Thierry and Boscher, Daniel (1995). A three-dimensional analysis of the electron radiation belt by the Salammbô code. *J. Geophys. Res.: Space Physics*, **100**(A8), 14853–14861.

Beutier, Thierry, Boscher, Daniel, and France, Martin (1995). Salammbô: A three-dimensional simulation of the proton radiation belt. *J. Geophys. Res.: Space Physics*, **100**(A9), 17181–17188.

Birmingham, Thomas J. (1992). Birkeland currents in an anisotropic, magnetostatic plasma. *J. Geophys. Res.: Space Physics*, **97**(A4), 3907–3917.

Bortnik, J., Thorne, R. M., O'Brien, T. P., Green, J. C., Strangeway, R. J., Shprits, Y. Y., and Baker, D. N. (2006). Observation of two distinct, rapid loss mechanisms during the 20 november 2003 radiation belt dropout event. *J. Geophys. Res.: Space Physics*, **111**(A12), A12216.

Boscher, D., Bourdarie, S., and Beutier, T. (1996). Dynamic modeling of trapped particles. *IEEE Trans. Nucl. Sci.*, **43**(2), 416–425.

Bourdarie, S., Blake, B., Cao, J.B., Friedel, R., Miyoshi, Y., Panasyuk, M., and Underwood, C. (2008), Panel on radiation belt environment modeling (PRBEM) data analysis procedure, v1.2.

Bourdarie, Sebastien, Boscher, Daniel, Beutier, Thierry, Sauvaud, Jean-Andre, and Blanc, Michel (1996). Magnetic storm modeling in the earth's electron belt by the Salammbô code. *J. Geophys. Res.: Space Physics (1978–2012)*, **101**(A12), 27171–27176.

Bourdarie, S., Boscher, D., Beutier, T., Sauvaud, J.-A., and Blanc, M. (1997). Electron and proton radiation belt dynamic simulations during storm periods: A new asymmetric convection-diffusion model. *J. Geophys. Res.: Space Physics*, **102**(A8), 17541–17552.

Bourdarie, S., Lazaro, D., Hands, A., Ryden, K., and Nieminen, P. (2009a). Electron environment specification models for navigation orbits. In *2009 European Conference on Radiation and Its Effects on Components and Systems (RADECS)*, pp. 364–368.

Bourdarie, S., Lazaro, D., Nieminen, P., and Evans, H. (2014). In-depth analysis of SREM count rate measurements for future use in EnKF Salammbô data assimilation tool. *IEEE Trans. Nucl. Sci.*, **61**(4), pp. 1679–1686, doi:10.1109/TNS.2013.2288938.

Bourdarie, S., Sicard-Piet, A., Friedel, R., O'Brien, T.P., Cayton, T., Blake, B., Boscher, D., and Lazaro, D. (2009b). Outer electron belt specification model. *IEEE Trans. Nucl. Sci.*, **56**(4), 2251–2257.

Brautigam, DH, Dichter, B, Woolf, S, Holeman, E, Ling, A, and Wrazen, D (2006). Compact environmental anomaly sensor (CEASE): response functions. *Air Force Research Laboratory Technical Report AFRL-VS-HA-TR-2006-1030*, **1**.

Brautigam, DH, Gussenhoven, MS, and Mullen, EG (1992). Quasi-static model of outer zone electrons. *Nuclear Science, IEEE Trans. Nucl. Sci.*, **39**(6), 1797–1803.

Brautigam, D. H. and Albert, J. M. (2000). Radial diffusion analysis of outer radiation belt electrons during the october 9, 1990, magnetic storm. *J. Geophys. Res.: Space Physics*, **105**(A1), 291–309.

Burch, J. L. (2000). IMAGE mission overview. *Space Sci. Rev.*, **91**(1-2), 1–14.

Buzulukova, N., Fok, M.-C., Pulkkinen, A., Kuznetsova, M., Moore, T. E., Glocer, A., Brandt, P. C., Tth, G., and Rasttter, L. (2010). Dynamics of ring current and electric fields in the inner magnetosphere during disturbed periods: CRCM-BATS-R-US coupled model. *J. Geophys. Res.: Space Physics*, **115**(A5), A05210.

Cain, Joseph C, Hendricks, Shirley J, Langel, Robert A, and Hudson, William V (1967). *A proposed model for the international geomagnetic reference field-1965*. NASA, Goddard Space Flight Center.

Cayton, T. E., Belian, R. D., Gary, S. P., Fritz, T. A., and Baker, D. N. (1989). Energetic electron components at geosynchronous orbit. *Geophys. Res. Lett.*, **16**(2), 147–150.

Chen, A. J. (1970). Penetration of low-energy protons deep into the magnetosphere. *J. Geophys. Res.*, 75(13), 2458–2467.

Chen, Margaret W., Liu, Shuxiang, Schulz, Michael, Roeder, James L., and Lyons, Larry R. (2006). Magnetically self-consistent ring current simulations during the 19 october 1998 storm. *J. Geophys. Res.: Space Physics*, **111**(A11), A11S15.

Chen, Margaret W., Roeder, James L., Fennell, Joseph F., Lyons, Larry R., and Schulz, Michael (1998). Simulations of ring current proton pitch angle distributions. *J. Geophys. Res.: Space Physics*, **103**(A1), 165–178.

Chen, Yue, Friedel, Reiner H. W., Henderson, Michael G., Claudepierre, Seth G., Morley, Steven K., and Spence, Harlan E. (2014). REPAD: an empirical model of pitch angle distributions for energetic electrons in the earth's outer radiation belt. *J. Geophys. Res.: Space Physics*, 2013JA019431.

Chen, Yue, Reeves, Geoffrey, Friedel, Reiner H. W., Thomsen, Michelle F., Looper, Mark, Evans, David, and Sauvaud, Jean-Andre (2012). LEEM: a new empirical model of radiation-belt electrons in the low-earth-orbit region. *J. Geophys. Res.: Space Physics*, **117**(A11), A11205.

Claflin, E. Scott (1970). Charge-exchange cross sections for hydrogen and helium ions incident on atomic hydrogen: 1 to 1000 keV,. Technical report.

Cornwall, John M, Coroniti, Ferdinand V, and Thorne, Richard M (1970). Turbulent loss of ring current protons. *J. Geophys. Res.*, 75(25), 4699–4709.

Cornwall, John M., Coroniti, F. V., and Thorne, R. M. (1971). Unified theory of SAR arc formation at the plasmapause. *J. Geophys. Res.*, **76**(19), 4428–4445.

Daglis, Ioannis A. (1997). The role of magnetosphere-ionosphere coupling in magnetic storm dynamics. In *Geophysical Monograph Series* (ed. B. T. Tsurutani, W. D. Gonzalez, Y. Kamide, and J. K. Arballo), Volume 98, pp. 107–116. American Geophysical Union, Washington, D. C.

De Zeeuw, Darren L., Sazykin, Stanislav, Wolf, Richard A., Gombosi, Tamas I., Ridley, Aaron J., and Tóth, Gabor (2004). Coupling of a global MHD code and an inner magnetospheric model: Initial results. *J. Geophys. Res.: Space Physics*, **109**(A12), A12219.

Ebihara, Y., Fok, M.-C., Wolf, R. A., Immel, T. J., and Moore, T. E. (2004). Influence of ionosphere conductivity on the ring current. *J. Geophys. Res.: Space Physics*, **109**(A8), A08205.

Ejiri, Masaki (1978). Trajectory traces of charged particles in the magnetosphere. *J. Geophys. Res.: Space Physics*, **83**(A10), 4798–4810.

Fälthammar, Carl-Gunne (1965). Effects of time-dependent electric fields on geomagnetically trapped radiation. *J. Geophys. Res.*, **70**(11), 2503–2516.

Fälthammar, Carl-Gunne (1968). Radial diffusion by violation of the third adiabatic invariant. In *Earth's Particles and Fields*, Volume 1, p. 157.

Fennell, J.F., Koons, H., Chen, M.W., and Blake, J.B. (2000). Internal charging: a preliminary environmental specification for satellites. *IEEE Trans. Plasma Sci.*, **28**(6), 2029–2036.

Finlay, C. C., et al. (2010). International geomagnetic reference field: the eleventh generation. *Geophys. J. Int.*, **183**(3), 1216–1230.

Fok, M.-C., Kozyra, J. U., Nagy, A. F., and Cravens, T. E. (1991). Lifetime of ring current particles due to coulomb collisions in the plasmasphere. *J. Geophys. Res.: Space Physics*, **96**(A5), 7861–7867.

Fok, Mei-Ching, Moore, Thomas E., Kozyra, Janet U., Ho, George C., and Hamilton, Douglas C. (1995). Three-dimensional ring current decay model. *J. Geophys. Res.: Space Physics*, **100**(A6), 9619–9632.

Fok, M.-C., et al. (2003). Global ENA IMAGE simulations, in *Magnetospheric Imaging The Image Prime Mission* (ed. J. L. Burch), pp. 77–103. Springer Netherlands.

Fok, M.-C., Wolf, R. A., Spiro, R. W., and Moore, T. E. (2001). Comprehensive computational model of earth's ring current. *J. Geophys. Res.: Space Physics*, **106**(A5), 8417–8424.

Friedel, R. H. W., Bourdarie, S., and Cayton, T. E. (2005). Intercalibration of magnetospheric energetic electron data. *Space Weather*, **3**(9), S09B04.

Gallagher, D. L., Adrian, M. L., and Liemohn, M. W. (2005). Origin and evolution of deep plasmaspheric notches. *J. Geophys. Res.: Space Physics*, **110**(A9), A09201.

Gannon, J. L., Li, X., and Heynderickx, D. (2007). Pitch angle distribution analysis of radiation belt electrons based on combined release and radiation effects satellite medium electrons a data. *J. Geophys. Res.: Space Physics*, **112**(A5), A05212.

Ganushkina, N. Yu., Pulkkinen, T. I., and Fritz, T. (2005). Role of substorm-associated impulsive electric fields in the ring current development during storms. *Ann. Geophys.*, **23**(2), 579–591.

Ginet, G.P., Huston, S.L., Roth, C. J., O'Brien, T.P., and Guild, T.B. (2010). The trapped proton environment in medium earth orbit (MEO). *IEEE Trans. Nucl. Sci.*, **57**(6), 3135–3142.

Ginet, G. P., et al. (2013). AE9, AP9 and SPM: new models for specifying the trapped energetic particle and space plasma environment. *Space Sci. Rev.*, **179**(1-4), 579–615.

Glauert, Sarah A. (2005). Calculation of pitch angle and energy diffusion coefficients with the PADIE code. *J. Geophys. Res.*, **110**(A4).

Goldstein, J., Spiro, R. W., Reiff, P. H., Wolf, R. A., Sandel, B. R., Freeman, J. W., and Lambour, R. L. (2002). IMF-driven overshielding electric field and the origin of the plasmaspheric shoulder of May 24, 2000. *Geophys. Res. Lett.*, **29**(16), 66–1.

Gombosi, TI, DeZeeuw, DL, Groth, CPT, Powell, KG, and Song, P (1998). The length of the magnetotail for northward IMF: Results of 3D MHD simulations. *Physics of Space Plasmas*, **15**, 121–128.

Gonzalez, Walter D., et al. (1989). Solar wind-magnetosphere coupling during intense magnetic storms (1978–1979). *J. Geophys. Res.: Space Physics*, **94**(A7), 8835–8851.

Guild, TB, O'Brien, TP, Mazur, JE, and Looper, MD (2009). On-orbit inter-calibration of proton observations during solar particle events. Volume TOR-2007 (3905)-22. The Aerospace Corporation, Houston, TX.

Gussenhoven, M.S., Mullen, E.G., Violet, M.D., Hein, C., Bass, J., and Madden, D. (1993). CRRES high energy proton flux maps. *IEEE Trans. Nucl. Sci.*, **40**(6), 1450–1457.

Gussenhoven, M. S., Mullen, E. G., Bell, J. T., Madden, D., and Holeman, E. (1997). APEXRAD: low altitude orbit dose as a function of inclination, magnetic activity and solar cycle. *IEEE Trans. Nucl. Sci.*, **44**(6), 2161–2168.

Hamilton, D. C., Gloeckler, G., Ipavich, F. M., Stdemann, W., Wilken, B., and Kremser, G. (1988). Ring current development during the great geomagnetic storm of february 1986. *J. Geophys. Res.: Space Physics*, **93**(A12), 14343–14355.

Harel, M., et al. (1981). Quantitative simulation of a magnetospheric substorm 1. model logic and overview. *J. Geophys. Res.: Space Physics*, **86**(A4), 2217–2241.

Heinemann, M. and Wolf, R. A. (2001). Relationships of models of the inner magnetosphere to the rice convection model. *J. Geophys. Res.: Space Physics*, **106**(A8), 15545–15554.

Hess, W. N., Canfield, E. H., and Lingenfelter, R. E. (1961). Cosmic-ray neutron demography. *J. Geophys. Res.*, **66**(3), 665–677.

Horne, RB, Meredith, SA, Glauert, SA, and Kersten, T (2014). Wave driven diffusion in radiation belt dynamics, (this volume).

Horne, Richard B., Thorne, Richard M., Glauert, Sarah A., Albert, Jay M., Meredith, Nigel P., and Anderson, Roger R. (2005). Timescale for radiation belt electron acceleration by whistler mode chorus waves. *J. Geophys. Res.: Space Physics*, **110**(A3), A03225.

Hultqvist, Bengt, Riedler, Willi, and Borg, Hans (1976). Ring current protons in the upper atmosphere within the plasmasphere. *Planet. Space Sci.*, **24**(8), 783–797.

Huston, S. L. (2002). Space environments and effects: Trapped proton model. *NASA STI/Recon Technical Report N*, **2**, 63598.

Huston, S. L. and Pfitzer, K. A. (1998). Space environment effects: Low-altitude trapped radiation model. Volume QC809.R3 H8 1998. Marshall Space Flight Center, National Aeronautics and Space Administration, Hunstville, AL.

Jaggi, R. K. and Wolf, R. A. (1973). Self-consistent calculation of the motion of a sheet of ions in the magnetosphere. *J. Geophys. Res.*, **78**(16), 2852–2866.

Jensen, DC and Cain, JC (1962). An interim geomagnetic field, *J. Geophys. Res.*, **67**, 3568.

Jordanova, Vania K. (2005). Sources, transport, and losses of energetic particles during geomagnetic storms. In *Geophysical Monograph Series* (ed. T. I. Pulkkinen, N. A. Tsyganenko, and R. H. Friedel), Volume 155, pp. 9–21. American Geophysical Union, Washington, D. C.

Jordanova, V. K., et al. (1999). Simulation of off-equatorial ring current ion spectra measured by polar for a moderate storm at solar minimum. *J. Geophys. Res.: Space Physics*, **104**(A1), 429–436.

Jordanova, V. K., Farrugia, C. J., Thorne, R. M., Khazanov, G. V., Reeves, G. D., and Thomsen, M. F. (2001). Modeling ring current proton precipitation by electromagnetic ion cyclotron waves during the May 14-16, 1997, storm. *J. Geophys. Res.: Space Physics*, **106**(A1), 7–22.

Jordanova, V. K., Kistler, L. M., Kozyra, J. U., Khazanov, G. V., and Nagy, A. F. (1996). Collisional losses of ring current ions. *J. Geophys. Res.: Space Physics*, **101**(A1), 111–126.

Jordanova, V. K., Kistler, L. M., Thomsen, M. F., and Mouikis, C. G. (2003). Effects of plasma sheet variability on the fast initial ring current decay. *Geophys. Res. Lett.*, **30**(6), 1311.

Jordanova, V. K., Kozyra, J. U., Khazanov, G. V., Nagy, A. F., Rasmussen, C. E., and Fok, M.-C. (1994). A bounce-averaged kinetic model of the ring current ion population. *Geophys. Res. Lett.*, **21**(25), 2785–2788.

Jordanova, V. K., Welling, D. T., Zaharia, S. G., Chen, L., and Thorne, R. M. (2012). Modeling ring current ion and electron dynamics and plasma instabilities during a high-speed stream driven storm. *J. Geophys. Res.: Space Physics*, **117**(A9).

Jordanova, V. K., Zaharia, S., and Welling, D. T. (2010). Comparative study of ring current development using empirical, dipolar, and self-consistent magnetic field simulations. *J. Geophys. Res.*, **115**.

Kelley, M. C., Fejer, B. G., and Gonzales, C. A. (1979). An explanation for anomalous equatorial ionospheric electric fields associated with a northward turning of the interplanetary magnetic field. *Geophys. Res. Lett.*, **6**(4), 301–304.

Kistler, L. M., Ipavich, F. M., Hamilton, D. C., Gloeckler, G., Wilken, B., Kremser, G., and Stüdemann, W. (1989). Energy spectra of the major ion species in the ring current during geomagnetic storms. *J. Geophys. Res.: Space Physics*, **94**(A4), 3579–3599.

Kistler, L. M., et al. (1999). Testing electric field models using ring current ion energy spectra from the equator-s ion composition (ESIC) instrument. *Annales Geophysicae*, **17**(12), 1611–1621.

Kistler, Lynn M. and Larson, Douglas J. (2000). Testing electric and magnetic field models of the storm-time inner magnetosphere. *J. Geophys. Res.: Space Physics*, **105**(A11), 25221–25231.

Kozyra, J. U., Fok, M.-C., Sanchez, E. R., Evans, D. S., Hamilton, D. C., and Nagy, A. F. (1998). The role of precipitation losses in producing the rapid early recovery phase of the great magnetic storm of february 1986. *J. Geophys. Res.: Space Physics*, **103**(A4), 6801–6814.

Landau, L and Lifchitz, E (1968). Mecanique. In *Physique theorique - Tome I, Editions MIR-French edition.*

Lejosne, Solène, Boscher, Daniel, Maget, Vincent, and Rolland, Guy (2012). Bounce-averaged approach to radial diffusion modeling: From a new derivation of the instantaneous rate of change of the third adiabatic invariant to the characterization of the radial diffusion process. *J. Geophys. Res.: Space Physics*, **117**(A8), A08231.

Lejosne, Solène, Boscher, Daniel, Maget, Vincent, and Rolland, Guy (2013). Deriving electromagnetic radial diffusion coefficients of radiation belt equatorial particles for different levels of magnetic activity based on magnetic field measurements at geostationary orbit. *J. Geophys. Res.: Space Physics*, **118**(6), 3147–3156.

Lenchek, A. M. and Singer, S. F. (1962). Geomagnetically trapped protons from cosmic-ray albedo neutrons. *J. Geophys. Res.*, **67**(4), 1263–1287.

Lennartsson, W., Shelley, E. G., Sharp, R. D., Johnson, R. G., and Balsiger, H. (1979). Some initial ISEE-1 results on the ring current composition and dynamics during the magnetic storm of december 11, 1977. *Geophy. Res. Lett.*, **6**(6), 483–486.

Liemohn, M. W., Kozyra, J. U., Clauer, C. R., and Ridley, A. J. (2001). Computational analysis of the near-earth magnetospheric current system during two-phase decay storms. *J. Geophys. Res.: Space Physics*, **106**(A12), 29531–29542.

Liemohn, M. W., Kozyra, J. U., Jordanova, V. K., Khazanov, G. V., Thomsen, M. F., and Cayton, T. E. (1999). Analysis of early phase ring current recovery mechanisms during geomagnetic storms. *Geophys. Res. Lett.*, **26**(18), 2845–2848.

Liemohn, Michael W., et al. (2005). Parametric analysis of nightside conductance effects on inner magnetospheric dynamics for the 17 april 2002 storm. *J. Geophys. Res.: Space Physics*, **110**(A12), A12S22.

Liemohn, Michael W., Ridley, Aaron J., Gallagher, Dennis L., Ober, Daniel M., and Kozyra, Janet U. (2004). Dependence of plasmaspheric morphology on the electric field description during the recovery phase of the 17 April 2002 magnetic storm. *J. Geophys. Res.: Space Physics*, **109**(A3), A03209.

Liemohn, Michael W., Ridley, Aaron J., Kozyra, Janet U., Gallagher, Dennis L., Thomsen, Michelle F., Henderson, Michael G., Denton, Michael H., Brandt, Pontus C., and Goldstein, Jerry (2006). Analyzing electric field morphology through data-model comparisons of the geospace environment modeling inner Magnetosphere/Storm assessment challenge events. *J. Geophys. Res.: Space Physics*, **111**(A11), A11S11.

Lingenfelter, R. E. (1963). The cosmic-ray neutron leakage flux. *J. Geophys. Res.*, **68**(20), 5633–5639.

Liu, S., Chen, M. W., Schulz, M., and Lyons, L. R. (2006). Initial simulation results of storm-time ring current in a self-consistent magnetic field model. *J. Geophys. Res.: Space Physics*, **111**(A4), A04225.

Lyons, J. G., Brecht, S. H., Huba, J. D., Fedder, J. A., and Palmadesso, P. J. (1981). Computer simulation of a geomagnetic substorm. *Phys. Rev. Lett.*, **46**(15), 1038–1041.

Lyons, Lawrence R. and Thorne, Richard Mansergh (1972). Parasitic pitch angle diffusion of radiation belt particles by ion cyclotron waves. *J. Geophys. Res.*, 77(28), 5608–5616.

Lyons, Lawrence R., Thorne, Richard Mansergh, and Kennel, Charles F. (1972). Pitch-angle diffusion of radiation belt electrons within the plasmasphere. *J. Geophys. Res.*, 77(19), 3455–3474.

Maget, V., et al. (2014). Unfolding JASON-2/ICARE-NG high-energy particles measurements using a singular value decomposition approach. *IEEE Trans. Nucl. Sci.*, **Early Access Online**.

Mauk, B. H., Fox, N. J., Kanekal, S. G., Kessel, R. L., Sibeck, D. G., and Ukhorskiy, A. (2014). Science objectives and rationale for the radiation belt storm probes mission, in *The Van Allen Probes Mission* (ed. N. Fox and J. L. Burch), pp. 3–27. Springer US.

McIlwain, Carl E. (1961). Coordinates for mapping the distribution of magnetically trapped particles. *J. Geophys. Res.*, **66**(11), 3681–3691.

McIlwain, C. E. (1972). Plasma convection in the vicinity of the geosynchronous orbit, in *Earths Magnetospheric Processes* (ed. B. M. McCormac), Number 32 in Astrophysics and Space Science Library, pp. 268–279. Springer Netherlands.

Northrop, Theodore G (1963). *The adiabatic motion of charged particles*. Volume 21. Interscience publishers New York.

O'Brien, TP (2012). Adding multiple time lags to AE9/AP9 V1.0. Volume TOR-2012 (1237)-3. The Aerospace Corporation, El Segundo, CA.

O'Brien, TP (2013). Data cleaning guidelines for AE-9/AP-9 data sets. Volume TOR-2012 (1237)-4. The Aerospace Corporation, El Segundo, CA.

O'Brien, T. P., Fennell, J. F., Roeder, J. L., and Reeves, G. D. (2007). Extreme electron fluxes in the outer zone. *Space Weather*, **5**(1), S01001.

Ogino, Tatsuki (1986). A three-dimensional MHD simulation of the interaction of the solar wind with the earth's magnetosphere: The generation of field-aligned currents. *J. Geophys. Res.: Space Physics*, **91**(A6), 6791–6806.

Olson, WP and Pfitzer, KA (1977). Magnetospheric magnetic field modeling, annual scientific report, AFOSR contract no. *F44620-75-c-0033*.

Ozeke, Louis G., et al. (2012). ULF wave derived radiation belt radial diffusion coefficients. *J. Geophys. Res.*, **117**(A4).

Parker, E. N. (1957). Newtonian development of the dynamical properties of ionized gases of low density. *Phys. Rev.*, **107**(4), 924–933.

Parker, E. N. and Stewart, H. A. (1967). Nonlinear inflation of a magnetic dipole. *J. Geophys. Res.*, 72(21), 5287–5293.

Pfitzer, KA, Olson, WP, and Mogstad, T (1988). A time dependent source driven magnetospheric magnetic field model. *Eos Trans. AGU*, **69**(16), 426.

Raeder, J., Walker, R. J., and Ashour-Abdalla, M. (1995). The structure of the distant geomagnetic tail during long periods of northward IMF. *Geophys. Res. Lett.*, **22**(4), 349–352.

Reeves, G. D., et al. (2003a). IMAGE, POLAR, and geosynchronous observations of substorm and ring current ion injection. In *Geophysical Monograph Series* (ed. A. Surjalal Sharma, Y. Kamide, and G. S. Lakhina), Volume 142, pp. 91–101. American Geophysical Union, Washington, D. C.

Reeves, G. D., McAdams, K. L., Friedel, R. H. W., and O'Brien, T. P. (2003b). Acceleration and loss of relativistic electrons during geomagnetic storms. *Geophys. Res. Lett.*, **30**(10), 1529.

Ridley, A. J. and Liemohn, M. W. (2002). A model-derived storm time asymmetric ring current driven electric field description. *J. Geophys. Res.: Space Physics*, **107**(A8), SMP 2–1.

Rodgers, DJ, Hunter, KA, and Wrenn, GL (2003). The FLUMIC electron environment model. *Proceedings 8th SCTC, Huntsville.*

Roeder, J. L., Chen, M. W., Fennell, J. F., and Friedel, R. (2005). Empirical models of the low-energy plasma in the inner magnetosphere. *Space Weather*, **3**(12), S12B06.

Roederer, J. G. (1970). Periodic drift motion and conservation of the third adiabatic invariant. In *Dynamics of Geomagnetically Trapped Radiation*, Number 2 in Physics and Chemistry in Space, pp. 72–83. Springer Berlin Heidelberg.

Sóraas, F., Lundblad, J. Å., and Hultqvist, B. (1977). On the energy dependence of the ring current proton precipitation. *Planet. Space Sci.*, **25**(8), 757–763.

Sandberg, et al. (2014). Development and validation of the electron slot region radiation environment model. *Nuclear Science, IEEE Transactions on*, **PP**(99), 1–1.

Sawyer, D. M. and Vette, J. I. (1976, December). AP-8 trapped proton environment for solar maximum and solar minimum. *NASA STI/Recon Technical Report N*, 77, 18983.

Sazykin, S., Wolf, R. A., Spiro, R. W., Gombosi, T. I., De Zeeuw, D. L., and Thomsen, M. F. (2002). Interchange instability in the inner magnetosphere associated with geosynchronous particle flux decreases. *Geophys. Res. Lett.*, **29**(10), 88–1.

Schulz, M. (1991). The magnetosphere, in *Geomagnetism*, Volume 1, pp. 87–293.

Schulz, M and Lanzerotti, LJ (1974). Particle diffusion in the radiation belts.

Selesnick, R. S., Looper, M. D., and Mewaldt, R. A. (2007). A theoretical model of the inner proton radiation belt. *Space Weather*, **5**(4).

Seltzer, Stephen M (1994). Updated calculations for routine space-shielding radiation dose estimates: Shieldose-2. *NIST Publication NISTIR*, **5477**.

Sergeev, V. A., Malkov, M., and Mursula, K. (1993). Testing the isotropic boundary algorithm method to evaluate the magnetic field configuration in the tail. *J. Geophys. Res.: Space Physics*, **98**(A5), 7609–7620.

Sicard-Piet, A., et al. (2008). A new international geostationary electron model: IGE-2006, from 1 keV to 5.2 MeV. *Space Weather*, **6**(7).

Smith, Paul H. and Hoffman, R. A. (1974). Direct observations in the dusk hours of the characteristics of the storm time ring current particles during the beginning of magnetic storms. *J. Geophys. Res.*, **79**(7), 966–971.

Spiro, R. W., Wolf, R. A., and Fejer, B. G. (1988). Penetrating of high-latitude-electric-field effects to low latitudes during SUNDIAL 1984. *Annales Geophysicae*, **6**, 39–49.

Stern, David P. (1975). The motion of a proton in the equatorial magnetosphere. *J. Geophys. Res.*, **80**(4), 595–599.

Subbotin, Dmitriy, Shprits, Yuri, and Ni, Binbin (2010). Three-dimensional VERB radiation belt simulations including mixed diffusion. *J. Geophys. Res.: Space Physics*, **115**(A3), A03205.

Summers, Danny, Thorne, Richard M., and Xiao, Fuliang (1998). Relativistic theory of wave-particle resonant diffusion with application to electron acceleration in the magnetosphere. *J. Geophys. Res.: Space Physics*, **103**(A9), 20487–20500.

Tóth, Góbor, et al. (2005). Space weather modeling framework: A new tool for the space science community. *J. Geophys. Res.*, **110**(A12).

Takahashi, Shun and Iyemori, Toshihiko (1989). Three-dimensional tracing of charged particle trajectories in a realistic magnetospheric model. *J. Geophys. Res.: Space Physics*, **94**(A5), 5505–5509.

Toffoletto, F. R., Sazykin, S., Spiro, R. W., Wolf, R. A., and Lyon, J. G. (2004). RCM meets LFM: initial results of one-way coupling. *J. Atmos. Sol.-Terr. Phys.*, **66**(1516), 1361–1370.

Tsyganenko, N. A. (1987). Global quantitative models of the geomagnetic field in the cislunar magnetosphere for different disturbance levels. *Planet. Space Sci.*, **35**(11), 1347–1358.

Tsyganenko, N. A. (1989). A magnetospheric magnetic field model with a warped tail current sheet. *Planet. Space Sci.*, **37**(1), 5–20.

Tsyganenko, N. A., Singer, H. J., and Kasper, J. C. (2003). Storm-time distortion of the inner magnetosphere: How severe can it get? *J. Geophys. Res.: Space Physics*, **108**(A5), 1209.

Tsyganenko, N. A. and Sitnov, M. I. (2005). Modeling the dynamics of the inner magnetosphere during strong geomagnetic storms. *J. Geophys. Res.: Space Physics*, **110**(A3), A03208.

Tsyganenko, Nikolai A. and Stern, David P. (1996). Modeling the global magnetic field of the large-scale birkeland current systems. *J. Geophys. Res.: Space Physics*, **101**(A12), 27187–27198.

Turner, D. L., et al. (2013). On the storm-time evolution of relativistic electron phase space density in earth's outer radiation belt. *J. Geophys. Res.: Space Physics*, **118**(5), 2196–2212.

Vampola, A.L. (1997). Outer zone energetic electron environment update, in *High Energy Radiation Background in Space, 1997 Conference on the*, pp. 128–136.

Vasyliunas, Vytenis M. (1970). Mathematical models of magnetospheric convection and its coupling to the ionosphere, in *Particles and Fields in the Magnetosphere* (ed. B. M. McCormac), Number 17 in Astrophysics and Space Science Library, pp. 60–71. Springer Netherlands.

Vette, James I. (1991). The AE-8 trapped electron model environment. *NASA STI/Recon Technical Report N*, **92**, 24228.

Violet, M. D., Lynch, K., Redus, R., Riehl, K., Boughan, E., and Hein, C. (1993). Proton telescope (PRO) on the CRRES spacecraft. *IEEE Trans. Nucl. Sci.*, **40**(2), 242–245.

Volland, H. (1973). A semiempirical model of large-scale magnetospheric electric fields. *J. Geophys. Res.*, **78**(1), 171–180.

Weimer, D. R. (1996). A flexible, IMF dependent model of high-latitude electric potentials having "space weather" applications. *Geophys. Res. Lett.*, **23**(18), 2549–2552.

Welling, D. T., Jordanova, V. K., Zaharia, S. G., Glocer, A., and Toth, G. (2011). The effects of dynamic ionospheric outflow on the ring current. *J. Geophys. Res.: Space Physics*, **116**(A2), A00J19.

Wentworth, R. C., MacDonald, W. M., and Singer, S. F. (1959). Lifetimes of trapped radiation belt particles determined by coulomb scattering. *Physics of Fluids (1958–1988)*, **2**(5), 499–509.

Williams, D. J. and Lyons, L. R. (1974a). Further aspects of the proton ring current interaction with the plasmapause: Main and recovery phases. *J. Geophys. Res.*, **79**(31), 4791–4798.

Williams, D. J. and Lyons, L. R. (1974b). The proton ring current and its interaction with the plasmapause: Storm recovery phase. *J. Geophys. Res.*, **79**(28), 4195–4207.

Wolf, R. A. (1970). Effects of ionospheric conductivity on convective flow of plasma in the magnetosphere. *J. Geophys. Res.*, **75**(25), 4677–4698.

Wolf, R. A. (1974). Calculations of magnetospheric electric fields. In *Magnetospheric Physics* (ed. B. M. McCormac), Number 44 in Astrophysics and Space Science Library, pp. 167–177. Springer Netherlands.

Wolf, R. A., Harel, M., Spiro, R. W., Voigt, G.-H., Reiff, P. H., and Chen, C.-K. (1982). Computer simulation of inner magnetospheric dynamics for the magnetic storm of july 29, 1977. *J. Geophys. Res.: Space Physics*, **87**(A8), 5949–5962.

Xapsos, M.A., O'Neill, P.M., and O'Brien, T.P. (2013). Near-earth space radiation models. *IEEE Trans. Nucl. Sci.*, **60**(3), 1691–1705.

Xiao, Fuliang, Su, Zhenpeng, Zheng, Huinan, and Wang, Shui (2010). Three-dimensional simulations of outer radiation belt electron dynamics including cross-diffusion terms. *J. Geophys. Res.: Space Physics*, **115**(A5), A05216.

Yu, Y., Jordanova, V.K., Welling, D., Larsen, B., Claudepierre, Seth G., and Kletzing, Craig (2014). The role of ring current particle injections: Global simulations and van allen probes observations during 17 March 2013 storm. *Geophys. Res. Lett.*, **41**, 1126–1132.

Zaharia, S., Cheng, C. Z., and Maezawa, K. (2004). 3-D force-balanced magnetospheric configurations. *Ann. Geophys.*, **22**(1), 251–265.

Zaharia, S., Jordanova, V. K., Welling, D., and Tóth, G. (2010). Self-consistent inner magnetosphere simulation driven by a global MHD model. *J. Geophys. Res.: Space Physics*, **115**(A12), A12228.

Zheng, Y., Lui, A. T. Y., Fok, M.-C., Anderson, B. J., Brandt, P. C., Immel, T. J., and Mitchell, D. G. (2006). Relationship between region 2 field-aligned current and the ring current: Model results. *J. Geophys. Res.: Space Physics*, **111**(A11), A11S06.

6

Monitoring ULF Waves from Low Earth Orbit Satellites

Georgios Balasis, Constantinos Papadimitriou, Eftyhia Zesta, and Viacheslav Pilipenko

■ Abstract

Recent magnetic field missions flying in a low Earth orbit (LEO), like Ørsted, ST5, and CHAMP, have enabled us to study in situ the occurrence of ultra low frequency (ULF) waves in the topside ionosphere. CHAMP satellite has been one of the most successful missions for the study of the Earth's magnetic field for more than a decade (July 2000–September 2010), allowing, for the first time, statistical studies on the occurrence of ULF wave events in the topside ionosphere. This chapter provides a review of recent studies of ULF waves using spacecraft in low Earth orbit. It combines a topical review of the various subclasses of ULF waves that are amenable to study from LEO and a detailed description of many recent observational studies supported with relevant theoretical efforts. The chapter concludes by focusing on further studies made possible by the Swarm mission. Finally, we summarize the most pressing open questions that current and future missions must address.

Waves, Particles, and Storms in Geospace. First Edition. Georgios Balasis, Ioannis A. Daglis, Ian R. Mann.

6.1 Introduction

Magnetospheric ultra low frequency (ULF) waves in the topside ionosphere are typically transmitted from magnetosperic and upstream solar wind sources. Just as is the case for ULF waves observed on the ground, the amplitude of the waves in the topside ionosphere is significantly smaller than that of the background magnetic field B_o, approximately 10^{-4} of B_o. Observations in the topside ionosphere therefore require magnetometers that are both extremely sensitive (<1 nT) and have a large dynamic range ($\pm$60,000 nT). ULF observations in the ionosphere were first reported in the late 80s during the MAGSAT era (Iyemori and Hayashi, 1989) when data from the mission were used to detect Pc1 waves (with frequency $f \simeq 0.2$–5 Hz). Recent magnetic and electric field missions flying in a low Earth orbit (LEO), like Ørsted, FAST, ST5, C/NOFS, Chibis-M, etc., have enabled us to study in situ the occurrence of ULF waves in the topside ionosphere. The CHAMP satellite has been one of the most successful missions for the study of the Earth's magnetic field, with high-sensitivity and accuracy magnetometer measurements orbiting within an altitude range 450–300 km for more than a decade (July 2000–September 2010). For the first time long-term statistical studies on the occurrence of ULF wave events in the topside ionosphere were possible. LEO observations of ULF waves can only be reliably done (i.e. without too much spatial aliasing) for the Pc1/Pi1 and Pc2/3 waves ($f \simeq 22$–100 mHz). Due to the fast motion through field lines in a LEO orbit the lower frequency Pc4-5 waves (1–10 mHz) cannot be accurately determined, their period being longer than the spacecraft transition time through the wave region. Though Pi2 waves have lower frequencies ($f \simeq 2$–25 mHz), thanks to their large spatial scales at low latitudes, they have also been detected by LEO satellites. However, the spatial scale of Pi2 waves along an orbit becomes is with the scale of lithospheric magnetic anomalies (Maus et al., 2002). Therefore, to reliably isolate Pi2 signatures from LEO the knowledge of a detailed magnetic field model of the Earth (e.g., Maus et al., 2005) is necessary.

Topside ionosphere observations of ULF waves with LEO satellites can offer unique understanding in the physics (i.e., excitation and propagation mechanisms) underlying magnetospheric pulsations recorded on the ground and, most importantly, how they relate to observations in the magnetosphere. The so far limited LEO satellite observations have already started filling the "ignorance gap" in ULF wave physics created by the fact that our notions have been solely based on ground and magnetospheric observations. LEO observations have indicated already that some existing theoretical views have to be revised. In the future, conjunctions of magnetospheric, LEO, and ground observations ought to offer good boundaries to ULF wave propagation models from the magnetosphere to the ground.

This chapter is structured as follows: in Section 6.2 we present methods and analysis techniques for monitoring ULF waves with LEO satellites. In Section 6.3 we discuss studies of ULF wave observations from LEO. Section 6.4 presents efforts to model the relationship between the ULF compressional disturbance above the ionosphere and its ground response. We conclude by discussing possible applications and extensions of these ULF wave studies to the Swarm era.

6.2 Methods and techniques

A Mean Field-Aligned (MFA) coordinate system is usually used in the analysis of satellite magnetic field observations in order to separate ULF field variations perpendicular to, as well as along the magnetic field direction. The unit vectors of the MFA coordinate system are defined as follows: the parallel component in the coordinate system, $\hat{p}$, is obtained from a running average of the instantaneous magnetic field. The other components are then chosen to be $\hat{\varphi} = (\hat{p} \times \vec{R})/|\hat{p} \times \vec{R}|$, where $\vec{R}$ is the radius vector of the satellite, and $\hat{r} = \hat{\varphi} \times \hat{p}$. Thus, $\hat{\varphi}$ is the azimuthal component and is positive eastward, while $\hat{r}$, completing the orthogonal system, is meridional and points radially outward at the magnetic equator. ULF wave components in the $\hat{p}$, $\hat{\varphi}$, and $\hat{r}$ directions are referred to as compressional $b_{\|}$, toroidal b_T, and poloidal b_P, respectively. The standard transformation of polar LEO satellite measurements into an Earth-oriented frame adds undesirable attitude noise to the high-latitude data (Heilig et al., 2007); thus, generating clean vector data requires a lot of manual intervention. However, when the compressional (field-aligned) component dominates over the transverse components, the wave signature can well be derived from the total field variations for a LEO satellite (Jadhav et al., 2001; Heilig et al., 2007), thus avoiding the attitude noise problems. It is, therefore, much easier and fruitful to conduct long-term studies from CHAMP ULF waves at polar latitudes for times when the compressional component dominates, such as the upstream source Pc3 waves. Field-line resonances (FLRs), widely studied by ground magnetometers, present mostly in the transverse component in space and can not be routinely identified in polar LEO observations, only at mid and low latitudes (Ndiitwani and Sutcliffe, 2009), where the azimuthal component is clean of attitude noise, introduced by the coordinate transformation. Where the component analysis can be accurately done without attitude noise, LEO observations can be powerful easily separating the contributions from upstream Pc3 waves and FLRs above the ionospheric layer that introduces a rotation to ground observed components (Hughes and Southwood, 1976).

ULF waves have been traditionally identified through visual inspection of spectrograms based on the Fast Fourier Transform (FFT). Motivated by the continuously increasing amount of data collected by space missions and ground-based instruments, algorithms have been developed based on FFT spectrograms to automatically identify ULF waves (Menk and Waters, 2013). Since the 1990s, the wavelet spectral analysis has become popular, as it allows the quantitative monitoring of localized variations of power within the time series data (Balasis et al., 2005; Mandea and Balasis, 2006; Balasis and Mandea, 2007). The wavelet techniques have been shown to be much more appropriate for the detection of ULF waves from LEO, particularly the lower frequency Pc3 waves

(Heilig et al., 2007). For sufficient frequency resolution the FFT would require a time window large enough that spatial aliasing in a fast moving satellite would be a problem. The wavelet transform is indicated instead because it provides frequency analysis at all scales simultaneously.

Balasis et al. (2005) performed wavelet spectral analysis of magnetic field magnitude data derived from CHAMP vector fluxgate magnetometer (FGM) measurements. The wavelet spectral analysis was able to separately detect, identify, and classify artificial noise sources – such as instrument problems and pre-processing errors – from geophysical sources – such as ULF waves or ionospheric plasma bubbles. Furthermore, wavelet techniques are powerful in that they are even able to detect signals in LEO satellite data associated with earthquakes or flares from distant neutron stars (Balasis and Mandea, 2007; Mandea and Balasis, 2006). We note that there are several parameters of the wavelet transform, such as frequency range and power spectral density amplification factor, which need to be correctly adjusted in order to capture different kinds of anomalous signals. As an example, Figure 6.1 presents CHAMP total magnetic field time series on October 31, 2003, during the dayside portion of the orbit derived from the 1 s FGM measurements after applying a 16 mHz high-pass filter along with its corresponding wavelet power spectrum in the Pc3 frequency band. The order of the filter is 5, but note that we perform zero-phase forward and reverse digital infinite impulse response (IIR) filtering. This means that after filtering in the forward direction, the filtered sequence is then reversed and run back through the filter. The result is the time reverse of this second filtering operation and has precisely zero phase distortion. It was empirically found that a cutoff of 16 mHz for the high-pass filter used at the preprocessing of the time series is able to reduce the amplitude of pulsations with $f \leq 10$ mHz by ~90%. This figure also includes electron density data derived from the 15 s Planar Langmuir Probe measurements. The inclusion of the electron density recordings helps to identify time segments of the signal that contain signatures of post-sunset equatorial spread F events (Stolle et al., 2006), and, therefore, discriminate between Pc3 wave and plasma depletion occurrence. A prominent Pc3 ULF wave is observed starting at ~14:44 UT and lasting ~15 minutes. The strong ionospheric currents' signatures near the poles covering lower frequencies can also be seen. The time period plotted is a descending dayside half orbit. It is worth noting that Pc3 waves are observed over the auroral zones in the 10–32 mHz range and in the dayside equator in the 40–100 mHz range, while wave power decreases significantly at mid-latitudes. This is a typical Pc3 wave profile of a high-inclination LEO orbit for quiet and moderate activity periods. During disturbed conditions (storms, compressions) Pc3 waves are detected at mid-latitudes and are significantly stronger in the auroral zone and equator.

Vellante et al. (2004) and Ndiitwani and Sutcliffe (2009) in their studies of conjunctions between CHAMP and ground measurements found the need for using the Maximum Entropy Spectral Analysis (MESA) technique. An inherent shortcoming of the FFT in analyzing short duration signals is the limited resolving power. MESA is a nonlinear method for computing spectral estimates, and it provides a greater resolution

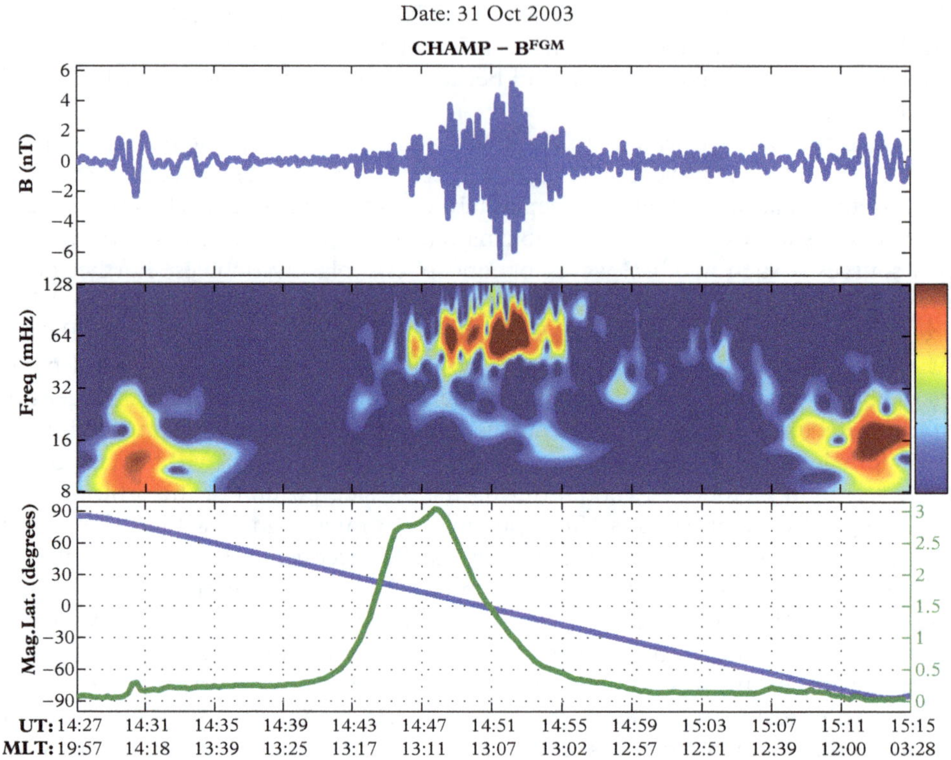

Figure 6.1 *The CHAMP track from 14:27 to 15:13 UT on October 31, 2003. From top to bottom are shown the time series of the CHAMP total magnetic field, calculated from the 1 Hz FGM data after applying a 16 mHz high-pass filter, its corresponding wavelet power spectrum as well as the temporal variation of the CHAMP electron density data along with its magnetic latidutinal dependence. The corresponding MLT values are given at the bottom of the graph. [Figure 6.1 is modified after Figure 4 of Balasis et al. (2012).]*

than linear methods and has the ability to analyze short data records. It is therefore more useful when precise conjunctions between space and ground need to be identified. MESA spectra could be very beneficial in the study of Pc1/Pi1 waves when brief bursts of activity need to be identified and accurately timed. Recently automated methods of deriving the characteristics of ULF waves in the magnetosphere have been developed (Balasis et al., 2012, 2013), which can be effectively applied to the huge datasets from a space mission in order to retrieve, on an operational basis, new information about the near-Earth electromagnetic environment. Automated techniques allow the simultaneous assembly and analysis of ground, magnetospheric, and topside ionospheric observations for a better understanding of the propagation of the ULF waves through the magnetosphere and down to the ionosphere illuminating pathways of energy transport. The database of automated wave identification from the European Commission's FP7 project MAARBLE—Monitoring Analyzing and Assessing Radiation Belt Loss and Energization (including observations from Cluster, CHAMP, ground) can now enable studies with a high data flow.

6.3 Studies of ULF wave observations from LEO

There have been only a handful of studies on ULF wave observations from LEO and we will outline here some of the key case studies.

6.3.1 Pc1 wave observations

Pc1-2 waves may occur during the main phase of some storms whereas they occur quite often during the recovery phase. Because these waves are expected to be highly effective in depleting relativistic electrons from the outer radiation belt and ring current protons, studies of the relative occurrence of these waves during magnetic storms became a high priority. Pc1 waves adjust the accumulation of energy in the radiation belt, control the duration of geospace magnetic storms, and may lead to the redistribution of plasma along geomagnetic field lines (Guglielmi and Kangas, 2007). The interest in Pc1 waves in Space Sciences is motivated by the opportunity to gain understanding on magnetospheric processes by using in situ and remote hydromagnetic wave diagnostics. In other fields, like Biology, Pc1 waves with frequency near the mammalian heart beating rate attract also the attention of biologists and medical specialists as a cosmic factor which is expected to affect the vital activity of organisms.

According to present day knowledge, Pc1 pulsations represent packets of electromagnetic ion-cyclotron (EMIC) waves which are excited as a result of cyclotron instability of energetic protons of the ring current with anisotropic temperature distribution ($T_\perp > T_\parallel$), and they propagate along geomagnetic field lines. It is commonly believed that the EMIC instability of the ring current protons is convective, that is, the region of instability at the apex of the field line works as an amplifier of running EMIC waves. The wave packets themselves at the same time oscillate between the conjugate ionospheres and are intensified at each passage through the equatorial region of the magnetosphere (the "bouncing wave packet" model), though recent observations have cast increasing doubts on this model (Mursula, 2007).

The studies of Pc1 waves from LEO are truly sparse. The amplitude of Pc1 waves on the ground was found to be more than two orders of magnitude smaller than that observed by MAGSAT (Iyemori and Hayashi, 1989) and ST5 (Engebretson et al., 2008). Engebretson et al. (2008) pointed out that such large variations in the satellite to ground amplitude ratio (~5–100) may be caused by (1) variations in the spatial extent of the wave source region (if the conjunction of space-ground is not close); (2) wave partial absorption/reflection by heavy ions in the magnetosphere or lower ionosphere; (3) ionospheric ducting of waves originating in a distant source region. Conjunction studies between space and ground Pc1 waves are not easily done as Pc1 waves are ducted within the ionosphere and appear on the ground over a much wider region than they exist in space. In a case study, well-defined Pc1 waves had been observed simultaneously from a ground array of five search-coil magnetometers (–62° to –87° magnetic latitude) in the morning sector of Antarctica on March 23, 2007 (Kim et al., 2010). A CHAMP satellite conjunction showed a transverse and nearly linearly polarized Pc1 ULF wave at the altitude of the ducting layer (350 km) over a limited latitudinal extent (–53° to –61° ILAT). The polarization analysis performed using the ground data supported the idea that CHAMP detected the wave activity near the wave injection region.

Park et al. (2013) presented a global climatology of Pc1 pulsations as observed by the CHAMP satellite from 2000 to 2010. They found that diurnal variation of Pc1 occurrences shows a primary maximum early in the morning and a secondary maximum during pre-midnight hours. The annual variations of the occurrence rates exhibit a clear preference for local summer. The solar cycle dependence of the occurrence rate reveals a maximum at the declining phase (2004–2005), whereas neither magnetic activity nor solar wind velocity controls the Pc1 occurrence rate significantly. Pc1 occurrence rate peaks at subauroral latitudes, but the steep cutoff toward higher latitudes is due to auroral field-aligned currents masking the Pc1 pulsations. An interesting feature found is that the global distribution of Pc1 exhibits highest occurrence rates near the longitude sector of the South Atlantic Anomaly.

EMIC Pc1 waves were successfully identified in a pearls-on-a-string configuration by the ST5 mission (Engebretson et al., 2008) in the topside ionosphere. ST5 consisted of three identical probes in almost identical orbits with distances between them from thousands to hundreds of km. The three probes crossing the same spatial region with a delay of $\sim$1–10 min provided the possibility of separating spatial and temporal structures of EMIC waves in the topside ionosphere. Detected EMIC wave packets manifested their narrow localization in latitude. In all events, where the burst of Pc1 waves was detected by two ST5 spacecraft, wave packets had a characteristic transverse dimension $\Delta\Phi \simeq 0.5$–$0.7°$ in latitude or $\Delta x \simeq 50$–90 km.

EMIC emissions were never detected with comparable amplitudes by all three ST5 probes. At the same time, in the cases when at the moment of registration of the wave packet in the ionosphere, the satellite orbit passed in the vicinity of a ground station, a prolonged series of emission at the same frequency was observed at the ground. In order to reconcile these observational facts, one may assume that the EMIC instability develops in the near-equatorial magnetosphere in the form of a series of irregular bursts of instability (alike "frying pop-corn"). At the same time, the EMIC waves are excited not in the regime of continuous emission, but in the form of relatively short (< 10 min) strongly localized bursts of emission distributed chaotically in time and space in a finite region with intense ring current. That is why at the flight of a LEO satellite, only short bursts of EMIC waves can be observed, while at ground stations prolonged emission is registered within the same frequency range collected from a large area. Thus, the EMIC instability of the ring current may work not as a convective amplifier of multiple oscillating wave packets but as a generator of wave bursts (absolute instability). Future studies need to explore this generation and propagation physics of Pc1 waves in more depth, as more and better observations are becoming available.

The real structure of the radial distribution of plasma is rather irregular, particularly closer to the plasmapause, which allows one to suggest the existence of more local waveguides for EMIC waves, not related directly to the plasmapause. A waveguide for EMIC waves can be formed due to the joint action of the transverse dispersion of the wave and transverse plasma inhomogeneity (Leonovich et al., 1985). The waveguide nature of EMIC waves is supported by the effect revealed by consideration of polarization features of the wave structures in the topside ionosphere (Pilipenko et al., 2012a). Figure 6.2 presents an example of a typical EMIC wave packet recorded by the ST5 probe at geomagnetic latitude of $\sim 48°$ with frequency of 1.2 Hz in the dusk sector. The polarization in the $b_P - b_T$ plane calculated by the dynamical wavelet method showed that ellipticity ϵ changes its sign in the region of maximum amplitude of the transverse component b_T. This polarization reversal may point to the standing-mode structure of EMIC waves in the transverse direction, which is characteristic for waveguide modes.

6.3.2 Pi1 bursts

Pi1B bursts are known to be a wave signature of sudden auroral intensifications in the midnight local time region where typical onsets occur. Lessard et al. (2006) suggested that Pi1B could be not just a marker, but even a driver of auroral activations. Using a good conjunction of ground stations with geosynchronous GOES observations and with a FAST crossing over the same region, they found that Pi1B waves propagated earthward past GOES as a compressional mode and coupled to transverse shear waves at LEO altitudes, as observed by FAST. Observations of Pi1B by Cluster and Polar spacecraft showed Pi1B pulsations associated with fast flows from deep in the tail. The implication is that fast flows carry compressional wave power that couples to shear mode waves that drive Alfvénic aurora at substorm onset. So far, only one such event was reported (Lessard et al., 2006), so this hypothesis awaits further validation and LEO satellite observations can be ideal for resolving physical sources of these pulsations.

6.3.3 Pc3 wave observations

The CHAMP observations motivated and enabled many studies examining ULF waves in the Pc3 frequency range. Pc3 waves are typically observed for quiet and moderate activity period over the auroral zones in the 8–30 mHz range and the dayside equator in the 60–80 mHz range, while wave power decreases significantly at mid-latitudes, a profile that we attribute to strong ionospheric currents in the auroral and dayside equatorial regions. During disturbed conditions Pc3 waves are also detected at mid-latitudes and are significantly stronger in the auroral zone and equator.

A case of simultaneous wave detection on July 6, 2002, from LEO and the ground was studied by Vellante et al. (2004). The event occurred when CHAMP flew almost

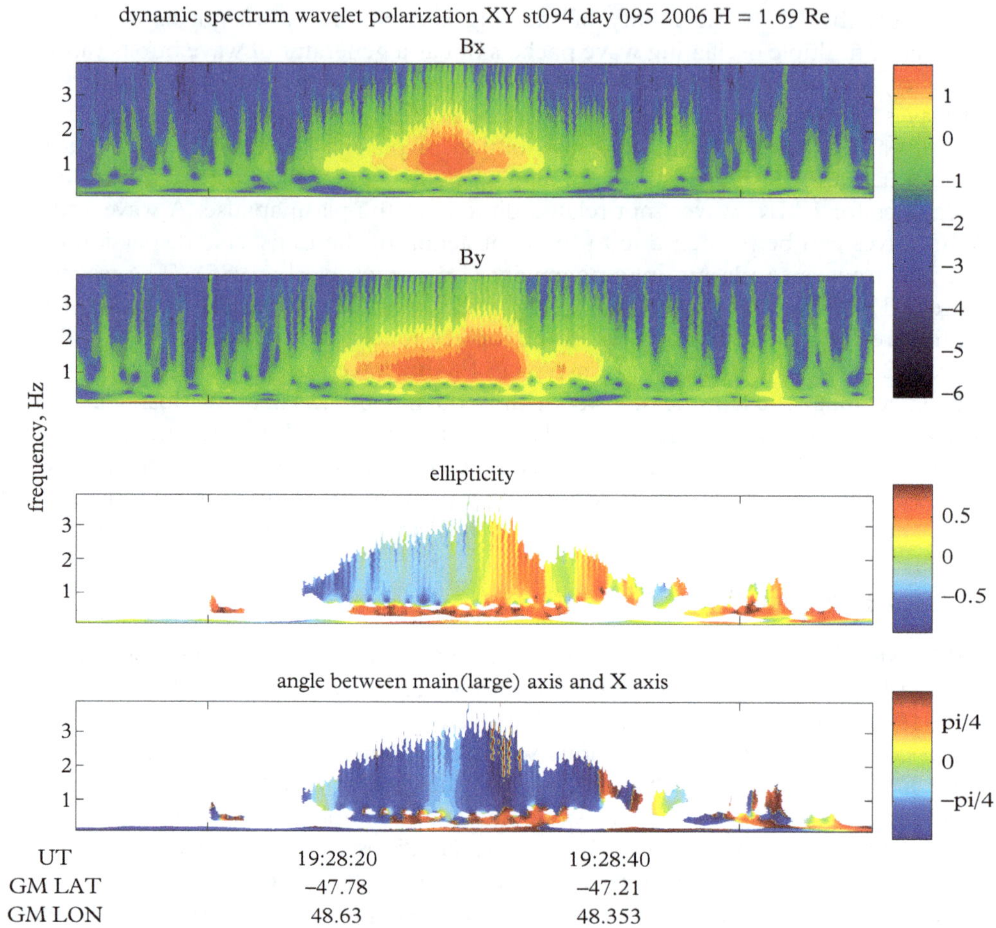

Figure 6.2 *Dynamic polarization spectrum estimated with the wavelet method in the transverse to B_o plane on April 5, 2006 (day 95) of Pc1 wave packet recorded by ST5-094 probe: wavelet spectra of B_x (poloidal, b_P) and B_y (toroidal, b_T) components (upper panels), ellipticity ϵ (middle panel), and ellipse orientation angle (bottom panel).*

directly above ground stations of the mid-latitude SEGMA network, simultaneously detecting Pc3 wave activity in space and on the ground. Clear activity in almost the entire range of the Pc3 waves could be seen in all three components of the magnetic field. The authors produced dynamic amplitude spectra that allowed a visual comparison between space and ground measurements. The spectra were computed by using the MESA method with a short data window, necessary because CHAMP was moving across magnetic shells very quickly. They succeeded in revealing the characteristic signatures of the Alfvén FLR in the spatial structure of Pc3 waves. The behavior of the

CHAMP azimuthal component showed amplitude-phase structure typical for a FLR and a $\pi/2$ rotation of the polarization ellipse through the ionosphere.

Sutcliffe et al. (2013) confirmed that a typical Pc3-4 pulsation frequency range contains a strong FLR contribution with latitude-dependent frequency. At the same time, wave frequency was proportional to the interplanetary magnetic field magnitude B_{IMF} both in the ground-based and LEO satellite data. This correspondence indicates that upstream waves are the likely source of these pulsations. In addition to contributions from FLRs the Pc3-4 wave band consistently contains signals at frequencies that are independent of latitude or magnitude of the interplanetary magnetic field (IMF), B_{IMF} which suggests that contributions in this additional frequency could be fast mode resonances related to cavity, waveguide, or virtual modes.

Heilig et al. (2007) and Ndiitwani and Sutcliffe (2009) found that the compressional power was unexpectedly large at LEO and that the source energy came from the upstream waves. In magnetic field measurements from the CHAMP satellite Pc3 waves rather surprisingly were seen clearly in the field-aligned $b_{\|}$ (compressional) component, whereas on the ground, their signatures were found in the North-South H (mostly due to the in situ transverse) component. The coherence between ground and satellite wave signatures was high over wide latitude and longitude ranges. Observations of Pc3 pulsations by the scalar magnetometer on the Ørsted satellite (h = 650–900 km) also showed the dominance of the compressional component (Jadhav et al., 2001). Pc3 wave packets were almost simultaneous at Ørsted and on ground magnetic stations. For nighttime events the Pc3 packets had about the same amplitude both in space and on the ground, but during the daytime Pc3 amplitudes at the satellite were larger than on the ground, especially at lower latitudes. A high transparency of the nighttime ionosphere to Pc3 waves was also shown in coordinated CHAMP–SAMBA satellite-ground observations at near-equatorial latitudes (Cuturrufo et al., 2015).

Many observations of compressional Pc3 waves in the near-equatorial regions of the magnetosphere were interpreted as fast mode inward transport of wave energy from the boundaries of the magnetosphere (Kim and Takahashi, 1999; Takahashi et al., 1994; Yumoto and Saito, 1983). However, the occurrence of a significant compressional component of the Pc3 wave structure in the upper ionosphere was unexpected because traditional notions assumed that Pc3 waves on the ground were mainly produced by Alfvén oscillations, excited by compressional waves in a FLR resonant region. The main features of the dayside compressional Pc3 activity at LEO were found to be controlled by interplanetary parameters (Heilig et al. 2007): Wave amplitudes are controlled by the solar wind speed and IMF cone angle, while the dominant frequency is determined by the IMF strength. All these findings support the upstream origin of the compressional waves observed at LEO. However, because fast waves are to be reflected from regions with a high Alfvén velocity V_A, they are expected to be localized in the near-equatorial plane of the magnetosphere, and they can reach the ionosphere at other latitudes only as an evanescent mode. Therefore, the unexpected phenomenon of the recent Ørsted and CHAMP observations of ample compressional Pc3 power in the topside ionosphere at all latitudes clearly showed that existing theoretical view, assuming that in effect only Alfvén waves can reach the ionosphere, has to be revised.

The key parameter for comparison of theoretical predictions with ground-satellite observations is the ratio κ of the compressional component above the ionosphere to the ground magnetic signal (H component). The global latitude-MLT dependence of the CHAMP compressional power demonstrated a near-noon (09–13 MLT) maximum at low latitudes ($< 15°$). An additional nighttime maximum at low latitudes and high-latitude maxima on the day- and nightside were probably produced by the contribution by spatial structures sampled by the fast moving satellite, namely the equatorial spread F phenomenon and field-aligned currents. The MLT dependence of the satellite/ground amplitude ratio κ showed that this ratio remains pretty much the same, varying in the range $\kappa = 1.0 \pm 0.5$ (Pilipenko et al., 2008).

6.3.4 Pi2 wave observations

Using data from CHAMP, Sutcliffe and Lühr (2003) were for the first time able to extract and clearly resolve Pi2 pulsations in vector magnetometer data at LEO. These observations showed the occurrence of a significant compressional component in Pi2 pulsations at LEO. The compressional component mapped directly to variations of the magnetic field H component on the ground. Simultaneous observations of night-side Pi2 waves at CHAMP and on the ground with the spectral power ratio $\kappa \simeq 1.0$ have been recently reported (Cuturrufo et al., 2015). The compressional component was in phase with the H component and the poloidal component was in anti-phase (Ndiitwani and Sutcliffe, 2009). Pi2 pulsations observed by the Ørsted satellite in a low-latitude region were also nearly pure compressional waves (Han et al., 2004), and the Pi2 compressional component mapped directly to the H-component on the ground. These observations were interpreted such that on the nightside the Pi2 waves at low latitude are due to cavity fast mode waves directly transmitted to the ground. The idea of mid-latitude Pi2 activity as a cavity mode oscillatory response of the inner magnetosphere to the substorm activation was firmly supported by numerous satellite observations in the nightside magnetosphere (Takahashi et al., 1995; Keiling et al., 2011).

6.3.5 Doppler effect on LEO observations

Analyzing ST5 data, Le et al. (2011) found a unique type of ULF waves in the range of ~30–200 mHz, or in the nominal Pc2-3 band. These Pc2-3 waves appeared as wave packets with durations in the order of 1–10 min. The coordinated ground-based magnetic observations at the spacecraft footprints, however, did not see waves in this band; instead, the waves appear to be the Pc4-5 waves associated with FLR. The ST5 Pc2-3 waves are not seen in the ground-based data. The signatures of large-scale field-aligned currents were absent, indicating that the spacecraft trajectories were all equatorward from the auroral zone and the polar cusp, i.e., in subauroral latitudes. The low-frequency variations in the magnetic field components as well as the magnetic field strength were the signatures of Earth's crustal magnetic fields (Purucker et al., 2007). They were mainly visible by the ST5 spacecraft near the orbit perigee at altitudes below ~600 km. The Pc2-3 wave signatures were seen only in the magnetic field components, not in the magnetic field strength and the wave oscillation fields were in the directions transverse to the background magnetic field. Le et al. (2011) suggested that these unique Pc2-3 waves are in fact Doppler-shifted poloidal (small-scale in transverse direction with $m \simeq 40$–250) Pc4-5 waves that result from the rapid traverse of the spacecraft azimuthally across the resonant field lines at low altitudes with $\Delta\omega \simeq \mathbf{k}_\perp \mathbf{V}_s$. These observations indicated that the high-m dont you mean Pc2-3 waves? waves, which cannot be detected on the ground, may coexist with the low-m waves. This fact has not been theoretically comprehended yet.

In contrast to Le et al. (2011), Balasis et al. (2012) showed Pc3 waves in CHAMP data without a Doppler shift in comparison to Cluster Pc3 wave observations. Moreover, Balasis et al. (2013) showed Pc3 waves in CHAMP data without a Doppler shift in

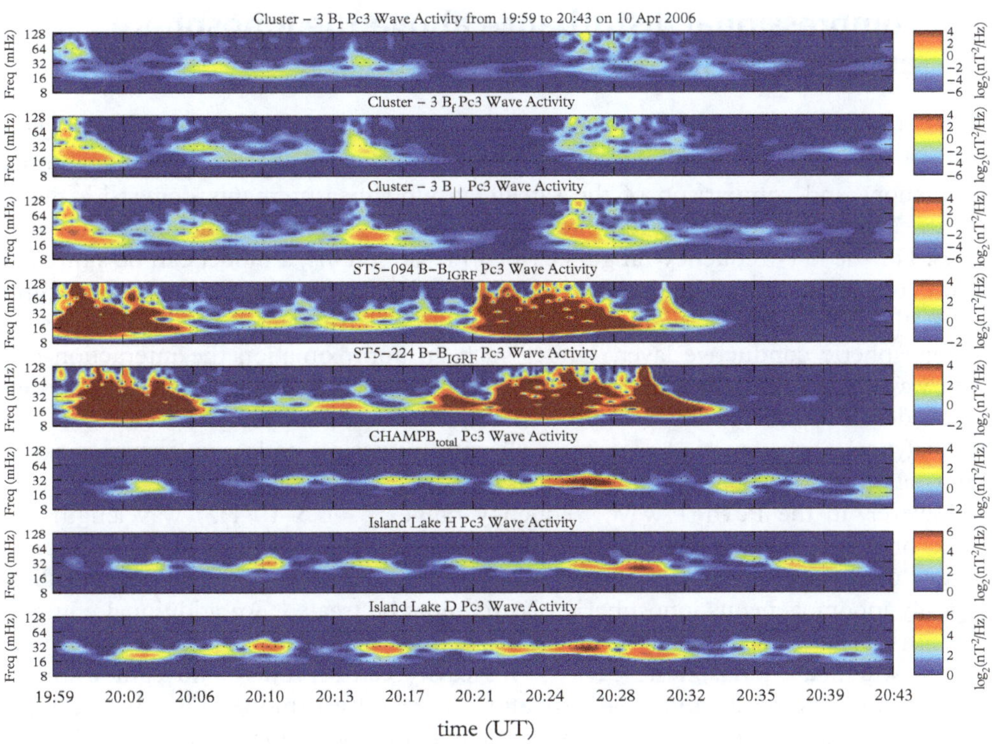

Figure 6.3 *From top to bottom, the wavelet power spectra of the Pc3 wave activity from Cluster-3 poloidal, toroidal and compressional components, ST5-094 and ST5-224 total field (after subtracting IGRF model), CHAMP total field, and Island Lake ground station H and D components, between 19:59 and 20:43 UT on April 10, 2006. [Figure 6.3 is modified after Figure 9 of Balasis et al. (2013).]*

comparison to ground Pc3 wave observations. For instance, Figure 6.3 presents wavelet dynamic spectra in the Pc3 range observed by the Cluster-3, ST5-094 and ST5-224, and CHAMP satellites as well as at the Island Lake ground station between 19:59 and 20:43 UT on April 10, 2006. This figure shows the wavelet power spectra of the poloidal, toroidal, and compressional components for Cluster-3 on the dayside magnetosphere and the total field power for ST5-094 and ST5-224, and CHAMP located at ~18 MLT and 15 MLT, respectively. CHAMP and Island Lake simultaneously observe Pc3 wave ($f \simeq 22 - 32$ mHz) with peak wave power seen at ~20:25 UT (LEO observations are in perfect match with the ground D component). Similar to the observations to Balasis et al. (2012), a recent study by Yagova et al. (2015) found examples of similar Pc2-3 waves on the ground, at CHAMP, and at high altitudes.

6.4 Modeling the relationship between the ULF compressional disturbance above the ionosphere and ground signal

The ULF wave pattern in the topside ionosphere is a complicated mixture of incident, reflected, and mutually converted waves. Therefore, interpretation of the low-altitude observations and comparison of the satellite/ground measurements demand a model that is capable of describing the wave reflection, absorption, and conversion processes. For the low-frequency range, i.e., Pc3/Pi2 pulsations, a model can be simplified by the use of the thin ionosphere approximation. This is a realistic approximation as the wavelength of such waves is typically much longer than the thickness of the ionospheric conductive layer. Under such an approximation the interaction of a plane wave harmonic with the magnetosphere–ionosphere–atmosphere–ground system can be described analytically (Alperovich and Fedorov, 2007). For periods $T > 20$ s the analytical approximation matches well with numerical code results presented by Waters and Sciffer (2008) and Waters et al. (2013).

However, in the Pc1/Pi1 wave frequency range (f ~0.5–10 Hz) wavelengths are of the order of the thickness of the ionospheric conducting layer and the approximation of a thin ionosphere is not valid, so the full wave equations in a realistic vertically inhomogeneous ionospheric plasma must be treated. An additional complication in the modeling of Pc1 waves is that a part of the Pc1 wave energy can be trapped within the ionospheric cavity. This cavity can serve as a waveguide for the fast magnetosonic mode and as a resonator for the Alfvén mode. The channeling of Pc1 wave energy in the ionospheric waveguide allows the propagation of the signal to large distances along the ionosphere (Fujita and Tamao, 1988). The partial trapping of the incident wave energy in the ionospheric Alfvén resonator results in oscillatory frequency-dependent transmission properties of the ionosphere in the Pc1 band (Lysak, 1997).

The occurrence of a ULF compressional component in the topside ionosphere can be caused by two possible mechanisms:

- an incident Alfvén wave generates an evanescent fast compressional mode upon interaction with the anisotropically conducting ionosphere (Hughes and Southwood, 1976; Kivelson and Southwood, 1988);
- transport of ULF wave energy from a distant source to the ionosphere occurs predominantly via the fast magnetosonic (FMS) mode.

The thin ionosphere model represents the magnetosphere as a half-space filled with a cold plasma immersed in a straight magnetic field $\mathbf{B}_0$, bounded on the bottom by a thin ionosphere—an anisotropically conducting layer at altitude h with height-integrated Pedersen and Hall conductances, Σ_P and Σ_H. The magnetospheric plasma is characterized by an Alfvén velocity V_A. In a Cartesian coordinate system the x-, y-, and z-axes are directed southward, eastward, and upward, respectively. The magnetic field inclination is denoted as I (vertical $\mathbf{B}_0$ corresponds to $I = \pm\pi/2$, in the equatorial ionosphere $I \rightarrow 0$). The atmosphere and ground are assumed to be isotropic conductors with conductivities σ_a and σ_g.

The wave electric (**e**) and magnetic (**b**) fields can be decomposed into two modes. The magnetospheric wave fields are the sum of:

- an Alfvén mode, where the disturbed magnetic field $\mathbf{b}_\perp$ is perpendicular to $\mathbf{B}_0$, whereas the longitudinal component is vanishing $b_\parallel = 0$; and
- an FMS mode, where the field-aligned component of the current vanishes $j_\parallel = 0$. The field-aligned (compressional) magnetic component $b_\parallel$ is dominant in the FMS mode and characterizes the plasma and magnetic field compression: $\delta N/N \simeq b_\parallel/B_o$.

The electromagnetic disturbance in the atmosphere and on the ground is composed of the magnetic H-mode, where the vertical component of the disturbed electric field is absent, $e_z = 0$; and the electric E-mode, where the vertical component of the disturbed magnetic field is absent, $b_z = 0$. The general set of Maxwell and ideal magnetohydrodynamic (MHD) equations for the magnetospheric plasma may be decomposed into two uncoupled sets of equations for the Alfvén and FMS modes. In a similar way, the decomposition into uncoupled equations for the E- and H-modes in the atmosphere/ground region can be done.

The electromagnetic field in the magnetosphere may be presented as a combination of incident (i) and reflected (r) waves, e.g., $\mathbf{b} = \mathbf{b}^{(i)} + \mathbf{b}^{(r)}$ and $\mathbf{e} = \mathbf{e}^{(i)} + \mathbf{e}^{(r)}$. The "thin ionosphere" theory provides analytical expressions for the matrix $\mathbf{R}$ of reflection coefficients, e.g., the ratio of the horizontal (transverse) magnetic wave components $\mathbf{b}_\tau$ after and before reflection. Similarly, to characterize the wave penetration to the ground, the matrix $\mathbf{T}$ of transmission coefficients for the ratio of the horizontal magnetic field wave amplitude at the ground to the horizontal magnetic field wave amplitude in the ionosphere is used. The elements of the reflection/transmission matrices are given in the theory of MHD wave interaction with the thin ionosphere (Alperovich and Fedorov, 2007), and take into account the Hall effect of the Alfvén wave reflection (Yoshikawa and Itonaga, 1996), excitation of the ionospheric surface mode (Pilipenko et al., 2000), and the finite conductivity of the atmosphere and ground.

In the following paragraphs, we consider separately the mechanisms by which incident harmonics of Alfvén and FMS waves ($\propto \exp(-i\omega t + ik_x x + ik_y y)$) interact with the ionosphere-atmosphere-ground system. Waves are assumed to be elongated in the azimuthal direction, so $k_y = 0$ and $k = k_x$.

Incident Alfvén waves. Upon interaction of an incident Alfvén wave with the anisotropic ionosphere, a compressional component $b_\parallel^{(A)}$ of the reflected evanescent mode can be produced. The ratio κ_A of this component to the ground magnetic signal $b_x^{(g)}$ (H-component) is (Pilipenko et al., 2008)

$$\kappa_A(z,k) = \frac{b_\parallel^{(A)}(z)}{b_x^{(g)}} = \exp(iI)kh_* \exp(-kz) \qquad h_* = h + (1+i)\delta_g/2. \tag{6.1}$$

Relation (6.1) is valid when the wave scale is much larger than both the skin-depth in the ground, $k\delta_g \ll 1$, and the height of the ionospheric conductive layer, $kh \ll 1$. The compressional component produced by an incident Alfvén wave becomes noticeable for wave scales in the range $k \simeq 10^{-2}$–10^{-3} km^{-1}. The ratio κ_A does not depend on the ionospheric conductance, but is determined mainly by the wave scale and the altitude of the space monitor. The latitude (or inclination I) determines the phase shift between the compressional and ground signals, but not the amplitude of their ratio.

Incident FMS mode. This mechanism assumes that the wave energy is transported from a source toward the ground by a FMS mode, without conversion into Alfvén waves. However, FMS waves cannot reach the ionosphere directly under reasonable horizontal wave numbers, because upon their propagation toward plasma with a higher V_A they encounter a non-propagation (opaque) region. However, thanks to its large horizontal scale (more than thousands of km), even an evanescent FMS mode can convey significant wave energy toward the Earth.

The factor κ_F characterizing the ratio of the compressional magnetic disturbance $b_\parallel^{(F)}$ above the ionosphere at a low altitude z to the ground magnetic disturbance b_x^g, induced by the fast mode, is determined by the transmission properties of the whole ionosphere-atmosphere-ground system. The thin ionosphere theory (Pilipenko et al., 2008, 2011) provides a remarkably simple relationship for a highly-conductive ground ($\sigma_g \to \infty$) as follows:

$$\kappa_F(z,k) = \frac{b_{\parallel}^{(F)}(z)}{b_x^{(g)}} = (1 - ip)\cos I. \tag{6.2}$$

The parameter $p = \omega h/V_C$ controls the penetration of the FMS mode through the ionosphere and to the ground. Here $V_C = (\mu\Sigma_C)^{-1}$ is the ionospheric Cowling velocity, determined by the Cowling conductance $\Sigma_C = \Sigma_P + \Sigma_H^2/\Sigma_P$. For the ULF waves in the nightside ionosphere this parameter is $|p| \ll 1$. Thus, the nightside ionosphere may be considered as transparent for the FMS mode, so the incident wave is reflected mainly by the Earth's surface, and $\kappa_F \simeq 1$. The component $b_{\parallel}$ is expected to be in phase with the H-component on the ground. On the dayside $|p| \simeq 1$, therefore, the dayside ionosphere can partially screen the magnetospheric signal from the ground, and introduce a noticeable phase delay. Relationship (6.2) predicts that the experimentally measured ratio κ should be proportional to $\cos I$, that is, it should decrease away from the dip equator. These theoretical predictions agree with the LEO/ground observations of Pi2 waves (Han et al., 2004; Cuturrufo et al., 2015).

Comparison of two mechanisms. Large-scale Alfvén and FMS modes both provide ground response mainly in the H-component, so just ground observations cannot resolve the contributions from the two modes. It is, however, possible to separate these two mode contributions by using conjugate and simultaneous observations from LEO and the ground. The theoretically predicted ratios between the compressional $b_{\parallel}$ (observed at LEO) and horizontal $b_x^{(g)}$ (observed on the ground) components for both modes may be compared with the results of satellite/ground observations. Comparison of the estimate of the factors $\kappa_A(z,k)$ and $\kappa_F(z,k)$ with CHAMP/ground mid-latitude observations of Pc3 waves was made in (Pilipenko et al., 2008). In general, the CHAMP observed ratio corresponds better to the scenario of direct fast mode transmission to the ground, $\kappa \simeq \kappa_F \gg \kappa_A$. The coordinated CHAMP/ground Pc3 pulsation observations resulted to a paradigm change—the Pc3 wave activity in the upper ionosphere has turned out to be dominated by compressional waves, whereas Alfvén waves dominate only in a narrow FLR region.

6.5 Discussion: prospects of further studies

Observations at CHAMP of the global distribution of dayside compressional Pc3 pulsation activity and ground Pc3 waves have shown that a significant part of the magnetospheric fast mode energy can leak in to the ionosphere, and eventually to the ground. The statistics of the Pc3 time intervals considered are probably dominated by events with a larger contribution of intervals when CHAMP was away from the local Alfvén resonant region. In these events a ground Pc3 signal is expected to be coherent over large distances, and resonant distortions of amplitude/phase behavior will be hardly noticeable. Observational results and theoretical modeling also have confirmed the idea of a fast mode cavity mechanism of mid-latitude Pi2 oscillations. The coordinated CHAMP-ground observations of Pc3/Pi2 waves are generally in a qualitative agreement with the predictions of the theory, assuming the incidence onto the upper ionosphere of a pure FMS mode.

However, in a general situation, coupled Alfvén and FMS modes may be present in an incident ULF wave. Therefore, a situation is possible in which the compressional component in the upper ionosphere is due to FMS mode, whereas the ground signal and ionospheric electric field are produced mainly by the Alfvén mode (Pilipenko et al., 2013). The theoretical model for this general situation is still to be developed. The demand for the elaboration of a reliable qualitative model of ULF transmission through the ionosphere to the ground comes not only from ULF wave studies, but from the wider space community. Such a model is necessary to evaluate how adequately ground observations of ULF waves correspond to magnetospheric wave activity, which is the important driver/loss factor for ring current protons and relativistic electrons.

The model of a thin ionosphere enables us to couple Pc3/Pi2 wave signatures between the upper ionosphere and the ground. For short-period Pc1 waves the problem becomes significantly more complicated due to the ionospheric Alfvén resonator excitation and mode trapping into the FMS waveguide. These effects make the ULF wave transmission properties strongly frequency and scale dependent, and demand more sophisticated numerical models for their description (e.g., Lysak, 1997; Prikner et al., 2004).

One of the gaps in the state-of-the-art knowledge of the wave background is how EMIC waves vary along a field line (see Chapter 10). The efficiency of EMIC waves in generating loss of relativistic electrons from the Van Allen belts to the atmosphere is determined not only by the resonance conditions and hence pitch angle diffusion rates, but also by their spatial extent, especially in L. A distinction must be made between waves triggered by external compressions, which can have a broad extent in L shell, and events triggered by substorm injections, which can have very narrow L-shell extents. The first category of these waves is often observed at higher L shells (Engebretson et al., 2002; Yahnina et al., 2008), which may make their observation from spacecraft in low Earth orbit difficult. However, the L-shell spatial extent of EMIC waves in a few multi-satellite events (Engebretson et al., 2008; Usanova et al., 2008) was found to be rather small ($\sim 10^2$ km). This fact has not been taken into account yet by modelers of the radiation belt and ring current dynamics.

The Swarm era in ULF wave observations has begun. Swarm is an Earth Explorer mission in ESA's Living Planet Programme (Friis-Christensen et al., 2006) launched on November 22, 2013. This mission, which comprises three satellites at low Earth polar orbits, with two spacecraft flying side by side in low altitude (~ 460 km) and one flying at a slightly higher altitude ($\sim$ 510 km), delivers data that will provide new insight into the Earth's system by improving our understanding of the Earth's interior as well as the near-Earth electromagnetic environment. Swarm is also expected to fill the gap in knowledge about ULF wave features at the spacecraft cruising altitudes (see the recent article on Swarm ULF wave observations in the topside ionosphere by Balasis et al., 2015). Conjunction studies with Van Allen Probes, Cluster, or THEMIS as well as with ground magnetometers can reveal how the spatial extent of ULF waves in situ correlates with their ionospheric propagation. The combination of Swarm and Cluster ULF wave observations could be a step forward in this direction. The ULF wave analysis methods presented by Balasis et al. (2012, 2013) can be applied in observations from the Swarm constellation, which is anticipated to enable joint studies with the Cluster mission. Conjunction studies from magnetospheric and LEO spacecraft with ground observations can help identify the extent of potential ducting and the way a narrow mangetospheric source maps to a larger area on the ionosphere and ground, but such studies have not yet been conducted.

Engebretson et al. (2008) noted that: "...insofar as Pc1 waves are also highly localized and evidently transient, a comprehensive understanding of their occurrence and role during geospace magnetic storms must also await the simultaneous deployment of a larger fleet of spacecraft, preferably with energetic particle instruments as well as magnetometers." Along these lines a combination of the electric and magnetic field measurements together with the plasma density measurements obtained by the instruments onboard each spacecraft of the fleet of the Swarm mission offers the greatest hope for a more in-depth understanding of EMIC waves at this point in time.

The bulk of information on Pi2 morphology is provided by ground-based magnetometers. While space-borne magnetometers provide important in situ information, they cannot match the spatio-temporal continuity of ground-based instruments. There are several sites where HF ionospheric radars and ground magnetometers are collocated. These facilities provide an opportunity to apply novel methods using simultaneously information on the wave electric field in the ionosphere and ground magnetic response (Pilipenko et al., 2012b; Ponomarenko and Waters, 2013). These observations can now be compared to measurements made onboard Swarm satellites, which fly at different altitudes in the topside ionosphere.

The Swarm mission will provide answers to some unresolved problems of ULF wave physics: How effectively can the FMS Pc3 and Pi2 waves illuminate a whole planet, in particular the polar, near equatorial, and mid-latitude regions? What fraction of compressional mode energy can tunnel toward the E-layer? Can upstream compressional waves refract on the nightside and produce "nightside" Pc3 pulsations? Is there any noticeable wave energy transfer to the auroral ionosphere by Pi1B wave bursts?

Significant progress in the ULF wave studies will be achieved, thanks to the availability of both electric and magnetic field data from the Swarm mission. It will allow calculation of the direction of the wave energy flow (Poynting vector). Additionally, Fedorov and Pilipenko (2014) suggested that an approach similar to the magnetotelluric sounding (MTS) technique can be used to find the planetary interior conductivity using the registration of variable electric and magnetic fields on a LEO space probe above a planetary ionosphere, while FMS waves in the planetary magnetosphere can play the role of sounding waves. The Swarm probe measures electric and magnetic wave components above the ionosphere, which enables one using the spectra of electric $E_\tau(\omega, k)$ and magnetic $B_\tau(\omega, k)$ variations tangent to the ionosphere to determine the matrix of an apparent spectral impedance $Z(\omega, k)$. Satellite MTS seems feasible because FMS waves feel the ionosphere only weakly, and mainly reflect from the Earth's surface.

Thus, the reflected FMS modes carry along information about the crustal conductivity. The theory developed by Fedorov and Pilipenko (2014) provides simple relationships to estimate the planet's surface impedance contribution to the apparent impedance of the entire magnetosphere–ionosphere–atmosphere–ground system measured onboard a probe. Such an approach can be tested based on electromagnetic observations on low orbiting satellites of the Swarm type with electric and magnetic sensors onboard in the Earth's topside ionosphere before it can be applied to sound other planets.

Acknowledgments

This work has received funding from the European Union Seventh Framework Programme (FP7-SPACE-2011-1) under grant agreement no 284520 for the MAARBLE (Monitoring, Analyzing and Assessing Radiation Belt Loss and Energization) collaborative research project, Russian Foundation for Basic Research (grant 16-55-16012 to VAP) and National Science Foundation (grant ATM 0636939 for the SAMBA magnetometer array).

REFERENCES

Alperovich, L.S. and Fedorov, E.N. (2007), Hydromagnetic Waves in the Magnetosphere and the Ionosphere, vol. 353 of Series: Astrophysics and Space Science Library, XXIV, ISBN:9781-402066368.

Balasis G., S. Maus, H. Lühr, and M. Rother (2005), Wavelet analysis of CHAMP flux gate magnetometer data, in Earth Observation with CHAMP, ed. by C. Reigber, H. Lühr, P. Schwintzer and J. Wickert, 347–352, *Springer*, New York.

Balasis, G., and M. Mandea (2007), Can electromagnetic disturbances related to the recent great earthquakes be detected by satellite magnetometers?, *Tectonophysics*, **431**, doi:10.1016/j.tecto.2006.05.038.

Balasis, G., I. A. Daglis, E. Zesta, C. Papadimitriou, M. Georgiou, R. Haagmans, and K. Tsinganos (2012), ULF wave activity during the 2003 Halloween superstorm: multipoint observations from CHAMP, Cluster and Geotail missions, *Ann. Geophys.*, **30**, 1751–1768, doi:10.5194/angeo-30-1751-2012.

Balasis, G., I. A. Daglis, M. Georgiou, C. Papadimitriou, and R. Haagmans (2013), Magnetospheric ULF wave studies in the frame of Swarm mission: a time-frequency analysis tool for automated detection of pulsations in magnetic and electric field observations, *Earth, Planets and Space*, **65**, 1385–1398.

Balasis, G., C. Papadimitriou, I. A. Daglis, and V. Pilipenko (2015), ULF wave power features in the topside ionosphere revealed by Swarm observations, *Geophys. Res. Lett.*, **42**, doi:10.1002/2015GL065424.

Cuturrufo F., V. Pilipenko, B. Heilig, M. Stepanova, H. Lühr, P. Vega, and A. Yoshikawa (2015), Near-equatorial Pi2 and Pc3 waves observed by CHAMP and on SAMBA/MAGDAS stations, *Adv. Space Res.*, **55**, 1180–1189.

Engebretson, M. J., W. K. Peterson, J. L. Posch, M. R. Klatt, B. J. Anderson, C. T. Russell, H. J. Singer, R. L. Arnoldy, and H. Fukunishi (2002), Observations of two types of

Pc 1-2 pulsations in the outer dayside magnetosphere, *J. Geophys. Res.*, **107**, 1451, doi:10.1029/2001JA000198.

Engebretson, M. J., J. L. Posch, A. M. Westerman, N. J. Otto, J. A. Slavin, G. Le, R. J. Strangeway, and M. R. Lessard (2008), Temporal and spatial characteristics of Pc1 waves observed by ST5, *J. Geophys. Res.*, **113**, A07206, doi:10.1029/2008JA013145.

Fedorov, E.N., and V.A. Pilipenko (2014), Electromagnetic sounding of planets from a low-orbiting probe, *Cosmic Research*, **52**, 46–51, DOI: 10.1134/S0010952514010031.

Friis-Christensen, E., H. Lühr, and G. Hulot (2006), Swarm: A constellation to study the Earth's magnetic field, *Earth Planets Space*, **58**, 351–358.

Fujita, S., and T. Tamao (1988), Duct propagation of hydromagnetic waves in the upper ionosphere, 1, Electromagnetic field distributions in high latitudes associated with localized incidence of a shear Alfvén wave, *J. Geophys. Res.*, **93**, 14665-14673.

Guglielmi, A., and J. Kangas (2007), Pc1 waves in the system of solar-terrestrial relations: New reflections, *J. Atm. Solar-Terrestr. Phys.*, **69**, 1635–1643.

Han, D. S., et al. (2004), A comparative analysis of low-latitude Pi 2 pulsations observed by Ørsted and ground stations, *J. Geophys. Res.*, **109**, A10209.

Heilig, B., H. Lühr, and M. Rother (2007), Comprehensive study of ULF upstream waves observed in the topside ionosphere by CHAMP and on the ground, *Ann. Geophys.*, **25**, 737–754.

Hughes, W. J., and D. J. Southwood (1976), An illustration of modification of geomagnetic pulsation structure by the ionosphere, *J. Geophys. Res.*, **81**, 3241–3247.

Jadhav, G., M. Rajaram, and R. Rajaram (2001), Modification of daytime compressional waves by the ionosphere: first results from Ørsted, *Geophys. Res. Lett.*, **28**, 103–106.

Iyemori T, and Hayashi K. (1989), Pc 1 micropulsations observed by Magsat in the ionospheric F region. *J. Geophys. Res.*, **94**, 93–100.

Keiling, A., and Takahashi K. (2011), Review of Pi2 models, *Space Sci. Rev.*, **161**, 63–148.

Kim, H., M.R. Lessard, M. J. Engebretson, and H. Lühr (2010), Ducting characteristics of Pc 1 waves at high latitudes on the ground and in space, *J. Geophys. Res.*, **115**, A09310, doi:10.1029/2010JA015323.

Kim, K.H., and K. Takahashi (1999), Statistical analysis of compressional Pc3–4 pulsations observed by AMPTE CCE at L = 2–3 in the dayside magnetosphere, *J. Geophys. Res.*, **104**, 4539–4558.

Kivelson, M.G., and D.J. Southwood (1988), Hydromagnetic waves and the ionosphere, *Geophys. Res. Lett.*, **15**, 1271–1274.

Le, G., P. J. Chi, R. J. Strangeway, and J. A. Slavin (2011), Observations of a unique type of ULF wave by low altitude Space Technology 5 satellites, *J. Geophys. Res.*, **116**, A08203.

Leonovich, A.S., et al. (1985), MHD waveguides in inhomogeneous plasma, Plasma Physics Reports (*Fiz. Plazmy)*, **11**, 1106–1115.

Lessard, M.R., et al. (2006), Nature of Pi1B pulsations as inferred from ground and satellite observations, *Geophys. Res. Lett.*, **33**, L14108, doi:10.1029/2006GL026411.

Lysak, R. L. (1997), Propagation of Alfvén waves through the ionosphere, *Phys. Chem. Earth*, **22**, 757–766.

Mandea M., and Balasis G. (2006), The SGR 1806-20 magnetar signature on the Earth's magnetic field, *Geophysical Journal International*, **167**, 586–591.

Maus, S., Rother, M., Holme, R., Lühr, H., Olsen, N. and Haak, V. (2002), First scalar magnetic anomaly map from CHAMP satellite data indicates weak lithospheric field, *Geophys. Res. Letters*, **29**, 10.1029/2001GL013685.

Maus, S., Lühr, H., Balasis, G., Rother, M., and Mandea, M. (2005), Introducing POMME, The Potsdam Magnetic Model of the Earth, in: Earth Observation with CHAMP, edited by: Reigber, C., Luehr, H., Schwintzer, P., and Wickert, J., pp. 293–298, Springer, Berlin-Heidelberg.

Menk, F. W. and Waters, C. L. (2013), Magnetoseismology: Ground-Based Remote Sensing of Earth's Magnetosphere, doi: 10.1002/9783527652051, Wiley-VCH Verlag GmbH & Co. KGaA, Weinheim, Germany.

Mursula, K. (2007), Satellite observations of Pc1 pearl waves: The changing paradigm, *J. Atm. Sol. Terr. Phys.*, **69**, 1623–1634.

Ndiitwani D. C. and P. R. Sutcliffe (2009), The structure of low-latitude Pc3 pulsations observed by CHAMP and on the ground, *Ann. Geophys.*, **27**, 1267–1277.

Park, J., Luhr, H., and Rauberg, J. (2013), Global characteristics of Pc1 magnetic pulsations during solar cycle 23 deduced from CHAMP data, *Ann. Geophys.*, **31**, 1507-1520, doi:10.5194/angeo-31-1507-2013.

Pilipenko, V., M. Vellante, and E. Fedorov (2000), Distortion of the ULF wave spatial structure upon transmission through the ionosphere, *J. Geophys. Res.*, **105**, 21225–21236.

Pilipenko, V., E. Fedorov, B. Heilig, and M.J. Engebretson (2008), Structure of ULF Pc3 waves at low altitudes, *J. Geophys. Res.*, **113**, A11208, doi:10.1029/2008JA013243.

Pilipenko, V., E. Fedorov, B. Heilig, M. J. Engebretson, P. Sutcliffe, and H. Lühr (2011), ULF waves in the topside ionosphere: Satellite observations and modeling, in: Dynamic Magnetosphere, IAGA Special Sopron Book Series 3, DOI 10.1007/978-94-007-0501-2_14, Springer.

Pilipenko, V. A., Fedorov, E. N., Teramoto, M., and Yumoto, K. (2013), The mechanism of mid-latitude Pi2 waves in the upper ionosphere as revealed by combined Doppler and magnetometer observations, *Ann. Geophys.*, **31**, 689-695.

Pilipenko V.A., Polozova T.L., and Engebretson M.J. (2012a), Spatial-temporal structure of ion-cyclotron waves in the upper ionosphere as observed by ST5 satellites, *Cosmic Research (Kosmicheskie issledovanija)*, **50**, 355–365.

Pilipenko V., V. Belakhovsky, A. Kozlovsky, E. Fedorov, and K. Kauristie (2012b), Determination of the wave mode contribution into the ULF pulsations from combined radar and magnetometer data: Method of apparent impedance, *J. Atmospheric and Solar-Terrestrial Physics*, 77, 85–95.

Ponomarenko P. V., and C. L. Waters (2013), Transition of Pi2 ULF wave polarization structure from the ionosphere to the ground, *Geophys. Res. Lett.*, **40**, 1474–1478.

Prikner K., K. Mursula, J. Kangas, R. Kerttula, and F. Z. Feygin (2004), An effect of the ionospheric Alfvén resonator on multiband Pc1 pulsations, *Annales Geophysicae,* **22**, 643–651.

Purucker, M., T. Sabaka, G. Le, J. A. Slavin, R. J. Strangeway, and C. Busby (2007), Magnetic field gradients from the ST5 constellation: Improving magnetic and thermal models of the lithosphere, *Geophys. Res. Lett.*, **34**, L24306, doi:10.1029/2007GL031739.

Stolle, C., H. Lühr, M. Rother, and G. Balasis (2006), Magnetic signatures of equatorial spread F as observed by the CHAMP satellite, *J. Geophys. Res.*, **111**, A02304, doi:10.1029/2005JA011184.

Sutcliffe, P. R. and H. Lühr (2015), A comparison of Pi2 pulsations observed by CHAMP in low Earth orbit and on the ground at low latitudes, *Geophys. Res. Lett.*, **30**, doi:10.1029/2003GL018270, 2003.

Sutcliffe, P. R., Heilig, B., and Lotz, S. (2013), Spectral structure of Pc3-4 pulsations: possible signatures of cavity modes, *Ann. Geophys.*, **31**, 725–743.

Takahashi, K., Anderson, B. J., Newell, P. T., Yamamoto, T., and N. Sato (1994), Propagation of compressional Pc3 pulsations from space to the ground: a case study using multipoint measurements, in: *Solar wind sources of magnetospheric ULF waves*, pp. 355–363, Geophysical Monogr. 81, AGU, Washington, D.C.

Takahashi,K., S. Ohtani, and B.J. Anderson (1995), Statistical analysis of Pi2 pulsations observed by the AMPTE CCE spacecraft in the inner magnetosphere, *J. Geophys. Res.*, **100**, 21929–21941.

Usanova, M. E., et al. (2008), Multipoint observations of magnetospheric compression-related EMIC Pc1 waves by THEMIS and CARISMA, *Geophys. Res. Lett.*, **35**, L17S25, doi:10.1029/2008GL034458.

Vellante, M., et al. (2004), Ground/satellite signatures of field line resonance: A test of theoretical predictions, *J. Geophys. Res.*, **109**, A06210, doi:10.1029/2004JA010392.

Waters, C. L. and M. D. Sciffer (2008), Field line resonant frequencies and ionospheric conductance: Results from a 2-D MHD model, *J. Geophys. Res.*, **113**, A05219, doi:10.1029/2007JA012822.

Waters CL, Lysak RL, Sciffer MD (2013), On the coupling of fast and shear Alfvén wave modes by the ionospheric Hall conductance, *Earth Planets and Space*, **65** 385–396.

Yagova, N., Heilig, B., and Fedorov, E. (2015), Pc2-3 geomagnetic pulsations on the ground, in the ionosphere, and in the magnetosphere: MM100, CHAMP, and THEMIS observations, *Ann. Geophys.*, **33**, 117–128.

Yahnina, T. A., H. U. Frey, T. Bsinger, and A. G. Yahnin (2008), Evidence for subauroral proton flashes on the dayside as the result of the ion cyclotron interaction, *J. Geophys. Res.*, **113**, A07209, doi:10.1029/2008JA013099.

Yoshikawa, A., and M. Itonaga (1996), Reflection of shear Alfvén waves at the ionosphere and the divergent Hall current, *Geophys. Res. Letters*, **23**, 101–104.

Yumoto, K., and T. Saito (1983), Relation of compressional HM waves at GOES 2 to low-latitude Pc 3 magnetic pulsations, *J. Geophys. Res.*, **88**, 10041–10052.

7

Monitoring Magnetospheric Waves from the Ground

Colin Waters and Fred Menk

■ Abstract

The electric fields of ULF plasma waves may energize charged particles in the magnetosphere, but many details of these interactions are unclear. Much of the present understanding of ULF wave properties has come from ground magnetometer arrays and associated data analysis techniques. In particular, the resonant properties of ULF waves may be used to remote sense the magnetospheric cold plasma reservoir. Electric field magnitudes in the equatorial plane may be estimated by tracing along the geomagnetic field to the ground where the ULF wave magnetic field is measured. However, the wide range of wave temporal and spatial scale sizes is a persistent challenge. Smaller spatial scale size waves involved in resonant interactions with particles are more readily seen in the ionosphere with HF radars and Doppler sounders. This chapter provides an overview of these aspects of ULF wave properties and their relationship to particle energization in the magnetosphere.

The magnetosphere is a dynamic environment, buffeted by the solar wind and in constant transition between different levels of energy storage and release. While the larger energy release processes are often the most spectacular and involve accelerations of particles to MeV levels, lower amplitude temporal perturbations of electric and magnetic

Waves, Particles, and Storms in Geospace. First Edition. Georgios Balasis, Ioannis A. Daglis, Ian R. Mann.

fields are an important mechanism for energy transport, enhancement and loss and are ubiquitous in the magnetosphere, occurring over all local times, latitudes and a wide frequency range.

Alfvénic electric and magnetic perturbations in the solar wind incident on the magnetosphere exhibit a spectrum that varies as f^{-1} to f^{-2} for frequencies less than the cyclotron frequency, f_c (Denskat and Neubauer, 1982). The interaction of solar wind kinetic energy and embedded waves with the magnetosphere and the effect of propagation and resonance processes alters the spectrum measured at the ground. For example, the power in the Pc3-4 band (Pc4; 7–22 mHz, Pc3; 22–100 mHz) at cusp latitudes appears to vary as f^{-4} in addition to an upstream related, band-limited portion (Ponomarenko et al., 2002). Given that the lower frequencies exhibit larger amplitudes, the Pc4-5 (Pc5; 1.6–7 mHz) ultra-low frequency (ULF) bands are thought to contain sufficient electric field magnitudes to significantly boost particle energies.

For Alfvén waves with frequencies less than about 100 mHz, the wavelengths are of similar dimensions to those of the magnetosphere. The shear Alfvén wave mode, with energy guided along the background magnetic field, forms field line resonances that may be used to remote sense the cold, core plasma mass density. Therefore, ULF waves contribute two important mechanisms involved in energetic particle dynamics in the magnetosphere: (i) the wave electric fields drive radial transport of charged particles, while (ii) field line resonances provide estimates of the cold, core plasma mass density reservoir from which high-energy particles may arise.

Advances in understanding relationships between particle dynamics and ULF waves are possible due to previous, basic research focused on ULF wave properties in the magnetosphere along with advances in suitable instrumentation. This chapter begins with a review of developments in instrumentation used to obtain ULF wave data in parallel with fundamental ULF wave properties gleaned from those early data. This is followed by results of the distribution of Pc5 power as seen by ground-based magnetometers. Applications of ULF waves to remote sensing, such as estimating the equatorial ULF electric fields and the cold, plasma mass density, are highlighted. The smaller spatial scale size, Pc5 waves involved in resonant interaction with particles are more readily seen in HF radar data. The chapter concludes with a discussion of both Pc5 and Pc3 indices and their use in space weather monitoring.

7.1 Overview of instrumentation and techniques

Perturbations in the geomagnetic field have been measured using ground magnetometer arrays for 170 years. In order to study these phenomena a network of British Colonial Observatories was established in 1840 under the direction of Edward Sabine. Using hourly measurements (recorded by hand) from Toronto, Canada and Hobart, Australia over 1843–1848, Sabine concluded that variations in magnetic declination "correspond precisely both in period and epoch with the variation in the frequency and magnitude of the solar spots" (Sabine, 1852). This may be regarded as the origin of solar-terrestrial physics (Cliver, 1984).

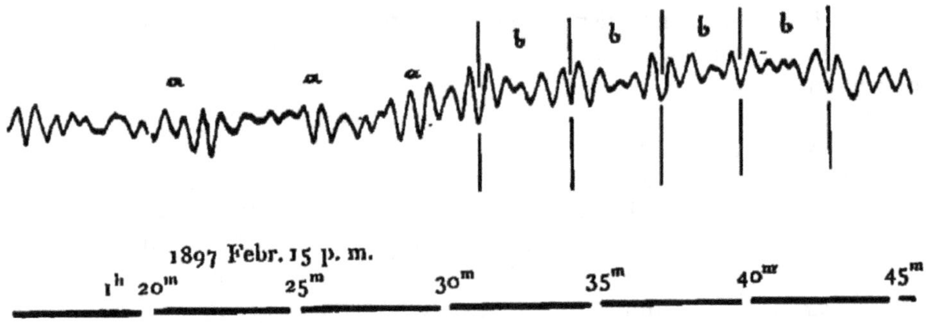

Figure 7.1 *Magnetic variations recorded at Potsdam Observatory on 14 Feb 1897 (Eschenhagen, 1897). Wave packets labelled aa, bb, etc. result from superposition of waves of period 43 and 34 s.*

The development of light beam magnetometers in 1857 allowed continuous recording of small, rapid geomagnetic perturbations on photosensitive paper (Stewart, 1859). By the late nineteenth century, perturbations with periods (at Potsdam) typically around 30 s, but as short as 12 s during large magnetic storms, were being recorded by Eschenhagen (1897). For one example, shown in Fig. 7.1, he determined that the observed pulsations were due to the superposition of two wavetrains of period 43 and 34 s, with resultant amplitude of 1.2 nT. In order to investigate the spatial extent of such signals he initiated coordinated observations every 5 s for selected hours at 15 globally distributed observatories. He also advocated the use of galvanometer sensors with large wire spools aligned with the components of the geomagnetic field to detect these perturbations, subsequently called geomagnetic pulsations by Bemmelen (1908). These pulsations are now recognized as the ground signature of ultra-low frequency (ULF; 1 mHz–1 Hz) Alfvén waves propagating through the magnetosphere.

Considerable early attention focused on characteristic, sinusoidal, “giant pulsations”, labelled Pg, observed in the auroral zones. Sucksdorff (1939) recorded 150 such events at Sodankylä over 1914–1938. Kato and Watanabe (1954) established that the periods of these were harmonically related to a fundamental period of 650 s (f=1.53 mHz) for storm-time events and 615 s (f=1.63 mHz) for the others. The majority of these Pg events occurred in the Pc4 band with periods between 1/9 and 1/6 of the fundamental. Following Dungey’s work Kato and Watanabe (1956) concluded that the pulsations were caused by “toroidal” oscillations of auroral field lines, with period depending on latitude and mass density at the apex of the field line.

An understanding of the Alfvénic properties (Alfvén, 1942) of ULF waves in space was developed by early researchers using ground-based measurements. Since Alfvén waves do not exist in the neutral atmosphere, attention also focused on the transition of the signal from space to ground-based sensors. The “shielding effect” which results in wave amplitude reduction as a function of wave spatial scale size and the physics of wave transition from Alfvénic in space, to electromagnetic in the atmosphere was discussed by

Sugiura (1950), Oguti and Nagata (1954) and Ashour and Price (1948). The role of the ionospheric Hall conductivity in ULF wave mode coupling was discussed by Kato and Tamao (1956), and has important consequences for remote sensing ULF wave electric field amplitudes using ground-based measurements of the wave magnetic field (Sciffer and Waters, 2011). Watanabe (1957) showed that the signal at the ground is reduced in amplitude when the ionospheric conductivity is high, and the wave polarization is rotated by $\pi/4$.

Some early researchers used detectors comprising large, multi-turn coil loops or antennas (e.g. Campbell, 1959). The detection of geomagnetic pulsations was facilitated by the development of ferromagnetic core induction coil sensors and slow-speed tape recording that allowed signals to be examined by spectrum analyses (e.g. Duffus et al. 1958). An array of four large, horizontal buried loop systems across Australia and spectrum analysis of data enabled the detection of sustained daytime bands of Pc oscillations (with frequency decreasing postdawn), short impulsive damped trains (i.e. Pi2) at night and more complex patterns including broadband noise with impulsive onset accompanying magnetic storms (Duncan, 1961). Wave periods were observed to decrease with decreasing latitude, as expected for standing Alfvén oscillations of field lines, as suggested by Obayashi and Jacobs (1958).

During the 1960s and 1970s magnetometer arrays with induction coil sensors, precise timing and digital data logging had been established for studies of the propagation of Pc1-2 (0.1–1 Hz) and Pc3 (~30 mHz) ULF waves (e.g. Fraser and Summers, 1972; Zelwer and Morrison, 1972; Ansari and Fraser, 1986). These sensors are relatively simple to manufacture and operate and their frequency response favours weaker, higher frequency signals. A recent review of induction coil sensors was presented by Tumanski (2007).

Instruments with a flat frequency response that measure magnetic variations in the orthogonal directions with suitable sensitivity for studies of pulsations include rubidium vapour and fluxgate (saturable core) magnetometers. Green (1976) investigated the longitudinal propagation of mid-latitude Pc3-4 pulsations using a rubidium magnetometer array with recordings digitized from chart paper.

The development of the EISCAT-magnetometer cross array in 1982 heralded the era of multi-national arrays using digitally sampled, triaxial fluxgate instruments (Lühr et al., 1984). In 1991 the EISCAT stations provided the foundation for the International Monitor for Auroral Geomagnetic Effects (IMAGE) array, which by its 20th anniversary numbered 30 stations covering northern Europe from subauroral to polar cap latitudes. Each station incorporates GPS timing and 10 s averaged sampling. A defining feature of the array is an open data policy through website access to quick-look and user-defined magnetograms. From 1987 IMAGE was supplemented by the SAMNET array that provides 1 s digital data and extends coverage to the United Kingdom.

A new quasi-meridional European Magnetometer Array (EMMA) commenced operation in 2011. Funded and operated as a consortium, the purpose of EMMA is to provide observations of ULF resonant oscillations of plasmaspheric field lines used to automatically derive near real-time estimates of equatorial mass density. These measurements, along with VLF-whistler derived electron densities, form input to a data assimilative model of the plasmasphere (Lichtenberger et al., 2013). In order to provide suitable coverage of mid-latitudes, EMMA incorporates stations from several magnetometer networks such as IMAGE, MM100 (Hungary), SEGMA (Italy) and stations on the same meridian in South Africa.

Many other multi-national, digitally sampled magnetometer networks have been established for studies of geomagnetic pulsations, especially field line resonances. An incomplete list includes the South American Meridional B Field Array (SAMBA; Boudouridis and Zesta, 2007) now operated through the African Meridian B Field Education and Research array (AMBER; Yizengaw and Moldwin, 2009), the Magnetic Data Acquisition System/Circum-pan Pacific Magnetometer Network (MAGDAS/CPMN; about 40 stations; Yumoto and the MAGDAS group, 2006), the Canadian Array for Realtime Investigations of Magnetic Activity (CARISMA) with 28 fluxgate and additional induction coil sensors at 8 sites, and the Mid-continent Magnetoseismic Chain (McMac), comprising 9 stations and extending CARISMA to form a large meridional array.

There now exist sufficient arrays that portals have been developed to facilitate integrated and coordinated access to global data sets. These include Ultima (Ultra Large Terrestrial International Magnetometer Array) and SuperMAG. The latter provides access to standardized magnetometer data for over 350 stations from multiple arrays, merged for providing some key tools, like equivalent current patterns (Gjerloev, 2012). SuperMAG incorporates and extends the Intermagnet array which provides online open access to a global network of 1-min mean standard observatory data. One wonders how Sabine and Eschenhagen would reflect on these developments.

7.2 Remote sensing geospace using data from ground magnetometer arrays

7.2.1 Data analysis techniques

Techniques for the analysis of ULF geomagnetic pulsation data have advanced in line with developments in data recording and access. The importance of understanding the spectral content of pulsation signals prompted the scaling by hand of wave periods from paper records (e.g. Maple, 1959) and hand digitization to produce frequency-time dynamic spectra. The development of the "Sona-Graph" sound spectrograph recorder by Kay Electric Co. under licence to Bell Laboratories permitted the production of dynamic spectra in 2.4 s samples over an 8 kHz frequency range from suitably sped-up tape recordings of pulsation data (e.g. Duncan, 1961; McNicol and Mainstone, 1963). The complex fine structure of Pc1 emissions was revealed for the first time (Tepley, 1961). Such technology involving electronic filtering was soon superseded by digital spectral analysis. One early study by Ness et al. (1961) tracked the dominant spectral peak in the 75–175 mHz range over 5 days using data from Fredericksburg (50^0 GM latitude) while Explorer X was aloft.

The wave nature of geomagnetic pulsations is best examined using time series techniques applied to multi-station analyses of spectral and correlation properties including cross-power, coherency, phase and cross-phase (e.g. Fraser, 1979; Glangeaud, 1982). Wave polarization, including ellipticity and azimuth, is also an important property (e.g. Olson and Samson, 1979; Samson and Olson, 1981). The polarization of pulsations at the ground has been measured at least since the International Geophysical Year in 1957 (e.g. Wilson and Sugiura, 1961), leading to the notion of global field line oscillations driven by Earthward propagating compressional and transverse hydromagnetic waves accompanying sudden commencements.

7.2.2 Remote sensing plasma mass density in space

The improved availability and increased number of ground-based magnetometer installations since the 1960s has prompted the development of techniques to monitor various magnetosphere and ionosphere space weather processes. The most developed methods exploit the latitude (radial distance) dependence of the resonant properties of the shear Alfvén ULF wave mode, readily identified using cross-phase spectral methods. These are often grouped together and described as *magnetoseismology* (Menk and Waters, 2013).

The coupling of a propagating fast mode wave of frequency ω_o to shear Alfvén field line oscillations can establish a field line resonance (FLR) when $\omega_o = k_{\parallel} V_A$, where $k_{\parallel}$ is the wavenumber parallel to the **B** field and V_A is the Alfvén velocity. FLRs exhibit characteristic variations in amplitude and polarization with radial distance in the magnetosphere (i.e. latitude across the ground), illustrated in Fig. 7.2.

Using a seven-station Canadian magnetometer array, Samson et al. (1971) suggested that the variation of polarized power and polarization with latitude for Pc5 signals at auroral latitudes was consistent with the excitation of toroidal mode field line

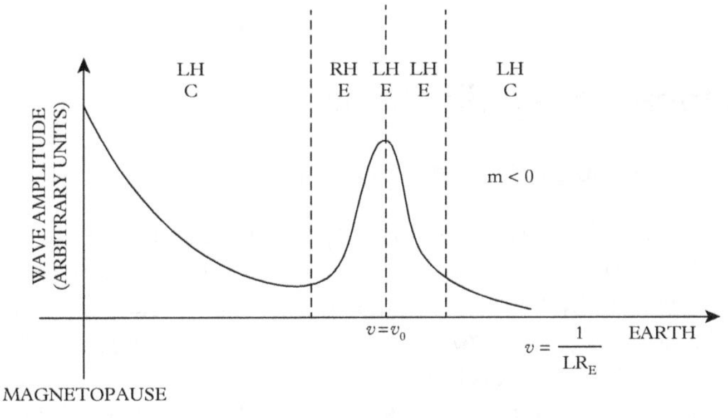

Figure 7.2 *Wave amplitude and polarization with radial distance in the equatorial plane for local morning, tailward propagating pulsations (Hasegawa and Chen, 1974). Wave ellipticity ("E") is large near the resonant field line at* $\nu = \nu_o$ *and reverses sense across the resonance.*

eigenoscillations by surface modes at the flanks of the magnetosphere. The variation with latitude of pure state power, phase and wave polarization (ellipticity and azimuth) for FLRs at low latitudes was described in detail by Ziesolleck et al. (1993). However, these properties are complicated by the rapid change in plasma mass density and hence Alfvén speed across a well-defined plasmapause (Lanzerotti et al., 1974).

At low and mid-latitudes, ULF field line resonances may not provide the dominant contribution to the pulsation spectrum. Determination of the FLR frequency is greatly facilitated using amplitude and phase difference measurements between two closely spaced magnetometers (Waters et al., 1991). Through its dependence on the Alfvén velocity, knowledge of the local resonant frequency affords an estimate of plasma mass density near the field line apex (Schulz, 1996; Troitskaya, 1997). The magnetometer networks mentioned previously may therefore be used to monitor spatial and temporal changes in magnetospheric plasma mass density (e.g. Waters et al. 1996; Menk et al. 2004; Menk and Waters, 2013), and automated FLR detection algorithms have been developed to facilitate this (e.g. Berube et al. 2003).

Storm-time plasma erosion from the plasmasphere and subsequent refilling from the underlying ionosphere are readily measured using FLR detection methods (e.g. Obana et al. 2010). Statistical analysis of FLR data has been used to develop a plasmaspheric density model and provide information on the annual variation in plasmaspheric density and the ion density in the plasmasphere and outer magnetosphere (e.g. Takahashi et al. 2004; Berube et al. 2005; Menk et al. 2012; Chi et al. 2013). The utility of ground magnetometer measurements is extended by comparison with other data sets. Information on heavy ion mass loading is obtained by comparing FLR-derived mass densities with electron densities obtained by ground-based VLF whistler or Doppler experiments. These can also be used to intercalibrate in situ particle measurements and imager observations (Clilverd et al., 2003) which allow estimates of the plasma composition (Grew et al., 2007).

7.3 Space weather applications

7.3.1 Remote sensing Pc5 electric fields

Large increases in the MeV electron flux appear to correlate with the appearance of large amplitude, 1–10 mHz ULF plasma waves throughout the magnetosphere (Baker et al., 1998; Rostoker et al., 1998). The drifting particles may resonantly interact with the radial electric field of global, toroidal mode, 1–10 mHz waves resulting in radial stochastic diffusion and hence acceleration (Parker, 1960; Elkington et al., 2003). The key parameter for particle energization appears to be the electric field of 1–10 mHz waves in the equatorial plane of the magnetosphere. In situ measurements of these fields are sparse and difficult to obtain so electric field values are "guessed" in order to estimate the maximum electron energy gain. However, ground magnetometer arrays routinely measure these plasma wave signatures.

Ozeke et al. (2009) described a simple method for estimating the ULF electric field in the equatorial plane using the measured perturbation magnetic field at the ground. The process involved ratios of electric and magnetic fields of the shear Alfvén ULF mode with east-west, magnetic signatures of b^i_ϕ in the ionosphere, north-south b^g_ν from the magnetometers at the ground and e^{eq}_ν, the electric fields in the equatorial plane, mapped into space along a dipole background magnetic field. Starting from the measured b^g_ν, the ratio b^i_ϕ/b^g_ν was estimated from ratios of the Pedersen and Hall conductance and the scale size of the field line resonance. Solutions to the "toroidal" mode ULF wave equations were then used to estimate e^{eq}_ν/b^i_ϕ.

Such an approach may provide an important ground-based remote sensing capability for the energization of "killer" particles. However, this method assumes *toroidal* and *poloidal* approximations of the ULF mode eigensolutions in the magnetosphere, the absence of Hall conductance effects, and a plane wave based, spatial scale estimate of the resonance region as seen at the ground. More realistic calculations of the resonance structure and inclusion of Hall and Pedersen conductance effects in the ionosphere significantly alter the ratio b^i_ϕ/b^g_ν (Sciffer and Waters, 2011). This is due to the generation of fast mode energy which appears in the topside ionosphere, as shown in Fig. 7.3.

The *toroidal* mode comes from an assumption of zero azimuthal spatial variation in a dipole field that allows separation of the Alfvén wave differential equation. At high

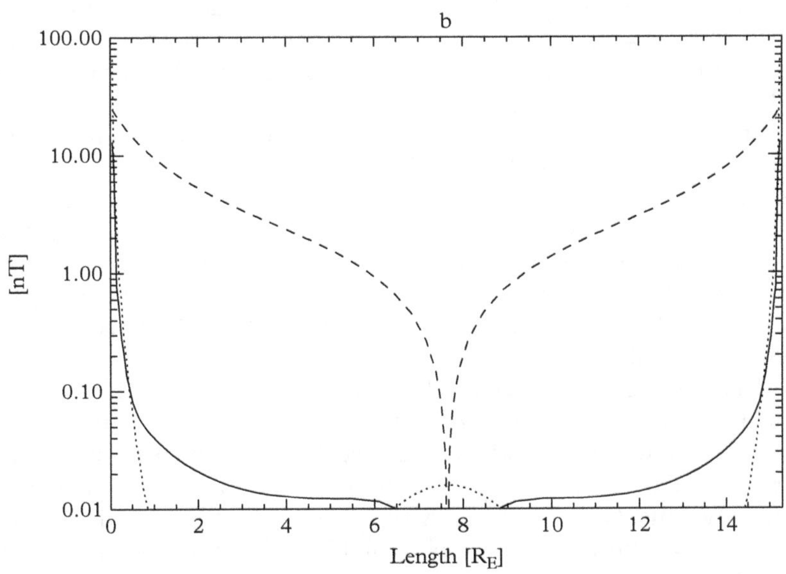

Figure 7.3 *Amplitude of the ULF magnetic field for a FLR on an $L = 6.3$ field line. The field components are ν (solid), ϕ (azimuthal; dashed) and μ (field-aligned; dotted). The amount of the μ component that appears in the topside ionosphere depends on the ionospheric Hall conductance and influences the ratio b^i_ϕ/b^g_ν.*

Adapted from Sciffer and Waters (2011).

latitudes the background magnetic field is non-dipolar, with associated magnetic tension and torsion terms that change the resonant frequency. Some of these effects have been quantified by Rankin et al. (2006), Kabin et al. (2007) and Degeling et al. (2007). Further studies of the parameter space, particularly the dependence on Hall conductance, are in progress.

7.3.2 Latitude and local time dependence of Pc5 power

The impact of ULF wave electric fields on the ionized particle population depends on the distribution of wave power in the magnetosphere. There have been a number of studies of the distribution of power, sorted by ULF wave mode, using spacecraft data (e.g. Anderson et al. 1990). These have provided a picture of the average spatial distribution of power over the Pc3-5 bands. Pc5 occurrence favours radial distances between 7 and 9 R_E with Pc4 waves more likely seen around 5 to 6 R_E. Such frequency-dependent penetration of ULF power into the magnetosphere is explained from the familiar refractive index, dispersion relation and wave cut-off physics. The fast mode dispersion relation has cutoff locations at larger radial distances for smaller frequencies. Toroidal mode Pc5 waves appear around dawn and dusk with FLR occurrence biased towards the morning. Poloidal modes are seen around noon and there is distinct storm-time Pc5 activity around midnight. Surveys that distinguish ULF mode and type (e.g. Anderson et al. 1990; Liu et al. 2009) are useful for radiation belt studies since the poloidal mode is thought to have the most influence on the radial diffusion of energetic particles, which influences particle loss processes (Elkington et al., 2003).

Spacecraft measurements often use magnetic field sensors located near the equatorial plane. For ULF waves that exhibit nodal structure, such as FLRs, measurements of ULF wave amplitude in the equatorial plane can be underestimated. Measurements of ULF wave activity on the ground avoid this bias and also provide improved spatio-temporal resolution compared with spacecraft observations. However, ULF activity with spatial scale sizes shorter than the ionosphere height is attenuated (e.g. Hughes and Southwood, 1976; Glassmeier, 1984; Ponomarenko et al. 2001). Furthermore, ground-based studies show a distinct afternoon decrease in Pc5 power at auroral latitudes while spacecraft data indicate Pc5 power is relatively constant over the daytime, as seen, for example, in Fig. 2 of Glassmeier and Stellmacher (2000). Given that the spacecraft observations also show that "toroidal" mode oscillations dominate during the morning, changing to "poloidal" modes during the afternoon, and local time changes in plasma mass density gradients, Glassmeier and Stellmacher (2000) proposed that a decrease in Pc5 spatial scale size during the afternoon inhibits ground-magnetometer detection of these waves. As discussed below, measurements of ULF activity in the ionosphere using HF radars can compensate for this disadvantage.

Pc4-5 activity measured at high latitudes also depends on solar wind properties, including effects from travelling convection vortices and dayside magnetopause processes such as wave generation by the Kelvin–Helmholtz instability (e.g. Engebretson et al. 1998). Given the extent and availability of (at least) 1 min ground magnetometer data, for example from the SuperMAG network (Gjerloev, 2012), it is now possible to systematically examine the variation of Pc5 power over a large range in latitude and all local times.

In order to examine the global distribution in Pc5 power we have analysed all available data from all stations in the SuperMAG database for 2004. A sliding Fast Fourier Transform (FFT) window of 0.5 hour duration was stepped through the north-south and east-west component magnetometer data with a 15 min step interval. A cosine taper was applied to each time series segment before the FFT process. The spectral amplitude estimates over 2–7 mHz were averaged to provide the Pc5 integrated amplitude. This process yields integrated Pc5 amplitude estimates for all stations and local times. For each station, the integrated Pc5 amplitudes were averaged over each local time segment using the data for all days in 2004. The results for the north-south components are shown in Fig 7.4.

The larger amplitude Pc5 occurs at high latitudes, where the north-south component data show the typical diurnal variation seen in previous studies. Some of these studies attempted to discriminate between Pc5 wave trains and broadband magnetic variations that happened to overlap the Pc5 band by calculating the ratio b^2_{Pc5}/b^2_{tot}, where b^2_{Pc5} is the integrated power over the Pc5 band and b^2_{tot} is power over a wider band, typically up to 100 mHz. This requires data sampled more often than 1 min. Plots of this ratio generally show the absence of the night-time, storm related activity (Engebretson et al., 1998) while the afternoon decrease in power remains.

The latitude of maximum amplitude generally moves higher around local noon. Waters et al. (1996) showed that if the input Pc5 spectrum is relatively constant and there is a significant "toroidal" mode population, then compression of the magnetosphere by the solar wind would result in a move of FLRs at a given frequency to higher latitudes near noon, as seen. The decrease in amplitude during the afternoon has been attributed to an ionosphere "screening" effect, as discussed previously and by Glassmeier and Stellmacher (2000). Furthermore, a daytime, equatorial enhancement appears, possibly related to the Sq system. The east-west component, averaged amplitude distribution showed much less variation with local time but similar variation in latitude.

7.3.3 Storm-time Pc5 activity and indices

During large magnetic storms, waves with the same frequency and wave structure across the globe are observed, including at very low latitudes (e.g. Trivedi et al. 1997; Marin et al. 2014). The signals may exhibit discrete spectral peaks (Kato and Watanabe, 1954; Lee et al., 2007) and are associated with high-speed solar wind streams (Potapov et al., 2009). In fact, Pc5 power seems to be strongly correlated with solar wind speed (Engebretson et al., 1998; Mathie and Mann, 2001) and in particular elevated irregular dynamic pressure (Engebretson et al., 1998; Mathie and Mann, 2001; Marin et al., 2014) at the same time as storm-time increases in MeV electron fluxes in the outer radiation belt, although a causal link is yet to be proven.

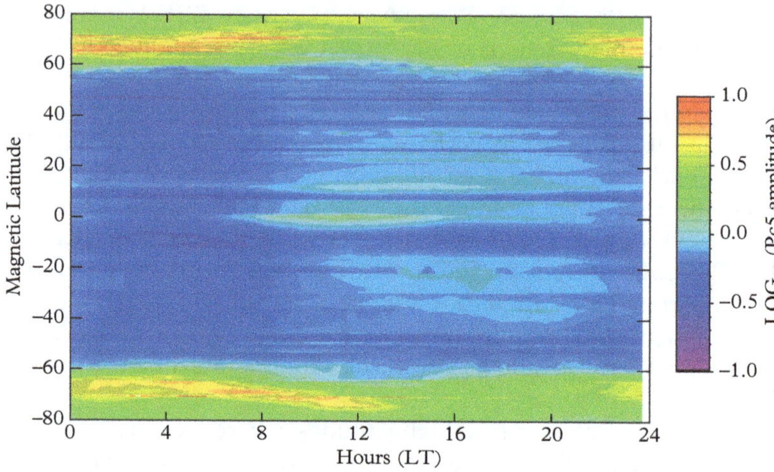

Figure 7.4 *Averaged integrated amplitude over 2–7 mHz for all north-south component SuperMAG data available during 2004.*

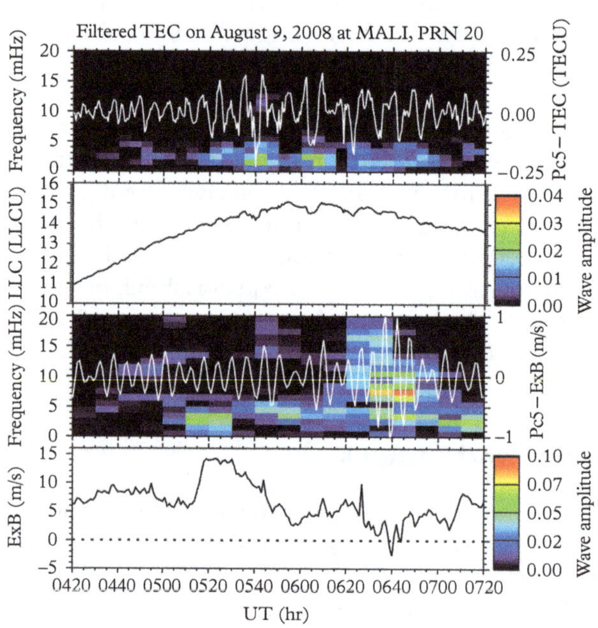

Figure 7.5 *Filtered time series, dynamic spectra and magnitude of total electron content (top two panels) and equatorial vertical ExB drift (bottom panels) observed at the same locations. Adapted from Yizengaw et al. (2013)*

The storm-time wave activity is most intense during storm main phase and in the morning and afternoon sectors of the auroral zone. The activity may result from oscillations of the entire MHD waveguide when the magnetosphere-magnetosheath boundary becomes over-reflecting under high solar wind speed conditions (Marin et al., 2014).

The modes couple to local Alfvén oscillations with a latitude dependence determined by the Alfvén speed profile, which is modified by increased heavy ion populations. Furthermore, the large Pc5 wave amplitudes at synchronous orbit may cause nonlinear effects such as ponderomotive forces that produce field-aligned redistribution of heavy ions in the high-latitude magnetosphere (Guglielmi and Lundin, 2001).

Global ULF wave activity is intense not only during typical Coronal Mass Ejection (CME) storms that exhibit large reduction in the Dst index, but also during High Speed Stream (HSS)/Corotating Interaction Region (CIR) disturbances which may produce only moderate Dst reduction but drive strong coupling to the magnetosphere. Yizengaw et al. (2013) showed that Pc5 may occur within the magnetosphere and ionosphere during such CIR storms in response to similar strong wave activity in the upstream solar wind. They also showed that Pc5 waves may penetrate to the equatorial ionosphere and modulate the equatorial electrojet (EEJ) there as well as the ionospheric density. Figure 7.5 shows filtered time series (white lines), dynamic spectra and magnitudes of the total electron content (TEC) and equatorial vertical ExB drift (which is directly proportional to the EEJ strength and calculated from ground magnetometer observations) observed at the same equatorial location. The time period shown is 3 hr during the CIR when wave power was high throughout the magnetosphere and ionosphere. Large 2–4 mHz Pc5 waves are present in both the TEC and vertical drift and were also recorded by widely spaced ground and space-based magnetometers. Yizengaw et al. (2013) concluded that the source of the TEC oscillations was the global magnetospheric Pc5 wave, which was in turn directly driven by upstream waves.

The potential importance of enhanced Pc5 wave power for particle energization and hence space weather effects has motivated the development of a wave power index quantifying planetary wave activity in the Pc5 range (Kozyreva et al. 2007). The index is based on pulsation signal amplitude (relative to background noise) determined over the 2–7 mHz range each hour window using 1-min data from several extensive ground magnetometer arrays, and is well correlated with storm-time increases in relativistic electron fluxes at geostationary orbit. A related ULF wave index has also been derived from magnetic field measurements by geostationary and interplanetary spacecraft. The ULF index database is presently available from 1991 onwards from the space.augsburg.edu anonymous FTP site.

7.3.4 ULF Waves in the ionosphere

The ionosphere is the window to space for ground-based observers. Much knowledge on the ionosphere has come from ground-based radars. A global, multi-national network of dual, over-the-horizon radars, called SuperDARN, operates as a consortium under agreed operational and data sharing protocols. Such radars transmit high-frequency ($\sim$ 10 MHz) signals which are then reflected back to the receivers either from decameter scale, field-aligned electron density irregularities mostly in the ionospheric F-region or by scattering from the ground. In the first case, Doppler frequency shifts due to nearly horizontal $\mathbf{E} \times \mathbf{B}_o$ plasma drifts are routinely observed and these are often modulated by the electric fields of ULF waves. In the second case, ULF wave fields cause vertical motions of the ionospheric plasma which also produce observed Doppler shifts.

Radars offer better spatial resolution compared to ground magnetometers and improved coverage over regions inaccessible to magnetometers. However, HF radar signals are strongly affected by radio propagation conditions in the ionosphere, often limiting their performance during disturbed times. Nevertheless, studies using the SuperDARN radars have provided new information on ULF waves, such as the existence and properties of FLRs at high latitudes (Walker et al., 1979), apparent discrete, fixed frequency waves and spatially limited Pc4-5 events generated by wave-particle instabilities. These "high-m" events are not readily detectable by magnetometers. Fig. 7.6 shows an illustrative example of ULF wave-related Doppler oscillations detected at mid-latitudes by the TIGER SuperDARN radar.

Comparison of ULF wave fields measured in the ionosphere by radars and on the ground with magnetometers allows the ionospheric transfer function to be determined, and has been achieved for Pi2 pulsations (Ponomarenko and Waters, 2013). In addition to providing information on the transition of ULF waves from space to the ground, these measurements provide rough estimates of the ionosphere conductance. Assuming an infinite, horizontally uniform sheet Pedersen current with conductance, Σ_P (no Hall conductance) over the detection region, the ground magnetic field may be approximated by $b_g = \mu_0 \Sigma_P E$ with the ULF electric field, E, given by the Doppler velocity measurement. For the Pi2 event in Ponomarenko and Waters (2013), $b_g \approx 2.5$ nT and $E \approx 5$ mV m^{-1} to give $\Sigma_P \approx 4S$.

An important class of Pc5 waves that interact with energetic particles in the ring current region are *high-m* number poloidal oscillations (Klimushkin et al., 2004). Large amplitude, monochromatic, pulsations (Pg) and storm-time Pc5 involve resonant interactions between spatially localized poloidal modes and energetic particles in drift

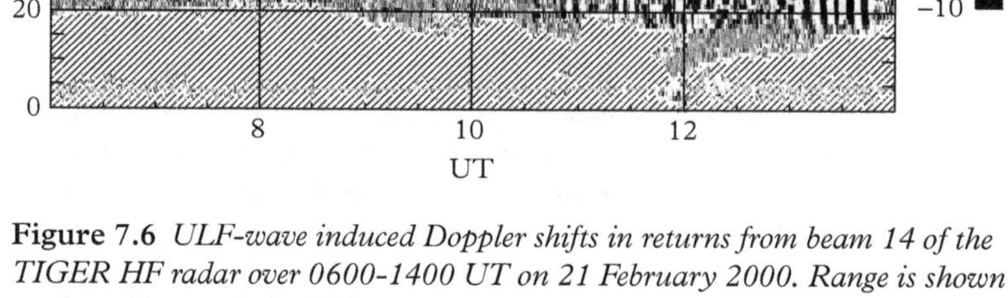

Figure 7.6 *ULF-wave induced Doppler shifts in returns from beam 14 of the TIGER HF radar over 0600-1400 UT on 21 February 2000. Range is shown on the ordinate axis in 45-km gates.*
(Ponomarenko et al., 2003).

and bounce resonance motions. There have been a number of theoretical treatments of these wave-particle interactions but the small spatial scale size hampers their detection in ground magnetometer data. Observations from ionospheric Doppler sounders and HF radars have allowed some experimental checks on the models (Fenrich et al., 1995; Yeoman et al., 2000; Baddeley et al., 2005). The energy transfer from the particles to the waves, which then channels into the ionosphere appears to be of order 10^{10}–10^{11} J. Many questions remain, particularly the details of energy transfer from the waves to particle energization.

7.3.5 Favoured frequencies in the Pc5 band

In addition to storm-time enhancement in amplitude, Pc5 pulsations appear to prefer specific discrete frequencies. This was first reported by Samson et al. (1992b) who analysed ground magnetometer and HF radar Doppler data to reveal discrete frequencies at 1.3, 1.9, 2.6 and 3.4 mHz. This has been confirmed by subsequent studies using radars (e.g. Fenrich et al., 1995) and magnetometers (e.g. Mathie et al., 1999). At low latitudes the discrete frequencies are more prominent under high solar wind pressure conditions (Villante and Lepidi, 2001).

A recent, comprehensive study by Norouzi-Sedeh et al. (2015) examined statistically the frequency composition of the Pc5 band by analysing over 3 years of magnetometer data from Macquarie Island (MQI; 54° 30′ S, 158° 57′ E) and Doppler data from the Tasmanian SuperDARN radar with observations over Macquarie. The Maximum Entropy Method (MEM) was used, with sufficient order to attain 0.05 mHz resolution. The histogram of frequencies seen in the radar data is shown in Fig. 7.7, with frequency peaks similar to those reported by other studies. Most of these Pc5 radar events occurred during 06-11 UT (16-21 LT), similar to results reported earlier by Keys (1965). The event occurrence with time of day is likely related to HF propagation conditions. This inhibits investigation of local time properties in the spectral composition of the Doppler data. However, examination of the MQI magnetometer data by Norouzi-Sedeh et al. (2015) showed that discrete, favoured frequencies were more evident in the spectra around local dawn and dusk, suggesting a link with magnetosphere flank processes.

The analysis used by Norouzi-Sedeh et al. (2015) was repeated for the north-south component of the MQI data in the SuperMAG data set for the year 2004. The significant spectral peaks over 3 hour intervals for each day of the year were counted to form the histograms shown in Fig. 7.8. There is a decrease in the counts for frequencies less than 0.5 mHz due to the pre-processing of the data by a high-pass, fourth-order Butterworth filter. Peaks appear at a number of discrete frequencies. Some explanations for the existence of such frequencies (or *Samson* frequencies) have been proposed. Resonant oscillations of the fast mode in the magnetosphere cavity have frequency spacing around 0.3 mHz (Samson et al., 1992a). The difficulty with this idea is that the dynamic cavity boundaries at high latitudes would average out favoured frequencies. However, this mechanism appears to operate in the inner magnetosphere (e.g. Samson

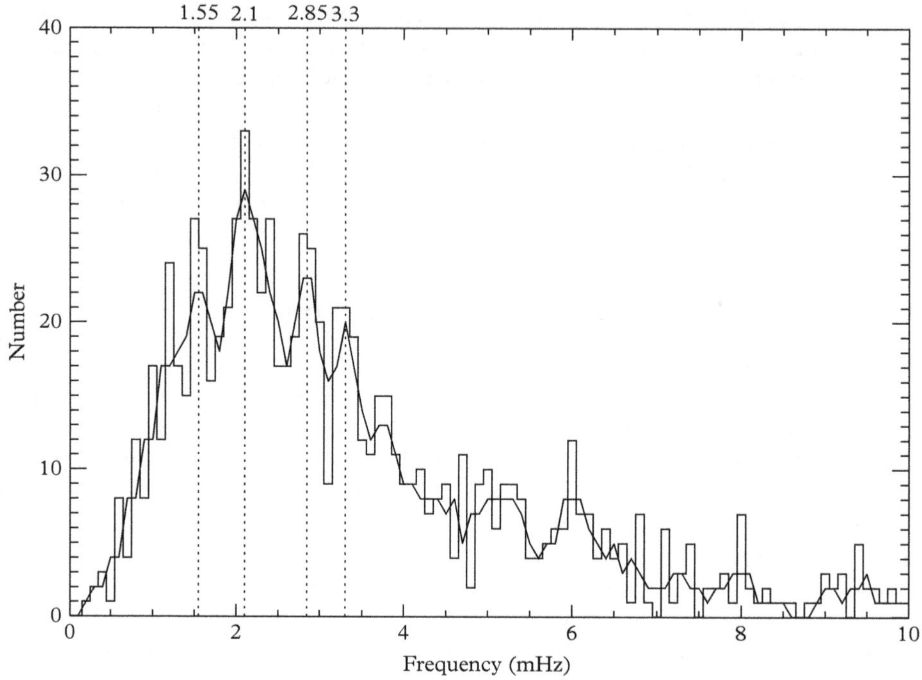

Figure 7.7 *Histogram of significant spectral peaks seen in the Doppler radar data from the Tasmanian SuperDARN radar, using a three-beam scan mode, over 2006–2009. A three-point, box-car smoothed line is also shown (from Norouzi-Sedeh et al., 2015).*

et al. 1995). The discrete, favoured frequencies might be directly driven by similar frequencies in the solar wind (Stephenson and Walker, 2002; Fenrich and Waters, 2008) or by solar wind driven magnetopause surface modes (Plaschke et al., 2009). Norouzi-Sedeh et al. (2015) showed that discrete, favoured Pc5 frequencies appear to have a local time dependence. Both the existence and local time dependence of these lack satisfactory explanations.

7.3.6 Pc3 waves and indices

Waves in the Pc3 range (approx. 10–45 s) are most likely due to ion-cyclotron resonance in proton beams reflected from Earth's bow shock, resulting in fast mode waves which may propagate into and through the magnetosphere, coupling energy to FLRs. Accordingly Pc3 waves are readily observed near and Earthward of the plasmapause (e.g. Menk and Waters, 2013). However, Pc3 are also detected in the outer magnetosphere by ground magnetometers and HF radars, where they are unlikely to result from local fundamental mode FLRs. A statistical study of high latitude, high coherence length Pc3-4 pulsations with the IMAGE magnetometer array (Howard and Menk, 2005)

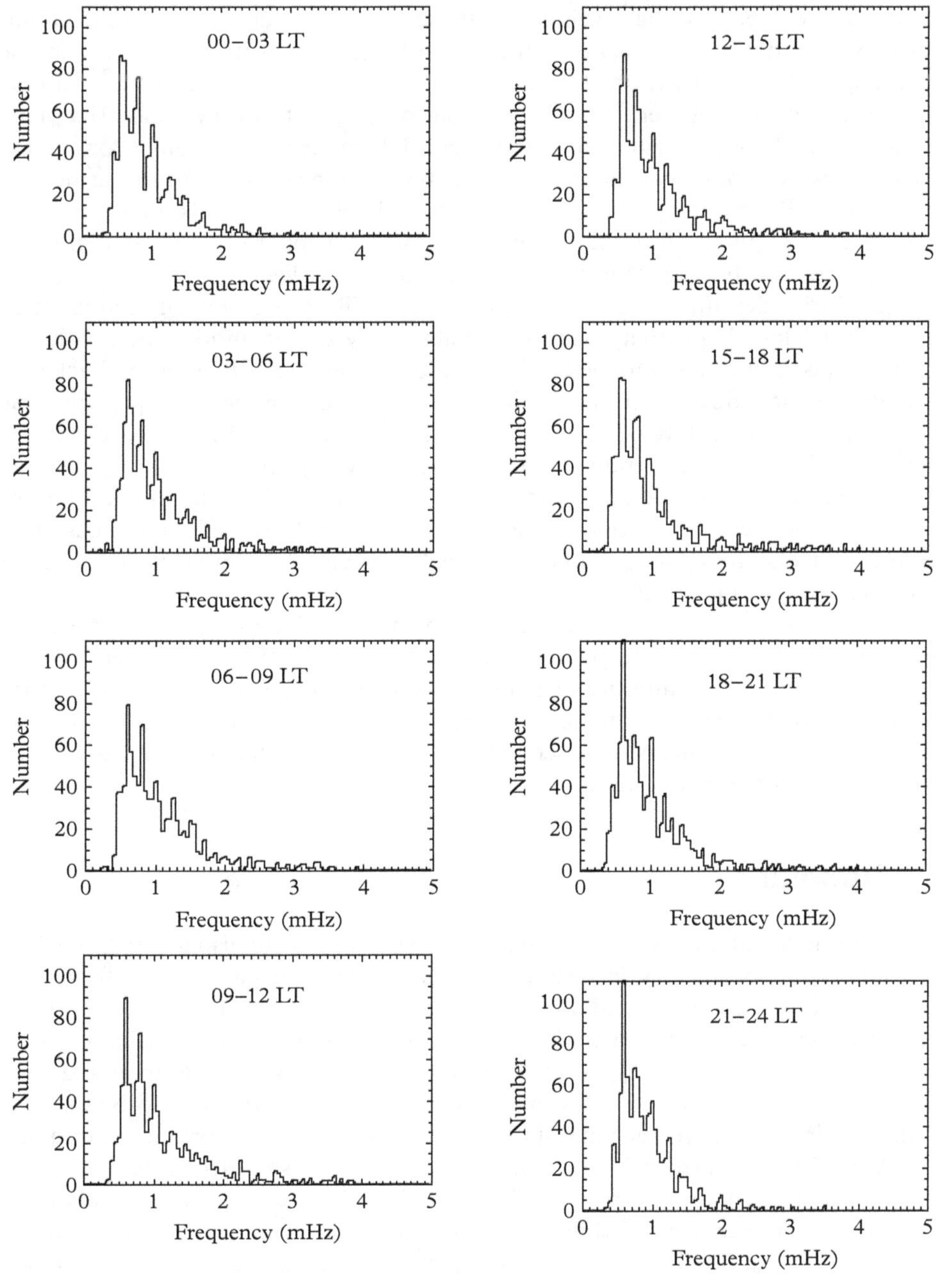

Figure 7.8 *Histogram of significant spectral peaks seen in the Macquarie Island north-south magnetometer component for all data during 2004.*

showed that the apparent phase velocity of these pulsations across the ground is mostly directed poleward and may relate to the time of flight for isotropic fast mode waves that propagate Earthward from the magnetopause in the equatorial plane and couple energy to field-aligned Alfvénic oscillations that propagate from the equatorial plane to the ionosphere. This scenario is consistent with HF radar observations (Ponomarenko et al., 2005) which show that the Pc3-4 waves probably result from third harmonic poloidal mode FLRs driven by the incoming fast mode wave energy. Nevertheless, Pc3-4 pulsations are ubiquitous throughout the outer magnetosphere and polar regions, and their generation mechanism(s) in these regions remain unclear.

While a Pc5 index may be used as a proxy of the likely presence of relativistic radiation belt particles, Pc3 activity occurs virtually every day at middle and low latitudes and a Pc3 pulsation index may be useful for identifying contamination by background activity in geomagnetic and aeromagnetic surveys used in geophysical exploration. Such Pc3 indices have been developed for the Australian region and are used by these industries. The indices are generated from 20-min intervals of two- and three-component magnetometer data at selected locations, filtered over the Pc3 pulsation band and multiplied by a scale factor to produce an integer value from 0 to 10. The indices are also used to produce activity contour maps and daily averages. The indices and maps may be accessed from `http://www.sws.bom.gov.au/`.

A similar Pc3 index has been used to investigate the relationship between Pc3 wave power at L=1.8 and solar wind parameters (Heilig et al., 2010). It has been found that Pc3 power depends on a number of parameters including solar wind speed and interplanetary magnetic field cone angle, density, pressure and Alfvénic and magnetosonic Mach number. There appears to be scope for further development of Pc3 activity indices as proxies of solar wind conditions.

7.4 Summary

Efforts to understand ULF wave source and propagation mechanisms in the Pc3-5 band have a relatively long history in near-Earth space physics research. This fundamental research has provided the foundation for developing remote sensing applications using these waves and for understanding wave-particle energy transfer. The basic wave propagation and resonance properties are well understood, with experimental data consistent with theoretical models. However, there is much that remains to be discovered and this requires extensive instrumentation and data access, both on the ground and in space. A persistent challenge is the ability of the natural system to operate over a wide range of spatial and temporal scales. The large spatial scale waves are readily seen by ground magnetometer arrays, while ionospheric Doppler sounders and HF radars are crucial for obtaining small-scale ULF wave data. These, in addition to improved modelling efforts and spacecraft data, provide the tools for advancing our understanding of the relationships between particle dynamics and ULF wave and their place in space weather processes.

Acknowledgments

We sincerely thank J.W. Gjerloev for the SuperMAG data, Geoscience Australia for the fluxgate magnetometer data from Macquarie Island, J. Devlin for TIGER radar data and E. Zesta for helpful discussions and suggestions.

REFERENCES

Alfvén, H. (1942). Existence of electromagnetic-hydromagnetic waves. *Nature*, **150**, 405–406.

Anderson, B. J., Engebretson, M. J., Rounds, S. P., Zanetti, L. J., and Potemra, T. A. (1990). A statistical study of Pc3-5 pulsations observed by the AMPTEE/CCE magnetic fields experiment, 1. occurrence distributions,. *J. Geophys. Res.*, **95**, 10495–10523.

Ansari, I. A. and Fraser, B. J. (1986). A multistation study of low latitude Pc3 geomagnetic pulsations. *Planet. Space Sci.*, **34**, 519–536.

Ashour, A. A. and Price, A. T. (1948). The induction of electric currents in a non-uniform ionosphere. *Proc. Roy. Soc.*, **A195**, 198–224.

Baddeley, L. J., Yeoman, T. K., Wright, D. M., Trattner, K. J., and Kellet, B. J. (2005). On the coupling between unstable magnetospheric particle populations and resonant high m ULF wave signatures in the ionosphere. *Ann. Geophys.*, **23**, 567–577.

Baker, D. N., et al. (1998). Coronal mass ejection, magnetic clouds and relativistic magnetopsheric electron events: Istp. *J. Geophys. Res.*, **103**, 17279–17291.

Bemmelen, W. Van (1908). On pulsations. *Obs. R. Mag. Met. Obs.*, **29**, 1–10.

Berube, D., Moldwin, M. B., Fung, S. F., and Green, J. L. (2005). A plasmaspheric density model and constraints on its heavy ion concentration. *J. Geophys. Res.*, **110**, A04212.

Berube, D., Moldwin, M. B., and Weygand, J. M. (2003). An automated method for the detection of field line resonance frequencies using ground magnetometer techniques. *J. Geophys. Res.*, **108**, 1348.

Boudouridis, A. and Zesta, E. (2007). Comparison of Fourier and wavelet techniques in the determination of geomagnetic field line resonances. *J. Geophys. Res.*, **112**, A08205.

Campbell, W. H. (1959). Studies of magnetic field micropulsations with periods of 5 to 30 seconds. *J. Geophys. Res.*, **64**, 1819–1826.

Chi, P. J., et al. (2013). Sounding of the plasmasphere by Mid-continent MAgnetoseismic Chain (McMAC) magnetometers. *J. Geophys. Res.*, **118**, 3077–3086.

Clilverd, M. A., et al. (2003). In situ and ground-based intercalibration measurements of plasma density at L = 2.5. *J. Geophys. Res.*, **108**, 1365.

Cliver, E. W. (1984). Solar activity and geomagnetic storms: the first 40 years. *EOS Trans. AGU*, 75.

Degeling, A. W., Rankin, R., Kabin, K., Rae, I. J., and Fenrich, F. R. (2007). Modeling ULF waves in a compressed dipole magnetic field. *Planet. Space Sci.*, **55**, 820–828.

Denskat, K. U. and Neubauer, F. M. (1982). Statistical properties of low-frequency magnetic field fluctuations in the solar wind from 0.29 to 1.0 au during solar minimum conditions: Helios 1 and helios 2. *J. Geophys. Res.*, **87**, 2215–2223.

Duffus, H. J., Nasmyth, P. W., Shand, J. A., and Wright, C. (1958). Sub-audible geomagnetic fluctuations. *Nature*, **181**, 1258–1259.

Duncan, R. A. (1961). Some studies of geomagnetic micropulsations. *J. Geophys. Res.*, **66**, 2087–2094.

Elkington, S. R., Hudson, M. K., and Chan, A. A. (2003). Resonant acceleration and diffusion of outer zone electrons in an asymmetric geomagnetic field. *J. Geophys. Res.*, **108**.

Engebretson, M., Glassmeier, K.-H., Stellmacher, M., and Hughes, W. J. (1998). The dependence of high-latitude Pc5 wave power on solar wind velocity and on the phase of high-speed solar wind streams. *J. Geophys. Res.*, **193**, 26271–26283.

Eschenhagen, M. (1897). On minute, rapid, changes of the Earth's magnetism. *Terr. Magn.*, **2**, 105–114.

Fenrich, F. R., Samson, J. C., Sofko, G., and Greenwald, R. A. (1995). Ulf high- and low-m field line resonances observed with the Super Dual Auroral Radar Network. *J. Geophys. Res.*, **100**, 21535–21547.

Fenrich, F. R. and Waters, C. L. (2008). Phase coherence analysis of a field line resonance and solar wind oscillation. *Geophys. Res. Lett.*, **35**, L20102.

Fraser, B. J. (1979). The use of multistation networks in geomagnetic pulsation studies. *Annals. Telecomm.*, **34**, 220–236.

Fraser, B. J. and Summers, W. R. (1972). Polarizations of pc1 micropulsations propagated through ionospheric ducts. *Nature*, **235**, 170–171.

Gjerloev, J. W. (2012). The SuperMAG data processing technique. *J. Geophys. Res.*, **117**, A09213.

Glangeaud, F. (1982). Analysis of pulsations. *Planet. Space Sci.*, **30**, 1249–1258.

Glassmeier., K. H. (1984). On the influence of ionospheres with non-uniform conductivity distribution on hydromagnetic waves. *J. Geophys.*, **54**, 125.

Glassmeier, K.-H. and Stellmacher, M. (2000). Concerning the local time asymmetry of Pc5 wave power at the ground and field line resonance widths. *J. Geophys. Res.*, **105**, 18847–18855.

Green, C. A. (1976). The longitudinal phase variation of mid-latitude pc3-4 micropul-sations. *Planet. Space Sci.*, **24**, 79–85.

Grew, R. S., Menk, F. W., Clilverd, M. A., and Sandel, B. R. (2007). Mass and electron densities in the inner magnetosphere during a prolonged disturbed interval. *Geophys. Res. Lett.*, **34**, L02108.

Guglielmi, A. and Lundin, R. (2001). Ponderomotive upward acceleration of ions by ion cyclotron and alfven waves over the polar regions. *J. Geophys. Res.*, **106**, 13219–13236.

Heilig, B., Lotz, S., Vero, J., Reda, J., Pajunpaa, K., and Raita, T. (2010). Empirically modelled pc3 activity based on solar wind parameters. *Ann. Geophys.*, **28**, 1703–1722.

Howard, T. A. and Menk, F. W. (2005). Ground observations of high-latitude pc3-4 ulf waves. *J. Geophys. Res.*, **110**, A04205.

Hughes, W. J. and Southwood, D. J. (1976). The screening of micropulsation signals by the atmosphere and the ionosphere. *J. Geophys. Res.*, **81**, 3234–3240.

Kabin, K., Rankin, R., Waters, C. L., Marchand, R., Donovan, E. F., and Samson, J. C. (2007). Different eigenproblem models for field line resonances in cold plasma: Effect on magnetospheric density estimates. *Planet. Space Sci.*, **55**, 820–828.

Kato, Y. and Tamao, T. (1956). Hydromagnetic oscillations in a conducting medium with hall conductivity under the uniform magnetic field. *Sci. Rept. Tohoku Univ., Ser. 5, Geophys.*, **3**, 147.

Kato, Y. and Watanabe, T. (1954). A possible explanation of the cause of giant pulsations. *Sci. Rept. Tohoku Univ., Ser. 5, Geophys.*, **6**, 95–104.

Kato, Y. and Watanabe, T. (1956). Further study on the cause of giant pulsations. *Sci. Rept. Tohoku Univ., Ser. 5, Geophys.*, **8**, 1–10.

Keys, J. G. (1965). Pulsating auroral radar echoes and their possible hydromagnetic association. *J. Atmos. Terr. Phys.*, **27**, 385–388.

Klimushkin, D. Y., Mager, P. N., and Glassmeier, K. H. (2004). Toroidal and poloidal alfvén waves with arbitrary azimuthal wavenumbers in a finite pressure plasma in the earth's magnetosphere. *Ann. Geophys.*, **22**, 267–287.

Kozyreva, O., Pilipenko, V., Engebretson, M. J., Yumoto, K., Watermann, J., and Romanova, N. (2007). In search of a new ULF wave index: Comparison of Pc5 power with dynamics of geostationary relativistic electrons. *Planet. Space Sci.*, **55**, 755–769.

Lanzerotti, L. J., Fukunishi, H., and Chen, L. (1974). ULF pulsation evidence of the plasmapause 3. interpretation of polarization and spectral amplitude studies of Pc3 and Pc4 pulsations near L=4. *J. Geophys. Res.*, **79**, 4648–4653.

Lee, E. A., Mann, I. R., Loto'aniu, T. M., and Dent, Z. C. (2007). Global pc5 pulsations observed at unusually low L during the great magnetic storm of 24 march 1991. *J. Geophys. Res.*, **112**.

Lichtenberger, J., Clilverd, M. A., Heilig, B., Vellante, M., Manninen, J., Rodger, C. J., Collier, A. B., Jorgensen, A. M., Reda, J., Holzworth, R. H., Friedel, R., and Simon-Wedlund, M. (2013). The plasmasphere during a space weather event: first results from the PLASMON project. *J. Space Weather Space Clim.*, **3**.

Liu, W., et al. (2009). Electric and magnetic field observations of Pc4 and Pc5 pulsations in the inner magnetosphere: A statistical study,. *J. Geophys. Res.*, **114**, A12206.

Lühr, H., Thürey, S., and Klöcker, N. (1984). The EISCAT-magnetometer cross. *Geophys. Surv.*, **6**, 305–315.

Maple, E. (1959). Geomagnetic oscillations at middle latitudes. *J. Geophys. Res.*, **64**, 1395–1404.

Marin, J., et al. (2014). Global pc5 pulsations during strong magnetic storms: excitation mechanisms and equatorward expansion. *Ann. Geophys.*, **32**, 319–331.

Mathie, R. A. and Mann, I. R. (2001). On the solar wind control of Pc5 ULF pulsation power at mid-latitudes: Implications for MeV electron accelearation in the outer radiation belt. *J. Geophys. Res.*, **106**, 29783–29796.

Mathie, R. A., Mann, I. R., Menk, F. W., and Orr, D. (1999). Pc5 ULF pulsations associated with waveguide modes observed with the IMAGE magnetometer array. *J. Geophys. Res.*, **104**, 7025–7036.

McNicol, R. W. E. and Mainstone, J. S. (1963). General characteristics of pc-type pulsations. *Aust. J. Phys.*, **16**, 507–512.

Menk, F. W., Ables, S. T., Grew, R. S., Clilverd, M. A., and Sandel, B. R. (2012). The annual and longitudinal variations in plasmaspheric ion density. *J. Geophys. Res.*, A03215.

Menk, F. W., Mann, I. R., Smith, A. J., Waters, C. L., Clilverd, M. A., and Milling, K. (2004). Monitoring the plasmapause using geomagnetic field line resonances. *J. Geophys. Res.*, A04216.

Menk, F. W. and Waters, C. L. (2013). *Magnetoseismology: Ground-based remote sensing of Earth's magnetosphere*. Wiley-VCH.

Ness, N. F., Skillman, T. L., Scearce, C. S., and Heppner, J. P. (1961). Magnetic field fluctuations on the Earth and in space. *Int. Conf. on Cosmic Rays and the Earth Storm, Kyoto.*

Norouzi-Sedeh, L., Waters, C. L., and Menk, F. W. (2015). Survey of ULF wave signatures seen in the Tasman International Geospace Environment Radars data, *J. Geophys. Res. Space Physics*, **120**, doi:10.1002/2014JA020652.

Obana, Y., Menk, F. W., and Yoshikawa, I. (2010). Plasma refilling rates for L=2.3-3.8 flux tubes. *J. Geophys. Res.*, **115**, A03204.

Obayashi, T. and Jacobs, J. A. (1958). Geomagnetic pulsations and the Earth's outer atmosphere. *Geophys. J. Roy. Astron. Soc.*, **1**, 53–63.

Oguti, T. and Nagata, T. (1954). Model experiments of the shielding effect of the ionosphere. *Rep. Ionos. Res. Japan*, **8**, 171–183.

Olson, J. V. and Samson, J. C. (1979). On the detection of the polarization states of pc micropulsations. *Geophys. Res. Lett.*, **6**, 413–417.

Ozeke, L. G., Mann, I. R., and Rae, I. J. (2009). Mapping guided alfven wave magnetic field amplitudes observed on the ground to equatorial electric field amplitudes in space. *J. Geophys. Res.*, **114**, A01214.

Parker, E. N. (1960). Geomagnetic fluctuations and the form of the outer zone of the van allen radiation belt. *J. Geophys. Res.*, **65**, 3117–3130.

Plaschke, F, Glassmeier, K.-H., Auster, H. U., Constantinescu, O. D., Magnes, W., Angelopoulos, V., Sibeck, D. G., and McFadden, J. P. (2009). Standing alfvén waves at the magnetopause. *Geophys. Res. Lett.*, **36**, L02104

Ponomarenko, P. V., Fraser, B. J., Menk, F. W., Ables, S. T., and Morris, R. J. (2002). Cusp-latitude Pc3 spectra: band-limited and power-law components. *Ann. Geophys.*, **20**, 1539–1551.

Ponomarenko, P. V., Menk, F. W., and Waters, C. L. (2003). Visualization of ULF waves in SuperDARN data. Geophys. Res. Lett., 30, 1926P

Ponomarenko, P. V., Menk, F. W., Waters, C. L., and Sciffer, M. D. (2005). Pc3-4 ULF waves observed by the SuperDARN TIGER radar. *Ann. Geophys.*, **23**, 1271–1280.

Ponomarenko, P. V. and Waters, C. L. (2013). Transition of pi2 ULF wave polarization structure from the ionosphere to the ground. *Geophys. Res. Lett.*, **40**, 1474–1478.

Ponomarenko, P. V., Waters, C. L., Sciffer, M. D., and Fraser, B. J. (2001). Spatial structure of ULF waves: Comparison of magnetometer and Super Dual Auroral Radar Network data. *J. Geophys. Res.*, **106**, 10509–10517.

Potapov, A. S., Tsegmed, B., and Polyushkina, T. N. (2009). Contribution of global pc5 oscillations to magnetic disturbance during geomagnetic storms. *Geomag. Aeron.*, **49**, 1182–1188.

Rankin, R., Kabin, K., and Marchand, R. (2006). Alfvénic field line resonances in arbitrary magnetic field topology. *Adv. Space Res.,*, **38**, 1720–1729.

Rostoker, G., Skone, S., and Baker, D. N. (1998). On the origin of relativistic electrons in the magnetopshere associated with some geomagnetic storms. *Geophys. Res. Lett.*, **25**, 3701–3704.

Sabine, E. (1852). On periodical laws discoverable in the mean effects of the larger magnetic disturbances. *Phil. Trans. R. Soc. Lond.*, **142**, 103–124.

Samson, J. C., Harrold, B. G., Ruohoniemi, J. M., Greenwald, R. A., and Walker, A. D. M. (1992a). Field line resonances associated with MHD waveguides in the magnetosphere. *Geophys. Res. Lett.*, **19**, 441–444.

Samson, J. C., Jacobs, J. A., and Rostoker, G. (1971). Latitude-dependent characteristics of long-period geomagnetic micropulsations. *J. Geophys. Res.*, **76**, 3675–3683.

Samson, J. C. and Olson, J. V. (1981). Data-adaptive polarization filters for multichannel geophysical data. *Geophys.*, **46**, 1423–1431.

Samson, J. C., Wallis, D. D., Hughes, T. J., Creutzberg, F., Ruohoniemi, J. M., and R. A., Greenwald (1992b). Substorm intensifications and field line resonances in the nightside magnetosphere. *J. Geophys. Res.*, **97**, 8495–8518.

Samson, J. C., Waters, C. L., Menk, F. W., and Fraser, B. J. (1995). Fine structure in the spectra of low latitude field line resonances. *Geophys. Res. Lett.*, **22**, 2111–2114.

Schulz, M. (1996). Eigenfrequencies of geomagnetic field lines and implications for plasma density modeling. *J. Geophys. Res.*, **101**, 17385–17397.

Sciffer, M. D. and Waters, C. L. (2011). Relationship between ULF wave mode mix, equatorial electric fields and ground magnetometer data. *J. Geophys. Res.*, **116**, A06202.

Stephenson, J. A. E. and Walker, A. D. M. (2002). Hf radar observations of Pc5 ULF pulsations driven by the solar wind. *Geophys. Res. Lett.*, **29**, 1297–1300.

Stewart, B. (1859). Construction of the self-recording magnetographs at present in operation at the Kew Observatory of the British Association. *29th Rept. Brit. Assoc. for the Advancement of Science*, 200–228.

Sucksdorff, E. (1939). Giant pulsations recorded at Sodankylä during 1914–1938. *Terr. Mag.*, **44**, 157–170.

Sugiura, M. (1950). The shielding effect of the ionosphere. *Rep. Ionos. Res. Japan*, **4**, 31–36.

Takahashi, K., Denton, R. E., Anderson, R. R., and Hughes, W. J. (2004). Frequencies of standing Alfvén wave harmonics and their implication for plasma mass density distribution along geomagnetic field lines: Statistical analysis of crres data. *J. Geophys. Res.*, **109**, A08202.

Tepley, L. R. (1961). Observations of hydromagnetic emissions. *J. Geophys. Res.*, **66**, 1651–1658.

Trivedi, N. B., et al. (1997). Global Pc5 geomagnetic pulsations of March 24, 1991, as observed along the American sector. *Geophys. Res. Lett.*, **24**, 1683–1686.

Troitskaya, V. A. (1997). Early ground based approach to hydromagnetic diagnostics of outer space. *Discovery of the Magnetosphere (eds. C.S. Gillmour and J.R. Spreiter), American Geophysical Union.*

Tumanski, S. (2007). Induction coil sensors - a review. *Meas. Sci. Tech.*, **18**, R31.

Villante, U., P. Francia and Lepidi, S. (2001). Pc5 geomagnetic field fluctuations at discrete frequencies at a low latitude station. *Annales Geophys.*, **19**, 321–325.

Walker, A. D. M., Greenwald, R. A., Stuart, W. F., and Green, C. A. (1979). Stare auroral radar observations of Pc5 geomagnetic pulsations. *J. Geophys. Res.*, **84**, 3373–3388.

Watanabe, T. (1957). Electrodynamical behaviour and screening effect of the ionosphere. *Sci. Rpt. Tohoku Univ. Series 5*, **9**, 81–98.

Waters, C. L., Menk, F. W., and Fraser, B.J. (1991). The resonance structure of low latitude Pc3 geomagnetic pulsations. *Geophys. Res. Lett.*, **18**, 2293–2296.

Waters, C. L., Samson, J. C., and Donovan, E. F. (1996). Variation of plasmatrough density derived from magnetospheric field line resonances. *J. Geophys Res.*, **101**, 24737–24745.

Wilson, C. R. and Sugiura, M. (1961). Hydromagnetic interpretation of sudden commencements of magnetic storms. *J. Geophys. Res.*, **66**, 4091–4111.

Yeoman, T. K., Wright, D. M., Chapman, P. J., and Stockton-Chalk, A. B. (2000). High-latitude observations of ULF waves with large azimuthal wavenumbers. *J. Geophys. Res.*, **105**, 5453–5462.

Yizengaw, E. and Moldwin, M. B. (2009). African meridian B-field education and research (AMBER) array. *Earth Moon Planet*, **104**, 237–246.

Yizengaw, E., et al. (2013). Observations of ULF wave related equatorial electrojet and density fluctuations. *J. Atmos. Solar-Terr. Phys.*, **103**, 157–168.

Yumoto, K. and the MAGDAS group (2006). MAGDAS project and its application for space weather. *International Living with a star (ILWS) Workshop Proceedings, Eds. N. Gopalswamy and A. Bhattacharyya*, **ISBN 81-87099-40-2**, 309–4–5.

Zelwer, R. and Morrison, H. F. (1972). Spatial characteristics of midlatitude geomagnetic micropulsations. *J. Geophys. Res.*, 77, 674–694.

Ziesolleck, C. W. S., Fraser, B. J., Menk, F. W., and McNabb, P. W. (1993). Spatial characteristics of low-latitude Pc3-4 geomagnetic pulsations. *J. Geophys. Res.*, **98**, 197–207.

8

Chorus Waves in Geospace and their Influence on Radiation Belt Dynamics

Jacob Bortnik, Richard M. Thorne, Wen Li, and Xin Tao

Abstract

This chapter explores the physics of the resonant interactions between whistler-mode chorus waves and energetic electrons, typical of the near-Earth space environment, and focuses specifically on those elements of the resonant interaction that are unique to chorus. After describing the general morphology and characteristics of chorus waves, we introduce the basic elements of the resonant wave-particle interaction by examining the regions of resonance, and trajectories of resonant electrons in velocity space. We then turn our attention to the different modes of scattering that resonant electrons can experience in response to chorus waves. First, the linear response is discussed and it is shown how multiple linear wave-particle interactions can cause diffusive spreading of the electron population. An example is given of a recent ultra-relativistic acceleration event that has been successfully modeled using a quasilinear diffusion approach. We then discuss the regimes of nonlinear phase bunching and phase trapping and show how these effects manifest under progressively more realistic conditions: first by including large amplitude waves, then subpacket structure, frequency drift, and finally a realistically modeled chorus element based on satellite observations. Throughout the chapter, particular emphasis is placed on developing a conceptual, intuitive understanding of the ideas presented, for

Waves, Particles, and Storms in Geospace. First Edition. Georgios Balasis, Ioannis A. Daglis, Ian R. Mann.

the worker who may be new to this field. We conclude by briefly outlining a few of the open areas where understanding is still incomplete, and which are ripe for discoveries in the coming years.

8.1 Introduction

The resonant interaction of energetic electrons with whistler-mode chorus waves has become a topic of extreme interest in recent years, mainly due to two reasons: the first is the steadily developing realization that chorus plays a key role in rebuilding radiation belt fluxes to very high levels after they have been depleted, by accelerating lower energy electrons ($\sim$100 keV) to relativistic energies on timescales that are comparable to a day [e.g., Horne et al., 2005; Thorne et al., 2013]. A second, and closely related reason, is the recent launch of the twin Van Allen Probes on August 30, 2012 [Mauk et al., 2012] whose primary objectives involve the understanding of the physical processes responsible for radiation belt dynamics, where chorus is thought to play a fundamental role [Thorne, 2010]. In addition to the electron radiation belt population, recent studies have demonstrated that chorus is the dominant driver of the pulsating aurora [Nishimura et al., 2010, 2011; Li et al., 2012] and the diffuse aurora [Thorne et al., 2010; Ni et al., 2011; Tao et al., 2011a], which involve the precipitation of electrons into the Earth's dense upper atmosphere, with energies of tens of keV, and a few keV, respectively. The diffuse aurora, in particular, acts as an essential feedback element in the coupled magnetosphere-ionosphere system, by modifying ionospheric conductivities, and hence 'shorting out' the electric field in regions of the ionosphere, which, when mapped out to the magnetosphere along magnetic field lines, tends to slow the rate of convection and transport. Thus, whistler-mode chorus waves form a key component of the dynamic and highly coupled inner magnetospheric environment, by transferring energy from the lower energy plasma sheet electron population to the high energy radiation belt population, by scattering and precipitating the lower energy electrons into the atmosphere and creating auroras, and by modifying the ionospheric conductivities and hence the macroscopic structure of the convection electric field that drives geomagnetic activity.

In this chapter, we aim to introduce and explore the resonant interaction of energetic electrons with whistler-mode chorus waves, and focus specifically on the unique aspects of chorus waves that give rise to these interactions. Chorus waves occur in a specific frequency band, at well-defined regions of geospace, and with unique wave normal and Poynting vector characteristics, which limit the morphology of the interaction to specific regions in velocity space. These chorus characteristics are described in Section 8.2 and the basic trajectories of resonant particles in velocity space are introduced and discussed in Section 8.3. Unlike the conceptual framework of quasilinear diffusion which assumes a band of small amplitude, uncorrelated waves, chorus occurs with widely varying amplitudes which can lead to qualitatively different responses of the resonantly interacting electrons. These different responses are discussed in Section 8.4, where simply increasing the amplitude of the wave results in scattering that progresses from linear, to phase bunched, to phase trapped. Section 8.4.1 discusses the linear response, and shows how it evolves from individual wave-particle interactions, and how it has been applied to ultra-relativistic electron acceleration. Section 8.4.2 then shows those aspects of the chorus wave-electron interaction that are not described by quasilinear theory, namely the finite (i.e., large) amplitude of the chorus wave leading to nonlinear phase bunching and trapping, the modulation of the chorus wave amplitude, how it can be modeled and understood in a simplified way, and finally a realistic chorus element with all its unique features, such as frequency variation, amplitude modulation, and propagation effects.

8.2 Characteristics of chorus waves

Whistler-mode chorus waves are electromagnetic electron cyclotron (i.e., whistler mode) waves, which are naturally excited near the Earth's geomagnetic equator due to cyclotron resonant interactions with anisotropic, plasma sheet electrons in the energy range of ~1–100 keV [e.g., Kennel and Petschek, 1966; Nunn, 1974; Nunn et al., 1997; Omura et al., 2008, 2009] that are injected into the inner magnetosphere during geomagnetically active conditions [e.g., Burtis and Helliwell, 1969; Tsurutani and Smith, 1977; Li et al., 2009; Meredith et al., 2003].

Figure 8.1a shows a cutaway view of the inner magnetosphere, where the drift of plasma sheet electrons from the magnetotail, through the dawn side, and into the day side of the Earth is illustrated using a red arrow. As these plasma sheet electrons drift closer to the Earth from the nightside, their anisotropy is increased due to the conservation of the first two adiabatic invariants [Roederer, 1970], leading to a source of free energy which is then released in the form of chorus waves. Consequently, chorus waves are predominantly observed over the Magnetic Local Time (MLT) range of ~2100–1500, coincident with the drift trajectories of plasma sheet electrons. However, the latitudinal extent to which chorus waves propagate is controlled by a more subtle effect.

A unique feature of chorus waves is the observation that they only propagate away from the geomagnetic equator [Lauben et al., 1998, 2002; Ledocq et al., 1998; Santolík et al., 2010; Li et al., 2013a], implying that their source region is located at, or very near to the equator [e.g., Helliwell, 1967; Nunn, 1974; Trakhtengerts, 1999; Omura et al., 2008, 2009]. Second, some mechanism extinguishes the wave energy in the course of their propagation, before the waves are able to reflect at high latitudes and return to the equator. The first implication follows directly from the source of free energy that creates the waves, namely the temperature anisotropy of the electron populations which is maximum at the equator (i.e., the region of minimum B along the field line, where ~90° electrons are trapped). The second implication, namely the necessity of a damping mechanism, is less obvious since the plasma is generally collisionless, and thus requires some sort of wave-particle interaction to play the role of collisions and remove the energy from the wave. An illustration of the chorus wave excitation at the

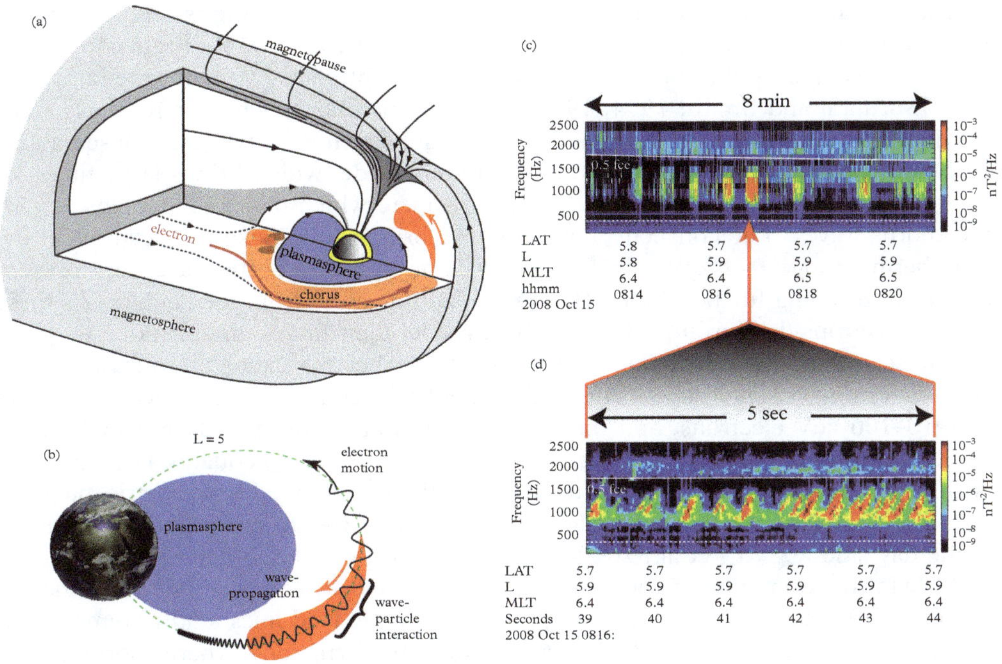

Figure 8.1 *Morphology of chorus waves in the Earth's inner magnetosphere. (a) A cutaway of the magnetosphere showing the drift of plasma sheet electrons from the magnetotail, through the dawn side and into the dayside. The equatorial projection of the chorus source region is shown in orange as well as a meridional projection of chorus propagation on the dayside. (b) A meridional projection illustrating the propagation of chorus waves from the equatorial region toward the southern hemisphere, and the traversal of an energetic electron through the chorus wave field. (c) An 8-minute observation of chorus waves from the THEMIS satellite, showing the large scale (~minute) modulation of the wave intensity. (d) A 5-second expanded view of a sequence of rising-tone chorus elements from (c). The white line represents half of the equatorial electron gyrofrequency.*

geomagnetic equator, propagation to high latitudes, and interaction with an energetic electron is given in Figure 8.1b where the orange color denotes the chorus wave packet. In a study that combined ray tracing with path-integrated linear growth rates calculated using statistical data from the CRRES satellite, Bortnik et al. [2007] showed that Landau resonant interactions of chorus with ~1 keV electrons resulted in damping rates that produced latitudinal propagation characteristics that were remarkably consistent with chorus observations, and that varied as a function of wave frequency, L-shell, and MLT. Nightside chorus was seen to propagate to latitudes of ~10°–15° which steadily increased with increasing MLT to latitudes of ~25°–30° (or more) on the dayside, consistent with observations of the chorus distribution. The modeled wave normal angle distribution of chorus power as a function of latitude was shown to be relatively field aligned [Chen et al., 2013], with most of the power contained within ~20°–30°, since

Landau interactions quickly damp more oblique wave normal angles, and is consistent with observations [Agapitov et al., 2013; Santolík et al., 2014]. Nightside chorus is very intense [Li et al., 2009] and more directly controlled by geomagnetic activity [e.g., Tsurutani and Smith, 1974, 1977] than dayside chorus which is relatively weak and does not seem to be strongly modulated by substorm activity. These observations are consistent with convective injection on the nightside, which brings large fluxes of ~10–100 keV electrons into the inner magnetosphere which then excite intense nightside chorus waves. The convective injection also brings large fluxes of electrons in the suprathermal range (~1 keV) which quickly damp oblique chorus waves and prevent them from traveling further than ~10°–15° in latitude away from the equator. As these injected electrons drift around dawn to the dayside, their fluxes diminish due to wave-particle scattering and precipitation into the atmosphere to cause the diffuse aurora. This scattering is energy dependent, such that ~keV electrons are depleted faster than the ~10–100 keV electrons, and the damping rates are consequently reduced on the dayside, allowing the waves to propagate further away from the equator along the field line. In addition, drift shell splitting due to the Earth's compressed, asymmetric magnetic field leads to an enhancement of the anisotropy of the ~10–100 keV electron population on the day side [e.g., Li et al., 2010], and hence ensures that sufficient free energy is available for the excitation of chorus waves. Studies have also demonstrated that there is a region on the dayside where the compressed magnetic field lines tend to remain uniform over extended distances, dubbed the 'dayside uniform zone' where chorus can be readily excited [Keika et al., 2012], and since Landau resonant electron fluxes are reduced, chorus waves propagate to high latitudes with little damping. Since there is a few hour time-lag between the electron injection on the nightside and the excitation of the chorus on the dayside (which is also largely controlled by magnetic-gradient induced anisotropy), the correlation of chorus to geomagnetic activity is much weaker than on the nightside.

The latitudinal distribution of the chorus wave power is thus largely shaped and controlled by the distribution of the ~keV suprathermal fluxes relative to the ~10–100 keV source population of electrons. The resulting latitudinal distribution, in turn, has profound implications on the net balance between chorus-driven loss and acceleration of radiation belt particles, since equatorially confined chorus waves tend to resonate with higher pitch-angle electrons, whereas chorus waves at high latitudes can more readily access the loss-cone [e.g., Horne and Thorne, 2003]. For a given electron energy, the net effect of the chorus waves on the particle population is critically dependent on the distribution of chorus wave power as function of latitude, and thus indirectly, on the ~keV electron population that controls this distribution.

In addition to the global distribution of chorus wave power as a function of L, MLT, latitude, and time (where time is taken as, for example, storm or substorm phase), chorus shows a remarkable degree of structure over a wide range of temporal and spatial scales. Figure 8.1c shows an 8-minute observation of chorus waves recorded on the Time History of Events and Macroscale Interactions during Substorms (THEMIS) E satellite on Oct 15, 2008, which reveals a periodic pulsation on the scale of ~10–100 s, which has been shown to be directly related to the pulsating aurora observed by ground-based all sky imagers [e.g., Nishimura et al., 2010, 2011]. When expanded further, as in Figure 8.1d over a 5 s timescale, the 'on' phase of the periodic pulsation is seen to consist of a multitude of distinct rising tones, with time scales on the order of ~0.1 s and separation on the order of ~0.5–1 s. It is these chorus elements that, when played over a loud speaker, resemble the sound of a rookery of birds in their 'dawn chorus' and give rise to the name of the emission [Storey, 1953; Isted and Millington, 1953]. When expanded further still (not shown) it is found that each individual chorus element consists of a sequence of finely structured sub-elements, with duration on the order of a few milliseconds [e.g., Santolík et al., 2004]. These temporal structures have also been shown to have corresponding spatial scales, for example the ~10–100 s periodic pulsation of the chorus shown in Figure 8.1c has been associated with regions of scale length ~1 Re [Nishimura et al., 2011] that pulsate coherently and seem to be associated with density enhancements and depletions [Li et al., 2011]. These regions are illustrated schematically on the nightside of Figure 8.1a and have the property that each patch pulsates with a unique pattern that is uncorrelated to any of its neighboring patches thus allowing a one-to-one correlation with a patch pulsating in the ionosphere [e.g., Nishimura et al., 2010, 2011]. The individual chorus elements shown in Figure 8.1d have spatial scales that are on the order of a wavelength (tens of km) in the direction perpendicular to the magnetic field line and roughly a factor of 10 larger in the parallel direction [Santolík and Gurnett, 2003].

At the present time, the distinctive spatial and temporal scales exhibited by chorus emissions are not well understood, although they are undoubtedly associated with the physics of the generation mechanism, and much progress has been made in recent years in replicating the chorus element structure in large-scale Particle-In-Cell computer simulations [Katoh and Omura, 2006, 2007, 2011; Omura et al., 2008, 2009; Hikishima et al., 2010; Omura and Nunn, 2011; Hikishima and Omura, 2012]. However, it is relatively straightforward to assess the effect that such structuring has on the electron scattering, as shown later in this chapter. To begin, we describe the essential elements of the wave-particle interaction process in the next section.

8.3 Introduction to resonant wave particle interactions

A charged particle moving in a background magnetic field $\mathbf{B}_0 = \mathbf{z}B_0$ will gyrate in the direction perpendicular to $\mathbf{z}$ with an angular frequency $\Omega = -qB_0/m$ where q, B_0, and m are the particle's (signed) charge, magnetic field magnitude, and non-relativistic mass, respectively. If we neglect the small radiative loss due to the particle's constant centripetal acceleration, it can be said that the particle will remain in the same trajectory forever without any alteration of its total energy E or equatorial pitch angle $\alpha_0 = \tan^{-1}(v_\perp/v_{||})$ (here, $v_\perp$ and $v_{||}$ are the components of the particle velocity perpendicular and parallel to $\mathbf{B}_0$, respectively). However, if a circularly polarized wave with angular frequency ω and wave number k is allowed to propagate along the field line and the charged particle moves through this wave, the situation changes considerably and the particle is able to respond to the electromagnetic wave field in a number of ways. For purposes of discussion, we shall assume that the wave has electric and magnetic field components denoted by E_w and B_w, respectively, and further that the particle dynamics are primarily governed by the background magnetic field, so that the wave field constitutes only a small perturbation, i.e., $B_w/B_0 \ll 1$. It might be intuitively expected that the wave field will generally appear to the particle as a minor perturbative noise whose effects will average to zero over several gyroperiods. However, if the particle moves through the wave packet with a particular parallel velocity such that the Doppler-shifted wave frequency $\omega + kv_{||}$ is exactly equal to the gyrofrequency of the particle Ω/γ (where $\gamma^2 = 1/(1 - v^2/c^2)$ is the relativistic correction factor), then the particle will experience the wave field as a quasi-stationary electromagnetic structure in its frame of reference, and may begin to deviate significantly from its unperturbed trajectory. This is the basis of a resonant wave-particle interaction.

The equation describing the wave-particle interaction process is the Lorentz force equation, which can be expressed as:

$$\frac{d\mathbf{p}}{dt} = q\left(\mathbf{E}_w + \frac{\mathbf{p}}{m\gamma} + [\mathbf{B}_0(\lambda) + \mathbf{B}_w]\right) \tag{8.1}$$

where $\mathbf{p} = \gamma m v$ is the particle momentum, and the magnetic field has been explicitly split into the wave component B_w and background field B_0, which varies as a function of latitude λ. Equation (8.1) is a vector equation consisting of 3 component equations in x, y, z, that can be readily transformed to a more convenient coordinate system with components parallel and perpendicular to the field line, and an angle η that represents phase angle between $v_\perp$ and B_w. For purposes of illustration, we assume that the wave is propagating strictly along the field line, and that the particle is non-relativistic, so that after gyro-averaging and neglecting second order terms, the equations result in the coupled set of ordinary differential equations:

$$\begin{aligned}
\frac{\mathrm{d}v_{\|}}{\mathrm{d}t} &= \left(\frac{qB_w}{m}\right) v_\perp \sin\eta = \frac{v_\perp^2}{2B}\frac{\partial B}{\partial z} \\
\frac{\mathrm{d}v_\perp}{\mathrm{d}t} &= -\left(\frac{qB_w}{m}\right)\left(v_{\|} + \frac{\omega}{k}\right) v_\perp \sin\eta + \frac{v_\perp v_{\|}}{2B}\frac{\partial B}{\partial z} \\
\frac{\mathrm{d}\eta}{\mathrm{d}t} &= \Omega - \omega - kv_{\|}
\end{aligned} \tag{8.2}$$

The set of equations (8.2) has been used extensively in past studies of wave-particle interactions [Dysthe, 1971; Inan et al., 1978; Bortnik et al., 2008] and only a few key features will be highlighted here. The first point to note is that if we set $B_w = 0$ then the equations reduce to purely adiabatic motion, and the particle simply travels periodically up and down a given magnetic field line, reflecting at the mirror points. In that case, the particle motion is fully described by the usual conservation of the adiabatic invariants [e.g., Roederer, 1970; Walt, 1994] and it is not necessary to use equations (8.2). If $B_w > 0$, then the wave begins to affect the particle motion through the first term on the

RHS of the top two equations of the set (8.2). However, this wave action usually averages to zero over several gyroperiods due to the rapid variation of η (and hence sin η), unless η remains constant over some period of time, that is, if $d\eta/dt = 0 = \Omega - \omega - kv_{||}$. This condition corresponds exactly to the resonance condition.

The resonance condition implies that any particle having a parallel velocity $v_{||} = (\Omega - \omega)/k$ will be resonant with the field-aligned wave described by the parameters (ω, k) regardless of its perpendicular velocity which would appear as a straight line when plotted in velocity-space (i.e., in the $v_{||}$-$v_{\perp}$ plane). If we generalize the resonance condition by including a relativistic mass correction, i.e., Ω is replaced with Ω/γ, the resonance condition becomes $\Omega/\gamma - \omega - kv_{||} = 0$, or $\omega + kv_{||} = \Omega(1 - v_{||}^2/c^2 - v_{\perp}^2/c^2)^{1/2}$, which now couples the perpendicular velocity into the resonance condition, and the formerly straight lines in velocity space which represented the resonance condition become resonant ellipses.

An example of two resonant ellipses is shown in Figure 8.2, where the parameters have now been specialized to represent chorus, with $\omega_1 = 0.1\Omega$ and $\omega_2 = 0.5\Omega$ representing the typical lower and upper frequency bounds of lower-band chorus, $\Omega = 2\pi(7\ \text{kHz})$ representative of the equatorial region at L$\sim$5, and $n_e = 10.23\ \text{cm}^{-3}$ giving $\omega_{pe}/\Omega = 4.1$. Together with the resonance curves (shown in solid) is a series of constant energy surfaces depicted using dashed lines, so that the intersection between these constant energy surfaces and the resonant ellipses indicates the resonant energies and approximate pitch angles that will be able to resonate with each of the waves. It should be noted that for positive values of the wave's phase velocity as depicted in the figure, i.e., $v_{ph1} = \omega_1/k_1 > 0$ and $v_{ph2} = \omega_2/k_2 > 0$, the first order cyclotron resonance will typically require that the respective resonance velocities be negative, i.e., $v_{||1}^{res} < 0$ and $v_{||2}^{res} < 0$. However, for sufficiently energetic electrons, γ can be large enough that $\Omega/\gamma = \omega$ and $v_{||}^{res} = 0$, or even $v_{||}^{res} > 0$ giving a relativistically shifted co-streaming interaction, or the previously described relativistic turning acceleration [e.g., Omura et al., 2007; Furuya et al., 2008].

Having determined the resonant energies involved in the wave-particle interaction, we next consider the approximate trajectory taken by the resonant particles, when they are acted upon by the wave. Reverting again to the non-relativistic limit for simplicity, and for purposes of illustration, the inhomogeneity is set to zero, representative of the equatorial region, and the set of equations (8.2) integrated once [e.g., Dysthe, 1971; Walker, 1993] to yield:

$$\left(v_{||} - \frac{\omega}{k}\right)^2 + v_{\perp}^2 = V_0^2 \qquad (8.3)$$

where V_0 is a constant of integration and defines a particular trajectory for resonant particles in velocity space. It should be noted that (8.3) can be alternatively obtained by performing a Lorentzian transformation to a coordinate system which is propagating with the phase velocity of the wave. In this reference frame, the wave electric field becomes zero, and hence no energy can be exchanged between the particle and the wave, resulting in constant radius circles in velocity space, centered on the phase velocity ω/k. The resonant particle trajectory including relativistic effects is more complicated and has been derived by Summers et al. [1998].

Returning now to Figure 8.2 where the relativistic resonant trajectories are shown as solid black curves confined between the $\omega_1 = 0.1\Omega$ and $\omega_2 = 0.5\Omega$ resonance ellipses, it is possible to visualize the energy exchange processes mediated by the chorus wave. In general, the population of electrons will be constrained to flow along the resonant trajectories shown in the figure, from regions of high phase space density to regions of low phase space density. For purposes of illustration, we have superimposed a typical phase space density onto Figure 8.2, represented by a normalized bi-Maxwellian distribution having a temperature anisotropy $A = T_\perp/T_{||} - 1 = 1$, and an empty loss-cone at pitch angles $\alpha < \alpha_{LC} = 3.7°$ (typical of L = 5). There are two regions indicated in the figure, denoted by R_1 and R_2. R_1 represents the low-energy region, where the

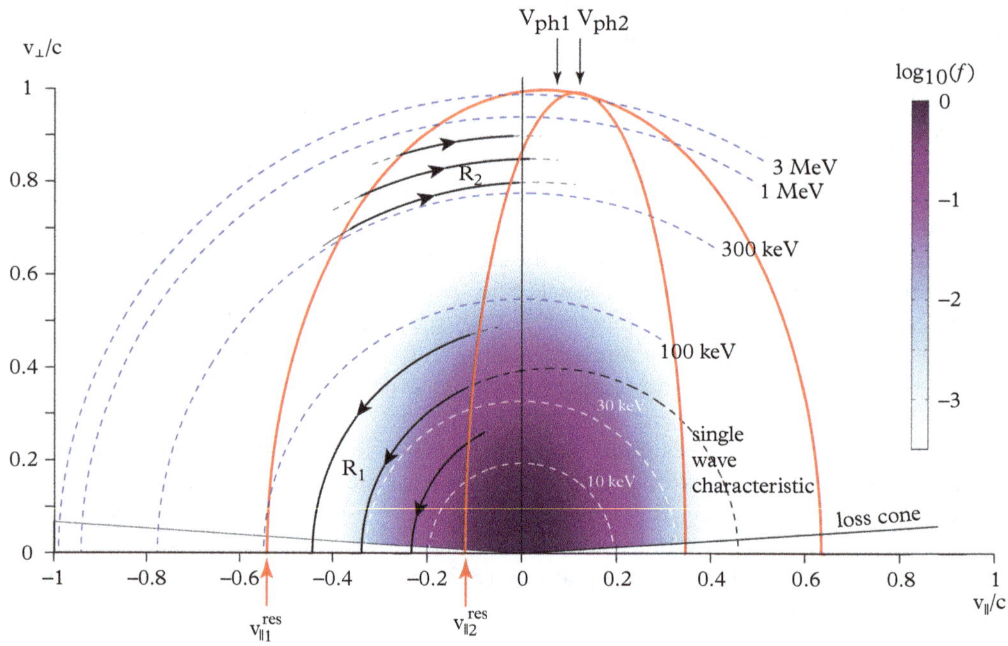

Figure 8.2 *Diffusive trajectory of electrons in velocity space. The dashed lines represent surfaces of constant energy. The solid lines represent resonance ellipses corresponding to $0.1f_{ce}$ and $0.5f_{ce}$ respectively, defining the region in velocity space where particles are able to diffuse. The solid lines with arrows depict the single wave characteristics of electrons resonant with $0.1f_{ce}$ waves, where* R_1 *represents the region of lower energy resonant electrons, preferentially diffusing toward the loss cone, and* R_2 *represents the region of higher energy electrons, preferentially diffusing to higher energies. The background shading depicts a typical phase space density distribution with a mild anisotropy ($A\sim1$) at lower energies.*

combination of a temperature anisotropy and loss cone results in a PSD which is large at high pitch angles, and decreases toward low pitch angles. Thus the net flow of particles is toward low pitch angles, as indicated by the arrows. When compared to the constant energy surfaces (e.g., the 30 keV white-dashed curve), it is apparent that the resonant particles lose energy on the whole, which is transferred to the chorus wave. From the figure, it is clear that the source population of electrons that excite chorus waves in the frequency range between $\omega_1 = 0.1\Omega$ and $\omega_2 = 0.5\Omega$ lies in the range $E \sim 10$–100 keV. The region R_2 represents the high energy electrons, and in this region the PSD decreases rapidly with increasing energy. Since the resonant trajectories increase in energy with increasing pitch angle, the lower pitch angle population will have a larger PSD and the net flow of electrons will be toward larger pitch angles. From the figure, it is apparent that electrons in the range $E \sim 300$ keV–3 MeV will absorb the wave energy and be preferentially accelerated to higher energies.

The process illustrated in Figure 8.2 depicts the essential elements of chorus-driven acceleration: the lower energy electrons (tens of keV) which contain free energy at medium to low pitch angles excite the chorus waves, and in the process get scattered toward the loss-cone and ultimately precipitated into the atmosphere, while the higher energy electrons (hundreds of keV) absorb the wave energy and get preferentially accelerated to higher pitch angles and energies. The chorus wave itself acts as a mediator of the energy flow between the different energy populations [e.g., Shklyar and Matsumoto, 2009; Thorne, 2010], and the transition energy between R_1 and R_2 occurs somewhere in the few hundred keV range [Summers et al., 2002; Horne and Thorne, 2003; Bortnik and Thorne, 2007; Turner et al., 2012].

The foregoing discussion relied on the basic assumption that electrons flow in the negative gradient direction of PSD. This is generally true but represents only one type of wave-particle interaction. Different modes of interaction are discussed next.

8.4 Modes of interaction

In order for the resonant electron population to be transported in pitch-angle and energy, the chorus wave must impart some non-adiabatic change to each electron, which can be described (in the simplest case) by the set of equations (8.2). Here, a situation is envisaged similar to Figure 8.1b, where an electron moves along its magnetic field line through the chorus wave packet, and the resonance condition, $d\eta/dt$ steadily approaches zero, at which point the particle is said to be in resonance with the wave, the wave-particle phase η (and hence sin η) remains approximately constant over a few gyroperiods, and the first two equations of (8.2) accumulate some permanent, finite changes in $v_\perp$ and $v_{||}$ (and hence also in α and E).

Intuitively, since the wave terms in equations (8.2) are proportional to $\sin\eta$, one might expect that the total scattering in α and E would be proportional to $\sin\eta$. In fact, this is true [e.g., Inan et al., 1978], but only when the parallel velocity of the electron through the wave is adequately approximated by the unperturbed, adiabatic parallel velocity, i.e., wave-induced changes in $v_{||}$ are negligible. In those instances when $v_{||}$ is significantly perturbed from its adiabatic trajectory, in such a way as to modify the evolution of the resonance condition (final equation in (8.2)), the scattering in α and E as a function of η can be quite different from that of a sinusoid.

As an example, Figure 8.3 shows the scattering of 24 electrons, all having the same initial α and E, but being uniformly distributed in η propagating through a chorus wave packet, and experiencing a wave-particle resonance [similar to Bortnik et al., 2008]. In Figure 8.3a, the background plasma is chosen to be representative of L = 5, the wave is assumed to be field-aligned, with $f = 2$kHz and $B_w \sim 1.4$ pT. The particles propagate from $\lambda = -9°$, experience a resonance at $\lambda \sim -5°$, and emerge at $\lambda \sim -2°$, with their pitch angles having been scattered sinusoidally as a function of the initial wave-particle phase η, as indicated in the adjoining panel on the right. Here, the scattering is small, with maximum pitch angle changes of ~0.03°–0.04°.

However, if we run the same simulation, but only increase the amplitude of the wave from $B_w \sim 1.4$ pT to $B_w \sim 1.4$ nT, the results look completely different as shown in Figure 8.3b. One might intuitively expect from (8.2) that increasing B_w will linearly increase the range of the scattering, which is true for small values of B_w, but the results of Figure 8.3b are qualitatively different, with all particles essentially behaving in a very similar way, and experiencing a rapid decrease of ~5° in pitch angle and ~5 keV in energy (shown in the right adjacent panel). In this regime, the initially uniform wave-particle phases η have all been bunched by the wave to be roughly the same, which is why all particles respond to the wave similarly. This regime is thus referred to as phase bunching.

In Figure 8.3c, the wave-particle interaction is changed slightly: the wave amplitude is kept constant at $B_w \sim 1.4$ nT, but the wave normal is changed to be ~60° and the initial α and E of the particles is modified so that they will resonate with the wave at $\lambda \sim -23°$. These parameters are discussed by Bortnik et al. [2008] and have been chosen to closely represent the recent observations of large-amplitude chorus waves [Cattell et al., 2008; Cully et al., 2008; Li et al. 2011]. Here, a combination of behaviors is observed: a large fraction of particles are phase bunched and transported to lower α and E, while some exhibit linear (and hence more symmetrical) scattering. One particle in particular, indicated by the red trajectory, is seen to rapidly increase in α, with a concomitant increase in E (not shown) of ~300 keV. This particle exhibits yet a third type of response known as phase trapping, where the particle's phase (relative to the wave) is trapped by the wave potential and is essentially forced to conform to the resonant velocity of the wave, as a function of latitude. This regime is known as phase trapping [e.g., Dysthe, 1971, Inan et al., 1978; Albert, 2000; Bortnik et al., 2008].

In order to determine which of the three particle responses should be expected from a particular wave-particle interaction, we return to the simplified set of equations (8.2). If the final equation is differentiated once, the equation for $dv_{||}/dt$ substituted and the

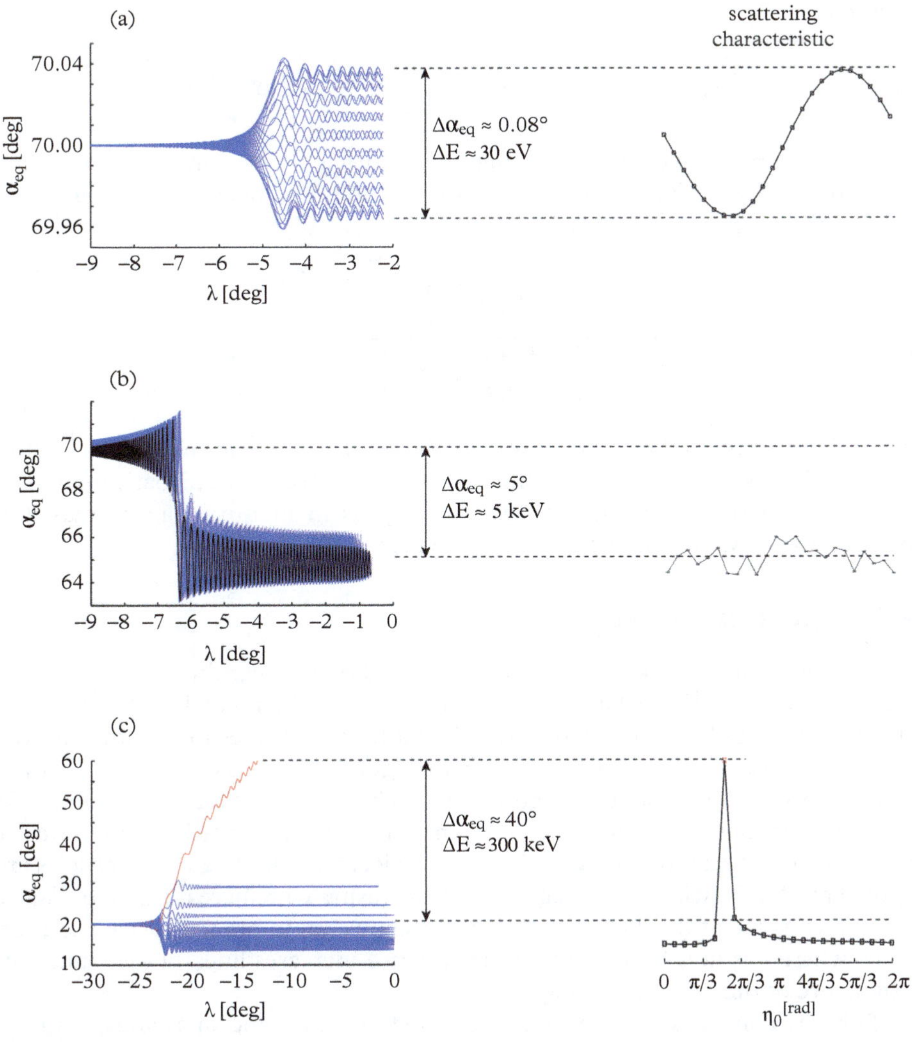

Figure 8.3 *An example of three types of electron responses to a resonance with a whistler wave. (a) Low amplitude wave (~1 pT) resonating at low latitude (~5°) resulting in linear scattering , (b) a large amplitude wave (~1 nT) resonating at low latitude (~5°) resulting in phase bunching and advection, and (c) a large amplitude wave (~1 nT) resonating at high latitudes (~23°), resulting in a combination of linear scattering, phase bunching, and phase trapping (shown rising to the dotted line). The attached panels on the right represent the pitch angle scattering characteristic as a function of initial wave-particle phase angle* η_0.

small order terms neglected, the result is a second order, oscillator equation which can be written as:

$$\frac{d^2\eta}{dt^2} + k\left(\frac{qB_w}{m}\right) v_\perp \sin\eta = \left[\frac{3}{2} + \frac{\Omega - \omega}{2\Omega}\tan^2\alpha\right] v_\parallel \frac{\partial\Omega}{\partial z} \tag{8.4}$$

It only remains to be determined whether the 'driving' force on the RHS of the equation will dominate the 'restoring' force of the pendulum. This is done using the ratio R, which compares these two terms:

$$R = \frac{k\left(\frac{qB_w}{m}\right) v_\perp}{\left[\frac{3}{2} + \frac{\Omega - \omega}{2\Omega}\tan^2\alpha\right] v_\parallel \frac{\partial\Omega}{\partial z}} \sim \frac{B_w}{\frac{\partial\Omega}{\partial z}} \tag{8.5}$$

Essentially, R shows that the dynamics of the particle are determined by a competition of the wave intensity B_w against the 'mirror' force, which causes adiabatic motion, $\partial\Omega/\partial z$. When $R \ll 1$, particles will be scattered linearly, as in Figure 8.3a, whereas if $R \gg 1$, nonlinear effects will result due to the dominance of B_w.

8.4.1 Quasilinear diffusion

One might legitimately ask whether the often-used quasilinear diffusion theory [Kennel and Petschek, 1966; Kennel and Engelman, 1966] can be related to the individual test particle scattering shown in Figure 8.3. To address this question in an intuitive way, we show the results of the simulation run in Figure 8.3a, but extended over the much longer interval of 60 seconds in Figure 8.4a such that a much larger number of wave-particle interactions are included (approximately 120 per particle). In contrast to the initial resonant interaction in Figure 8.3a, the trajectories of the 24 test particles are seen to spread randomly, with the average spread increasing with increasing time. Figure 8.4b shows the variance of the 24 test particles at each instant in time, displayed as a function of time, where it is now apparent that the variance increases linearly with time, which is the signature of diffusive spreading.

In fact, the comparison of test particle based diffusion and quasilinear diffusion has been done in a number of recent studies [Tao et al. 2011b, 2012] and shown to be equal. More formally, Albert [2010] has demonstrated that the analytical expressions for the quasilinear diffusion coefficients in the narrowband limit reduce precisely to the spreading derived from single-wave test particle interactions, when averaged over multiple, incoherent interactions similar to Figures 8.4a and 8.4b. Suitably averaged over a full wave distribution, the test particle based diffusion equations reproduce the quasilinear diffusion coefficients precisely.

To examine the evolution of individual test particle scattering interactions into diffusion we show the first resonant interaction again in Figure 8.4c, which is identical to Figure 8.3a but displayed as a function of time, not latitude. Since the scattering in α is sinusoidal as a function of initial η, it can be shown analytically that the results of

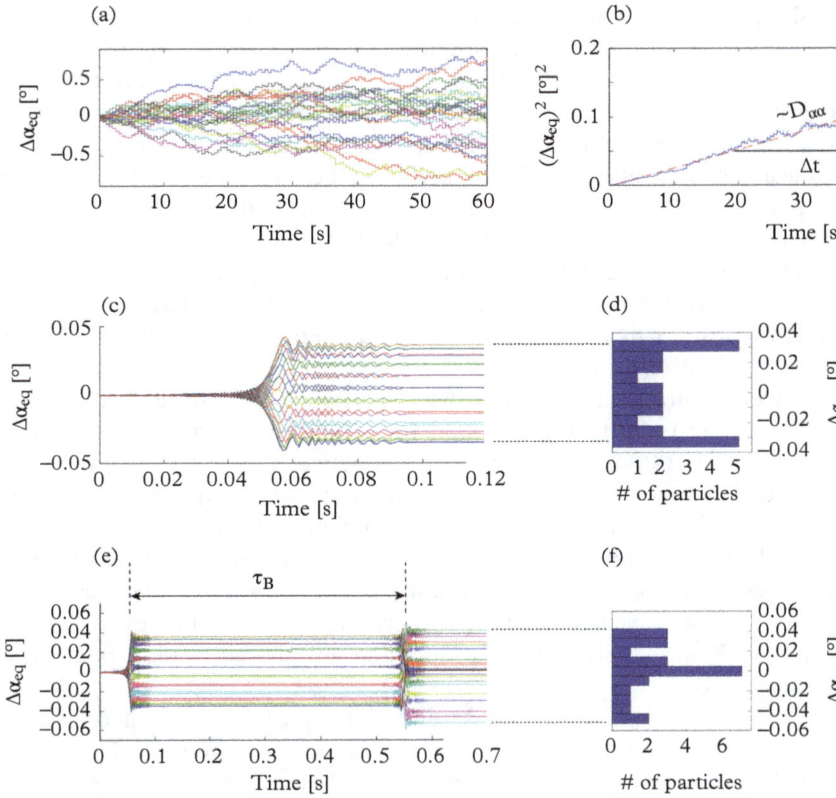

Figure 8.4 *A long-run simulation of 24 electrons experiencing cumulative linear scattering interactions (similar to Figure 8.3a), resulting in quasilinear diffusive behavior. (a) Change in equatorial pitch angle of all particles as a function of time. (b) The variance of all particles, shown to be increasing linearly as a function of time, consistent with diffusive scattering. (c) An expanded view of the first 0.12 sec of the simulation, showing the first linear resonant-scattering interaction (similar to Figure 8.3a), and (d) the resultant distribution of particles at 0.12 sec. (e) An expanded view showing the first two linear resonant-scattering interactions, and (f) the resultant distribution at 0.7 sec, showing that the total scattering signature is now more normally distributed, consistent with diffusive scattering.*

the first scattering take on a distribution of the form $h(\alpha) = [\pi((\Delta\alpha_{max})^2 - (\Delta\alpha)^2)^{1/2}]^{-1}$ [Bortnik et al., 2006] where $\Delta\alpha_{max}$ is readily determined from the equations of motion (8.2). A histogram showing the particle distribution of pitch angles following the resonant interaction is shown in Figure 8.4d, and is seen to indeed resemble $h(\alpha)$. If the time of the simulation is now extended to include the first two resonant interactions, as shown in Figure 8.4e, it is seen that the distribution of pitch angles following the second scattering no longer resembles $h(\alpha)$, but is the result of the convolution $h(\alpha)*h(\alpha)$, leading to a more 'bell shaped' distribution. Subsequent resonant interactions each lead to a further convolution of the existing particle distribution with $h(\alpha)$ which is essentially the solution of the diffusion equation.

Having established the basic equivalence of test particle scattering with quasilinear diffusion, we can now explore the limits of applicability of quasilinear diffusion. Figure 8.5 shows an example of the results of Tao et al. [2012], where the scattering of 400 electrons has been calculated in a whistler wave packet consisting of 100 individual, incoherent components spanning the range 0.2Ω–0.4Ω, with assumptions that satisfy those of quasilinear theory as closely as possible. Figures 8.5a and 8.5b show the pitch-angle and energy diffusion coefficients calculated using both the quasilinear theory and the test particle approach for a specific pitch angle and energy (E = 200 keV, $\alpha = 60°$), as a function of the integrated wave intensity, B_w^{rms}. For low wave amplitudes, i.e., $B_w^{rms} < 300$ pT, the diffusion coefficients agree very closely, but for large amplitudes i.e., $B_w^{rms} > 500$ pT, the test particle based diffusion coefficients saturate, and no longer agree with the theoretical diffusion coefficients.

We define the wave amplitude at which the two approaches are different by a factor of 2 as the 'hinge' point, and plot this hinge point for seven logarithmically spaced energies between 10 keV and 1000 keV, and pitch angles between $\alpha_0 = 10°$ and $\alpha_0 = 80°$ in Figure 8.5c. Using L = 5 as a convenient reference point, the approximate chorus amplitudes at which quasilinear diffusion coefficients begin to saturate can be as low as 100–300 pT for electron energies with $E < \sim 100$ keV, which is the typical range of chorus wave intensities used in calculations [Meredith et al., 2003; 2012]. However, the limit of applicability increases well beyond typical, averaged observable levels for energies larger than 1 MeV, which should make diffusion simulation at ultra-relativistic energies quite accurate.

An example of the application of quasilinear diffusion to a remarkable acceleration event occurring at ultra-relativistic (i.e., several MeV) energies is shown in Figure 8.6 [Thorne et al., 2013]. Here, a sequence of observations of electrons with energies E = 2.3–7.15 MeV made during the October 8–9, 2012, geomagnetic storm with the Van Allen Probes showed a PSD increase of 3–4 orders of magnitude within a span of several hours, together with characteristic 'tophat' shaped pitch angle distributions, shown in Figures 8.6a–8.6d.

By using a newly developed technique to infer the global, time-varying chorus wave field [Li et al., 2013b], and using these chorus amplitudes to calculate diffusion coefficients that were then used to solve a Fokker–Planck diffusion equation, the theoretical phase space density distributions shown in Figures 8.6e–8.6h were obtained. The theoretical curves show remarkable consistency with observed fluxes, replicating the increase of PSD as a function of time and energy, and showing the development of the tophat distributions in pitch angle, as a function of time. These simulations are particularly significant in that they demonstrate conclusively that it is chorus waves that cause the rapid acceleration of the ultra-relativistic electrons in this event, as opposed to any of the other candidate acceleration mechanisms [e.g., Friedel et al., 2002; Thorne, 2010].

8.4.2 Nonlinear wave-particle interactions

In the previous section, we showed that when the chorus waves can be considered 'weak' (as defined by the ratio R, for example), leading to linear scattering, that the

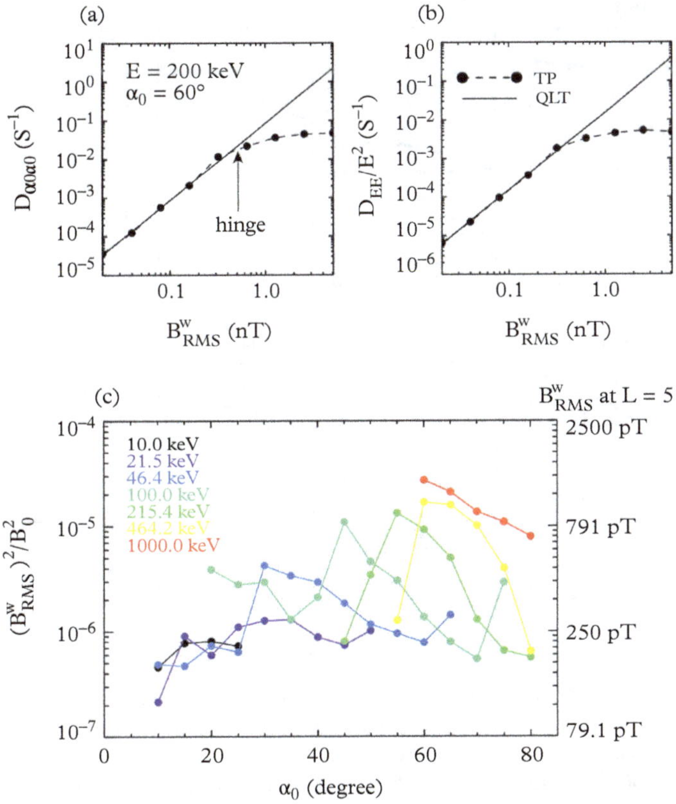

Figure 8.5 *The saturation of quasilinear diffusion coefficients. (a) A comparison of the pitch-angle diffusion coefficients of 200 keV, 60° electrons calculated numerically using test particle scattering, and by quasi-linear theory, showing their agreement at low amplitudes, and deviation at large amplitudes. (b) Similar to (a) but for energy diffusion coefficients. (c) A plot of the 'hinge' points at which test particles no longer respond diffusively to wave scattering, as a function of initial pitch angle and energy. The scale on the right of the figure shows the corresponding wave amplitudes at* L = 5 *(i.e.,* B_0 = *250 nT).*

result of multiple resonant interactions—even for a completely coherent monochromatic wave—reduced to the quasilinear diffusion formalism, and further that this formalism was applicable and accurate for resonant electron energies in the MeV range. However, when the waves cannot be considered 'weak', the resulting scattering can exhibit significant nonlinear behavior as shown in Figures 8.3b and 8.3c, and the result of (similar) multiple incoherent interactions is no longer diffusive but will contain strong advective

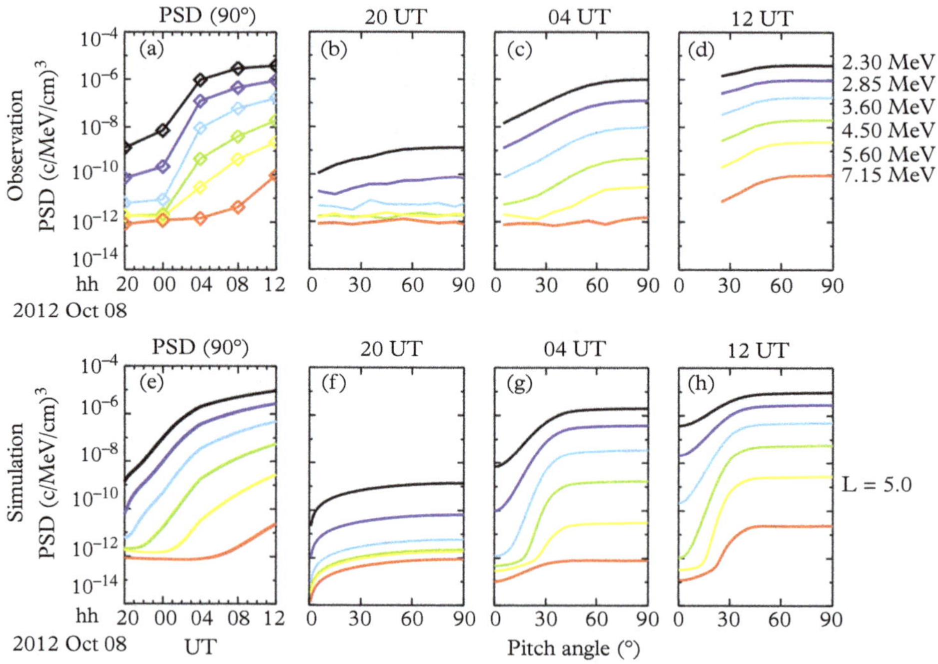

Figure 8.6 *Simulation of the October 9, 2012, acceleration event of ultra-relativistic electrons. (a) Observed phase space densities on the Van Allen Probes as a function of time together with pitch angle distributions parameterized by energy at (b) October 8, 20 UT, (c) October 9, 04 UT, (d) October 9, 12 UT. Simulated results showing the progression of the phase space densities and pitch angle distributions in (e)–(h), similar to the observations in (a)–(d).*

terms. In this section, we discuss such nonlinear behavior and describe how it can be understood and modeled.

The nonlinear scattering of test particles due to a single, coherent, monochromatic wave was shown in Figure 8.3 and discussed in Section 8.4.1. But realistic chorus elements contain more complexity than this simple picture, for example, consisting of a steadily rising frequency, and containing amplitude modulation within a single chorus element, which was briefly discussed in Section 8.2.

In order to quantify the effects of the chorus wave's amplitude modulation, we model the wave packet as a superposition of two sinusoids with closely spaced frequencies $\cos(\omega_1 t)$ and $\cos(\omega_2 t)$, such that the resultant wave is described by the standard formula:

$$g(t) = 2\cos\left(\frac{\omega_1 - \omega_2}{2}t\right)\cos\left(\frac{\omega_1 + \omega_2}{2}t\right) \tag{8.6}$$

Here, the small difference in frequencies $\delta\omega = (\omega_1 - \omega_2)/2$ can be interpreted as a slow modulation of a rapidly oscillating wave with frequency $<\omega> = (\omega_1 + \omega_2)/2$. While this model clearly does not capture all the intricacies of the chorus element, it is nevertheless extremely useful because it represents the amplitude modulated chorus wave in a way that can be analytically understood and analyzed [Zaslavsky, 1985, p.133]. Each component wave frequency will have its own resonant island [Lichtenberg and Lieberman, 1983], with a half-width given by:

$$\frac{\Delta p_{||}}{mc} = 2\sqrt{\frac{p_{\perp}}{mc}\frac{|\Omega|}{\omega}\frac{B_y^w}{B_0}\frac{1}{n}}$$

where the field is assumed to be uniform for simplicity, and the refractive index $n \gg 1$. Using the non-relativistic resonance condition, the separation between the centers of the resonant islands can be written as:

$$\frac{\delta p_{||}}{mc} = \frac{1}{c}\left|\frac{\omega_1 + \Omega}{k_1} - \frac{\omega_2 + \Omega}{k_2}\right|$$

In order to infer what the response of the particles would be to the superposed waves, it is useful to determine whether the resonant islands of the two waves overlap by comparing the width of the islands to the separation between their midpoints as shown in Figure 8.7c. This is accomplished with the overlap parameter K, defined as $K = (\Delta p_{||1} + \Delta p_{||2})/\delta p_{||}$, which is $\ll 1$ when the resonant islands are far apart and do not overlap, and >1 when resonant islands overlap and interfere with each other's trapping dynamics.

Figure 8.7 shows the results of a simulation that demonstrates the effects of varying the modulation frequency of the wave, or alternatively stated, the period between subelements defined as $\delta\tau = 1/\delta\omega$ following the work of Tao et al. [2013]. Here, the trajectories of 400 test particle electrons were calculated through a superposition of two cosine waves with a varying frequency of separation, indicated on the abscissa by δf. For low values of δf, between $\sim$1 and 7 mHz, the resonant islands overlap to a great degree as shown schematically in Figure 8.7a, corresponding to a very slow modulation (large value of $\delta\tau$) and behavior that closely resembles that of a single wave, exhibiting trapping and bunching behavior as in Figure 8.3c. As δf is increased further, the resonant islands separate further and the electron scattering becomes more stochastic, illustrated in Figure 8.7b. Finally, when δf is separated further still, $K > 1$, and the resonant islands no longer overlap, resulting in a response which is better described by two independent, non-interfering waves that act on the particles, illustrated in Figure 8.7c. Also shown in Figure 8.7d is a probability distribution of subelement separations, scaled from Figure 5 of Santolík et al. [2004], which shows that the most frequent values of δf are in the range of $K \sim 1$, bordering between the response of two separate waves and two overlapping resonant islands which result in stochastic scattering. We note for completeness that Figures 8.7a–8.7c only serve to illustrate the topology of the resonance schematically, but additional complexity develops when particles are allowed to propagate through the phase portrait, as shown in Lichtenberg and Lieberman [1983, pp. 252–256].

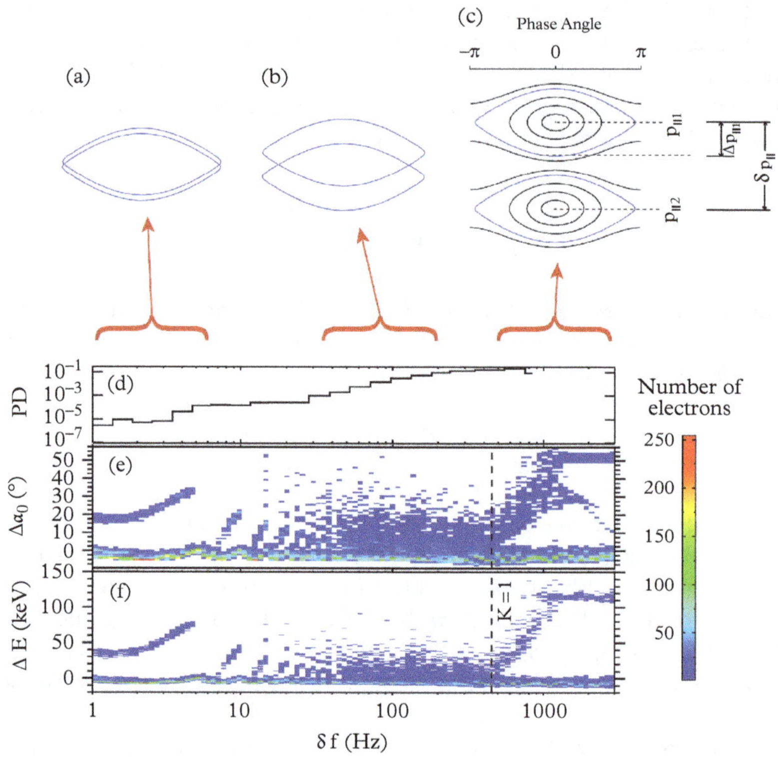

Figure 8.7 *A two-wave model of the modulated wave amplitude, and particle scattering results. (a) The degenerate region, where waves have very closely spaced frequencies, (b) the stochastic region, where resonant islands overlap, and (c) the non-overlap region, where particles respond to each wave independently. (d) The distribution of subpacket frequency separations, (e) test particle scattering distributions as a function of frequency separation between the two component waves (f) similar to (e) but for particle energies.*

These results suggest that the particle response in a typical chorus element will involve a complex interplay between the characteristic length of the subpacket structure (which controls $\delta p_{||}$), and the amplitude of the subpacket which controls the width of the resonant island and hence $\Delta p_{||}$. In general, we expect that nonlinear phase bunching and phase trapping effects will be somewhat reduced by the inclusion of subpacket structure. To demonstrate that this is the case, we show in Figure 8.8 the results of a simulation based on the work of Tao et al. [2012]. Here, a realistic chorus element, complete with frequency variation, subpacket structure, and amplitude modulation is modeled to accurately represent an observation of a chorus element made on THEMIS D on October 23, 2008, at L = 6.4, near the geomagnetic equator. Figure 8.8a shows the observed chorus element and Figure 8.8b shows the modeled chorus element, both

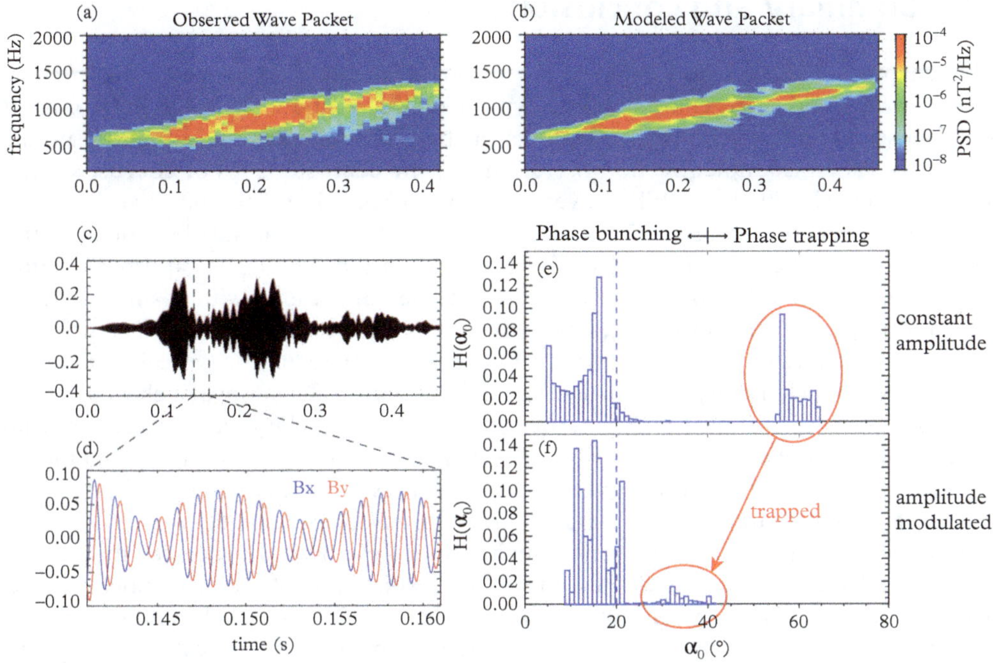

Figure 8.8 *A realistic simulation of test particle scattering by an amplitude modulated chorus element. (a) An observed chorus element, and (b) the modeled chorus element representing the observation in panel (a), through which the test particles are propagated. (c) The amplitude of the Bx and By components of the chorus element for the duration of the element, and (d) an expanded view at a particular time, which is similar for both the observed and the simulated wave. (e) Test particle scattering in response to a constant amplitude wave and (f) in response to the amplitude modulated wave, showing that the trapped particles do not travel as far as in (e) but nevertheless still exhibit evidence of trapping effects.*

having a time structure as shown in Figures 8.8c and 8.8d. A total of 1600 test particle electrons are run through the modeled wave packet, and a similar wave packet having constant amplitude and frequency to highlight the differences, as shown in Figures 8.8e and 8.8f. As expected, the population of particles that are trapped in the realistic wave packet (Figure 8.8f) has been substantially diminished compared to those in the constant amplitude wave, and the degree of scattering has been reduced, but qualitatively the test particle response still exhibits the signature of nonlinear scattering with a majority of particles advected to lower pitch angles and energies due to phase bunching, and a small fraction being advected to much larger pitch angles and energies due to phase trapping. The effect of multiple wave packets over an extended period of time on the test particles (e.g., Figure 8.4a) has not yet been determined but is the subject of current and ongoing research. Some early results suggest that the net effect is a very rapid acceleration of the tail of energetic electrons to high energies [Yoon, 2011].

8.5 Summary and conclusions

This chapter introduces and explores the resonant interaction between whistler mode chorus waves and energetic electrons, typical of the near-Earth space environment, and focuses specifically on those elements of the resonant interaction that are unique to chorus. As an example, Section 8.2 discussed the characteristics of chorus, and the unique structuring that it exhibits over a variety of spatial and temporal scales. Temporal structuring on the finest timescales, typically referred to as subelement structure, was shown in Section 8.4 to produce dynamics that are not typically captured in either the quasilinear diffusion framework or test particle simulations with a single wave of constant frequency and amplitude. These dynamics can be understood as a complex interplay of the chorus wave's characteristic intensity, compared to the typical periodicities of the chorus subelements. Test particle simulations with realistic wave packets show that while the classical nonlinear effects associated with constant, large-amplitude waves are reduced, the overall particle scattering is nevertheless far from linear, and global models of radiation belt variability will need to find novel ways of including the realistic particle scattering responses into the traditionally diffusion-based approaches.

In addition, many open questions remain about the physical processes that control the morphological features of chorus: the $\sim$1 Re spatial scale of magnetospheric chorus patches, and the associated $\sim$minute timescales over which these patches pulsate, the $\sim$1 s timescale of the chorus elements themselves, what controls this timescale, the frequency drift, and separation between individual chorus elements. Finally, what is the physical process that controls the subelement structure within each chorus element, and how can it be modeled and understood. The unprecedented volume and quality of data currently being generated, including those data from the recently launched Van Allen Probes, together with the intense activity in this field at the moment, will undoubtedly answer many of these outstanding questions in the near future.

REFERENCES

Agapitov, O., et al. (2013), Statistics of whistler-mode waves in the outer radiation belt: Cluster STAFF-SA measurements, *J. Geophys. Res. Space Physics*, **118**, 3407–3420, doi:10.1002/jgra.50312.

Albert, J.M. (2000), Gyroresonant interactions of radiation belt particles with a monochromatic electromagnetic wave, *J. Geophys. Res.*, **105** (A9), 21191–21209.

Albert, J. M. (2010), Diffusion by one wave and by many waves, *J. Geophys. Res.*, **115**, A00F05, doi:10.1029/2009JA014732.

Bortnik, J., U. S. Inan, and T. F. Bell (2006), Temporal signatures of radiation belt electron precipitation induced by lightning-generated MR whistler waves: 1. Methodology, *J. Geophys. Res.*, **111**, A02204, doi:10.1029/2005JA011182.

Bortnik, J., and R. M. Thorne (2007), The dual role of ELF/VLF chorus waves in the acceleration and precipitation of radiation belt electrons, *J. Atmos. Solar Terr. Phys.*, **69**, 378–386.

Bortnik, J., R. M. Thorne, and N. P. Meredith (2007), Modeling the propagation characteristics of chorus using CRRES suprathermal electron fluxes, *J. Geophys. Res.*, **112**, A08204, doi:10.1029/2006JA012237.

Bortnik, J., Thorne, R.M., and Inan, U.S. (2008). Nonlinear interaction of energetic electrons with large amplitude chorus. *Geophys. Res. Lett.*, **35**, L21102, http://dx.doi.org/10.1029/2008GL035500.

Burtis, W. J., and R. A. Helliwell (1969), Banded chorus—A new type of VLF radiation observed in the magnetosphere by OGO 1 and OGO 3, *J. Geophys. Res.*, **74** (11), 3002–3010, doi:10.1029/JA074i011p03002.

Cattell, C., et al. (2008), Discovery of very large amplitude whistler-mode waves in Earth's radiation belts. *Geophys. Res. Lett.*, **35**, L01105, http://dx.doi. org/10.1029/2007GL032009.

Chen L., R. M. Thorne, W. Li, and J. Bortnik (2013), Modeling the wave normal distribution of chorus waves, *J. Geophys. Res. Space Physics*, **118**, 1074–1088, doi:10.1029/2012JA018343.

Cully, C.M., J.W. Bonnell, and R.E. Ergun (2008), THEMIS observations of long-lived regions of large-amplitude whistler waves in the inner magnetosphere. *Geophys. Res. Lett.*, **35**, L17S16, http://dx.doi.org/10.1029/2008GL033643.

Dysthe, K.B. (1971), Some studies of triggered whistler emissions, *J. Geophys. Res.* **76** (28), 6915–6931.

Friedel, R. H. W., G. D. Reeves, and T. Obara (2002), Relativistic electron dynamics in the inner magnetosphere—A review, *J. Atmos. Sol. Terr. Phys.*, **64**, 265.

Furuya, N., Y. Omura, and D. Summers (2008), Relativistic turning acceleration of radiation belt electrons by whistler mode chorus, *J. Geophys. Res.*, **113**, A04224, doi: 10.1029/2007JA012478.

Helliwell, R. A. (1967), A theory of discrete emissions from the magnetosphere, *J. Geophys. Res.*, **72**, 4773–4790.

Hikishima, M., Y. Omura, and D. Summers (2010), Microburst precipitation of energetic electrons associated with chorus wave generation, *Geophys. Res. Lett.*, **37**, L07103, doi:10.1029/2010GL042678.

Hikishima, M., and Y. Omura (2012), Particle simulations of whistler-mode rising-tone emissions triggered by waves with different amplitudes, *J. Geophys. Res.*, **117**, A04226, doi: 10.1029/2011JA017428.

Horne, R. B., and R. M. Thorne (2003), Relativistic electron acceleration and precipitation during resonant interactions with whistler-mode chorus, *Geophys. Res., Lett.*, **30** (10), 1527, doi:10.1029/2003GL016973.

Horne, R. B., R. M. Thorne, S. A. Glauert, J. M. Albert, N. P. Meridith, and R. R. Anderson (2005), Timescales for radiation belt electron acceleration by whistler mode chorus waves, *J. Geophys. Res.*, **111**, A03225, doi:10.1029/2004JA010811.

Inan, U.S., T. F. Bell, and R. A. Helliwell (1978), Nonlinear pitch angle scattering of energetic electrons by coherent VLF waves in the magnetosphere, *J. Geophys. Res.*, **83** (A7), 3235–3253.

Isted, G. A., and G. Millington (1957), The "dawn chorus" in radio observations, *Nature*, **180**, 716.

Katoh, Y., and Y. Omura (2006), A study of generation mechanism of VLF triggered emission by self-consistent particle code, *J. Geophys. Res.*, **111**, A12207, doi: 10.1029/2006JA011704.

Katoh, Y., and Y. Omura (2007), Relativistic particle acceleration in the process of whistler-mode chorus wave generation, *Geophys. Res. Lett.*, **34**, L13102, doi: 10.1029/2007GL029758.

Katoh, Y., and Y. Omura (2011), Amplitude dependence of frequency sweep rates of whistler mode chorus emissions, *J. Geophys. Res.*, **116**, A07201, doi: 10.1029/2011JA016496.

Keika, K., M. Spasojevic, W. Li, J. Bortnik, Y. Miyoshi, and V. Angelopoulos (2012), PENGUIn/AGO and THEMIS conjugate observations of whistler mode chorus waves in the dayside uniform zone under steady solar wind and quiet geomagnetic conditions, *J. Geophys. Res.*, **117** (A7), A07212.

Kennel, C. F., and F. Engelman (1966), Velocity space diffusion from weak plasma turbulence in a magnetic field, *Phys. Fluids*, **9**, 2377.

Kennel, C. F., and H. E. Petschek (1966), Limit on stably trapped particle fluxes, *J. Geophys. Res.*, **71** (1), 1–28, doi:10.1029/JZ071i001p00001.

Lauben, D. S., U. S. Inan, T. F. Bell, D. L. Kirchner, G. B. Hospodarsky, and J. S. Pickett (1998), VLF chorus emissions observed by Polar during the January 10, 1997, magnetic cloud, *Geophys. Res. Lett.*, **25** (15), 2995–2998.

Lauben, D. S., U. S. Inan, T. F. Bell, and D. A. Gurnett (2002), Source characteristics of ELF/VLF chorus, *J. Geophys. Res.*, **107** (A12), 1429, doi:10.1029/2000JA003019.

LeDocq, M. J., D. A. Gurnett, and G. B. Hospodarsky (1998), Chorus source locations from VLF Poynting flux measurements with the Polar spacecraft, *Geophys. Res. Lett.*, **25** (21), 4063–4066.

Lichtenberg, A., and M. Lieberman (1983), *Regular and Chaotic Dynamics*, second ed. Springer Verlag, New York.

Li, W., et al. (2009), Evaluation of whistler-mode chorus intensification on the nightside during an injection event observed on the THEMIS spacecraft, *J. Geophys. Res.*, **114**, A00C14, doi:1029/2008JA013554.

Li, W., et al. (2010), THEMIS analysis of observed equatorial electron distributions responsible for the chorus excitation, *J. Geophys. Res.*, **115**, A00F11, doi: 10.1029/2009JA014845.

Li, W., J. Bortnik, Y. Nishsimura, R. M. Thorne, and V. Angelopoulos (2012), The origin of pulsating aurora: modulated whistler mode chorus waves, in *Auroral Phenomenology and Magnetospheric Processes: Earth and Other Planets, Geophysical Monograph Series 197*, AGU, 10.1029/2011GM001164.

Li, W., J. Bortnik, R. M. Thorne, Y. Nishimura, V. Angelopoulos, and L. Chen (2011), Modulation of whistler mode chorus waves: 2. Role of density variations, *J. Geophys. Res.*, **116**, A06206, doi:10.1029/2010JA016313.

Li, W., et al. (2013a), Characteristics of the Poynting flux and wave normal vectors of whistler-mode waves observed on THEMIS, *J. Geophys. Res. Space Physics*, **118**, 1461–1471, doi: 10.1002/jgra.50176.

Li, W., et al. (2013b), Constructing the global distribution of chorus wave intensity using measurements of electrons by the POES satellites and waves by the Van Allen Probes, *Geophys. Res. Lett.*, **40**, 4526–4532, doi: 10.1002/grl.50920.

Mauk, B. H., et al. (2012), Science objectives and rationale for the radiation belt storm probes mission, *Space Sci. Rev.*, 1–15, doi:10.1007/s11214-012-9908-y.

Meredith, N. P., R. B. Horne, R. M. Thorne, and R. R. Anderson (2003), Favored regions for chorus-driven electron acceleration to relativistic energies in the Earth's outer radiation belt, *Geophys. Res. Lett.*, **30**, 1871, doi:10.1029/2003GL017698, 16.

Meredith, N. P., et al. (2012), Global model of lower band and upper band chorus from multiple satellite observations, *J. Geophys. Res.*, **117**, A10225.

Ni, B., R. M. Thorne, N. P. Meredith, R. B. Horne, and Y. Shprits (2011), Resonant scattering of plasma sheet electrons leading to diffuse auroral precipitation: 2. Evaluation for whistler-mode chorus waves, *J. Geophys. Res.*, **116**, A04219, doi:10.1029/2010JA016233.

Nishimura, Y., et al. (2010), Identifying the driver of pulsating aurora, *Science*, **330**, 81–84, doi:10.1126/science.1193186.

Nishimura, Y, et al. (2011), Multi-event study of the correlation between pulsating aurora and whistler-mode chorus emissions, *J. Geophys. Res.*, **116**, A11221, doi:10.1029/2011JA016876.

Nunn, D. (1974), A self-consistent theory of triggered VLF emissions, *Planet. Space Sci.*, **22**, 349.

Nunn, D., Y. Omura, H. Matsumoto, I. Nagano, and S. Yagitani (1997), The numerical simulation of VLF chorus and discrete emissions observed on the Geotail satellite using a Vlasov code, *J. Geophys. Res.*, **102** (A12), 27,083–27,098, doi:10.1029/97JA02518.

Omura, Y., N. Furuya, and D. Summers (2007), Relativistic turning acceleration of resonant electrons by coherent whistler mode waves in a dipole magnetic field. *J. Geophys. Res.*, **112**, A06236, http://dx.doi.org/10.1029/ 2006JA012243.

Omura, Y., Y. Katoh, and D. Summers (2008), Theory and simulation of the generation of whistler-mode chorus, *J. Geophys. Res.*, **113**, A04223, doi: 10.1029/2007JA012622.

Omura, Y., M. Hikishima, Y. Katoh, D. Summers, and S. Yagitani (2009), Nonlinear mechanisms of lower-band and upper-band VLF chorus emissions in the magnetosphere, *J. Geophys. Res.*, **114**, A07217, doi: 10.1029/2009JA014206.

Omura, Y., and D. Nunn (2011), Triggering process of whistler mode chorus emissions in the magnetosphere, *J. Geophys. Res.*, **116**, A05205, doi:10.1029/2010JA016280.

Roederer, J. G. (1970), *Dynamics of Geomagnetically Trapped Radiation*, Springer, New York.

Santolík, O., and D. A. Gurnett (2003), Transverse dimensions of chorus in the source region, *Geophys. Res. Lett.*, **30** (2), 1031, doi:10.1029/2002GL016178.

Santolík, O., D. A. Gurnett, J. S. Pickett, M. Parrot, and N. Cornilleau-Wehrlin (2004), A microscopic and nanoscopic view of storm-time chorus on 31 March 2001, *Geophys. Res. Lett.*, **31**, L02801, http://dx.doi.org/10.1029/2003GL018757.

Santolík, O., J. S. Pickett, D. A. Gurnett, J. D. Menietti, B. T. Tsurutani, and O. Verkhoglyadova (2010), Survey of Poynting flux of whistler mode chorus in the outer zone, *J. Geophys. Res.*, **115**, A00F13, doi: 10.1029/2009JA014925.

Santolík, O., E. Macúšová, I. Kolmašová, N. Cornilleau-Wehrlin, and Y. de Conchy (2014), Propagation of lower-band whistler-mode waves in the outer Van Allen belt: Systematic analysis of 11 years of multi-component data from the Cluster spacecraft, *Geophys. Res. Lett.*, **41**, 2729–2737, doi:10.1002/2014GL059815.

Shklyar, D., and H. Matsumoto (2009), Oblique whistler mode waves in the inhomogeneous magnetospheric plasma: resonant interactions with energetic charges particles, *Surv. Geophys.*, **30**, 55–104, doi: 10.1007/s10712-009-9061-7.

Storey, L. R. O. (1953), An investigation of whistling atmospherics, *Phil. Trans. Roy. Soc. London*, **A246**, 113–141.

Summers, D., R. M. Thorne, and F. Xiao (1998), Relativistic theory of wave-particle resonant diffusion with application to electron acceleration in the magnetosphere, *J. Geophys. Res.*, **103**, 20,487–20,500.

Summers, D., et al. (2002), Model of the energization of outer-zone electrons by whistler-mode chorus during the October 9, 1990 geomagnetic storm, *Geophys. Res. Lett.*, **29** (24), 2174, doi:10.1029/2001GL016039.

Tao, X., J. Bortnik, J. M. Albert, K. Liu, and R. M. Thorne (2011b), Comparison of quasilinear diffusion coefficients for parallel propagating whistler mode waves with test particle simulations, *Geophys. Res. Lett.*, **38**, L06105, doi:10.1029/2011GL046787.

Tao, X., R. M. Thorne, W. Li, B. Ni, N. P. Meredith, and R. B. Horne (2011a), Evolution of electron pitch-angle distributions following injection from the plasma sheet, *J. Geophys. Res.*, **116**, A04229, doi:10.1029/2010JA016245.

Tao, X., J. Bortnik, J. M. Albert, and R. M. Thorne (2012), Comparison of bounce-averaged quasi-linear diffusion coefficients for parallel propagating whistler mode waves with test particle simulations, *J. Geophys. Res.*, **117**, A10205, doi:10.1029/2012JA017931.

Tao, X., J. Bortnik, J. M. Albert, R. M. Thorne, and W. Li (2013), The importance of amplitude modulation in nonlinear interactions between electrons and large amplitude whistler waves, *J. Atmos. Sol.-Terr. Phys.*, **99**, 67–72.

Thorne, R. M. (2010), Radiation belt dynamics: The importance of wave-particle interactions, *Geophys. Res. Lett.*, **37**, L22107, doi:10.1029/2010GL044990.

Thorne, R. M., B. Ni, X. Tao, R. B. Horne, and N. P. Meredith (2010), Scattering by chorus waves as the dominant cause of diffuse auroral precipitation, *Nature*, **467**, 943–946, doi:10.1038/nature09467.

Thorne, R. M., et al. (2013), Rapid local acceleration of relativistic radiation belt electrons by magnetospheric chorus, *Nature*, Paper #2013-08-11553B.

Trakhtengerts, V.Y. (1999), A generation mechanism for chorus emissions. *Ann. Geophys.*,**17**, 95–100.

Tsurutani, B. T., and E. J. Smith (1974), Postmidnight chorus: A substorm phenomenon, *J. Geophys. Res.*, **79** (1), 118–127, doi:10.1029/JA079i001p00118.

Tsurutani, B. T., and E. J. Smith (1977), Two types of magnetospheric ELF chorus and their substorm dependencies, *J. Geophys. Res.*, **82** (32), 5112– 5128, doi:10.1029/JA082i032p05112.

Turner, D. L., V. Angelopoulos, Y. Shprits, A. Kellerman, P. Cruce, and D. Larson (2012), Radial distributions of equatorial phase space density for outer radiation belt electrons, *Geophys. Res. Lett.*, **39**, L09101, doi: 10.1029/2012GL051722.

Walt, M. (1994), *Introduction to geomagnetically trapped radiation*, Cambridge University Press, Cambridge.

Walker, A. D. M. (1993), *Plasma waves in the magnetosphere*, Springer-Verlag, New York, 1993.

Yoon, P.H. (2011), Large-amplitude whistler waves and electron acceleration. *Geophys. Res. Lett.*, **38**, L12105, http://dx.doi.org/10.1029/2011GL047893.

Zaslavsky, G.M. (1985), *Chaos in Dynamic Systems*, Harwood Academic Publishers, New York.

9

Wave-Driven Diffusion in Radiation Belt Dynamics

Richard B. Horne, Nigel P. Meredith, Sarah A. Glauert, and Tobias Kersten

■ Abstract

The discovery of the van Allen radiation belts in 1958 posed two important problems: how can charged particles be accelerated to energies of several MeV or more, and what is responsible for the variability of the belts? Here we discuss the electron radiation belts. It has now been established that electron acceleration to form the belts must occur inside geostationary orbit by a local acceleration process. We describe electron acceleration and loss as a diffusion process driven by wave-particle interactions and present some of the most important types of magnetospheric plasma wave that can contribute to radiation belt variations. We present the results of simulations which solve the 3-d Fokker Planck equation where the diffusion rates are based on careful modelling of satellite wave data. The results show that radial diffusion alone cannot reproduce the variability of electron radiation belts, but losses due to plasmaspheric hiss and lightning generated whistlers are also required. Even so, the flux is too low. We show that whistler mode chorus waves increase the electron flux much closer to the observed levels and are a very important local electron acceleration process. We also show that EMIC waves cause electron loss at energies of several MeV and predict that EMIC waves should result in an anisotropic pitch angle distribution peaked at large pitch angles at energies of several MeV. Finally we describe our current understanding in

Waves, Particles, and Storms in Geospace. First Edition. Georgios Balasis, Ioannis A. Daglis, Ian R. Mann.

terms of a conceptual model for the formation of the Earth's electron radiation belts where local acceleration is provided by wave-particle interactions.

9.1 Introduction

The Earth's radiation belts were discovered in 1958 by James van Allen using Explorer I, the first U.S. satellite [Van Allen, 1958]. They consist of high energy charged particles trapped inside the Earth's magnetic field and which drift around the Earth. There is one proton belt which lies between $L = 1.2$ and $L = 2$ with a peak near $L = 1.4$ (where L is the equatorial crossing of the dipole magnetic field line in Earth radii). Proton energies can exceed 400 MeV. The proton belt is generally quite stable and is believed to be formed by the decay of cosmic rays backscattered from the atmosphere, a process known as cosmic ray albedo neutron decay or CRAND [Singer, 1958]. There are two electron belts consisting of an inner belt between $1.2 < L < 2.5$ and an outer belt which extends from approximately $L = 3.5$ to beyond geostationary orbit. Electron energies range from a few hundred keV to several MeV. Between the two electron belts is the slot region where the high energy electron flux is depleted due to wave-particle interactions [Lyons and Thorne et al., 1973]. The inner belt is relatively stable but the flux in the outer belt can vary rapidly, by up to five orders of magnitude on a timescale of several hours to a few days, but which can be as fast as 2 minutes. In this chapter we focus on the electron belts. For work on the proton belt see, for example, Selesnick et al. [2007] and for more information on the history of the radiation belts see Chapter 1 by Baker.

Two of the most important problems in radiation belt physics are to understand how electrons are accelerated up to such high energies, and what is responsible for the variations in the electron flux and the location of the belts. Research shows that the acceleration must take place inside the Earth's magnetosphere [Li et al., 1997] in a multi-step process [Baker et al., 1998], and that in effect, the magnetosphere acts as a giant particle accelerator.

The high altitude nuclear tests conducted by the U.S.A. and the U.S.S.R. in the late 1950s and early 1960s provided some of the earliest evidence that wave-particle interactions play a major role in radiation belt dynamics. Following a high altitude nuclear detonation fission fragments were injected across the magnetic field and decayed by beta decay to form a new electron radiation belt. The energy spectrum extended from less than 0.25 MeV to more than 7 MeV [Hess, 1963]. For $L < 1.25$ and energies of 2 MeV, the new radiation belt decayed with a lifetime consistent with losses due to collisions with atmospheric gasses [Walt, 1964]. The lifetime increased to approximately a year near $L = 1.5$ reducing to a month between $2.5 < L < 3.5$ [Roberts, 1969]. At these higher altitudes the lifetimes were much shorter than that due to collisions, indicating that other loss processes must be acting. As radial transport is very slow at low L electron diffusion into the loss cone via wave-particle interactions was established as the major loss process [e.g. Dungey, 1963; Kennel and Petschek, 1966]. The artificial belt formed by the starfish test was still observable some 6 years later for $L < 1.7$ [West and Buck, 1976].

Several theories were put forward to explain the origin of the electron radiation belts but the one that emerged in the 1970s involved electrons gaining access into the magnetosphere, possibly via magnetic reconnection in the magnetotail, and then being transported towards the planet via radial diffusion. Fluctuations in the Earth's magnetic field and electrostatic potential field would drive the diffusion [Falthammar, 1965]. As the electrons were diffused towards the planet, and hence higher magnetic field, conservation of the first two adiabatic invariants would result in an increase in the electron energy. Closer to the Earth, and inside the high density plasmasphere between $L = 2.5$–3.5, wave-particle interactions with very low frequency waves known as plasmaspheric hiss would break the first two adiabatic invariants causing pitch angle scattering into the loss cone and loss to the atmosphere. This model was able to explain the quiet time structure of the radiation belts [Lyons and Thorne, 1973].

It is now recognized that the outer radiation belt is highly dynamic, for example, the relativistic electron flux can vary by a factor of 10^5 on a timescale of a few minutes [Baker et al., 2007] and in 1991 a new radiation belt was formed in less than 2 minutes during the passage of an interplanetary shock [Blake et al., 1992]. However, more common are the variations observed during geomagnetic storms and fast solar wind streams. During a geomagnetic storm, the electron flux near geosynchronous orbit usually drops rapidly over a few hours and then increases by more than a factor of 2 above the pre-storm value over a period of 1–2 days in about 53% of cases. In contrast 19% cause a net loss by more than a factor of 2, and in the rest of cases the flux varies by less than a factor of 2 [Reeves et al., 2003]. Some of the largest increases in the radiation belts are associated with fast solar wind streams [Paulikas and Blake, 1979] and periods of long duration substorm activity lasting for several days associated with the passage of co-rotating interaction regions [Tsurutani et al., 2006].

The early ideas on the formation of the radiations belts have now been revised in view of the highly dynamic nature of the outer belt. Satellite measurements [Green and Kivelson, 2004; Iles et al., 2006; Chen et al., 2007], and now data from the NASA Van Allen Probes mission [Reeves et al., 2013] show evidence for a peak in the electron phase space density inside geostationary orbit, typically near $L = 5.5$. It is therefore difficult to see how the idea of inward radial diffusion from a source in the outer magnetosphere can account for the peak since the observed gradient in phase space density should transport electrons outwards from $L = 5.5$. Furthermore, the time evolution of the peak is such that it cannot be formed by a mixture of inward diffusion and loss in the outer region. Thus the formation of the electron radiation belts requires local acceleration inside geostationary orbit.

Radial diffusion is still a very important transport process and must be included in radiation belt dynamics. Coronal mass ejections and fast solar wind streams flowing along the boundary of the magnetosphere can excite ULF waves inside the magnetosphere and increase radial diffusion substantially [Elkington et al., 1999; Mathie and Mann, 2000].

However, radial diffusion may act as a net loss process by diffusing high energy electrons outwards from the peak in phase space density towards the magnetopause boundary where they are lost on to open drift paths [Shprits et al., 2006]. Inward radial diffusion from $L = 5.5$ is still possible and may also be very important for filling the slot region during magnetic storms.

While wave-particle interactions were generally accepted as an important loss process for the radiation belts, it was only in the late 1990s that it was shown that they could also cause substantial electron acceleration via Doppler-shifted cyclotron resonance [Summers et al., 1998; Horne et al., 1998]. In this process waves with frequencies less than the electron cyclotron frequency are Doppler shifted up to the electron cyclotron frequency by the relative motion of the waves and electrons travelling along the magnetic field. Resonance enables the efficient exchange of energy between the waves and particles and, for a sufficiently broad band of waves, diffusion of electrons into the loss cone at small pitch angles and acceleration of electrons at large pitch angles which remain trapped in space. Using the average observed wave power whistler mode chorus waves can increase the relativistic electron flux $>$ 1 MeV by a factor of 10 or more on a timescale of 18–24 hours [Horne et al., 2005a,b]. Furthermore, global simulations showed that these waves could produce a peak in the electron phase space density inside geostationary orbit [Varotsou et al., 2005, 2008; Shprits et al., 2009; Albert et al., 2009; Fok et al., 2008]. The discovery of a third radiation belt, or storage ring, by the Van Allen Probes satellites [Baker et al., 2013] could only be explained by a mixture of acceleration and loss, by wave-particle interactions and transport by radial diffusion [Shprits et al., 2013; Thorne et al., 2013].

Wave-particle interactions have now been established as a very important local acceleration process that operates in the heart of the radiation belts and which fulfils many of the observations. However, there are still many unknown factors regarding wave-particle interactions. For example, there are at least five types of waves that could contribute to electron acceleration and six types that could contribute to electron loss [Horne and Thorne, 1998]. Acceleration and loss processes depend on the spatial distribution of the waves in latitude, magnetic local time and L, their spectral properties and direction of propagation. These in turn depend on plasma properties such as the plasma density, magnetic field strength and the source electron and ion velocity distributions which govern the growth and damping of the waves and depend critically on geophysical conditions driven by the solar wind interaction with the magnetosphere. Thus understanding the radiation belts is very difficult. It has often been said that no two magnetic storms are the same: the complex chain of dependences may partly explain the reason why. To make progress one must study many different events where some processes dominate over others so that we can build up an understanding of the detailed physical processes and the relative contribution of each.

In this chapter we present a survey of some of the most important types of waves that cause electron acceleration and loss in the radiation belts. We describe how they can interact with high energy electrons causing loss and acceleration, and how the average wave properties can be used in global radiation belt models to simulate the variations in the outer electron belt over a period of several days. We illustrate some of the effects of electron acceleration, transport by radial diffusion and loss caused by whistler mode chorus, plasmaspheric hiss and EMIC waves.

9.2 Magnetospheric plasma waves

High energy radiation belt electrons drift around the Earth under the influence of gradient and curvature drift caused by the Earth's magnetic field. Typically it may take 15 minutes or so for a 1 MeV electron at $L = 4$ to complete an orbit. As they do so they encounter a variety of different plasma waves that can lead to wave-particle interactions resulting in electron acceleration and or loss. Four of the most important types of waves are illustrated in Figure 9.1. As the figure illustrates, the waves are usually only present for part of the electron drift orbit. We now discuss each wave mode in turn where the frequencies are related to the local electron gyro-frequency $f_{ce} = |e|B/(2\pi m_e)$, the local electron plasma frequency $f_{pe} = (ne^2/(2\pi m_e \epsilon_0))^{1/2}$ and the lower hybrid resonance frequency $f_{LHR} \approx f_{ce}/43$, and where the location is given in terms of L^* rather than L, which is a coordinate system based on the drift path of electrons around the Earth [Roederer, 1970]. In a dipole magnetic field $L^* = L$.

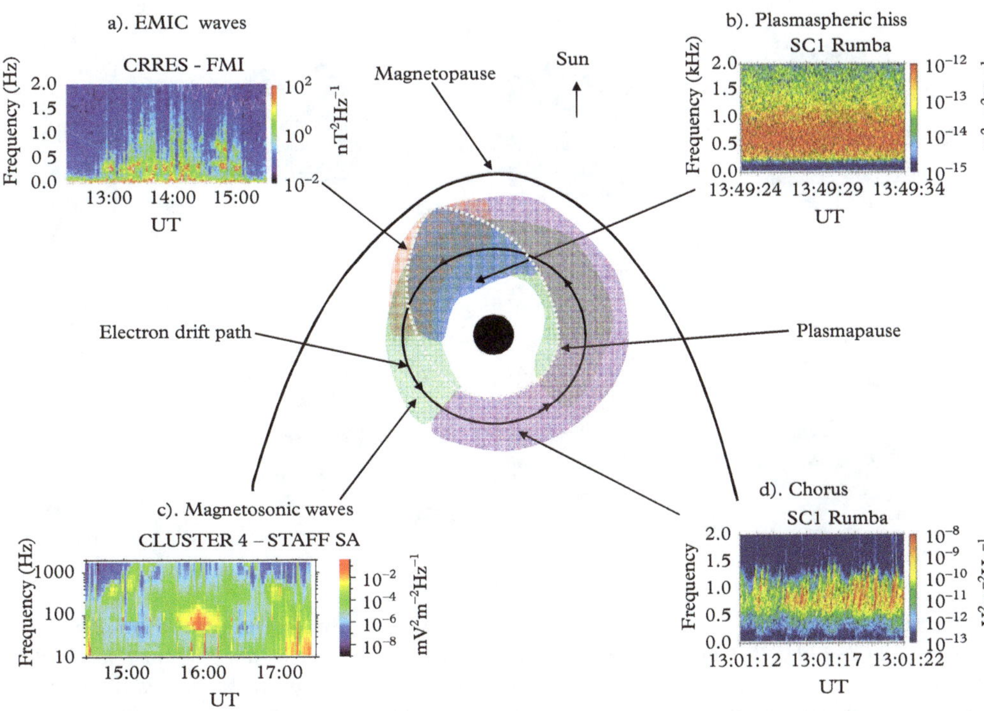

Figure 9.1 *Schematic diagram of the spatial distribution of magnetospheric waves that contribute to electron acceleration and loss in the radiation belts.*

9.2.1 Plasmaspheric hiss

Plasmaspheric hiss (Figure 9.1b) is a broadband, structureless, electromagnetic emission that occurs below f_{ce} in the frequency range from approximately 100 Hz to several kHz, with peak power near a few hundred Hz. The waves, as their name implies, are generally observed throughout the higher density regions associated with the Earth's plasmasphere and plasmaspheric plumes where $f_{pe}/f_{ce} > 1$.

The global distribution of equatorial plasmaspheric hiss in the frequency range 100 Hz to 2 kHz inside the plasmasphere is shown in the bottom panels of Figure 9.2 as a function of L^* and MLT for, from left to right, quiet (AE < 100 nT), moderate (100 < AE < 300 nT) and active (AE > 300 nT) conditions. These maps, which were constructed from a database of VLF waves from seven satellites using techniques described by Meredith et al. [2012], extend linearly out to L^* = 10 with noon at the top and dawn to the right. The emissions persist during quiet conditions (bottom left) and

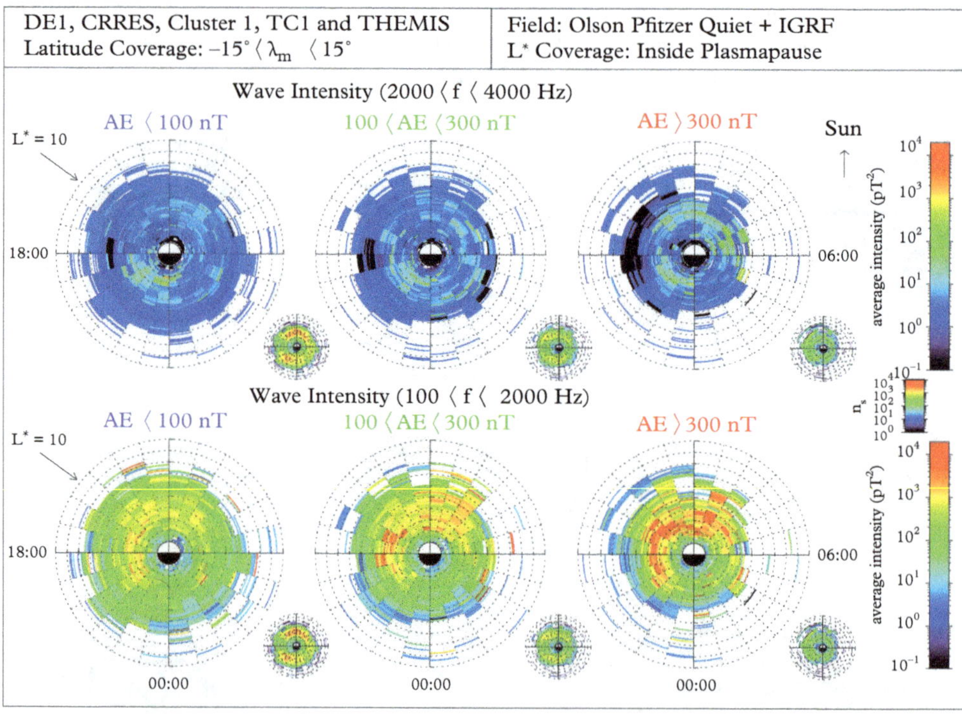

Figure 9.2 *Global model of equatorial plasmaspheric hiss in the frequency range 100 Hz–2 kHz (bottom) and 2–4 kHz (bottom) as a function of L* and MLT, for, from left to right, quiet, moderate and active conditions. The plots extend linearly out to L* = 10 with noon at the top and dawn to the right. The average intensities are shown in the large panels and the corresponding sampling distributions in the small panels.*

extend over most local times. The strongest emissions have band integrated intensities of the order 2000 pT^2 and occur during active conditions from 09–21 MLT over a range of L^* shells ($2 < L^* < 4$). At higher frequencies, $2 < f < 4$ kHz (top panels), the average intensity of plasmaspheric hiss is much weaker and typically about two orders of magnitude lower.

There have been a number of different theories to explain the origin of plasmaspheric hiss. Until recently the two leading candidates were *in situ* growth of wave turbulence in space [Thorne et al., 1973] and lightning-generated whistlers [Sonwalkar and Inan, 1989]. However, the former could not achieve the observed power levels [Church and Thorne, 1983] and the latter could not explain the lack of correlation between plasmaspheric hiss below 2 kHz and land mass [Meredith et al., 2006a], where more lightning occurs, or the geomagnetic control of plasmaspheric hiss [Meredith et al., 2004]. More recently, Bortnik et al. [2008], using ray tracing modelling, showed that plasmaspheric hiss could originate from bursts of short duration (~0.1s) chorus emissions which are excited outside the plasmapause and then propagate into and become trapped inside the plasmapause. Many of the observational characteristics of plasmaspheric hiss are reproduced by this new model, including the observed frequency band, the geomagnetic control and the day-night asymmetry in intensity. On further investigation, the new model was found to underestimate the typical plasmaspheric hiss power by 10–20 dB [Chen et al., 2012a; 2012b] but this deficiency was resolved when cyclotron resonant wave growth inside the plasmapause was included in the modelling [Chen et al., 2012c]. A number of observational studies have recently been conducted which support the new theory including case studies of a strong correlation between the amplitude modulation of chorus and plasmaspheric hiss on two THEMIS spacecraft [Bortnik et al., 2009] and two Cluster spacecraft [Wang et al., 2011] and dayside observations of downward propagating chorus and upward propagating plasmaspheric hiss by the Polar spacecraft [Tsurutani et al., 2012]. There is also statistical evidence for chorus extending to high latitudes in the pre-noon sector and, in the equatorial region, there is a clear gap of the order 1–2 Earth radii between plasmaspheric hiss at $L < 4$ and chorus further out [Meredith et al., 2013], consistent with ray tracing modelling from a chorus source.

Plasmaspheric hiss emissions above 2 kHz have been attributed to lightning-generated whistlers [e.g. Green et al., 2005; Meredith et al., 2006a] and, indeed, the emissions in the frequency range 2–4 kHz on the nightside, between dusk and midnight show no substorm dependence (Figure 9.2 top panels) and coincide with the region where lightning rates are high and D-region absorption is low, suggesting that these emissions are generated by lightning. However, on the dayside, the emissions in the frequency range 2–4 kHz are substorm dependent and have a similar distribution to plasmaspheric hiss which suggests that these emissions may be the weaker, higher frequency component of hiss from chorus.

Plasmaspheric hiss, ground-based VLF transmitters and lightning-generated whistlers cause substantial electron loss from the radiation belts inside the plasmapause [Abel and Thorne, 1998a, b]. Figure 9.3 (taken from Meredith et al. [2009]) shows a comparison between the electron lifetime for $L = 2$–3 and 2–6 MeV electrons measured by

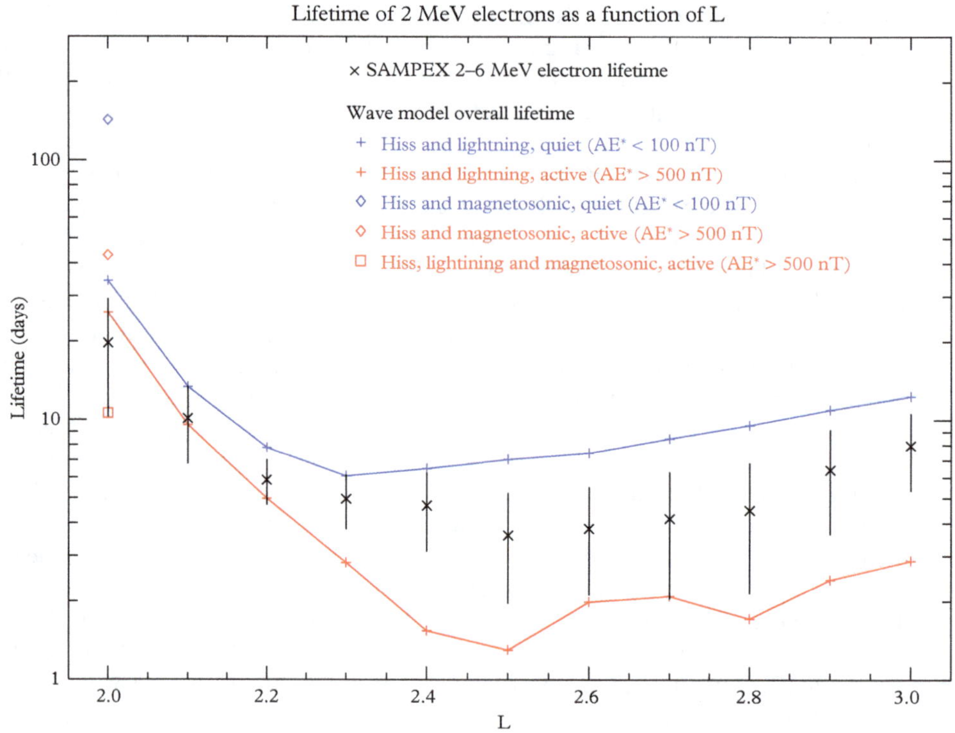

Figure 9.3 *The observed lifetimes (black) of 2–6 MeV electrons measured by SAMPEX as a function of L with the loss timescales for the combined spectrum of hiss and guided whistlers for (gray) quiet and (light gray) active conditions. The lifetimes for the combined spectrum of hiss and magnetosonic waves are shown at L = 2 for quiet and active conditions, and for the combined spectrum of hiss, lightning and magnetosonic waves for active conditions.*
Meredith et al., 2009. © John Wiley & Sons 2009

SAMPEX and the lifetime computed from pitch angle diffusion into the loss cone using the observed wave power. The lifetime is much shorter during active geomagnetic conditions as the wave power is higher. Note also that magnetosonic waves, discussed below, can also help contribute to loss when they are present with plasmaspheric hiss. The effects of magnetosonic waves are only shown at $L = 2$.

Resonant pitch angle scattering by plasmaspheric hiss largely accounts for the formation of the slot region between the inner and outer radiation belts [Lyons and Thorne, 1973; Meredith et al., 2007; 2009]. It also contributes to the loss of outer radiation belt electrons during the main and recovery phases of geomagnetic storms [Summers et al., 2007] and the quiet time decay of energetic electrons in the outer radiation belt [Meredith et al., 2006b; Summers et al., 2007].

9.2.2 Chorus

Chorus waves (Figure 9.1d) are usually characterized by short duration bursts of radiation below f_{ce} which rise or fall rapidly in frequency. They may only last a few milliseconds but they often overlap and recur repeatedly for many hours [Santolík et al., 2003]. Chorus is often observed in two frequency bands with a gap at $0.5f_{ce}$ [Tsurutani and Smith, 1974], separating the emissions into so-called lower band ($0.1f_{ce} < f < 0.5f_{ce}$) and upper band ($0.5f_{ce} < f < f_{ce}$) chorus. The waves are generated outside the plasmapause near the magnetic equator [LeDocq et al., 1998; Lauben et al., 2002] by cyclotron resonant interaction with suprathermal electrons [Katoh and Omura, 2007; Omura et al., 2009] injected into the inner magnetosphere during storms and substorms. While the initial phase of chorus excitation is linear the discrete nature of the emissions and the short duration rising frequency tones indicate that the generation is highly nonlinear. Observations show that some of the emissions can reach peak amplitudes of 240 mV m^{-1} or more [Cattell et al., 2008] which also suggests that nonlinear effects are very important.

Measurements of the wave normal angle, i.e. the angle between the k vector of the wave and the Earth's magnetic field, show that the wave normal angle is very small close to the equator in the source region, typically less than 15° [Agapitov et al., 2013; Li et al., 2011]. This suggests that the chorus waves tend to propagate along the magnetic field. However, the distribution of angles can be much broader at higher latitudes indicating propagation across the magnetic field. The Poynting energy flux is directed away from the magnetic equator [Santolík et al., 2010] indicating that the source is close to the magnetic equator. Chorus waves are sometimes observed on the ground in the polar regions [e.g. Smith et al., 2004] indicating the propagation along the whole field line is possible.

The global distribution of chorus during active conditions (AE> 300 nT) is shown as a function of frequency, magnetic latitude, MLT and L^* in Figure 9.4. These maps were constructed from a database of VLF waves using data from Dynamics Explorer 1, Cluster 1, Double Star TC1 and THEMIS A, D & E (see Meredith et al., [2012] for further details). Chorus is strongest and most extensive in the lower band in the equatorial region (Figure 9.4d), where band integrated waves intensities of the order 2000 pT^2 are observed in the region $4 < L^* < 9$ from 2300 to 1200 MLT. Strong lower band chorus is also seen at mid latitudes (Figure 9.4e) but these emissions are restricted to the dayside between 0700 and 1400 MLT, where Landau damping is weaker. Upper band chorus (Figure 9.4c–e) is confined to the equatorial region and is both weaker and less extensive than equatorial lower band chorus with peak intensities of the order a few hundred pT^2 between midnight and 1100 MLT in the region $3 < L^* < 7$. Near the magnetic equator there is little chorus power below $0.1f_{ce}$ compared to that in the lower band range $0.1f_{ce} < f < 0.5f_{ce}$. However, as lower band chorus generated near the equator in the region 0700–1400 MLT propagates to higher latitudes significant wave power appears below $0.1f_{ce}$ due to the increasing magnetic field strength (Figure 9.4j). Indeed, at mid to high latitudes significant wave power, with intensities of the order a few hundred pT^2, is observed in the band $f_{LHR} < f < 0.1f_{ce}$ from 07 to 14 MLT in the

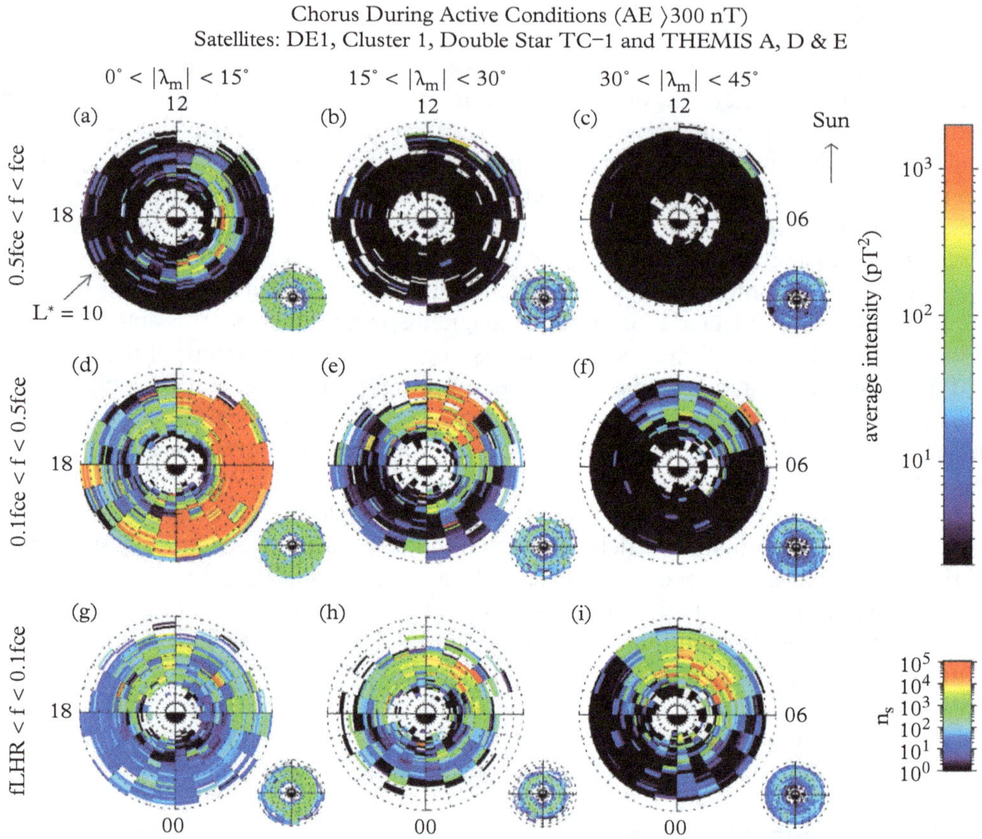

Figure 9.4 *Global model of (top row) upper band chorus, (middle row) lower band chorus and (bottom row) chorus between $f_{LHR} < f < 0.1f_{ce}$ as a function of L^* and MLT, for, from left to right, equatorial, mid and high latitudes. Each plot extends linearly out to $L^* = 10$ with noon at the top and dawn to the right. The average intensities are shown in the large panels and the corresponding sampling distributions in the small panels.*

region $30 < |\lambda_m| < 45°$ from $L^* = 4-8$ [Meredith et al., 2014]. This additional power, which is typically not included in radiation belt models, may have a significant effect on radiation belt dynamics.

Whistler mode chorus plays a major role in radiation belt dynamics contributing to both the acceleration and loss of relativistic electrons. For example, chorus waves are thought to be largely responsible for the gradual build-up of radiation belt electrons that occurs on a timescale of 1–2 days during the recovery phase of geoeffective storms [e.g. Horne et al., 2005a, b; 2006]. In contrast, storm-time chorus at high latitudes causes microburst precipitation and may lead to losses of radiation belt electrons on the timescale of the order of a day [Lorentzen et al., 2001; Thorne et al., 2005b]. At lower electron energies, pitch angle scattering by whistler mode chorus is largely responsible for both the diffuse aurora [Thorne et al., 2010] and the pulsating aurora [Nishimura et al., 2010].

9.2.3 EMIC waves

Electromagnetic ion cyclotron waves (EMIC) are waves which propagate at frequencies below the hydrogen ion gyrofrequency (Figure 9.1a). The frequency range of interest in the radiation belts varies according to the hydrogen ion gyrofrequency but is typically (0.1–5 Hz). The waves are generated by medium energy (1–100 keV) ring current ions injected into the inner magnetosphere during enhanced Earthward convection during geomagnetic storms and substorms [e.g. Cornwall et al., 1970; Kozyra et al., 1984]. EMIC waves are observed in three bands below the hydrogen, helium ion and oxygen ion gyrofrequencies, the excited band being controlled by the ion composition and anisotropy of the proton distribution [Kozyra et al., 1984; Horne and Thorne, 1994], the level of geomagnetic activity [Braysy et al., 1998] and the location with respect to the plasmapause [Fraser and Nguyen, 2001]. Hydrogen band EMIC waves, which are observed between the hydrogen and helium ion gyrofrequencies, are mainly found outside the plasmapause. In contrast, helium band EMIC waves, which are observed between the helium and oxygen ion gyrofrequencies, are observed both inside and outside the plasmapause. EMIC waves below the oxygen ion gyrofrequency, referred to as oxygen band EMIC waves, are rarely observed.

The global distribution of the number of helium band EMIC wave events for which the band integrated wave intensity is > 0.1 nT^2 is shown as a function of L^*, MLT and geomagnetic activity in the top panels of Figure 9.5. The number of events is shown in the large panels and the corresponding number of samples is shown in the smaller panels. Each plot extends linearly out to $L^* = 8$ with noon at the top and dawn to the right. The global distribution of the fractional occurrence of these events is in the same format in the bottom panels. The helium band EMIC wave activity is substorm dependent, as measured by AE, with the greatest frequency of events during active conditions with waves present approximately 4% of the time in the noon to dusk sector over a range of L* shells ($4 < L^* < 7$).

EMIC waves are able to resonate with highly relativistic electrons [Lyons and Thorne, 1972; Horne and Thorne, 1998; Summers et al., 1998] causing pitch angle scattering and loss to the atmosphere [Kersten et al., 2014; Thorne and Andreoli, 1980]. In a survey of over 800 EMIC wave events observed by CRRES, Meredith et al. [2003b] found that while the minimum resonant energy was above 2 MeV for the majority of the events, it could fall below 2 MeV in localized regions of high plasma density and or low magnetic field ($f_{pe}/f_{ce} > 10$) for wave frequencies just below the hydrogen or helium ion gyrofrequencies. Since the average wave spectral intensity of these lower energy scattering events (4–5 nT^{2}Hz^{-1}) was sufficient to cause strong diffusion scattering it was inferred that EMIC waves could play an important role in the loss of radiation belt electrons.

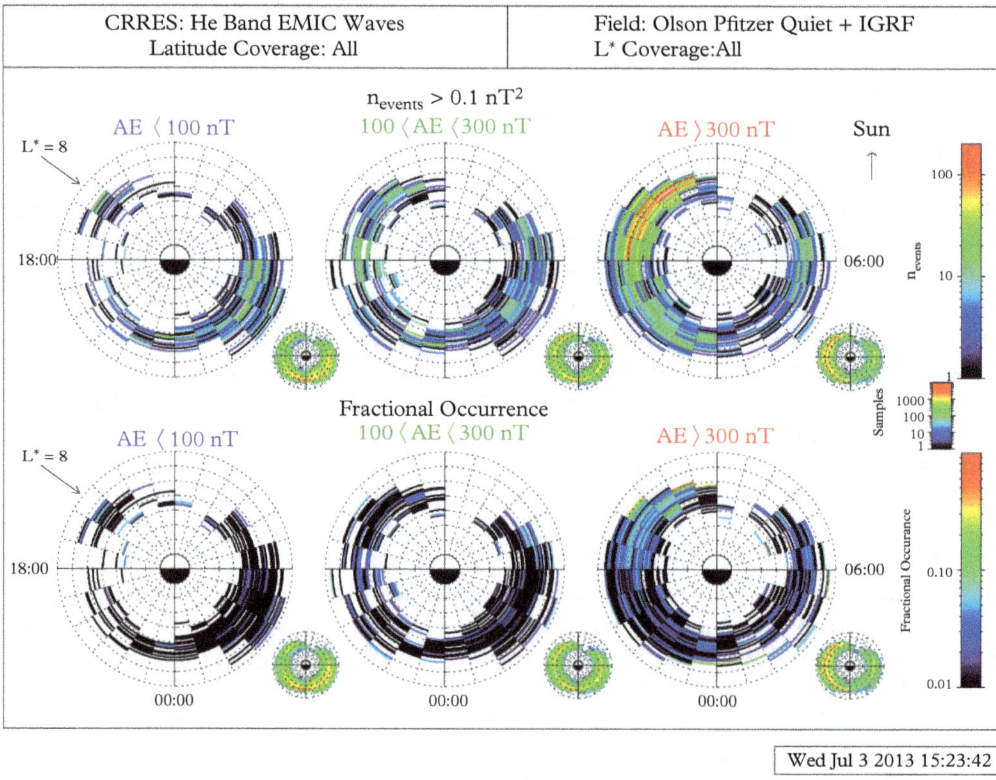

Figure 9.5 *Global model of the number of helium band EMIC wave events with band integrated intensities greater than 0.1 nT^2 (top) and their fractional occurrence (bottom) as a function of L^* and MLT, for, from left to right, quiet, moderate and active conditions. The corresponding sampling distributions are shown in the small panels. Each plot extends linearly out to $L^* = 8$ with noon at the top and dawn to the right.*

9.2.4 Magnetosonic waves

Magnetosonic waves (Figure 9.1c) are a natural, often intense electromagnetic emission observed in the inner magnetosphere at frequencies between the proton gyrofrequency and the lower hybrid resonance frequency. The waves are observed between $L = 2$ to 8, both inside and outside the plasmapause, and are largely confined to the magnetic equatorial region [Meredith et al., 2008]. Magnetosonic waves are thought to be generated by proton ring distributions at energies of the order 10 keV [Boardsen et al., 1992; Horne et al., 2000; Chen et al., 2011]. It has recently been suggested that these waves can lead to local acceleration and, in particular, may energise electrons from ~10 keV up to a few MeV in the outer radiation belt [Horne et al., 2007]. However, more work needs to be done to fully assess the role of these waves in radiation belt dynamics.

9.3 Global simulations

The dynamics of the Earth's radiation belts can be described by an equation based on the three adiabatic invariants [Schulz and Lanzerotti, 1974]. If there were no wave activity and no changes in the magnetic field then the adiabatic invariants would be conserved. However, for wave frequencies of the order of the cyclotron, bounce and drift frequencies of the electrons, which are typically kHz, Hz and mHz respectively, the invariants can be broken and the evolution of the particle distribution can be described by a diffusion equation.

As wave-particle interactions occur on time scales of milliseconds, some approximation is required to incorporate these effects into radiation belt models to simulate storms which may last several days. One of the most important approximations is quasi-linear theory, as used here. In this approximation the particles are assumed to be scattered by a series of small amplitude wave packets which have random phase so that the process is stochastic and can be treated as a diffusion problem. Particle trapping effects by large amplitude waves, or very narrow band waves, are omitted. Fully nonlinear wave-particle interactions are computationally expensive to simulate and can only be done for short periods of time. It is not yet possible to apply full nonlinear theory on a global scale lasting days. As shown below, quasi-linear theory provides a surprising amount of agreement with observations and a basic physical understanding of what controls the evolution of the radiation belts during geomagnetic storms.

Since plasmaspheric hiss is broad-band and has no frequency structure, quasi-linear theory is a good approximation. Quasi-linear theory is more questionable for chorus waves due to their bursty nature and high amplitudes. However, even here the non-uniformity of the magnetic field and the fact that electrons may interact with many chorus elements as they bounce between mirror points suggest that quasi-linear theory may not be as bad as one might expect. A comparison between quasi-linear and nonlinear theory and the regimes where they agree is given by Albert [2010].

The time evolution of the electron distribution function for radiation belt electrons can be written as [Subbotin and Shprits, 2009; Glauert et al., 2014]

$$\begin{aligned}\frac{\partial f}{\partial t} &= \frac{1}{G}\frac{\partial}{\partial \alpha_{eq}}\left[G\left(D_{\alpha_{eq}\alpha_{eq}}\frac{\partial f}{\partial \alpha_{eq}} + D_{\alpha_{eq}E}\frac{\partial f}{\partial E}\right)\right] \\ &+ \frac{1}{G}\frac{\partial}{\partial E}\left[G\left(D_{EE}\frac{\partial f}{\partial E} + D_{E\alpha_{eq}}\frac{\partial f}{\partial \alpha_{eq}}\right)\right] \\ &+ L^2\frac{\partial}{\partial L}\left[\frac{1}{L^2}D_{LL}\frac{\partial f}{\partial L}\right] - \left(\frac{f}{0.5\tau_b}\right)_{losscone}\end{aligned} \tag{9.1}$$

where

$$G = T(\alpha_{eq})\sin 2\alpha_{eq}(E+E_0)\sqrt{E(E+2E_0)} \tag{9.2}$$

$f(\mathcal{J}_1, \mathcal{J}_2, \mathcal{J}_3, t)$ is the bounce and drift averaged distribution function, $\mathcal{J}_1$, $\mathcal{J}_2$ and $\mathcal{J}_3$ are the three adiabatic invariants, E and E_0 are the electron energy and rest mass energy, α_{eq} is the equatorial pitch angle and τ_b is the electron bounce period. The derivatives with respect to L are for fixed $\mathcal{J}_1$ and $\mathcal{J}_2$ whereas the derivatives with respect to α_{eq} and E are for fixed L.

The first four terms on the right-hand side of (9.1) represent diffusion in pitch angle and energy due to wave-particle interactions, the fifth term represents transport across the magnetic field as a result of radial diffusion driven by ULF waves and the last term represents losses into the atmosphere for particles that are diffused into the loss cone. The timescale for loss is half the electron bounce period and is energy dependent.

9.3.1 Diffusion rates

As in any diffusion problem, a considerable amount of effort goes into calculating the diffusion coefficients, which here are pitch angle $D_{\alpha_{eq}\alpha_{eq}}$, energy D_{EE}, mixed pitch angle-energy $D_{\alpha_{eq}E}$ and radial diffusion D_{LL}. The radial diffusion coefficients have been calculated from theory [Brautigam and Albert, 2000], ground-based magnetic field measurements [Ozeke et al., 2012] and satellite observations [Lejosne et al., 2013]. They are scaled by the Kp index and have a remarkable level of agreement for the outer radiation belt. Only diffusion coefficients due to magnetic field component of ULF waves are used here [Brautigam and Albert, 2000].

Calculating the pitch angle and energy diffusion coefficients is more challenging. The formulation includes an integration over the power spectrum and the angular distribution of wave normal angles of waves in Doppler-shifted cyclotron resonance with the particles [Lyons and Williams, 1984]. This also requires a sum over all the most important cyclotron harmonic resonances. The diffusion rates must also be integrated over the bounce orbit of the particles and take into account the change in wave properties with latitude. Finally the diffusion rates must be averaged over the drift orbit of the particles and take into account the change in wave properties with MLT. This must be repeated for each type of wave and for different levels of geomagnetic activity.

One approach is to parameterize the wave properties and calculate the diffusion rates [e.g. Shprits et al., 2009]. A more detailed analysis based on multiple satellite observations has been given by Horne et al. [2013a]. In this method the diffusion rates were obtained by analysing wave data from 7 different satellites, identifying the different wave modes, converting the data into L^* using a magnetic field model and then binning the wave intensity in each frequency band into 18 bins in L^* between 1.5 and 10, 10 latitude bins between 0° and 60*, 8 three hour bins in MLT and 5 levels of Kp. The wave intensities were then fitted to obtain the characteristic frequency spectrum in every bin. The frequency spectra, along with a plasma density and a distribution of wave normal angles, peaked in the field aligned direction but with a Gaussian spread [Agapitov et al., 2013], were used to calculate the bounce and drift averaged diffusion rates. In this case every diffusion rate depends on geomagnetic activity through the Kp index. Thus a time sequence of Kp can be used to look up the diffusion rates and solve equation (9.1).

Figure 9.6 shows an example of the drift and bounce averaged pitch angle and energy diffusion rates for three types of waves. The diffusion rates are shown for Kp > 4 and selected values of L^*. At $L^* = 3$ (left column) pitch angle diffusion into the loss cone due to hiss increases for energies greater than 30 keV, and is generally larger than energy diffusion (bottom panel). Pitch angle diffusion is large near 90° and corresponds to Landau resonance. The strongest diffusion rates are for chorus waves (second column). Large pitch angle diffusion extends from 1 to 100 keV in energy and from small to large pitch angles. Thus chorus can diffuse electrons into the loss cone and contribute to electron loss. However, energy diffusion is also very significant particularly at energies of a few keV near the loss cone and at higher energies at large pitch angles. The usual interpretation is that as electrons are diffused into the loss cone they lose energy to the waves which provides wave growth. Equation (9.1) shows that the change in the distribution function with time depends on the gradients in the distribution function as well as the diffusion rates. Usually the distribution falls rapidly with energy and this enables chorus to accelerate a fraction of the electron population at large pitch angles to high energies and hence increase the flux at high energies in the radiation belts. The accelerated particles then remain trapped in space as pitch angle diffusion at high energies is small near the loss cone. Energy diffusion is much more efficient when the ratio of f_{pe}/f_{ce} is small, typically 3–4 [Horne et al., 2003; Thorne et al., 2013].

Figure 9.6 also shows the results for EMIC waves where the diffusion rates are due only to Helium band waves in the CRRES data which were found to be the strongest waves. They also depend on magnetic activity. EMIC waves cause significant pitch angle diffusion for energies greater than a few MeV whereas energy diffusion can be neglected. They cause loss of high energy electrons that may be accelerated to energies of several MeV. However, since the diffusion rates do not extend to large pitch angles it seems unlikely that they can remove all the electrons at high energies. Thus one might expect a peak in the electron distribution to remain at large pitch angles.

9.3.2 Comparison with data

A series of simulations were performed with the BAS radiation belt model to investigate the role of different diffusive processes. The simulation boundary conditions were $\partial f/\partial\alpha = 0$ at $\alpha = 0°$ and 90° and $f = 0$ at 10 MeV and above. The minimum energy boundary was set to 153 keV at $L^* = 5.55$, and followed a line of constant first invariant with L^*. At the minimum energy, and at $L^* = 2$ and 5.55, f was set according to data from CRRES. More details are given by Glauert et al. [2014].

Figure 9.7 shows a comparison between data from the CRRES satellite over a period of 100 days in 1990, and simulations results. The CRRES data (9.7a) are for an differential electron flux at 976 keV and a pitch angle of 90°. Only the slot region and the outer radiation belt are shown. The data illustrate some of the variability of the electron flux, in particular the large increase in flux near day 284. The Dst index (not shown) dropped to −135 nT indicating a magnetic storm at this time and the Kp and AE indices (9.7e) were both elevated. Note that there was a rapid drop in flux just before the increase.

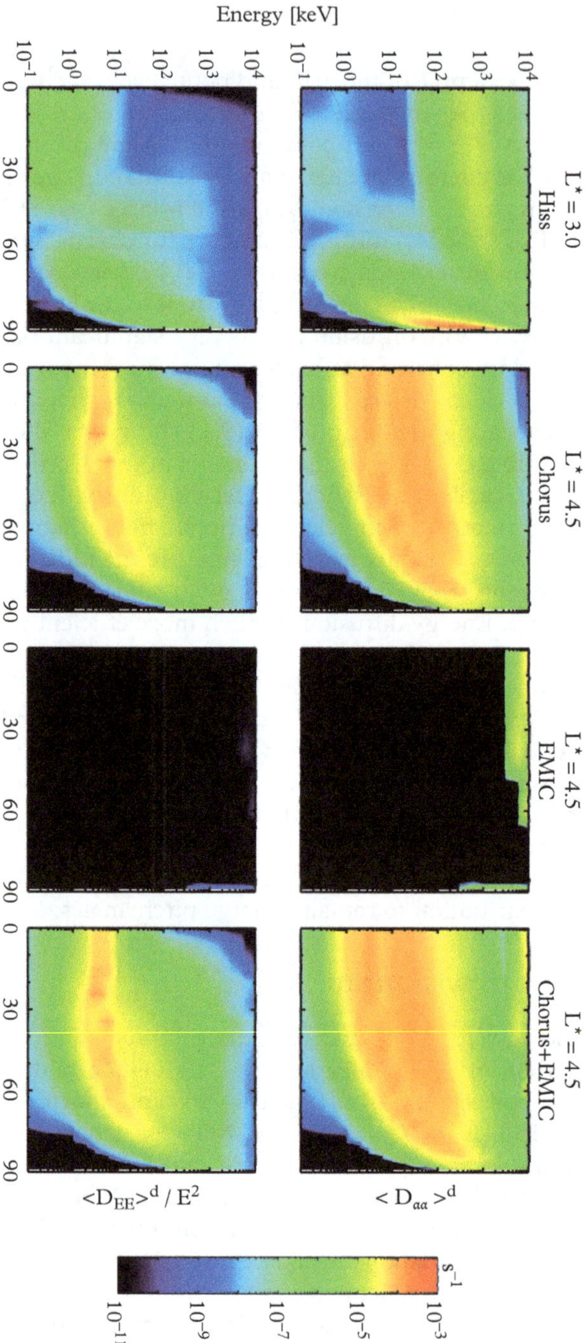

Figure 9.6 *Bounce and drift averaged electron diffusion rates as a function of energy and pitch angle for (top row) pitch angle diffusion and (bottom row) energy diffusion. The diffusion rates are shown for hiss, chorus, EMIC and combined chorus and EMIC waves at selected values of L^*. The diffusion rates are colour coded as shown on the right.*

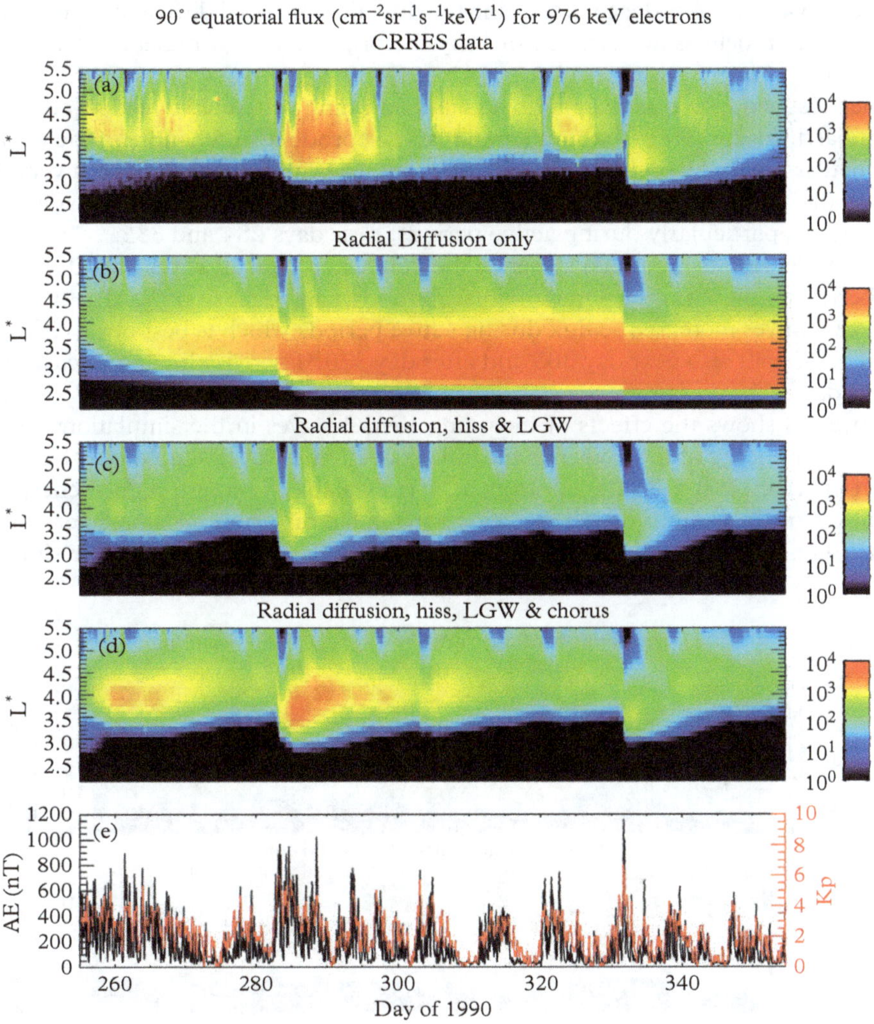

Figure 9.7 *Comparison between the electron flux measured by CRRES and simulation results from the BAS model as a function of L* and time at 976 keV and a pitch angle of 90°, for (a) CRRES data, (b) simulations for radial diffusion only, (c) simulations for radial diffusion, hiss and lightning generated whistlers, (d) simulations for radial diffusion, hiss, lightning generated whistlers and chorus, and (e) AE and Kp indices. The electron flux is colour coded as shown on the right.*

Simulations for radial diffusion alone (9.7b) could reproduce the increase in flux on day 284 but did not capture the variability of the radiation belts. Note that although radial diffusion produces a peak in the flux at fixed energy, this is not the same as a peak in electron phase space density which must be calculated for all L^* for constant adiabatic invariants in order to determine the direction of radial diffusion. The variability of the radiation belts is another reason why the original idea of radial diffusion alone is inadequate.

When losses due to plasmaspheric hiss and lightning generated whistlers are included with radial diffusion (9.7c) the simulation results reproduce much more of the variability. In particular the decay of the inner edge of the outer belt after day 305 has a very similar form to the observations. However, the electron flux is generally much smaller than observations, particularly during active periods, near days 284 and 335.

The model results are much closer to the data when chorus waves are added to plasmaspheric hiss, lightning-generated whistlers and radial diffusion (9.7d). In particular the chorus waves increase the flux during active periods which is best observed between day 285–295, but also near day 305 and after day 320. There is also an increase, but less so, near day 335.

Figure 9.8 shows the effects of including EMIC waves in the simulations, but for a much higher energy of 6 MeV since the effects of EMIC waves should only be observed at high energies. No CRRES electron data is available for these higher energies and so two sets of simulations are shown, one with radial diffusion, plasmaspheric hiss and chorus (9.8a), and the other with the addition of EMIC waves (9.8b). The simulations

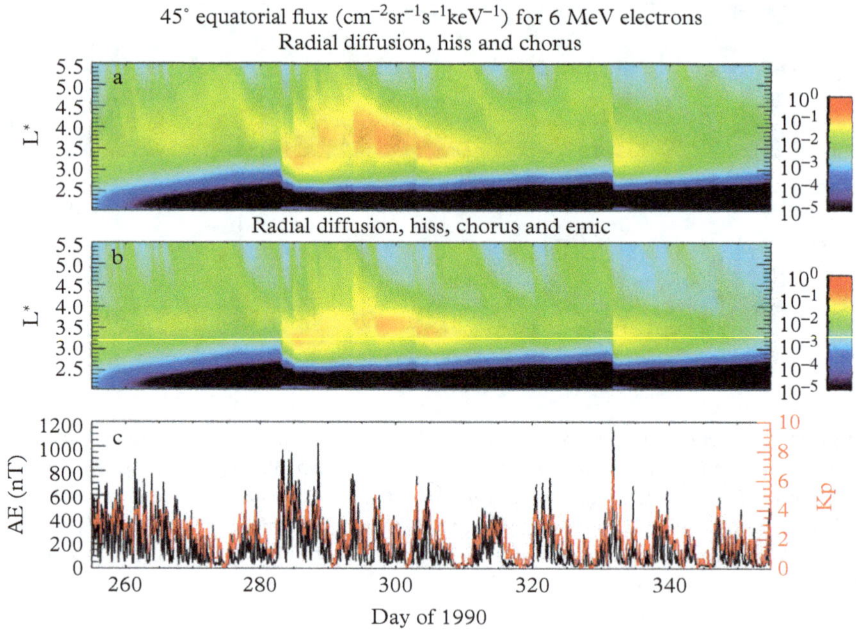

Figure 9.8 *Same as Figure 9.7 for the bottom three panels, but for 6 MeV and a pitch angle of 45°.*
Kersten et al. 2014. © John Wiley & Sons 2014.

are shown for a pitch angle of 45°. The EMIC waves are only present in the model for $L^* > 4$ due to the limited sensitivity of the wave instrument on CRRES from which these measurements are derived and the higher background field at lower L^*. However, a comparison between the results shows that the electron flux is significantly reduced by EMIC waves for $L^* > 4$ during active periods, just after day 260, between 285 and 294 and 310–330. There are little or no reductions in the flux at 90° (not shown). The results show that EMIC waves can be very effective in electron loss into the atmosphere at high energies.

Figure 9.9 illustrates the evolution of the pitch angle distribution at $L^* = 4.2$ at 1 and 6 MeV. At 1 MeV (top) the flux is depleted near the loss cone but there are no discernible effects of EMIC waves. At 6 MeV (bottom) EMIC waves deplete the electron flux between 0° and 50° leaving behind an anisotropic pitch angle distribution peaked near 90°. Thus we predict that EMIC waves should be correlated with anisotropic pitch angle distributions at energies of several MeV.

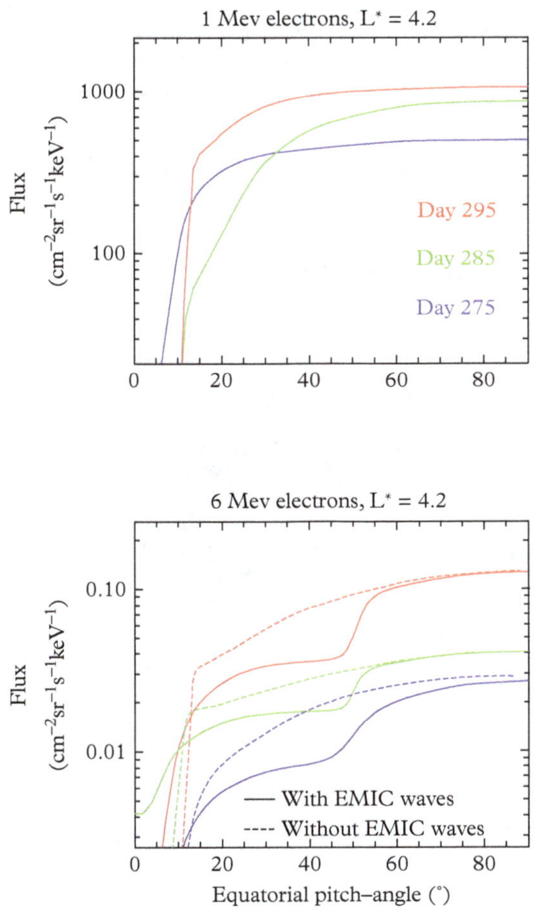

Figure 9.9 *Electron pitch angle distributions at $L^* = 4.2$ for (solid line) simulations with and (dashed) without the presence of EMIC waves at (top) 1 MeV and (bottom) 6 MeV.*

9.4 Discussion and conclusions

Our concept of how the radiation belts are formed and what causes the variability in the belts has changed considerably since the early 1970s. Figure 9.10, taken from Horne [2007], illustrates the new concept. Low energy electrons are transported towards the Earth from the magnetotail and plasma sheet via convective and inductive electric fields. Transport is particularly strong during substorms. These low energy electrons form a seed population of 1 to a few hundred keV electrons which becomes unstable due to an increasing loss cone and increasing temperature anisotropy as electrons are transported closer to the Earth. The unstable distribution excites whistler mode chorus waves which grow by diffusing low energy electrons into the loss cone but accelerate a fraction of the population to energies of several MeV. This process takes place from the plasmapause to geosynchronous orbit and beyond where the ratio of f_{pe}/f_{ce} is low, and forms a peak in electron phase space density near $L = 5.5$ at higher energies where wave acceleration is more effective. Radial diffusion diffuses the accelerated electrons away from the peak and works in concert with wave acceleration to fill up the whole radiation belts. During very active conditions and large magnetic storms whistler mode chorus and ULF waves are significantly enhanced. These waves can then drive enhanced diffusion and acceleration to fill up the slot region and transport electrons into the inner radiation belt. The slot region would subsequently decay by wave-particle interactions [e.g. Meredith et al., 2009].

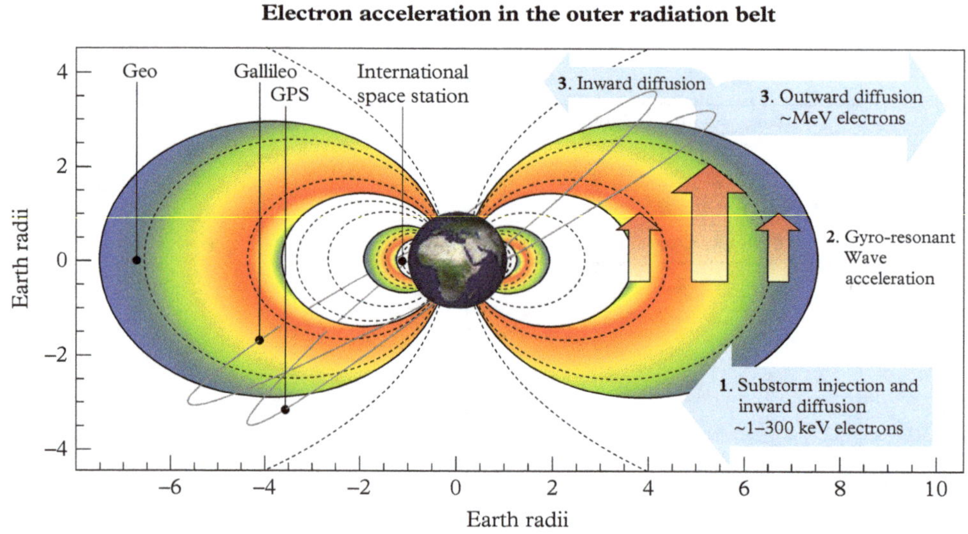

Figure 9.10 *General concept on the formation of the electron radiation belts (after Horne,* Nature Physics, *2007).*

Wave-particle interactions clearly play an important role in the loss and acceleration of radiation belt electrons, and contribute significantly to the variability of the belts. However, there are still several important areas where we need better understanding. For example, at the onset of a storm the electron flux is often observed to drop rapidly, known as a flux drop-out. It has been suggested that this is due to magnetopause compression and losses due to outward radial diffusion [Morley et al., 2010], or due to enhanced loss by EMIC waves [Thorne, 2010]. More generally understanding how the solar wind is coupled to the radiation belts is very important as ultimately the solar wind is the main driver. The formation of a radiation belt in 2 minutes by a fast CME is a clear example. Substorms play an important role leading to the excitation of wave-particle interactions, and there are still large uncertainties over how ULF waves interact with electrons, and how effective these waves are in the inner region. There are other questions over the importance of nonlinear wave-particle interactions and whether fully nonlinear effects would change wave scattering significantly so as to affect the radiation belts on a global scale. There are other types of waves that have hardly been assessed in terms of their effects, such as magnetosonic waves. There is also the question of how the ring current may be coupled to the radiation belts, and the importance of changes in the magnetic field, such as dipolarization effects. Missions such as the NASA Van Allen Probes and Japanese ERG satellite may help to answer these questions.

Finally, the radiation belts are not just an academic interest. High energy electrons are known to cause radiation damage to satellites and have been linked to internal charging and the loss of service, and occasional total loss of a satellite [Wrenn, 1995]. The prospect of a superstorm as big as the Carrington storm could significantly affect the radiation belts [Shprits et al., 2011] and lead to significant disruption to satellite services. During the 2003 Halloween storm about 10% of the entire satellite fleet was malfunctioning [Cannon et al., 2013]. Radiation belt models are now being used to forecast the radiation belts for space weather applications so as to help protect satellites on orbit [Horne et al., 2013b] (see for example the SPACECAST website www.fp7-spacecast.eu). Space weather forecasting is at an early stage, similar to that of weather forecasting in the 1960s. A better understanding of the basic physical processes will help improve forecasting and help protect space assets.

Acknowledgments

We thank Brian Fraser for providing data on EMIC waves from the CRRES satellite. The research leading to these results has received funding from the European Union Seventh Framework Programme (FP7/2007–2013) under grant agreement no. 262468 (SPACECAST), 606716 (SPACESTORM), and 284520 (MAARBLE). We also acknowledge financial support from the Natural Environment Research Council.

REFERENCES

Abel, B., and R. M. Thorne (1998a), Electron scattering loss in Earth's inner magnetosphere 1. Dominant physical processes, *J. Geophys. Res.,* **103**, 2385–2396.

Abel, B., and R. M. Thorne (1998b), Electron scattering and loss in Earth's inner magnetosphere 2. Sensitivity to model parameters, *J. Geophys. Res.,* **103**, 2397–2407.

Agapitov, O., et al. (2013), Statistics of whistler-mode waves in the outer radiation belt: Cluster STAFF-SA measurements, *J. Geophys. Res.,* **118**, 3407–3420, doi:10.1002/jgra.50312.

Albert, J. M., N. P. Meredith and R. B. Horne (2009), Three-dimensional diffusion simulation of outer radiation belt electrons during the October 9, 1990, magnetic storm, *J. Geophys. Res.,* **114**, A09214, doi:10.1029/2009JA014336.

Albert, J. M. (2010), Diffusion by one wave and by many waves, *J. Geophys. Res.,* **115**, A00F05, doi:10.1029/2009JA014732.

Baker, D. N., et al. (1998), Coronal mass ejections, magnetic clouds, and relativistic magnetospheric electron events: ISTP, *J. Geophys. Res.,* **103**, 17279–17291.

Baker, D. N., S. G. Kanekal, R. B. Horne, N. P. Meredith, and S. A. Glauert (2007), Low-altitude measurements of 2–6 MeV electron trapping lifetimes at $1.5 \leq L \leq 2.5$, *Geophys. Res. Lett.*, **34**, L20110, doi:10.1029/2007GL031007.

Baker, D. N., et al. (2013), A Long-Lived Relativistic Electron Storage Ring Embedded in Earth's Outer Van Allen Belt, *Science,* **340**, 186–190, doi: 10.1126/science.1233518.

Blake, J. B., W. A. Kolasinski, R. W. Fillius, and E. G. Mullen (1992), Injection of electrons and protons with energies of tens of MeV into $L < 3$ on 24 March 1991, *Geophys. Res. Lett.,* **19**, 821–824.

Boardsen, S. A., D. L. Gallagher, D. A. Gurnett, W. K. Petersen, and J. L. Green (1992), Funnel-shaped low-frequency equatorial waves, *J. Geophys. Res.,* **97**, 14,967–14,976.

Bortnik, J., R. M. Thorne, and N. P. Meredith (2007), Modeling the propagation characteristics of chorus using CRRES suprathermal electron fluxes, *J. Geophys. Res.,* **112**, A08204, doi:10.1029/2006JA012237.

Bortnik, J., R. M. Thorne, and N. P. Meredith (2008), The unexpected origin of plasmaspheric hiss from discrete chorus emissions, *Nature,* **452**, 62–66, doi:10.1038/nature06741.

Bortnik, J., et al. (2009), An observation linking the origin of plasmaspheric hiss to discrete chorus emissions, *Science,* **324**, 5928, 775–778, doi:10.1126/science.1171273.

Brautigam, D. H. and J. M. Albert (2000), Radial diffusion analysis of outer radiation belt electrons during the October 9, 1990, magnetic storm *J. Geophys. Res.,* **105**, 291–309.

Braysy, T., K. Mursula and G. Markland (1998), Ion cyclotron waves during a great geomagnetic storm observed by Freja double-probe electric field instrument, *J. Geophys. Res.,* **103**, 4145.

Cannon P., et al. (2013), Extreme space weather: impacts on engineered systems and infrastructure, Royal Academy of Engineering, Carlton House Terrace, London.

Cattell, C., et al. (2008), Discovery of very large amplitude whistler-mode waves in Earth's radiation belts, *Geophys. Res. Lett.,* **35**, L01105, doi:10.1029/2007GL032009.

Chen, Y., G. D. Reeves, and R. H. W. Friedel (2007), The energization of relativistic electrons in the outer Van Allen radiation belt, *Nature Physics,* **3**, 614–617.

Chen, L., R. M. Thorne, V. K. Jordanova, M. F. Thomsen, and R. B. Horne (2011), Magnetosonic wave instability analysis for proton ring distributions observed by the LANL magnetospheric plasma analyzer, *J. Geophys. Res.,* **116**, A03223, doi:10.1029/2010JA016068.

Chen, L., J. Bortnik, W. Li, R. M. Thorne, and R. B. Horne (2012a), Modeling the properties of plasmaspheric hiss: 1. Dependence on chorus wave emission, *J. Geophys. Res.,* **117**, A05201, doi:10.1029/2011JA017201.

Chen, L., J. Bortnik, W. Li, R. M. Thorne, and R. B. Horne (2012b), Modeling the properties of plasmaspheric hiss: 2. Dependence on the plasma density distribution, *J. Geophys. Res.,* **117**, A05202, doi:10.1029/2011JA017202.

Chen, L., W. Li, J. Bortnik, and R. M. Thorne (2012c), Amplification of whistler-mode hiss inside the plasmasphere, *Geophys. Res. Lett.,* **39**, L08111, doi:10.1029/2012GL051488.

Church, S., and R. M. Thorne (1983), On the origin of plasmaspheric hiss: ray path integrated amplification, *J. Geophys. Res.,* **88**, 7941.

Cornwall, J. M., F. V. Coroniti and R. M. Thorne (1970), Turbulent loss of ring current protons, *J. Geophys. Res.,* **75**, 4699.

Dungey, J. W. (1963), Loss of van Allen electrons due to whistlers, *Planet and Space Sci.,* **11**, 591–595.

Elkington, S. R., M. K. Hudson, and A. A. Chan (1999), Acceleration of relativistic electrons via drift resonant interactions with toroidal-mode Pc-5 ULF oscillations, *Geophys. Res. Lett.,* **26**, 3273–3276.

Falthammar, C-G. (1965), Effects of time dependent electric fields on geomagnetically trapped radiation, *J. Geophys. Res.,* **70**, 2503–2516.

Fok, M.-C., R. B. Horne, N. P. Meredith, and S. A. Glauert (2008), Radiation Belt Environment model: Application to space weather nowcasting, *J. Geophys. Res.,* **113**, A03S08, doi:10.1029/2007JA012558.

Fraser, B. J. and T. S. Nguyen (2001), Is the plasmapause a preferred source region of electromagnetic ion cyclotron waves in the magnetosphere ?, *J. Atmos. Terr. Phys.,* **63**, 1225.

Glauert, S. A., R. B. Horne, and N. P. Meredith (2014), Three-dimensional electron radiation belt simulations using the BAS Radiation Belt Model with new diffusion models for chorus, plasmaspheric hiss, and lightning-generated whistlers, *J. Geophys. Res. Space Physics,* **119**, doi:10.1002/2013JA019281.

Green, J. C., and M. G. Kivelson (2004), Relativistic electrons in the outer radiation belt: Differentiating between acceleration mechanisms, *J. Geophys. Res.,* **109**, A03213, doi:10.1029/2003JA010153.

Green, J. L., S. Boardsen, L. Garcia, W. W. Taylor, S. F. Fung, and B. W. Reinisch (2005), On the origin of whistler mode radiation in the plasmasphere, *J. Geophys. Res.,* **110**, A03201, doi:10.1029/2004JA010495.

Hess, W. N., (1963), The artificial radiation belt made on July 9, 1962, *J. Geophys. Res.,* **68**(3), 667–683.

Horne, R. B. and R. M. Thorne (1994), Convective instabilities of electromagnetic ion cyclotron waves in the outer magnetosphere, *J. Geophys. Res.,* **99**, 17,259.

Horne, R. B., and R. M. Thorne (1998), Potential waves for relativistic electron scattering and stochastic acceleration during magnetic storms, *Geophys. Res. Lett.,* **25**, 3011–3014.

Horne, R. B. (2002), The contribution of wave-particle interactions to electron loss and acceleration in the Earth's radiation belts during geomagnetic storms, *in URSI Review of Radio Science 1999–2002,* edited by W.R. Stone, pp. 801–828, Wiley.

Horne, R. B., G. V. H. Wheeler, and C. K. St. Alleyne (2000), Proton and electron heating by radially propagating fast magnetosonic waves, *J. Geophys. Res.,* **105**, 27,597–27,610.

Horne, R. B., S. A. Glauert, and R. M. Thorne (2003), Resonant diffusion of radiation belt electrons by whistler-mode chorus, *Geophys. Res. Lett.,* **30**, (9), 1493, doi:10.1029/2003GL016963.

Horne, R. B., R. M. Thorne, S. A. Glauert, J. M. Albert, N. P. Meredith, and R. R. Anderson (2005a), Timescale for radiation belt electron acceleration by whistler mode chorus waves, *J. Geophys. Res.,* **110**, A03225, 10.1029/2004JA010811.

Horne, R. B., et al. (2005b), Wave acceleration of electrons in the Van Allen radiation belts, *Nature,* **437**, 227–230, doi:10.1038/nature03939.

Horne, R. B., et al. (2006), Mechanisms for the acceleration of radiation belt electrons, in Co-rotating Solar Wind Streams and Recurrent Geomagnetic Activity, *in Geophys. Monogr. Ser.*, **167**, AGU, Washington, D. C.

Horne, R. B. (2007), Plasma astrophysics: Acceleration of killer electrons, *Nature Physics*, **3**, 590–591, doi: 10.1038/nphys703.

Horne, R. B., R. M. Thorne, S. A. Glauert, N. P. Meredith, D. Pokhotelov, and O. Santolík (2007), Electron acceleration in the Van Allen radiation belts by fast magnetosonic waves, *Geophys. Res. Lett.,* **34**, L17107, doi:10.1029/2007GL030267.

Horne, R. B., et al. (2013a), A new diffusion matrix for whistler mode chorus waves, *J. Geophys. Res. Space Physics,* **118**, 6302–6318, doi:10.1002/jgra.50594.

Horne, R. B., et al. (2013b), Space Weather effects on satellites and forecasting the Earth's electron radiation belts with SPACECAST, *Space Weather,* **11**, doi:10.1002/swe.20023.

Iles, R. H. A., N. P. Meredith, A. N. Fazakerley, and R. B. Horne (2006), Phase space density analysis of the outer radiation belt energetic electron dynamics *J. Geophys. Res.,* **111**, A03204, doi:10.1029/2005JA011206.

Katoh, Y., and Y. Omura (2007), Computer simulation of chorus wave generation in the Earth's inner magnetosphere, *Geophys. Res. Lett.,* **34**, L03102, doi:10.1029/2006GL028594.

Kennel, C. F., and H. E. Petschek (1966), Limit on stably trapped particle fluxes, *J. Geophys. Res.,* **71**, 1–28.

Kersten, T., R. B. Horne, S. A. Glauert, N. P. Meredith, B. J. Fraser, and R. S. Grew (2014), Electron losses from the radiation belts caused by EMIC waves, *J. Geophys. Res. Space Physics,* **119**, doi:10.1002/2014JA020366.

Kozyra, J. U., T. E. Cravens, A. F. Nagy, E. G. Fontheim, and R. S. B. Ong (1984), Effects of energetic heavy ions on electromagnetic ion cyclotron wave generation in the plasmapause region, *J. Geophys. Res.,* **89**, 2217.

Lauben, D. S., U. S. Inan, T. F. Bell, and D. A. Gurnett (2002), Source characteristics of ELF/VLF chorus, *J. Geophys. Res.,* **107**, 1429, doi:10.1029/2000JA003019.

LeDocq, M. J., D. A. Gurnett, and G. B. Hospodarsky (1998), Chorus source locations from VLF Poynting flux measurements with the Polar spacecraft, *Geophys. Res. Lett.*, **25**(21), 4063–4066, doi:10.1029/1998GL900071.

Lejosne, S., D. Boscher, V. Maget, and G. Rolland (2013), Deriving electromagnetic radial diffusion coefficients of radiation belt equatorial particles for different levels of magnetic activity based on magnetic field measurements at geostationary orbit, *J. Geophys. Res. Space Physics,* **118**, 3147–3156, doi:10.1002/jgra.50361.

Li, W., J. Bortnik, R. M. Thorne, and V. Angelopoulos (2011), Global distribution of wave amplitudes and wave normal angles of chorus waves using THEMIS wave observations, *J. Geophys. Res.,* **116**, A12205, doi:10.1029/2011JA017035.

Li, X., et al. (1997), Are energetic electrons in the solar wind the source of the outer radiation belt? *Geophys. Res. Lett.,* **24**, 923–926.

Lorentzen, K. R., J. B. Blake, U. S. Inan, and J. Bortnik (2001), Observations of relativistic electron microbursts in association with VLF chorus, *J. Geophys. Res.,* **106**, 6017–6027, doi:10.1029/2000JA003018.

Lyons, L. R., and R. M. Thorne (1972), Parasitic pitch angle diffusion of radiation belt particles by ion cyclotron waves, *J. Geophys. Res.,* 77, 5608.

Lyons, L. R., and R. M. Thorne (1973), Equilibrium structure of radiation belt electrons, *J. Geophys. Res.,* 78, 2142–2149.

Lyons, L. R., and D. J. Williams (1984), Quantitative aspects of magnetospheric physics, D. Reidel, Norwell, Mass..

Mathie, R. A., and I. R. Mann (2000), A correlation between extended intervals of ULF wave power and storm-time geosynchronous relativistic electron flux enhancements, *Geophys. Res. Lett.,* 27, 3261.

Meredith, N. P., R. M. Thorne, R. B. Horne, D. Summers, B. J. Fraser and R. R. Anderson (2003b), Statistical analysis of relativistic electron energies for cyclotron resonance with EMIC waves observed on CRRES, *J. Geophys. Res.,* **108**, (A6), 1250, doi:10.1029/2002JA009700.

Meredith, N. P., R. B. Horne, R. M. Thorne, D. Summers, and R. R. Anderson (2004), Substorm dependence of plasmaspheric hiss, *J. Geophys. Res.,* **109**, A06209, doi:1029/2004JA010387.

Meredith N. P., R. B. Horne, M. A. Clilverd, D. Horsfall, R. M. Thorne, R. R. Anderson (2006a), Origins of plasmaspheric hiss, *J. Geophys. Res.,* **111**, A09217, doi:10.1029/2006JA011707.

Meredith, N. P., R. B. Horne, S. A. Glauert, R. M. Thorne, D. Summers, J. M. Albert, and R. R. Anderson (2006b), Energetic outer zone electron loss timescales during low geomagnetic activity, *J. Geophys. Res.,* **111**, A05212, doi:10.1029/2005JA011516.

Meredith, N.P., R. B. Horne, S. A. Glauert, and R. R. Anderson (2007), Slot region electron loss timescales due to plasmaspheric hiss and lightning generated whistlers, *J. Geophys. Res.,* **112**, A08214, doi:10.1029/2006JA012413.

Meredith, N. P., R. B. Horne, and R. R. Anderson (2008), Survey of magnetosonic waves and proton ring distributions in the Earth's inner magnetosphere, *J. Geophys. Res.,* **113**, A06213, doi:10.1029/2007JA012975.

Meredith N.P., R. B. Horne, S. A. Glauert, D. N. Baker, S. G. Kanekal, and J. M. Albert (2009), Relativistic electron loss timescales in the slot region, *J. Geophys. Res.,* **114**, A03222, doi:10.1029/2008JA013889.

Meredith, N. P., et al. (2012), Global model of lower band and upper band chorus from multiple satellite observations, *J. Geophys. Res.,* **117**, A12209, doi:10.1029/2012JA017978.

Meredith, N. P., et al. (2013), Global statistical evidence for chorus as the embryonic source of plasmaspheric hiss, *Geophys. Res. Lett.,* **40**, 2891–2896, doi:10.1002/grl.50593.

Meredith, N. P., R. B. Horne, W. Li, R. M. Thorne, and A. Sicard-Piet (2014), Global model of low-frequency chorus ($f_{LHR} < f < 0.1 f_{ce}$) from multiple satellite observations, *Geophys. Res. Lett.,* **41**, doi:10.1002/2013GL059050.

Morley, S. K., R. H. W. Friedel, T. E. Cayton, and E. Noveroske (2010), A rapid, global and prolonged electron radiation belt dropout observed with the Global Positioning System constellation, *Geophys. Res. Lett.,* **37**, L06102, doi:10.1029/ 2010GL042772.

Nishimura et al. (2010), Identifying the driver of the pulsating aurora, *Science,* **330**, 81–84, doi:10.1126/science.1193186.

Omura, Y., M. Hikishima, Y. Katoh, D. Summers, and S. Yagitani (2009), Nonlinear mechanisms of lower-band and upper-band VLF chorus emissions in the magnetosphere, *J. Geophys. Res.,* **114**, A07217, doi:10.1029/2009JA014206.

Ozeke, L. G., et al. (2012), ULF wave derived radiation belt radial diffusion coefficients, *J. Geophys. Res.,* **117**, A04222, doi:10.1029/2011JA017463.

Paulikas, G. A., and J. B. Blake (1979), Effects of the solar wind on magnetospheric dynamics: Energetic electrons at the synchronous orbit, in *Quantitative modelling of Magnetospheric processes, Geophys. Mongr. Ser.*, vol. 21, edited by W. P. Olsen, pp. 180–202, AGU, Washington D.C.

Reeves, G. D., K. L. McAdams, R. H. W. Friedel, and T. P. O'Brien (2003), Acceleration and loss of relativistic electrons during geomagnetic storms, *Geophys. Res. Lett.,* **30**, doi10.1029/2002GL016513.

Reeves, G. D., et al. (2013), Electron Acceleration in the Heart of the Van Allen Radiation Belts, *Science express,* **340**, doi: 10.1126/science.1237743

Roberts, C. S. (1969), Pitch-angle diffusion of electrons in the magnetosphere, *Rev Geophys.*, 7, 305–337.

Roederer, J. G. (1970), Dynamics of Geomagnetically Trapped radiation, Springer, New York.

Santolík, O., D. A. Gurnett, and J. S. Pickett, (2003), Spatio-temporal structure of storm-time chorus, *J. Geophys. Res.,* **108**, (A7), 1278, doi:10.1029/2002JA009791.

Santolík, O., J. S. Pickett, D. A. Gurnett, J. D. Menietti, B. T. Tsurutani, and O. Verkhoglyadova (2010), Survey of Poynting flux of whistler mode chorus in the outer zone, *J. Geophys. Res.,* **115**, A00F13, doi:10.1029/2009JA014925

Schulz, M., and L. Lanzerotti (1974), Particle diffusion in the radiation belts, Springer-Verlag, New York.

Selesnick, R. S., M. D. Looper, and R. A. Mewaldt (2007), A theoretical model of the inner proton radiation belt, *Space Weather*, **5**, S04003, doi:10.1029/2006SW000275.

Shprits, Y. Y., et al. (2006), Outward radial diffusion driven by losses at magnetopause, *J. Geophys. Res.,* **111**, A11214, doi:10.1029/2006JA011657.

Shprits, Y. Y., D. Subbotin, and B. Ni (2009), Evolution of electron fluxes in the outer radiation belt computed with the VERB code, *J. Geophys. Res.,* **114**, A11209, doi:10.1029/2008JA013784.

Shprits, Y. Y., D. Subbotin, B. Ni., R. Horne, D. Baker, and P. Cruce (2011), Profound change of the near-Earth radiation environment caused by solar superstorms, *Space Weather*, **9**, S08007, doi:10.1029/2011SW000662.

Shprits, Y. Y., et al. (2013) Unusual stable trapping of the ultrarelativistic electrons in the Van Allen radiation belts, *Nature Phys.*, DOI: 10.1038/NPHYS2760.

Singer, S. F. (1958), Trapped albedo theory of the radiation belt, *Phys. Rev. Lett.*, **1**, 181–183.

Smith, A. J., R. B. Horne, and N. P. Meredith (2004a), Ground observations of chorus following geomagnetic storms, *J. Geophys. Res.*, **109** A02205, doi:10.1029/2003JA010204.

Sonwalkar, V. S., and U. S. Inan (1989), Lightning as an embryonic source of VLF hiss, *J. Geophys. Res.,* **94**, 6986.

Subbotin, D. A., and Y. Y. Shprits (2009), Three-dimensional modeling of the radiation belts using the Versatile Electron Radiation Belt (VERB) code, *Space Weather*, 7, S10001, doi:10.1029/2008SW000452.

Summers, D., R. M. Thorne, and F. Xiaio (1998), Relativistic theory of wave-particle resonant diffusion with application to electron acceleration in the magnetosphere, *J. Geophys. Res.,* **103**, 20, 487–20,500.

Summers, D., B. Ni, and N. P. Meredith (2007), Timescales for radiation belt electron acceleration and loss due to resonant wave-particle interactions: 2. Evaluation for VLF chorus, ELF hiss, and electromagnetic ion cyclotron waves, *J. Geophys. Res.,* **112**, A04207, doi:10.1029/2006JA011993.

Thorne, R. M., and L. J. Andreoli (1980), Mechanisms for intense relativistic electron precipitation, in *Explortion of the Polar Upper Atmosphere*, edited by C. S. Deehr, and J. A. Holtet, pp. 381–394, Springer, New York.

Thorne, R. M., E. J. Smith, R. K. Burton, and R. E. Holzer (1973), Plasmaspheric hiss, *J. Geophys. Res.*, 78(10), 1581–1596.

Thorne, R. M., et al. (2005a), The influence of wave-particle interactions on relativistic electrons during storms, in Inner Magnetosphere Interactions: New Perspectives From Imaging, *in Geophys. Monogr. Ser., vol.* **159**, edited by J. Burch, M. Schulz, and H. Spence, AGU, Washington D.C.

Thorne, R. M., T. P. O'Brien, Y. Y. Shprits, D. Summers, and R. B. Horne (2005b), Timescale for MeV electron microburst loss during geomagnetic storms, *J. Geophys. Res.*, **110**, A09202, doi:10.1029/2004JA010882.

Thorne, R. M. (2010), Radiation belt dynamics: The importance of wave-particle interactions, *Geophys. Res. Lett.*, **37**, L22107, doi:10.1029/2010GL044990.

Thorne, R. M., B. Ni, X. Tao, R. B. Horne, and N. P. Meredith (2010), Scattering by chorus waves as the dominant cause of the Earth's diffuse aurora, *Nature*, **467**, 943–946, doi:10.1038/nature09467.

Thorne, R. M. et al. (2013), Rapid local acceleration of relativistic radiation-belt electrons by magnetospheric chorus, *Nature*, **504**, 401, doi:10.1038/nature12889.

Tsurutani, B. T., and E. J. Smith (1974), Postmidnight chorus: A substorm phenomenon, *J. Geophys. Res.*, **79**, 118–127.

Tsurutani, B. T., N. Gopalswamy, R. L. McPherron, W. D. Gonzalez, G. Lu, and F. L. Guarnieri (2006), Magnetic storms caused by corotating solar wind streams, in *Recurrent Magnetic Storms: Corotating Solar Wind Streams, Geophys. Monogr. Ser.*, vol. 167, edited by B. T. Tsurutani, R. L. McPherron, W. D. Gonzalez, G. Lu, J. H. A. Sobral, and N. Gopalswamy, pp. 1–17, AGU, Washington, D. C.

Tsurutani, B. T., B. J. Falkowski, O. P. Verkhoglyadova, J. S. Pickett, O. Santolík, and G. S. Lakhina (2012), Dayside ELF electromagnetic wave survey: A Polar statistical study of chorus and hiss, *J. Geophys. Res.*, **117**, A00L12, doi:10.1029/2011JA017180.

Van Allen, J. A., G. H. Ludwig, E. C. Ray, and C. E. McIlwain (1958), Observation of high intensity radiation by Satellites 1958 Alpha and Gamma, *Jet Propulsion*, **28**, 588–592.

Varotsou, A., D. et al. (2005), Simulation of the outer radiation belt electrons near geosynchronous orbit including both radial diffusion and resonant interaction with whistler-mode chorus waves, *Geophys. Res. Lett.*, **32**, L19106, doi:10.1029/2005GL023282.

Varotsou, A., et al. (2008), Three-dimensional test simulations of the outer radiation belt electron dynamics including electron-chorus resonant interactions, *J. Geophys. Res.*, **113**, A12212, doi:10.1029/2007JA012862.

Walt, M. (1964), The effects of atmospheric collisions on geomagnetically trapped electrons, *J. Geophys. Res.*, **69**, 3947–3958.

Wang, C., Q. Zong, F. Xiao, Z. Su, Y. Wang, and C. Yue (2011), The relations between magnetospheric chorus and hiss inside and outside the plasmasphere boundary layer: Cluster observation, *J. Geophys. Res.*, **116**, A07221, doi:10.1029/2010JA016240.

West, H. I., and R. M. Buck (1976), A study of electron spectra in the inner belt, *J. Geophys. Res.*, **81**, 4696–4700.

Wrenn, G. L. (1995), Conclusive Evidence for Internal Dielectric Charging Anomalies on Geosynchronous Communications Spacecraft, *J. Spacecraft and Rockets*, **32**, 514–520.

10

Understanding the Role of EMIC Waves in Radiation Belt and Ring Current Dynamics: Recent Advances

Maria E. Usanova and Ian R. Mann

Abstract

Electromagnetic ion cyclotron (EMIC) waves are believed to be important for influencing the dynamics of energetic particles in the inner magnetosphere, especially in relation to ring current and radiation belt dynamics. In relation to the ring current, both the dynamical evolution of ion distributions along their drift paths and effects from the solar wind can result in EMIC wave generation from unstable ion distributions. EMIC waves often grow as a result of perpendicular temperature anisotropy, with the ion dynamics and energy and pitch angle distributions in the ring current being modified significantly by the growth of the EMIC instability. EMIC waves are also hypothesized to influence higher energy electrons in the Van Allen belts, through a Doppler-shifted cyclotron resonance, including stimulating potential scattering loss into the atmosphere. This review presents some of the latest results from selected studies of EMIC wave characteristics and EMIC wave impacts on energetic particles in the inner magnetosphere. It draws on results from recent satellite missions including the NASA THEMIS and ESA-NASA Cluster missions, as well as some of the very latest results from the NASA Van Allen Probes mission launched on August 31,

Waves, Particles, and Storms in Geospace. First Edition. Georgios Balasis, Ioannis A. Daglis, Ian R. Mann.

2012. We also highlight the value of data from networks of modern ground-based magnetometers in providing continuous monitoring over local, continental, and even global scales, especially in conjunction with in situ measurements from satellites. Such coordinated ground–satellite conjunction studies represent a powerful tool for understanding the self-consistent and cross-energy coupling between ions and electrons in the ring current and radiation belt via the intermediary of EMIC waves.

10.1 Introduction

Electromagnetic ion cyclotron (EMIC) waves are transverse plasma waves generated in the magnetosphere by energetic (a few keV to hundreds keV) protons, with perpendicular temperature anisotropy ($T_\perp > T_\|$). EMIC waves are usually, but not always, excited in the equatorial plane of the magnetosphere and can travel large distances from the source region both inside the magnetosphere and, if they reach the ionosphere, across large perpendicular distances inside the ionospheric duct [see, e.g., Fraser, 1975; Mann et al., 2014]. Often the waves are left-hand (LH) polarized and Alfvénic in the magnetosphere, corresponding to the sense of ion rotation around the magnetic field, although in some cases they can be excited with a significant compressional component. During wave propagation from the source region toward the ionosphere there can also be significant polarization rotation, including changes in polarization rotation at the crossover frequencies or due to interactions at local heavy ion gyrofrequencies and at the bi-ion frequency [e.g., Horne and Thorne, 1997]. Similarly, waves in the ionospheric duct observed by networks of ground-based magnetometers can also demonstrate complex patterns of polarization and polarization sense with respect to distance from the source [see, for example, Mann et al., 2014, and references therein, and the modeling by Woodroffe and Lysak, 2012].

Commonly, EMIC waves which are observed on the ground are referred to as Pc1-2 pulsations, often being seen either as structured emissions or in the form of IPDP (Intervals of Pulsations with Diminishing Periods) [e.g., Hayakawa et al., 1992 and references therein] in the frequency range between 0.1 and 5 Hz. Structured Pc1-2 pulsations are characterized by relatively slow variations in frequency, while IPDPs exhibit relatively fast frequency fluctuations on a timescale of tens of minutes. Early theoretical studies showed that the EMIC wave growth leads to the isotropization of the initially unstable ion (usually proton) distribution and consequent pitch-angle scattering and loss of protons into the ionosphere [e.g., Cornwall, 1965], this concept being further supported by conjugate observations of localized precipitation of energetic protons and the occurrence of EMIC waves [Søraas et al., 1980; Yahnin and Yahnina, 2007; Usanova et al., 2010a].

EMIC waves are believed to be important for influencing the dynamics of energetic particles in the inner magnetosphere, especially in relation to ring current ion dynamics and their effects are included in global ring current numerical simulation models [e.g.,

Khazanov et al., 2007; Jordanova, 2012]. Gyroresonant interaction with EMIC waves is also hypothesized to be an important potential internal loss mechanism for higher energy radiation belt electrons [see, e.g., Thorne and Kennel, 1971; Lyons and Thorne, 1972; Summers and Thorne, 2003]. The action of internal radiation belt loss processes resulting in the scattering of Van Allen belt electrons into the atmosphere is believed to be required to explain radiation belt flux variability during various magnetic storms [Reeves et al., 2003; Millan and Thorne, 2007] and in particular, radiation belt flux dropouts during the storm main phase [e.g., Turner et al., 2013].

In order to incorporate EMIC wave-related loss processes into global magnetospheric and space weather models, it is important to know the solar wind and magnetospheric conditions favorable for EMIC wave growth as well as the occurrence distribution of the waves in the magnetosphere. Early studies identified the plasmapause as a preferred region for EMIC wave generation and explained it by the fact that an increase in the number of cold ions would lead to a drastic enhancement in the convective EMIC wave growth rate (which is inversely proportional to the Alfvén speed, V_A) [e.g., Cornwall et al., 1970; Gendrin, 1975]. In many papers examining the potential interactions between radiation belt electrons and different plasma wave modes, EMIC waves are shown schematically to be preferentially excited along the duskside plasmapause during geomagnetically disturbed periods [see, e.g., Summers et al., 1998]. In this paradigm, storm-time enhancements in the convection electric field may lead to the overall intensification of the ring current and also to the formation of anisotropic particle distributions. The paradigm suggests that EMIC waves would be generated along the ion drift path in the dusk-to-noon sector at locations where the ring current overlaps with the plasmasphere. More recently, this model has been modified by extending the region of intense but localized EMIC waves onto the westward boundary of a plasmaspheric plume—region of the outer plasmasphere transported sunward during magnetic storms [see, e.g., Li et al., 2007]. In support of this paradigm, there are several case studies [see, e.g., Fraser et al., 2005; Morley et al., 2009; Yuan et al., 2010] showing EMIC wave observations in plumes. However, from a statistical perspective, a link between EMIC waves and plasmaspheric plumes had not been definitively established and we review the results of our recent analyses here.

Interestingly, the relationship between geomagnetic storms and EMIC waves, especially in regard to prevalence of EMIC waves during the storm main phase, remains controversial. For example, statistical studies by Engebretson et al. [2008a, 2008b], using both ground-based and in situ observations, did not support the hypothesis that EMIC waves were dominant during the storm main phase. Conversely, CRRES and AMPTE/CCE satellite surveys by Halford et al. [2010] and Keika et al. [2013] did show enhanced EMIC wave occurrence in the inner magnetosphere during the main phase.

Another important process that leads to enhanced energetic ion temperature anisotropies is magnetospheric compression by the solar wind. Anisotropic proton distributions, and hence EMIC waves, can be generated on high L-shells in the subsolar equatorial magnetosphere close to the magnetopause by solar wind compressions [Olson and Lee, 1983; Anderson and Hamilton, 1993; Denton et al., 2002]. This is also where EMIC wave occurrence rates reach their maximum [see, e.g., Anderson et al., 1992; Usanova et al., 2012; Keika et al., 2013]. A number of papers [e.g., Usanova et al., 2008, 2010a and references therein] have reported observations of EMIC waves at much lower L-shells in the inner magnetosphere, for example EMIC waves occurring just inside the plasmapause, and which are also associated with intense magnetospheric compression due to enhanced solar wind dynamic pressure. However, statistically these events are not very common and may only happen when the solar wind dynamic pressure is very significantly increased. For example, during the event considered by Usanova et al. [2008] the solar wind dynamic pressure reached ~12 nPa. We also review selected studies of the occurrence of EMIC waves in the inner magnetosphere, and their connection to enhancements in solar wind dynamic pressure.

This chapter is organized as follows. First, we discuss EMIC wave excitation mechanisms in the inner magnetosphere. Then we present in situ statistical observations of EMIC waves onboard the THEMIS and Cluster satellites and analyze their occurrence distributions. We examine the role of magnetospheric compression in EMIC wave generation and present the probability of EMIC wave occurrence in plasmaspheric plumes. Further we discuss the role of EMIC waves in energetic particle loss (both ions and relativistic electrons) and in particular present recent simultaneous observations of EMIC waves and radiation belt electron dynamics seen by the NASA Van Allen Probes. A discussion and conclusion section ends our chapter.

10.2 EMIC wave excitation in the inner magnetosphere

EMIC wave excitation is related to generation of temperature anisotropy ($T_\perp > T_\parallel$) in energetic ion, usually proton, distributions. These anisotropic distributions may form along ion drift trajectories from the magnetospheric tail to the dayside magnetosphere, for example as a result of the energy dispersion of injected populations, as a result of the dynamics of the convecting population, or as a result of the action of the solar wind. In principle, depending upon the location of the unstable distribution, EMIC waves may be generated at any location along their drift path.

For example, the convection electric field can generate ion anisotropy as particles injected from the tail are transported radially inward due to $E \times B$ drift during magnetospheric convection while conserving their first and second adiabatic invariants [e.g., Summers et al., 1998]. In this case, ion temperature anisotropy is generated since the perpendicular energization due to conservation of the first adiabatic invariant via betatron acceleration exceeds that arising in the parallel direction due to Fermi acceleration via conservation of the second invariant. This mechanism may lead to especially strong EMIC wave generation along duskside plasmapause in regions where ion drift paths overlap dense regions of plasmaspheric plasma, especially during disturbed geomagnetic conditions [e.g., Jordanova et al., 2001]. In addition to the expected westward drifting ion paths, a fraction of lower energy protons (with energies below ~10 keV) may instead reach the dayside along injection paths which traverse the morning side magnetosphere (instead of drifting westward through dusk) and give rise to morning side EMIC waves [Anderson et al., 1996a and references therein; see, e.g., also the modeling of ion trajectories presented by Ozeke and Mann, 2001]. Finally, dayside events in particular are often observed during periods of magnetospheric compression due to enhanced solar wind dynamic pressure, which may be related to the generation of increased ion temperature anisotropy at the equator due to drift-shell splitting or perhaps at off-equatorial locations related to Shabansky orbits [see, e.g., McCollough et al., 2010, and references therein]. We will consider these mechanisms for EMIC wave generation in more detail below.

10.2.1 Physical generation mechanisms for compression-related EMIC waves

The prevalence of dayside EMIC activity during intervals of enhanced solar wind dynamic pressure can be explained by three possible mechanisms, which can lead to the generation of anisotropic proton distributions.

10.2.1.1 Bursty EMIC wave events

Short-term magnetospheric compressions can increase the $T_{\perp}/T_{\parallel}$ ratio through adiabatic heating [see Olson and Lee, 1983] and can generate EMIC wave bursts during periods within which the magnetic field strength is increasing (i.e., $\frac{dB}{dt} > 0$). In a dipole field, the resulting anisotropy, A, generated if the magnetic field magnitude changes from the initial value, B_0, to the final value, B,

$$A = \frac{B^{1/3}}{B_0}(A_0 + 1) - 1,$$

and where A_0 is the initial anisotropy.

As an example, a bursty EMIC event from December 16, 2006, observed by the GOES 12 magnetometer at geosynchronous orbit is shown in Figure 10.1. Figure 10.1a shows the OMNI [King and Papitashvili, 2004] solar wind dynamic pressure time-shifted to the subsolar magnetopause (red curve), as well as the magnetic field magnitude observed by the GOES 12 satellite in the dayside magnetosphere (blue curve). Enhancements in solar wind dynamic pressure caused simultaneous enhancements in the magnetic field during which GOES also observed bursts of EMIC waves as shown in the dynamic spectrogram in Figure 10.1b. Using the above equation, we estimated the additional amount of anisotropy produced during this compression event: where the GOES magnetic field magnitude increased from 110 to 135 nT (first compression spike in Figure 10.1), and from 120 to 135 nT (second compression spike in Figure 10.1), this produces a 12% and a 7% increase in the temperature anisotropy, respectively. This observation is consistent with Anderson and Hamilton [1993] who suggested that an adiabatic increase in anisotropy of this magnitude may increase the anisotropy from an initial marginally stable level to one of significant growth, and hence lead to strong EMIC wave generation. This may explain the bursty structure of EMIC waves observed at GOES 12 in Figure 10.1, and which are seen simultaneously with increases of the background magnetic field strength. Nevertheless, this mechanism cannot explain why EMIC wave activity can last for many hours while the magnetic field remains compressed during continuous periods of enhanced solar wind dynamic pressure, and we consider this below.

10.2.1.2 Continuous EMIC wave events

Continuous EMIC wave events may be seen to last for several hours in ground-based magnetometer data, often occurring during intervals while the magnetosphere remains compressed in response to a continuous period of enhanced solar wind dynamic pressure. An example of a continuous and long-lasting EMIC event seen to span ~4.5 hours by ground-based magnetometers on June 29, 2007, is shown in Figure 10.2. The EMIC waves occurred coincident with a strong solar wind dynamic pressure enhancement (red curve in Figure 10.2a) following several days of sustained quiet geomagnetic conditions. This increased solar wind dynamic pressure produced an ~40%

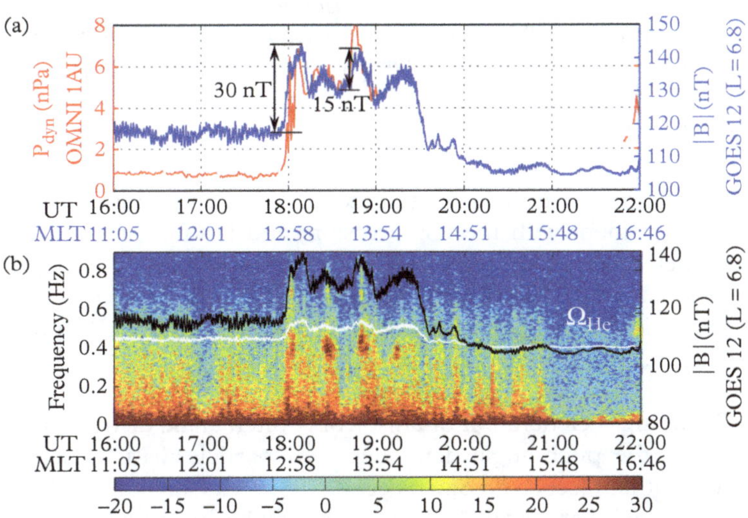

Figure 10.1 *Example of a bursty EMIC wave event from December 16, 2006. (a) Solar wind dynamic pressure time-shifted to the magnetopause (red) and GOES 12 magnetic field (blue) between 16 and 22 UT; (b) dynamic spectrogram of the power of the GOES 12 Hp (parallel to the spin axis) magnetic field component, overplotted with the GOES 12 total magnetic field (black; same as in panel a) and helium gyrofrequency (white). (Adapted from Usanova, 2010b.)*

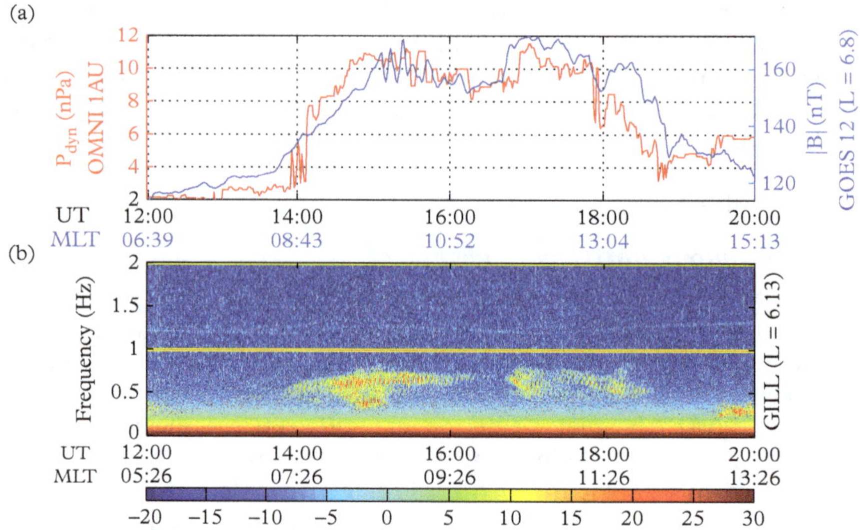

Figure 10.2 *Example of continuous EMIC wave event from June 29, 2007. (a) Solar wind dynamic pressure time-shifted to the magnetopause (red) and GOES 12 magnetic field (blue) between 12 and 20 UT; (b) dynamic spectrogram of the Y-component (geographic east-west) magnetic field perturbation from the Gillam (GILL) CARISMA station for the same period.*
(Adapted from Usanova et al., 2008.)

enhancement in the magnetic field magnitude at geosynchronous orbit (blue curve in Figure 10.2a). Coincident with the compression, and lasting throughout its duration, the ground magnetometers saw clear, nearly continuous, and long-lasting EMIC waves with frequencies below 0.7 Hz (Figure 10.2b). The physical mechanisms which may result in the generation of this type of long-lasting and continuous EMIC wave event are discussed further below.

Drift-shell Splitting Generation of anisotropic pitch angle distributions in the magnetosphere due to the pitch-angle dependence of particle drift trajectories has been studied by many authors [see, e.g., Sibeck et al., 1987; McCollough et al., 2010]. Sibeck et al. (1987) concluded that unstable distributions can be generated due to particles with different pitch angles following different drift paths, such that particles from different spatial locations can be brought together to create an unstable distribution. While in a symmetric dipolar magnetic field all particles drift around the Earth at the same (initial) equatorial distance, in a compressed magnetic field particles with near 90° pitch angles follow lines of constant magnetic field (conserving magnetic moment), and particles with smaller pitch angles move along more circular orbits. This is known as drift-shell splitting, which is illustrated by the schematic in Figure 10.3a.

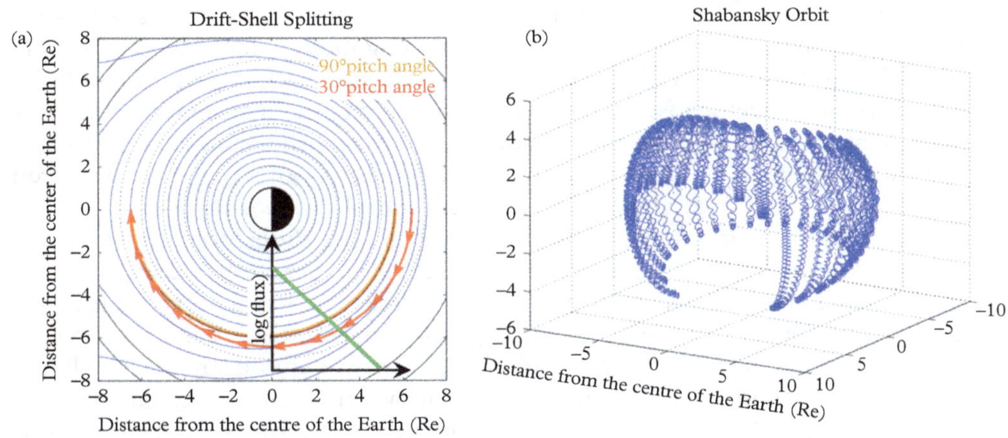

Figure 10.3 *Schematic illustrating mechanisms for continuous EMIC wave event: (a) Drift shell splitting—drift path of 90° (orange) and 30° (red) pitch angle protons from the nightside to the dayside magnetosphere; a schematic of the decreasing logarithm of proton flux (green) as a function of radial distance from the Earth is also shown. (b) Shabansky orbit—trajectory of a proton whose bounce trajectory changes from trapping between mirror points spanning the equatorial plane to trapping between two off-equatorial minimum B regions, resulting in trajectories which do not cross the equatorial plane in the dayside magnetosphere (also known as Shabansky orbits).*
(Adapted from Usanova, 2010b.)

Figure 10.3a shows example drift paths of particles with pitch angles of 30° and 90° as they drift from the nightside to the dayside in the magnetosphere. The drift paths of these two particles intersect at the dayside subsolar magnetosphere, while the initial starting point of their trajectories was located in the nightside magnetosphere in two different locations: for a 30° pitch angle particle further away from the Earth and for a 90° pitch angle particle closer to the Earth. In a situation where the flux of particles decreases with radial distance from the Earth as shown in Figure 10.3a, the 90° pitch-angle particles drift closer to the Earth, as compared to smaller pitch-angle particles, will have a larger flux. This leads to an enhancement of the perpendicular temperature with respect to the parallel temperature and makes the distribution temperature anisotropic (with $T_\perp > T_\parallel$), and thereby potentially unstable to EMIC waves. The drift trajectory splitting is more prominent for particles at higher L-shells and for higher degrees of compression. Therefore, for the inner magnetosphere this effect may be important only during intervals of highly increased solar wind dynamic pressure.

Shabansky Orbits The outer L-shells of the compressed dayside magnetosphere can have two off-equatorial minima in magnetic field strength, which are more pronounced during strong compressions, and can be regarded as a compression-related feature of the reconfigured dayside magnetospheric magnetic field. In such a field topology, particles with 90° pitch angles continue to drift through the equator, but particles with lower pitch angles can mirror at high latitudes without passing through the equator, executing so-called Shabansky orbits [Shabansky, 1971]. As particles reach the $\partial B/\partial S < 0$ region near the dayside magnetopause, where S is distance along the magnetic field, they escape off-equator and execute repeated bounces and are trapped at mid-latitudes. This is demonstrated in Figure 10.3b which shows the full trajectory of a proton drifting from the nightside to the dayside magnetosphere. These so-called *Shabansky regions* can lead to the formation of enhanced temperature anisotropy in two different locations: first, in the equatorial plane, in the absence of the lower pitch angle particles; and second, in the trapped off-equatorial regions where particles' pitch angle undergoes a net shift toward a higher pitch angle. As a result, the ion perpendicular temperature increases relative to the parallel temperature, with the resulting temperature anisotropy potentially giving rise to the growth of EMIC waves, especially in the off-equatorial regions [see, e.g., McCollough et al., 2010; Liu et al., 2012]. Similar to drift shell splitting, when magnetosphere is strongly compressed by the solar wind the regions supporting Shabansky orbits may move further inward, reaching locations inside geosynchronous orbit, to become an important source of dayside temperature anisotropy in the inner magnetosphere.

10.2.1.3 Discussion

The temporal duration of an EMIC wave event must be considered when trying to identify the appropriate physical mechanism which acts as the source of the waves. For example, magnetic drift-shell splitting, Shabansky orbits, or a combination thereof can act in isolation as the source of temperature anisotropy for a specific EMIC wave emission if the driving particle's drift time around the Earth is longer than the observed EMIC event duration, which is about 5 hours for the event of June 29, 2007. In such a case, a new population of particles can drift to the region characterized by sufficient anisotropy to drive the waves for the whole of the period where EMIC waves are observed. For example, low energy protons (with energy ~10 keV) have closed drift paths with periods in excess of 10 hours and could hence provide an energy source for the observed waves. However, higher energy protons also on closed orbits drift around the Earth faster such that they may encounter the source region twice (or more) during the interval of the EMIC wave emission. After the first crossing of the EMIC wave excitation region, the anisotropy will likely decrease due to the EMIC instability most likely being reduced to the level of marginal stability. On subsequent passes through the subsolar magnetosphere, these higher energy particles might not have enough remaining anisotropy to continue to excite the waves—presenting a challenge for explaining the duration of the emission.

It is reasonable to assume that a compression-related reconfiguration of the magnetic field can sustain wave activity for a timescale of less than one particle drift period, but that additional processes may be required to explain the continuous (re-)formation of anisotropic distributions. For example, the convection electric field may provide a source of sustained anisotropy through the supply of "fresh" particles by injection from the nightside magnetosphere onto open drift trajectories which reach the dayside magnetosphere. There are two possible effects of the convection electric field on the particle dynamics.

First, due to $E \times B$ drift, particles may end up on open drift trajectories which cross the dayside magnetosphere offering a potential energy source for EMIC waves there but which will eventually escape from the magnetosphere by crossing the magnetopause. Therefore to feed continuous EMIC wave excitation a steady supply of new injected particles is required. In such a case the convection electric field may provide both a source and sink of energetic particles, from the plasma sheet and out across the magnetopause, respectively, but with temperature anisotropy developing due to the fact that at least in a dipole field betatron acceleration, conserving the first adiabatic invariant, is always more efficient than Fermi acceleration, conserving the second adiabatic invariant, along the injection path from the nightside plasma sheet into the magnetosphere [note however the situation is more complex in other geometries, e.g., Wolf et al., 2009; Sergeev et al., 2001]. Second, both the pre-existing particles on closed drift trajectories and the newly injected particles will be mixed together. This may, nevertheless, sustain sufficient anisotropy for ongoing EMIC wave generation. Note also that under relaxation from active conditions, particles which are initially injected onto open trajectories can be captured onto closed trajectories as the separatrix between open and closed drift paths for particles of a given fixed energy moves outward [see, for example, Ozeke and Mann, 2001].

10.2.2 Effect of solar wind dynamic pressure and enhanced cold plasma density on EMIC wave generation

From a statistical perspective, the effect of enhanced solar wind dynamic pressure on EMIC wave generation was examined by Usanova et al. [2012]. These authors used magnetic field measurements from four THEMIS spacecraft between May 2007 and December 2011 and identified EMIC waves with an automated wave detection algorithm developed by Bortnik et al. [2007] which selects discrete spectral peaks that stand out above the background noise. Figure 10.4a shows an example of such an automated EMIC wave detection using magnetic field data from the THEMIS C satellite from July 4, 2007. The wave occurrence as a function of L and MLT based on ~26,000 EMIC spectral peaks (and a total of 1130 days) selected with this algorithm is plotted in Figure 10.4b. Consistent with earlier AMPTE CCE [Anderson et al., 1992] and CRRES [Fraser and Nguyen, 2001] studies, EMIC occurrence observed by THEMIS maximized at high L-shells in the dayside magnetosphere, while the occurrence in the inner magnetosphere did not exceed ~2%.

Usanova et al. [2012] also investigated dependence of the EMIC occurrence probability distribution on solar wind dynamic pressure, as shown in Figure 10.5. As demonstrated in Figure 10.5, EMIC waves in the noon sector are highly responsive to increases in solar wind dynamic pressure. The probability of EMIC occurrence increases with P_{dyn} most clearly at high L-shells, close to the magnetopause; however, it also increases at lower L-shells as well, though less dramatically. Also, the maximum probability region moves inward with enhanced P_{dyn}, which may be indicative of inward magnetopause motion.

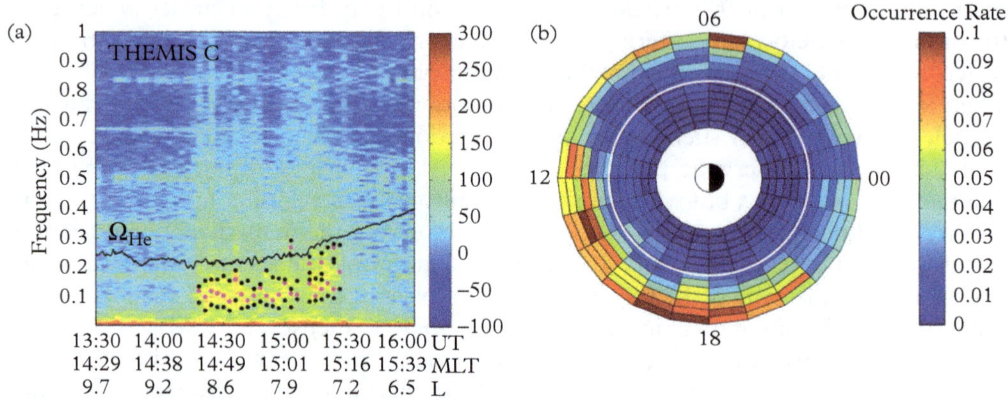

Figure 10.4 *(a) Example of EMIC wave identification on the THEMIS C satellite using the automated Bortnik et al. [2007] detection algorithm. The wave activity was observed between 14:20 and 14:30 MLT on July 4, 2007. Superposed on each spectrogram is the local helium gyrofrequency (black curve) and spectral peaks detected with the automated algorithm. The black and purple dots define the bandwidth and the maximum power, respectively. (b) EMIC wave occurrence distribution observed by THEMIS, plotted in L-MLT coordinates between L = 3–10 and obtained using the Bortnik et al. [2007] algorithm. The white solid line shows geosynchronous orbit.*
(Adapted from Usanova et al., 2012.)

Keika et al. [2013] examined the global distribution of EMIC waves from the AMPTE/CCE satellite between 1984 and 1989 during solar maximum conditions. Their conclusions about the occurrence rate of EMIC waves in the inner magnetosphere are consistent with those of Usanova et al. [2012]. Specifically, these authors found low EMIC wave occurrence (<5%) in the nightside magnetosphere and similar <5% occurrence in the dayside at $L < 6$—but with, and up to, 25% occurrence rates at $L > 6$. Keika et al. [2013] also investigated the occurrence rates in the hydrogen and helium bands separately and found that the helium-band waves often occur in the inner duskside magnetosphere with probability being >10% during disturbed geomagnetic conditions (Dst < 50 nT). Conversely, the hydrogen band waves occur frequently in the outer magnetosphere regardless of the level of geomagnetic activity. This lead Keika et al. [2013] to the conclusion that the helium band waves were associated with ion injections, while the hydrogen mode waves were most likely driven by solar wind/magnetospheric compression.

Anderson et al. [1990] explained the prevalence of high-L EMIC waves by examining the dependence of maximum EMIC convective growth rate on plasma parameters. For simplicity, they considered a proton-electron plasma with hot and cold protons immersed in a background magnetic field to approximate conditions appropriate for both the outer and inner magnetosphere. They showed that the wave growth rate is mostly strongly affected by the background magnetic field, such that a stronger magnetic field suppresses the EMIC wave growth. In addition, both the perpendicular temperature

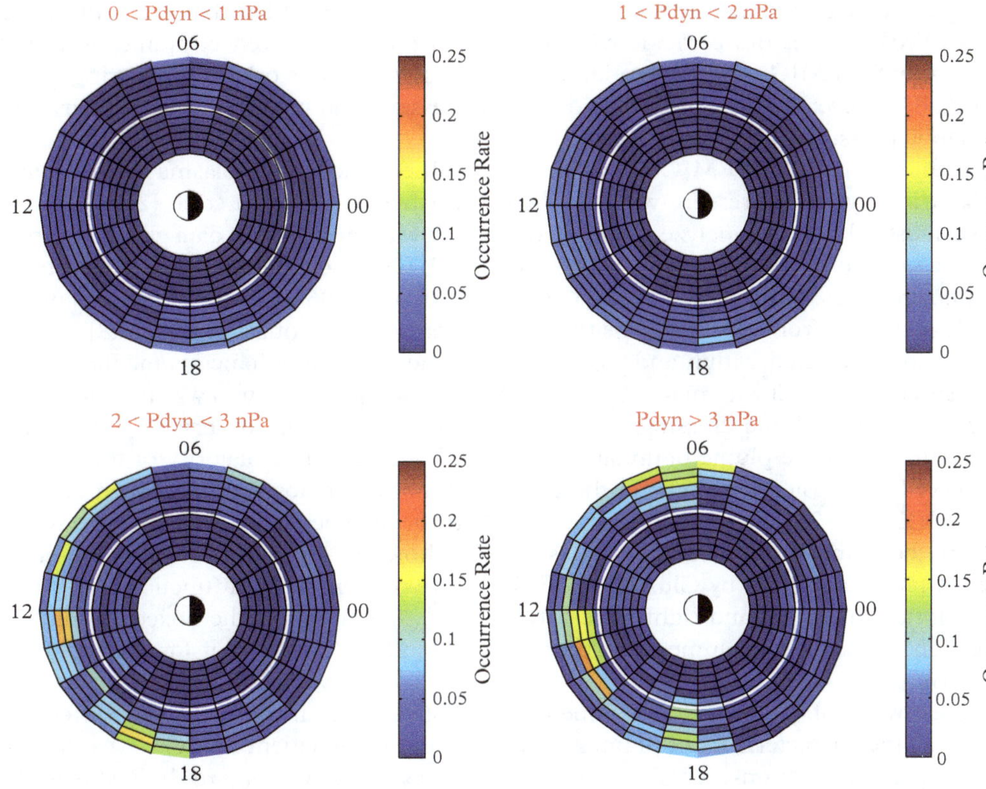

Figure 10.5 *EMIC wave occurrence distribution (in L-MLT) as observed by the THEMIS satellites for multiple intervals of solar wind dynamic pressure:* $0 < Pdyn < 1$, $1 < Pdyn < 2$, $2 < Pdyn < 3$, *and* $Pdyn > 3$ *nPa. The EMIC wave occurrence rate is color coded from 0 (blue) to 0.25 (red). The white solid line in each panel shows geosynchronous orbit.*
(Adapted from Usanova et al., 2012.)

and temperature anisotropy of hot protons have a profound effect on the growth rate. Conversely, the cold proton density has the least effect on the convective growth rate. Hence these authors concluded that the maximum growth rate would be expected to be reached close to the magnetopause, consistent with observations. Their analysis also suggested that cold plasma density gradients in the inner magnetosphere (for example at the plasmapause) would be more important than just enhanced cold proton density for aiding EMIC wave growth.

Posch et al. [2010] compared EMIC wave occurrence on the ground at Antarctic high latitude (L = 6.28–8.07) magnetometer stations with conjugate geosynchronous measurements of cold plasma density between 1996 and 2003. They found only a weak correspondence between EMIC wave occurrence and cold plasma density, either from plasmaspheric plumes or from intervals of expanded plasmasphere where the plasmapause crossed geosynchronous orbit. They concluded that even though at times EMIC waves did coincide with intervals of plume occurrence, plumes were not necessary for EMIC wave generation around geosynchronous orbit, even during storm times. Their events were observed most often in association with increases in solar wind dynamic pressure.

The dependence of EMIC occurrence on both enhanced cold plasma densities and solar wind dynamic pressure was investigated further by Usanova et al. [2013]. In the first statistical study to analyze simultaneous in situ wave and plasma data globally, across a wide range of L-shells (L = 4–11), Usanova et al. [2013] examined 993 plasmaspheric plume crossings observed by Cluster between 2001 and 2006 and which had been previously identified from electron density measurements by Darrouzet et al. [2008]. EMIC waves were detected within and close to the plumes from simultaneous magnetic field measurements by Cluster magnetometers. An interesting observation was that only during 10% of the plume crossings by Cluster were EMIC waves observed either inside the plume or at the plume boundary (inside and outside). The majority of the plumes observed by Cluster were seen in the duskside inner magnetosphere (between L = 5–7 and MLT = 16–22); however, an increased EMIC occurrence was not seen in this sector where the plume occurrence dominates. This is shown in Figure 10.6, where the number of plume crossings by Cluster and EMIC wave occurrence as a function of L/MLT are plotted in the left and right panels, respectively. As shown in the boxed region, the location of maximum plume occurrence corresponds to a region of low EMIC wave occurrence.

Usanova et al. [2013] also examined the statistics to see if there was any difference between the characteristics of plumes with and without coincident EMIC waves (tagged as "events" and "non-events" in Figure 10.7, respectively). Figure 10.7 shows the

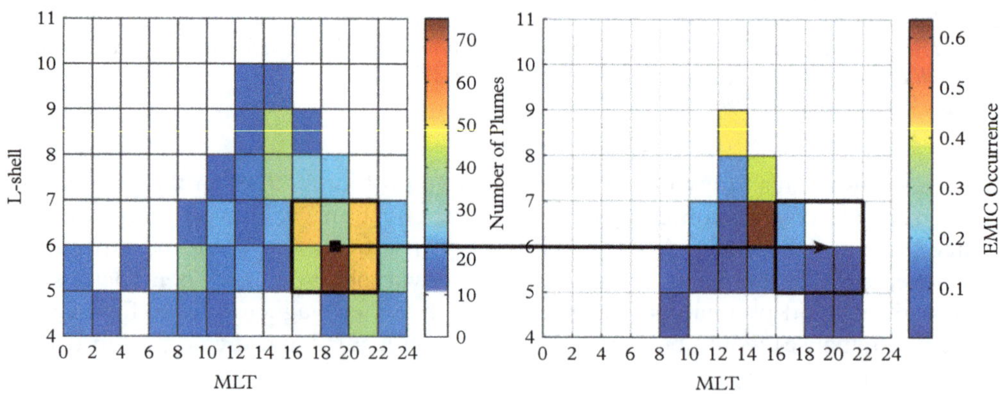

Figure 10.6 *(left) L/MLT distribution of plasmaspheric plumes and (right) L/MLT EMIC wave occurrence rate in plumes. The black rectangle shows the region with maximum plume occurrence, which corresponds to a region of low EMIC wave occurrence. Bins with less than 10 plume crossings are not colored.*

(From Usanova et al., 2013.)

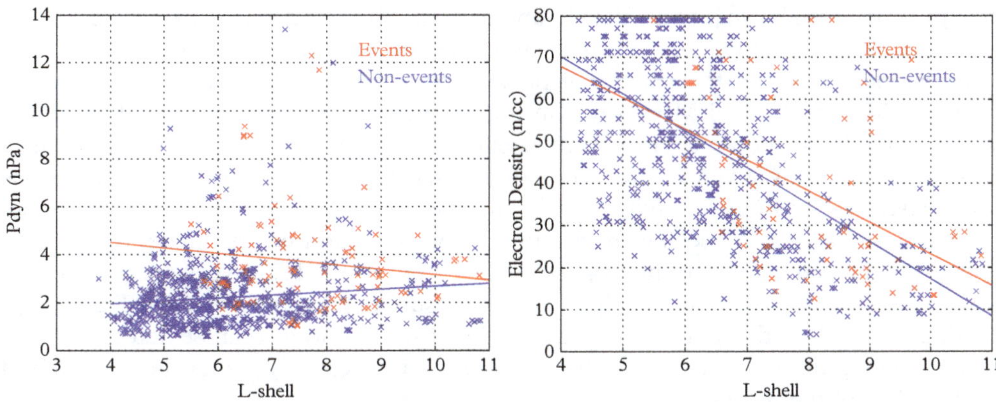

Figure 10.7 *(left) Solar wind dynamic pressure and (right) electron density of plasmaspheric plumes identified by Darrouzet et al. [2008] as a function of L-shell. Plumes with coincident EMIC waves (events) are shown in red, and those without coincident EMIC waves (non-events) are shown in blue. Linear regression lines are over-plotted for each population of events and non-events.*
(From Usanova et al., 2013.)

maximum value of electron density observed along one plume transit (right panel) and the value of solar wind dynamic pressure (left panel) for each individual plume crossing as a function of L-shell. In general, as shown in Figure 10.7, Usanova et al. [2013] found no significant difference in the plume electron density between those with EMIC waves (events) and those without EMIC waves (non-events). Interestingly, the average dynamic pressure during the events was 1.7 times higher than during the non-events, this difference being more significant at low L-shells. This is consistent with the concept that at higher L-shells, closer to magnetopause, the plasma is often marginally stable such that even small enhancements in solar wind dynamic pressure can lead to generation of EMIC waves [e.g., Anderson and Hamilton, 1993]. Finally, Usanova et al. [2013] concluded overall that solar wind dynamic pressure exerts the strongest control over EMIC occurrence, and that enhanced plasma density by itself is not sufficient for EMIC wave generation inside the plumes. This challenges the assumption that the westward plasmaspheric plume edge is a preferred location for EMIC wave generation and consequent scattering loss of energetic particles to the atmosphere as shown schematically by Li et al. [2007].

10.2.3 Radial and MLT extent of EMIC waves

Information about the radial and MLT extent of EMIC activity is extremely important for estimation of energetic particle (especially ring current ion and outer radiation belt electron) loss rates into the atmosphere. Such spatial extents will control the efficiency with which EMIC waves can cause large-scale changes in the global morphology of the ring current and radiation belt, as well as determine if there is any strong spatial structuring in the energy input into the atmosphere as a result of energetic particle precipitation. Ideally, to obtain details about both MLT and radial extent, simultaneous ground and conjugate in situ measurements are required. Satellite measurements from elliptical orbits in the equatorial plane provide an excellent basis for estimating the radial localization of EMIC waves in the magnetosphere [cf. Usanova et al., 2008; Mann et al., 2014]; however since such satellite orbits typically traverse a limited range of local time on any given L-shell, they only offer limited capacity for estimating the MLT extent of the EMIC activity seen in any specific single event. On the other hand, ground observations, especially from multiple locations around the globe, offer an alternative perspective for determining the MLT extent of EMIC wave events. While this is an improvement over single satellite in situ measurements in terms of MLT coverage, ducting of signals in the ionospheric waveguide means that waves can be detected on the ground at long distances away from the source—not necessarily just at the footpoint of the field-line of the source.

Using simultaneous ground–satellite data from the Polar satellite and high-latitude Antarctic stations, Engebretson et al. [2002] observed spatially localized EMIC waves, where continuous wave emissions were seen on the ground for extended periods of time. In space, the waves were observed on the Polar satellite for only a few minutes occurring only in radially narrow regions outside the plasmapause from L = 5–11. The majority of these events (75%) were associated with significant compressions of the magnetosphere. Usanova et al. [2008] reported a similar observation of compression-related EMIC that occurred inside the plasmapause, which is more unusual. The waves were observed on the ground for ~4.5 hours (see Figure 10.2) but for less than 30 minutes while three THEMIS spacecraft crossed the inner magnetosphere in a "string-of-pearls" configuration and observed EMIC waves on L-shells ranging from 5 to 6.5. On each satellite, the radial extent of EMIC activity was found to be ~1.3 Re and confined to a region just inside the plasmapause. EMIC wave spectrograms from the THEMIS magnetometer data for the event studied by Usanova et al. [2008] are shown in Figure 10.8a–c. In each panel, the waves appear to be confined within the high-density side of the plasmapause (the electron density being monitored via the proxy of spacecraft potential, and shown by the red line in each panel). The thick lines in panel (d) show the intervals where EMIC waves were seen by each of the three THEMIS satellites. Liu et al. [2013] presented a similar observation from the Cluster satellites, EMIC waves being detected just inside the plasmapause and also having a radial width < 1Re. Another EMIC wave case study from the Van Allen Probes mission (considered in more detail in Section 10.3.2), simultaneous with conjugate ground-based CARISMA magnetometer observations, also showed a very nice example of narrow-width (<0.5 Re) EMIC waves which were only observed for ~15 minutes in situ but were seen continuously on the ground for more than 18 hours (see also Usanova et al. [2014] and Section 10.3.2 for more details). All these case studies suggest that the plasma density and its gradient play a significant role in confining the waves to a narrow region in space which only spans a small region of L-shell. Such gradients may influence the field-guided nature of the wave propagation and also affect the EMIC wave growth rate. Indeed, simulation of EMIC wave propagation and growth with the HOTRAY ray tracing code by Chen et al. [2009] showed that density structures guide waves keeping the wave normal angle small, which can lead to an order of magnitude enhancement in net wave gain.

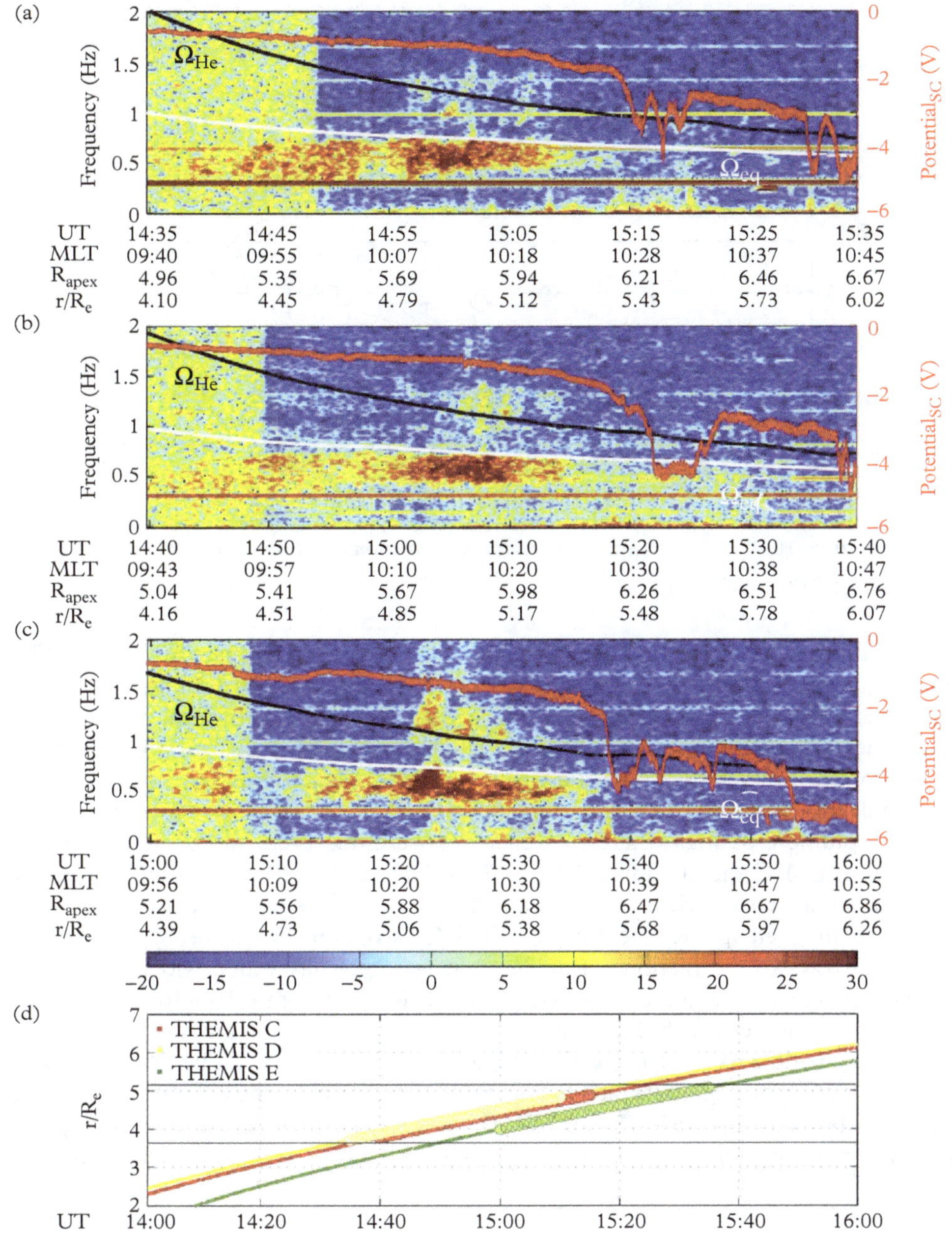

Figure 10.8 *(a–c) Fourier spectral density of the perpendicular (azimuthal) magnetic field component in field-aligned coordinates for one hour windows at THEMIS D, C, and E; over-plotted are the spin-plane spacecraft potential (red line), the local helium gyrofrequency (black line), and the helium gyrofrequency mapped to the equatorial plane (white line). The apexes of the geomagnetic field lines (see x-axis labels) were obtained by tracing field lines from THEMIS to the minimum |B| using the T96 model. (d) Geocentric radial distance of THEMIS C, D, and E as a function of UT. Thick lines show the intervals where each spacecraft sees EMIC waves.*

(Adapted from Usanova et al., 2008.)

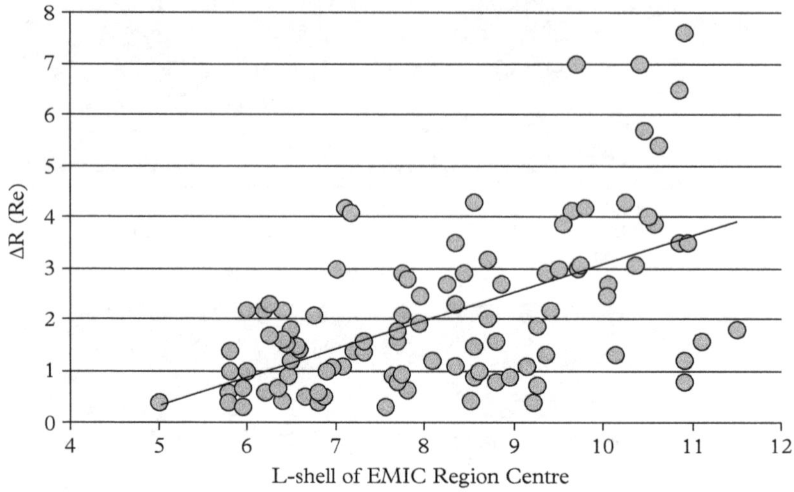

Figure 10.9 *EMIC region radial width as a function of L from Cluster plume crossings.*

In a further study, we also used a prior Cluster plume survey of Darrouzet et al. [2008] to investigate the relationship between EMIC wave occurrence in plasmaspheric drainage plumes. Since plasmaspheric plumes were identified based on the multiple spacecraft measurements available from the Cluster satellite constellation, this survey provided an excellent opportunity to examine the width of EMIC elements in the magnetosphere, and to look for evidence of radial confinement of EMIC wave elements within plumes. From 106 plume crossings where EMIC waves were observed, we found a significant dispersion in EMIC radial width (2.1 ± 1.7 Re). Despite the broad range of observed radial widths, there was a clear trend toward the width increasing with increasing L-shell. In the inner magnetosphere, the region of EMIC wave activity was usually observed to be quite narrow, a few fractions of an Re, but was typically much wider in the outer magnetosphere, where it could be as wide as a few Re (see Figure 10.9 for details). This conclusion is consistent with the results of Lee et al. [2013], who analyzed EMIC wave characteristic widths using data from multiple THEMIS satellites. They concluded that EMIC elements span regions which become wider at higher L-shells. Note however that the EMIC characteristic scale from THEMIS was found to lie between 0.2 and 1.4 Re, which is somewhat narrower than that in our Cluster survey and shown in Figure 10.9.

10.3 Role of EMIC waves in energetic particle loss

As described earlier, EMIC waves are believed to be potentially important for influencing energetic particle dynamics in the magnetosphere. This can occur as a result of the changing ion dynamics during wave growth, or from resonant interactions with other ion or electron populations in the ring current and radiation belts.

Energetic particles (both ions and electrons) can undergo cyclotron resonance with EMIC waves and consequent pitch–angle scattering into the atmosphere if:

$$\omega - k_{\parallel} v_{\parallel} = n\,|\Omega|\,/\gamma, \quad n = 0, \pm 1 \ldots,$$

where ω is the wave frequency, $k_{\parallel}$ is the parallel (to the ambient magnetic field) wave number, $v_{\parallel}$ is the particle parallel velocity, Ω is the particle gyrofrequency, and γ is the relativistic correction [see, e.g., Thorne et al., 2013]. In order for the ions to be in resonance with the waves they must move in a direction opposite to the waves. The resonant energy of ions for EMIC wave interactions is thought to lie between a few keV to a few hundred keV. Electrons, despite their large gyrofrequency (~kHz), can also interact with EMIC waves (with frequencies ~Hz), but only if their parallel speed is sufficiently high such that $k_{\parallel} v_{\parallel}$ reaches the order of the electron gyrofrequency. In this case there is a Doppler shift of the wave frequency and the sense of rotation in the electron reference frame changes from left-handed to right-handed.

In terms of the EMIC wave dispersion relation, in a cold multi-ion-component magnetospheric plasma, there are forbidden band gaps for EMIC wave generation and propagation, generally close to the heavy ion gyrofrequencies and which split the wave spectrum into multiple branches (H^+, He^+, O^+), the growth rates on each branch being controlled by (typically proton) temperature anisotropy, minor ion composition, and cold plasma density [e.g., Kozyra et al., 1984; Horne and Thorne, 1997]. The resonance energy of relativistic electrons interacting with EMIC waves is very sensitive to the ratio of local plasma density to the magnetic field strength and, in the cold multi-ion plasma approximation, to the proximity of the wave frequency to ion gyrofrequencies [e.g., Summers and Thorne, 2003; Ukhorskiy et al., 2010]. In this case the minimum resonance energy is given by the equation:

$$E_{min} = \frac{B^2}{8\pi N_i}\frac{m_i}{m_e}\frac{\Omega_i - \omega}{\Omega_i},$$

where ω is the wave frequency, Ω_i is the ion gyrofrequency, N_i is the ion density, B is the magnitude of the background magnetic field, and m_i and m_e are the ion and electron mass, respectively [Ukhorskiy et al., 2010]. Interestingly, the importance of the ion gyrofrequency in controlling the minimum energy is not maintained under a warm plasma approximation [Silin et al., 2011]. Indeed, it is the value of the k-vector, which controls the interaction. In the cold plasma approximation, k becomes very large close to the local ion gyrofrequency allowing resonance with lower energy electrons. However, in a warm plasma such behavior is not maintained, the behavior is more complex and depends on the wave frequency ω only implicitly via the dispersion relation. In fact the regions of minimum energy are the regions of the dispersion relation where k is large, which in general do not occur at the heavy ion gyrofrequency. Consequently the formula for the minimum resonant energy in Ukhorskiy et al. [2010] may not apply—see Silin et al. [2011] for more details.

In the cold plasma approximation, EMIC waves can only interact with electrons below ~1–2 MeV (that constitute a bulk distribution of radiation belt electrons) in regions of high plasma density and low magnetic field, e.g., outer regions of the plasmasphere or plasmaspheric plumes. For ultra-relativistic (>2 MeV) electrons, high background plasma density together with the proximity of the wave frequency to the local ion gyrofrequency may not be necessary to satisfy the resonance condition. Therefore, EMIC waves are expected to have a much more pronounced effect on ultra-relativistic (>2 MeV) electron pitch-angle scattering. Meredith et al. [2003] performed a statistical survey of EMIC waves using the magnetic field data from the CRRES mission and found that typically, EMIC waves can only interact with >2 MeV energy electrons. This was later confirmed by numerical simulations of a third ultra-relativistic radiation belt formation [e.g., Shprits et al., 2013]. Nevertheless, in 11% of the events from Meredith et al. [2003] statistics the cold plasma resonance energy was below 2 MeV, and the majority of these EMIC wave events were generated on the helium branch.

As described earlier, recent theoretical and experimental studies have also extended EMIC wave analysis beyond the cold plasma approximation and focused on thermal plasma effects on the resonance energy. In a warm plasma, the wave energy can tunnel through a stop band close to the ion gyrofrequency, and the warm plasma dispersion relation leads to the modification of the equation for the minimum resonance energy. The lowest resonance electron energies are achieved for largest wave numbers; however at large wave numbers the waves may also be significantly damped [Silin et al., 2011; Chen et al., 2013]. As a result EMIC waves may still be restricted to resonant interactions with > a few MeV energy electrons [Chen et al., 2013]. Interestingly, Lee et al. [2012] provided observational evidence of EMIC wave tunneling through a stop band, as predicted by the warm plasma theory, using THEMIS satellite instrumentation and questioned the effectiveness of MeV electron loss caused by EMIC waves.

The effectiveness of relativistic electron scattering by EMIC waves also strongly depends on the wave normal angle distribution. Despite the fact that in most numerical models EMIC waves are assumed field aligned, there is experimental evidence that EMIC waves can sometimes be highly oblique and the wave normal angle can reach ~70° [see, e.g., Anderson et al., 1996b; Denton et al., 1996]. Khazanov and Gamayunov [2007] studied the effect of the wave normal angle characteristics on the EMIC wave-radiation belt electron scattering process using self-consistent numerical simulations. They found that oblique EMIC wave distributions reduce the pitch-angle range of electrons that can interact with EMIC waves, and strongly decrease the scattering rate for lower energy electrons (< 2 MeV).

10.3.1 Role in ion precipitation into the atmosphere

The theory that EMIC waves produce scattering of energetic protons into the atmosphere in the course of EMIC instability development and wave growth was supported by observations of localized precipitation of energetic protons (LPEP) by the ESRO satellite, and ground EMIC waves conjugate to the satellite, reported for the first time by Søraas et al. [1980]. Further connections between LPEP and EMIC waves have been examined using the low Earth orbit polar NOAA satellites [see, e.g., Yahnin and

Yahnina, 2007 and references therein]. Currently EMIC waves are considered to be one of the important loss mechanisms affecting ring current dynamics, and can now be included self-consistently in global ring current numerical simulation models [e.g., Jordanova, 1997; Jordanova et al., 2012].

Sandanger et al. [2007] reported simultaneous observations of collocated LPEP and relativistic electron precipitation in space and suggested that the scattering of both protons and relativistic electrons might have occurred due to interactions with the same EMIC waves—even though these authors did not report any direct evidence of the presence of EMIC waves. A later work by Miyoshi et al. [2008] presented an event where EMIC waves were observed by a ground-based magnetometer and the proton precipitation was detected by both NOAA and an all-sky camera as an isolated proton aurora in the hydrogen Balmer β line. This event was also accompanied by precipitation of relativistic electrons also observed by NOAA.

Proton aurora is produced by Doppler-shifted emissions of neutral hydrogen as a result of proton precipitation into the atmosphere and consequent charge exchange [see, e.g., Sakaguchi et al., 2007 and references therein]. Global images of the proton aurora were obtained by the IMAGE satellite mission. A large number of works identified proton aurora existing equatorward of the main auroral oval, and have characterized them as "proton spots," "proton arcs," and "proton flashes" [see, e.g., Frey et al., 2007]. These aurora are believed to be associated with LPEP due to interaction with EMIC waves [Yahnin and Yahnina, 2007]. Yahnin et al. [2013] performed a multi-event study of sub-oval proton aurorae using IMAGE data and found that 65% of proton auroral spots were located in the vicinity of the model plasmapause (within 0.5 Re). These authors suggested that the origin of these proton spots, and the proton precipitation that produced, is interaction with monochromatic EMIC waves which are often observed close to the plasmapause below the equatorial helium gyrofrequency.

An example of EMIC wave activity and energetic proton precipitation during the recovery phase of a moderate geomagnetic storm on October 11, 2012, is shown in Figure 10.10 [see also Usanova et al., 2014]. The EMIC wave activity was observed in situ on the Van Allen Probes and conjugately on the ground across the CARISMA array (`http://www.carisma.ca`) throughout an extended 18-hour interval. The Van Allen Probes were conjugate to the region of EMIC activity in Canada and observed helium-band EMIC waves on a narrow range of L-shells from L~4.0–4.5. EMIC waves were bounded at high L by a density drop, which is indicative of the plasmapause (red line in Figure 10.10c).

During the interval of EMIC activity on the ground, NOAA POES (the National Oceanic and Atmospheric Administration Polar Orbiting Environmental Satellites) observed a loss of energetic protons in low Earth orbit. For example, at ~14:08 UT, POES registered a localized enhancement of precipitating protons with energies >30 keV close to the ground magnetometer station PINA at L~4. The panels in Figure 10.10 (d–g) show POES proton data from 30 to 80 keV, 80 to 250 keV, and 250 to 800 keV, and the channel sensitive to >0.7 MeV electrons, respectively. The blue (red) trace in Figure 10.10 (d–g) shows the flux of precipitating (locally mirroring, trapped) particles observed by the Medium Energy Proton and Electron Detector MEPED [Evans and Greer, 2000]. Clearly, the precipitation of energetic protons associated with the narrow

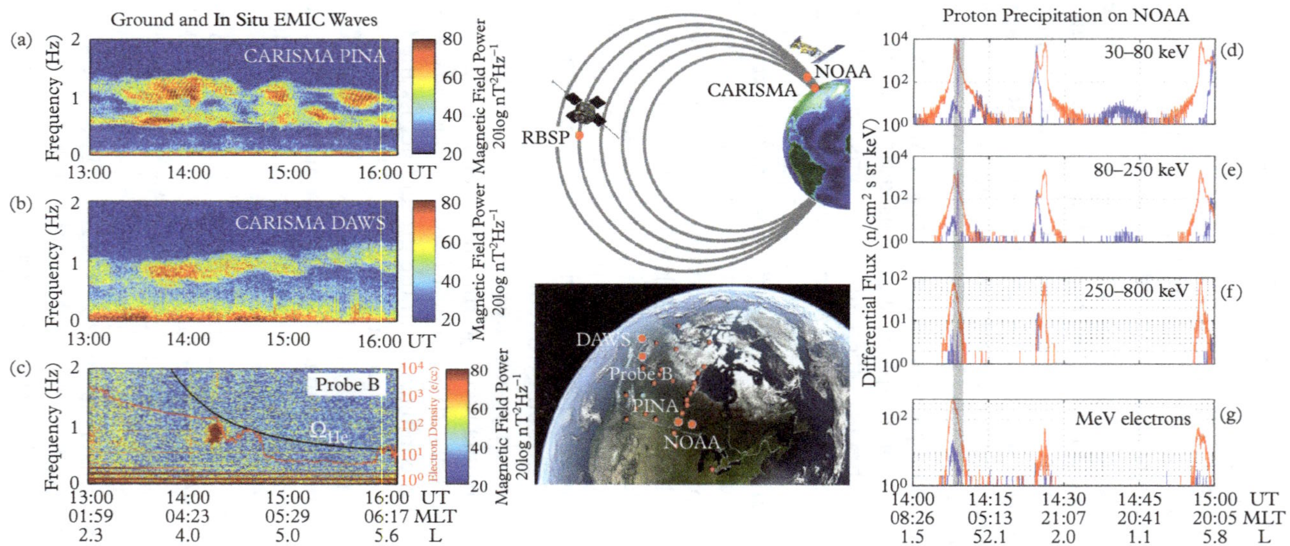

Figure 10.10 *EMIC wave observations from the CARISMA PINA (a) and DAWS (b) magnetometers, from GSM Bx component from Van Allen Probe B (c), together with loss of energetic protons on the POES 16 satellite (d–f) on October 11, 2012. Note that >0.7 MeV electron loss (panel g) was not observed on POES 16 at the same time. The central panel shows the schematic locations of the satellites and ground magnetometers.*

(From Usanova et al., 2014.)

L-range of EMIC waves is not accompanied by precipitating >0.7 MeV energetic electrons in this case at this time. Following the classification of Yahnin and Yahnina, [2007], we identify this proton event as LPEP of type 1, i.e., characterized by localization of proton precipitation in the morning MLT sector during quiet geomagnetic conditions, without the presence of a precipitating energetic electron population, and in conjunction with structured Pc 1 pulsations observed on the ground. A similar precipitation event, associated with magnetospheric compression and observed on the ground by CARISMA and in space by Cluster, was reported by Usanova et al. [2010a].

10.3.2 Role in radiation belt electron precipitation

The role of EMIC waves in the loss of radiation belt electrons is still a controversial subject. As discussed in the introduction, EMIC waves can only resonate with ~MeV energy electrons under specific conditions and they most effectively interact with low pitch angle particles. The direct resonance interaction of relativistic and even ultra-relativistic electrons of high pitch angle is very weak, and hence other additional wave modes are required to also interact with the particles at the same time in order for EMIC waves to contribute to the loss of the entire pitch angle distribution, including high pitch angle particles and affecting the core of the distribution up to 90° pitch angle.

The potential importance of EMIC waves as a radiation belt electron loss mechanism was recently highlighted in some theoretical/modeling papers [e.g., Li et al., 2007; Ukhorskiy et al., 2010; Shprits et al., 2013]. For example, Shprits et al. [2013] simulated the formation of the third radiation belt consisting of >2 MeV electrons observed by Baker et al. [2013] and concluded that EMIC waves were a necessary component of the plasma environment required to produce loss of the ultra-relativistic electrons and hence to explain the observed configuration of a narrow third radiation belt spatially separated from the higher-L outer belt. Observationally, Sakaguchi et al. [2013] reported EMIC waves in the radiation belt slot region seen by the Akebono satellite. This event happened during a recovery phase of a severe geomagnetic storm, on September 21, 1989, and where Akebono observed a decrease of >2.5 MeV electron flux. These authors suggested that the spatial collocation of EMIC waves and the slot region might be due to a contribution from EMIC waves to the loss of relativistic electrons and hence to the formation of the slot region. Turner et al. [2013] examined storm-time evolution of the outer radiation belt using THEMIS instrumentation and concluded that EMIC waves might have contributed to the loss of relativistic electrons observed during the recovery phase of selected magnetic storm; however, there was no evidence of EMIC waves being present during the main phase where the major loss occurred.

From an observational perspective, there are a limited number of studies showing direct spatiotemporal correlation between energetic electron precipitation and EMIC waves. Miyoshi et al. [2008] reported observations of simultaneous proton aurora and precipitation of energetic 30–80 keV protons and >3 MeV electrons on the POES satellite together with EMIC waves on the ground in a single event study and concluded that these EMIC waves produced the observed precipitation. Sandanger et al. [2009] observed spatially collocated proton and relativistic electron precipitation on NOAA POES during the recovery phase of geomagnetic storms and suggested that they were the result of interactions with EMIC waves inside the plasmapause. However, their observations did not include any simultaneous observations of EMIC waves which were assumed to be the cause of the precipitation. Carson et al. [2013] performed a statistical study of NOAA POES data between 1998 and 2010 (using the same approach as Sandanger et al. [2009] but in an automated manner) and found that only a subset of EMIC-driven proton precipitation events simultaneously produced any relativistic electron precipitation. Adding further complexity, however, the majority of the events showing both proton and electron precipitation in the Carson et al. [2013] survey occurred in the midnight/dawn sector. If such loss is to be associated with EMIC waves this is puzzling since such a spatial distribution is very dissimilar to the known distribution of EMIC-wave occurrence probability, which maximizes in the dusk-to-noon sector.

Usanova et al. [2014] investigated interactions between EMIC waves and ultra-relativistic electrons from both observational and modeling perspectives by examining the effect of electromagnetic ion cyclotron (EMIC) waves on the loss and pitch-angle scattering of relativistic (<2 MeV) and ultra-relativistic (≥2 MeV) electrons. They did not observe precipitation of relativistic electrons in the region where both EMIC waves and co-located proton precipitation observed by NOAA POES occurred (see Figure 10.10, adapted from Usanova et al. [2014] and the discussion in Section 10.3.2). In addition, the Van Allen Probe data did not show any significant radiation belt flux reduction in the equatorial magnetosphere. Figure 10.11 (a–c) shows spin averaged differential electron flux in the 2.3, 4.5, and 5.6 MeV energy channels as a function of L between October 9 and November 29, 2012, as observed by the Relativistic Electron Proton Telescope (REPT) [Baker et al., 2012] on the Van Allen Probes (the EMIC event on October 11 is tagged by the green arrow). Overall, the entire interval can be characterized by a gradual and continuous decrease in flux, interspersed with a sharp step decrease on November 1, and a similar sharp increase around November 15, 2012.

Figure 10.11 (d–f) shows electron differential flux as a function of pitch angle in the same 2.3, 4.5, and 5.6 MeV energy channels at L∼4.5, during the entire 51-day interval. In each panel, the flux is normalized by the 90° pitch-angle flux to highlight the changing width of the pitch-angle distributions: being narrow (i.e., more strongly peaked at 90°) at the beginning, then widening, and experiencing repeated fluctuations later in the interval. The EMIC wave occurrence as seen on the ground at two magnetometer stations (both at L∼4–4.5) is shown in Figure 10.11g and tagged as "No" for days of absent EMIC waves and "Yes" for days where EMIC waves are present. Figure 10.11 demonstrates a clear correlation and connection between rapid changes in ultra-relativistic pitch-angle distributions and the occurrence of EMIC waves. It provides good evidence that EMIC waves can generate bite-outs in flux at low pitch angles, which can last for extended intervals; however, especially given the lack of evidence for significant overall losses in the radiation belts suggests a limited role for generating drop-outs in flux across the entire pitch angle distribution which have effects right into the core of the peaked-at -90° pitch angle distributions.

To test this hypothesis, Usanova et al. [2014] computed bounce-averaged pitch-angle diffusion coefficients of electrons due to EMIC waves using a quasi-linear approach [e.g., Kennel and Engelmann, 1966; Summers et al., 2007]. Using the Van Allen Probe

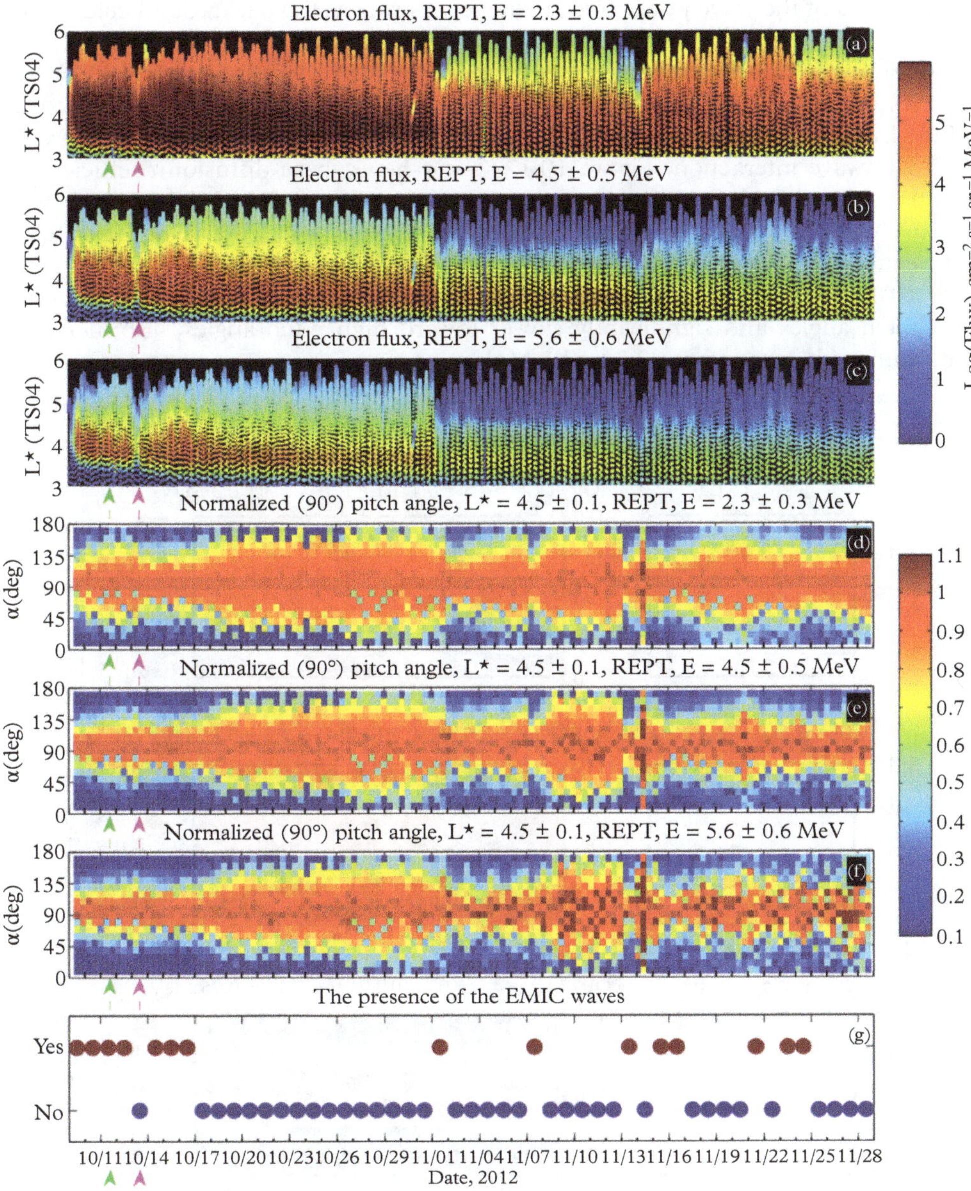

Figure 10.11 *Spin-averaged differential electron flux as a function of L* (a–c), and differential flux as a function of pitch angle, normalized by the 90° pitch-angle flux, at L*=4.5 (d–f) in the 2.3, 4.5, and 5.6 MeV energy channels, respectively, from REPT, and EMIC wave occurrence from L~4–4.5 on the ground (g) between October 9 and November 29, 2012. The green arrow shows the EMIC event on October 11. The purple arrow indicates the time of the minimum Dst at 11 UT on October 13, 2012.* *(From Usanova et al., 2014.)*

measurements of the wave power and frequency, as well as the background cold plasma density complemented by the assumption that the plasma is cold, and that the ion composition was 70% H^+, 20% He^+, and 10% O^+ (as in Meredith et al. [2003]), and the waves are left-handed and circular polarized. Usanova et al. [2014] estimated the likelihood of the observed electron pitch-angle distributions being produced through this EMIC wave interaction. Figure 10.12 (a–c) shows these diffusion coefficients as a function of equatorial pitch angle for electron energies of 2.3, 4.5, and 7.15 MeV, respectively. A particle loss timescale can be estimated using the inverse of the diffusion coefficient. Loss timescales in Figure 10.12 (a–c) lie in the interval of approximately 0.1 to 10 hours, depending on the electron energy and pitch angle, being very fast at low pitch angles and significantly slower toward high pitch angles, also increasing significantly with energy from 2.3 to 7.1 MeV.

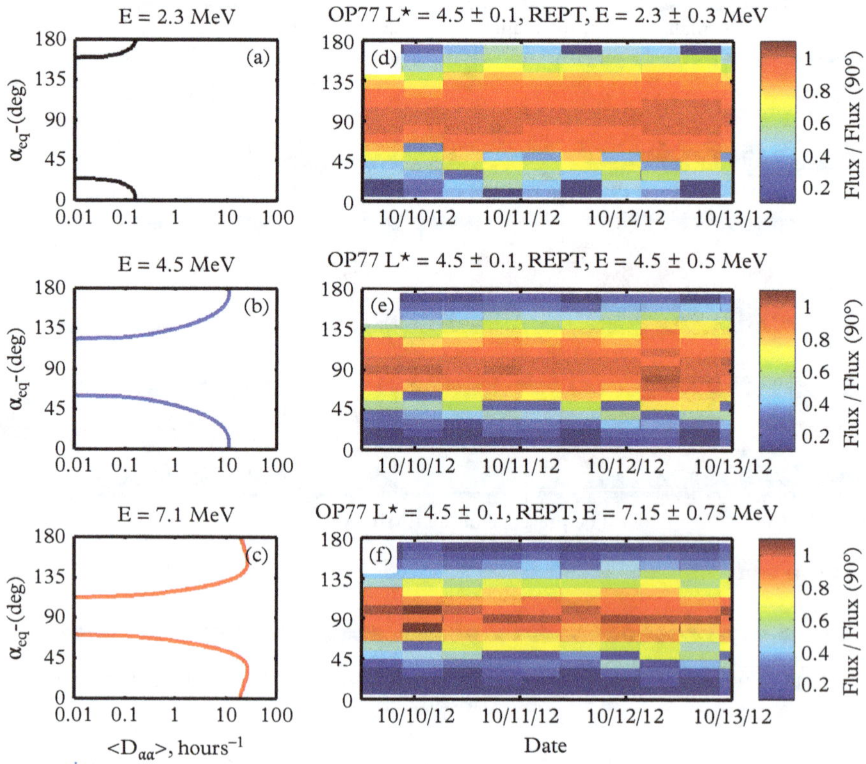

Figure 10.12 *Computed pitch-angle diffusion coefficients (a–c) and observed normalized electron flux as a function of pitch angle (d–f) in the 2.3, 4.5, and 7.15 MeV REPT energy channels, respectively, from Van Allen Probes for October 9–13, 2012.*
(From Usanova et al., 2014.)

Figure 10.12 demonstrates that with increasing energy, electrons with a larger range of pitch angles are scattered by EMIC causing the confinement of fluxes closer to equatorial pitch angles of ~90°, especially at higher (close to ~8 MeV) energies. This is in excellent agreement with the REPT instrument energy/pitch-angle spectrograms in Figure 10.12 (d–f) that show normalized electron flux as a function of pitch angle (d–f) for 2.3, 4.5, and 7.15 MeV during the times where EMIC waves were present. The fact that these EMIC waves do not affect the core of the electron distribution is consistent with the absence of significant precipitation observed by POES. The fraction of precipitating electrons during this event was ~1% of the >0.7 MeV detectable electron flux, which is below the background precipitating flux (5%) and hence most likely cannot be resolved by POES.

A number of theoretical works [e.g., Summers et al., 2007; Albert, 2008; Ukhorskiy et al., 2010] have previously indicated that EMIC waves alone are unable to resonate with close to 90° pitch angle electrons, and other mode waves (e.g., magnetosonic waves) are required to be present simultaneously with EMIC waves in order to remove the core of the higher energy radiation belt population. The study by Usanova et al. [2014] was the first clearly presenting direct observational evidence supporting this theory, significantly casting some doubt on the potential importance of EMIC waves for controlling relativistic electron loss from the outer Van Allen radiation belt. More work is clearly needed to investigate this further.

10.4 Discussion and conclusions

The modern era of satellite missions has provided an excellent data set with which to improve our understanding not only of the processes responsible for the excitation of EMIC waves, but also to examine their potential impacts for the scattering of energetic ions and electrons in the ring current and outer radiation belt into the atmosphere. Recent missions such as the NASA THEMIS mission, ESA-NASA Cluster mission, and most recently the NASA Van Allen Probes missions—combined with conjugate particle precipitation measurements from low Earth orbit as well as extensive ground-based magnetometer arrays—have led to significant advances in our understanding of the dynamical evolution and impacts of EMIC waves in the terrestrial magnetosphere.

In terms of EMIC wave excitation, recent observations have highlighted the importance of solar wind dynamic pressure for exciting EMIC waves, especially on the dayside, in addition to more traditional excitation due to the effects of drift dispersion. As has been long known, solar wind compression-related events are important at high-L and their excitation off-the-equatorial plane can be associated with Shabansky orbits. However, recent studies have also shown solar compression to be an important excitation mechanism even in the inner magnetosphere. The region just inside the plasmapause appears to be a preferential region for EMIC wave excitation, where it overlaps with an available unstable energetic ion source population; however, an active understanding of the coupling between the ring current, radiation belt, and the dense cold plasma population of the plasmasphere needs further work.

Interestingly, statistical studies have challenged some of the former perceived wisdoms about EMIC wave excitation. For example, the concept that EMIC waves are prevalent along a westward edge of plasmaspheric drainage plumes is not accurate—and there is little difference in the density of plumes with and without EMIC waves as observed by satellites. Although there is a slight increase in occurrence rates of EMIC waves in plumes compared to occurrence rates over all times, the general occurrence rate of EMIC waves remains rather low. Certainly solar wind dynamic pressure appears to be the best geomagnetic parameter with which to organize EMIC wave occurrence rates. Given that both energetic electrons and ion distributions in the equatorial plane are expected to develop increased anisotropy during solar wind compression events, an interesting future study would be to look for similar increased occurrence rates on the electron branch such as for lower band whistler modes to that seen for EMIC waves.

The low rate of occurrence of EMIC waves may also challenge inferences that the EMIC waves have an important role to play in radiation belt electron loss. Perhaps most important is the recent observation that while EMIC waves appear to be able to scatter low pitch angle ultra-relativistic electrons into the loss cone and hence into the atmosphere, their impact on higher pitch angle particles is very weak. Consequently, at least by themselves, they cannot cause loss which reaches the core of the electron pitch-angle distribution. Both the very slow diffusion rates and the lack of observed overall loss of entire relativistic electron distributions during EMIC-wave interactions very recently presented by Usanova et al. [2014] hence suggest that EMIC waves might not ordinarily be the dominant electron loss process from the outer Van Allen belts in the magnetosphere—but more studies are definitely needed. Conversely, there is much more evidence that EMIC wave excitation plays a critical role not only in the dynamical evolution of the ion ring current, but also in both ion precipitation loss to the atmosphere and in the generation of the proton aurorae.

Overall, since the morphology of the radiation belts is controlled by a delicate balance between competing acceleration and loss processes, it is clear that understanding the dominant processes controlling the loss of energetic particles from the radiation belts is at least as important as understanding those dominating acceleration. This is complicated by the fact that multiple and different acceleration and loss processes may also act with different efficiencies at different locations around an electron drift path—such that the capability to monitor wave-particle interactions in the radiation belts from a global perspective remains very important. Since true global constellation class missions are probably still decades away from being realized, at least in highly elliptical and high altitude orbits, combining wave and precipitation coverage from extensive networks of ground-based arrays such as magnetometers remains an important complement to the single, or few, in situ point measurements. Of course the location of the satellite is not necessary a preferred location for wave-particle interactions; hence by definition most of the time satellites are observing the fingerprints of wave-particle interactions being driven by wave populations which are non-local to the satellite.

In terms of the goal of developing accurate physics-based models of radiation belt dynamics, and hence accurate forecasts of energetic particle doses and charging impacts on Earth-orbiting satellites, understanding the conditions under which specific wave-particle interactions dominate remains critical but poorly understood. Similarly, it has recently been proposed that energetic particle precipitation into the atmosphere can allow changes in thermospheric NO_x [e.g., Randall et al., 2009] to have a significant impact on global temperatures [e.g., Seppala et al., 2009] and hence have a role to play in understanding the pathways influencing global climate change. Clearly a better understanding of the potential linkage between energetic particle precipitation and climate will benefit from an improved understanding of the processes controlling the scattering of energetic particles into the loss cone and their ultimate precipitation into the atmosphere.

EMIC waves are also often observed, especially on the ground, in the form of chirped and dispersing discrete wave packets. They are also often observed to have large amplitudes, and have polarization and mode characteristics which are expected to change significantly during their growth, saturation, and propagation through the inhomogeneous plasma of the inner magnetosphere. Even now we are only beginning to understand the effects of inhomogeneity, and non-linear effects in wave-particle interactions remain in their relative infancy. Certainly there has been great progress in developing an improved understanding of the role of EMIC waves in affecting ring current and radiation belt energetic particle dynamics. However, what is certain is that much still remains to be done if the processes through which EMIC waves influence the energetic particle populations in the inner magnetosphere are to be fully understood.

Acknowledgments

This work is supported in part by the MAARBLE (Monitoring, Analyzing and Assessing Radiation Belt Loss and Energization) consortium. MAARBLE has received funding from the European Community's Seventh Framework Programme (FP7-SPACE-2010-1, SP1 Cooperation, Collaborative project) under grant agreement number 284520. This paper reflects only the authors' views and the Union is not liable for any use that may be made of the information contained herein. MEU was partly supported by the Canadian Space Agency and by NASA award NAS5-01072. IRM is supported by a Discovery Grant from Canadian NSERC.

REFERENCES

Albert, J. M. (2008), Efficient approximations of quasi-linear diffusion coefficients in the radiation belts, *J. Geophys. Res.*, **113**, A06208, doi:10.1029/2007JA012936.

Anderson, B. J., K. Takahashi, R. E. Erlandson, and L. J. Zanetti (1990), Pc1 pulsations observed by AMPTE/CCE in the Earth's outer magnetosphere, *Geophys. Res. Lett.*, 17, doi: 10.1029/GL017i011p01853.

Anderson, B. J., R. E. Erlandson, and L. J. Zanetti (1992), A statistical study of Pc 1-2 magnetic pulsations in the equatorial magnetosphere. 1. Equatorial occurrence distributions, *J. Geophys. Res.*, **97** (A3), 3075–3088, doi:10.1029/91JA02706.

Anderson, B. J. and D. J. Hamilton (1993), Electromagnetic ion cyclotron waves stimulated by modest magnetospheric compressions, *J. Geophys. Res.*, **98**, 11369–11382, doi:10.1029/93JA00605.

Anderson, B. J., R. E. Denton, G. Ho, D. C. Hamilton, S. A. Fuselier, and R. J. Strangeway (1996a), Observational test of local proton cyclotron instability in the Earth's magnetosphere, *J. Geophys. Res.*, **101** (A10), 21527–21543, doi:10.1029/96JA01251.

Anderson, B. J., R. E. Denton, and S. A. Fuselier (1996b), On determining polarization characteristics of ion cyclotron wave magnetic field fluctuations, *J. Geophys. Res.*, **101**, 13195.

Baker, D. N., et al. (2012), The Relativistic Electron-Proton Telescope (REPT) instrument on board the Radiation Belt Storm Probes (RBSP) spacecraft: characterization of Earth's radiation belt high-energy particle populations, *Space Sci. Rev.*, **179**, 337–381, doi:10.1007/s11214-012-9950-9.

Baker, D. N., et al. (2013), A long-lived relativistic electron storage ring embedded in Earth's outer Van Allen belt, *Science*, **340**, 6129, doi 10.1126/science.1233518.

Borovsky, J. E., R. H. W. Friedel, and M. H. Denton (2014), Statistically measuring the amount of pitch angle scattering that energetic electrons undergo as they drift across the plasmaspheric drainage plume at geosynchronous orbit, *J. Geophys. Res. Space Physics*, **119**, doi:10.1002/2013JA019310.

Bortnik, J., J. W. Cutler, C. Dunson, and T. E. Bleier (2007), An automatic wave detection algorithm applied to Pc1 pulsations, *J. Geophys. Res.*, **112**, A04204, doi:10.1029/2006JA011900.

Carson, B. R., C. J. Rodger, and M. A. Clilverd (2013), POES satellite observations of EMIC-wave driven relativistic electron precipitation during 1998–2010, *J. Geophys. Res. Space Physics*, **118**, 232–243, doi:10.1029/2012JA017998.

Chen, L., R. M. Thorne, and R. H. Horne (2009), Simulation of EMIC excitation in a model magnetosphere including structured high-density plumes, *J. Geophys. Res.*, **114**, A07221, doi:10.1029/2009JA014204.

Chen, L., R. M. Thorne, Y. Shprits, and B. Ni (2013), An improved dispersion relation for parallel propagating electromagnetic waves in warm plasmas: application to electron scattering, *J. Geophys. Res. Space Physics*, **118**, 2185–2195, doi:10.1002/jgra.50260.

Cornwall, J. M. (1965), Cyclotron instabilities and electromagnetic emissions in the ultra low frequency and very low frequency ranges, *J. Geophys. Res.*, **70**, 61–69.

Cornwall, J. M., F. W. Coroniti, and R. M. Thorne (1970), Turbulence loss of ring current protons, *J. Geophys. Res.*, **75**, 4699–4709.

Darrouzet, F., J. De Keyser, P.M.E. Décréau, F. El Lemdani-Mazouz, and X. Vallières (2008), Statistical analysis of plasmaspheric plumes with CLUSTER/WHISPER observations, *Ann. Geophys.* **26** (8), 2403–2417.

Denton, R. E., B. J. Anderson, G. Ho, and D. C. Hamilton (1996), Effects of wave superposition on the polarization of electromagnetic ion cyclotron waves, *J. Geophys. Res.*, **101**, 24869.

Denton, R. E., J. LaBelle, and X. Zhu (2002), Location of Pc 1–2 waves relative to the magnetopause, *Ann. Geophys.*, **20**, 1763–1767.

Engebretson, M. J., W. K. Peterson, J. L. Posch, M. R. Klatt, B. J. Anderson, C. T. Russell, H. J. Singer, R. L. Arnoldy, and H. Fukunishi (2002), Observations of two types of Pc 1–2 pulsations in the outer dayside magnetosphere, *J. Geophys. Res.*, **107** (A12), 1451, doi:10.1029/2001JA000198.

Engebretson, M. J., J. L. Posch, A. M. Westerman, N. J. Otto, J. A. Slavin, G. Le, R. J. Strangeway, and M. R. Lessard (2008a), Temporal and spatial characteristics of Pc1 waves observed by ST5, *J. Geophys. Res.*, **113**, A07206, doi:10.1029/2008JA013145.

Engebretson, M. J., et al. (2008b), Pc1–Pc2 waves and energetic particle precipitation during and after magnetic storms: superposed epoch analysis and case studies, *J. Geophys. Res.*, **113**, A01211, doi:10.1029/2007JA012362.

Evans, D. S. and M. S. Greer (2000), Polar orbiting environmental satellite space environment monitor 2: instrument description and archive data documentation, *NOAA Tech.Memo.OAR SEC-93*, NOAA, Boulder, Colorado.

Fraser, B. J. (1975), Ionospheric duct propagation and Pc 1 pulsation sources, *J. Geophys. Res.*, **80** (19), 2790–2796, doi:10.1029/JA080i019p02790.

Fraser, B. J. and T. S. Nguyen (2001), Is the plasmapause preferred source region of electromagnetic ion cyclotron waves in the magnetosphere? *J. Atm. Solar-Terr. Phys.*, **63**, 1225–1247.

Fraser, B. J., H. J. Singer, M. L. Adrian, D. L. Gallagher, and M. F. Thomsen (2005), The relationship between plasma density structure and EMIC waves at geosynchronous orbit, in *Inner Magnetosphere Interactions: New Perspectives from Imaging*, *Geophys. Monogr. Ser.*, vol. 159, edited by J. Burch, M. Schulz, and H. Spence, pp. 55–70, AGU, Washington, D. C., doi:10.1029/159GM04.

Frey, H. U. (2007), Localized aurora beyond the auroral oval. *Rev. Geophys.*, **45**, RG1003, http://dx.doi.org/10.1029/2005RG000174.

Gendrin, R. (1975). Is the plasmapause a preferential region for proton precipitation? *Ann. Geophys.*, **31**, 127.

Halford, A. J., B. J. Fraser, and S. K. Morley (2010), EMIC wave activity during geomagnetic storm and nonstorm periods: CRRES results, *J. Geophys. Res.*, **115**, A12248, doi:10.1029/201.

Hayakawa, M., S. Shimakura, T. Kobayashi, and N. Sato (1992), A study of polarization of irregular pulsations of diminishing period and their generation mechanism, *Planet. Space Sci.*, **40**, 1081–1091.

Horne, R. B. and R. M. Thorne (1997), Wave heating of He+ by electromagnetic ion cyclotron waves in the magnetosphere: heating near the H^+–He^+ bi-ion resonance frequency, *J. Geophys. Res.*, **102** (A6), 11457–11471, doi:10.1029/97JA00749.

Jordanova, V. K., J. U. Kozyra, A. F. Nagy, and G. V. Khazanov (1997), Kinetic model of the ring current-atmosphere interactions, *J. Geophys. Res.*, **102** (A7), 14279–14291, doi:10.1029/96JA03699.

Jordanova, V. K., C. J. Farrugia, R. M. Thorne, G. V. Khazanov, G. D. Reeves, and M. F. Thomsen (2001), Modeling ring current proton precipitation by electromagnetic ion cyclotron waves during the May 14–16, 1997, storm, *J. Geophys. Res.*, **106** (A1), 7–22, doi:10.1029/2000JA002008.

Jordanova, V. K., D. T. Welling, S. G. Zaharia, L. Chen, and R. M. Thorne (2012), Modeling ring current ion and electron dynamics and plasma instabilities during a high-speed stream driven storm, *J. Geophys. Res.*, **117**, A00L08, doi:10.1029/2011JA017433.

Khazanov, G. V. and K. V. Gamayunov (2007), Effect of electromagnetic ion cyclotron wave normal angle distribution on relativistic electron scattering in outer radiation belt, *J. Geophys. Res.*, **112**, A10209, doi:10.1029/2007JA012282.

Khazanov, G. V., K. V. Gamayunov, D. L. Gallagher, J. U. Kozyra, and M. W. Liemohn (2007), Self-consistent model of magnetospheric ring current and propagating electromagnetic ion cyclotron waves. 2. Wave-induced ring current precipitation and thermal electron heating, *J. Geophys. Res.*, **112**, A04209, doi:10.1029/2006JA012033.

Keika, K., K. Takahashi, A. Y. Ukhorskiy, and Y. Miyoshi (2013), Global characteristics of electromagnetic ion cyclotron waves: occurrence rate and its storm dependence, *J. Geophys. Res. Space Physics*, **118**, 4135–4150, doi:10.1002/jgra.50385.

Kennel, C. F. and F. Engelmann (1966), Velocity space diffusion from weak plasma turbulence in a magnetic fields, *Phys. Fluids*, **9**, 2377–2388.

King, J. H. and N. E. Papitashvili (2004), Solar wind spatial scales in and comparisons of hourly Wind and ACE plasma and magnetic field data, *J. Geophys. Res.*, **110** (A2), A02209, 10.1029/2004JA010804.

Kozyra, J. U., T. E. Cravens, F. Nagy, and E. G. Fontheim (1984), Effects of energetic heavy ions on electromagnetic ion cyclotron wave generation in the plasmapause region, *J. Geophys. Res.*, **89** (A4), 2217–2233.

Lee, J. H., L. Chen, V. Angelopoulos, and R. M. Thorne (2012), THEMIS observations and modeling of multiple ion species and EMIC waves: implications for a vanishing He^+ stop band, *J. Geophys. Res.*, **117**, A06204, doi:10.1029/2012JA017539.

Lee J., K. Min, and K.-S. Kim (2013), Characteristic dimension of electromagnetic ion cyclotron wave activity in the magnetosphere, *J. Geophys. Res. Space Physics*, **118**, 1651–1658, doi:10.1002/jgra.50242.

Li, W., Y. Y. Shprits, and R. M. Thorne (2007), Dynamic evolution of energetic outer zone electrons due to wave-particle interactions during storms, *J. Geophys. Res.*, **112**, A10220, doi:10.1029/2007JA012368.

Liu, Y. H., B. J. Fraser, and F. W. Menk (2012), Pc2 EMIC waves generated high off the equator in the dayside outer magnetosphere, *Geophys. Res. Lett.*, **39**, L17102, doi:10.1029/2012GL053082.

Liu, Y. H., B. J. Fraser, and F. W. Menk (2013), EMIC waves observed by Cluster near the plasmapause, *J. Geophys. Res. Space Physics*, **118**, 5603–5615, doi:10.1002/jgra.50486.

Lyons, L. R., and R. M. Thorne (1972), Parasitic pitch-angle diffusion of radiation belt particles by ion-cyclotron waves, *J. Geophys. Res.*, 77, 5608–5616.

Mann, I. R., M. E. Usanova, K. Murphy, M. T. Robertson, D. K. Milling, A. Kale, C. Kletzing, J. Wygant, S. Thaller, and T. Raita (2014), Spatial localization and ducting of EMIC waves: Van Allen Probes and ground-based observations, *Geophys. Res. Lett.*, **41**, 785–792, doi:10.1002/2013GL058581.

McCollough, J. P., S. R. Elkington, M. E. Usanova, I. R. Mann, D. N. Baker, and Z. C. Kale (2010), Physical mechanisms of compressional EMIC wave growth, *J. Geophys. Res.*, **115**, A10214, doi:10.1029/2010JA015393.

Meredith, N. P., R. M. Thorne, R. B. Horne, D. Summers, B. J. Fraser, and R. R. Anderson (2003), Statistical analysis of relativistic electron energies for cyclotron resonance with EMIC waves observed on CRRES, *J. Geophys. Res.*, **108**, 1250, doi:10.1029/2002JA009700.

Millan, R. M. and R. M. Thorne (2007), Review of radiation belt relativistic electron losses, *JASTP*, **69**, 362.

Miyoshi, Y., K. Sakaguchi, K. Shiokawa, D. Evans, J. Albert, M. Connors, and V. Jordanova (2008), Precipitation of radiation belt electrons by EMIC waves, observed from ground and space, *Geophys. Res. Lett.*, **35**, L23101, doi:10.1029/2008GL035727.

Morley, S. K., S. T. Ables, M. D. Sciffer, and B. J. Fraser (2009), Multipoint observations of Pc1-2 waves in the afternoon sector, *J. Geophys. Res.*, **114**, A09205, doi:10.1029/2009JA014162.

Olson, J. V. and L. C. Lee (1983), Pc1 wave generation by sudden impulses, *Planet. Space Sci.*, **31**, 295–302.

Ozeke, L. G. and I. R. Mann (2001), Modeling the properties of high-m Alfvén waves driven by the drift-bounce resonance mechanism, *J. Geophys. Res.*, **106** (A8), 15583–15597, doi:10.1029/2000JA000393.

Posch, J. L., M. J. Engebretson, M. T. Murphy, M. H., M. R. Lessard, and R. B. Horne (2010), Probing the relationship between electromagnetic ion cyclotron waves and plasmaspheric plumes near geosynchronous orbit, *J. Geophys. Res.*, **115**, A11205, doi:10.1029/2010JA015446.

Randall, C. E., V. L. Harvey, D. E. Siskind, J. France, P. F. Bernath, C. D. Boone, and K. A. Walker (2009), NOx descent in the Arctic middle atmosphere in early 2009, *Geophys. Res. Lett.*, **36**, L18811, doi:10.1029/2009GL039706.

Reeves, G. D., K. L. McAdams, R. H. W. Friedel, and T. P. O'Brien (2003), Acceleration and loss of relativistic electrons during geomagnetic storms, *Geophys. Res. Lett.*, **30**, 1529, doi:10.1029/2002GL016513.

Sakaguchi, K., K. Shiokawa, A. Ieda, Y. Miyoshi, Y. Otsuka, T. Ogawa, M. Connors, E. F. Donovan, and F. J. Rich (2007), Simultaneous ground and satellite observations of an isolated proton arc at subauroral latitudes, *J. Geophys. Res.*, **112**, A04202, doi:10.1029/2006JA012135.

Sakaguchi, K., Y. Kasahara, M. Shoji, Y. Omura, Y. Miyoshi, T. Nagatsuma, A. Kumamoto, and A. Matsuoka (2013), Akebono observations of EMIC waves in the slot region of the radiation belts, *Geophys. Res. Lett.*, **40**, 5587–5591, doi:10.1002/2013GL058258.

Sandanger, M., F. Søraas, K. Aarsnes, K. Oksavik, and D. S. Evans (2007), Loss of relativistic electrons: evidence for pitch angle scattering by electromagnetic ion cyclotron waves excited by unstable ring current protons, *J. Geophys. Res.*, **112**, A12213, doi:10.1029/2006JA012138.

Sandanger, M., F. Søraas, M. Sørbø, K. Aarsnes, K. Oksavik, and D. Evans (2009), Relativistic electron losses related to EMIC waves during CIR and CME storms, *JASTP*, **71** (10–11), 1126–1144, doi:10.1016/j.jastp.2008.07.006.

Seppälä, A., C. E. Randall, M. A. Clilverd, E. Rozanov, and C. J. Rodger (2009), Geomagnetic activity and polar surface level air temperature variability, *J. Geophys. Res.*, **114**, A10312, doi: 10.1029/2008JA014029.

Sergeev, V. A., W. Baumjohann, and K. Shiokawa (2001), Bi-directional electron distributions associated with near-tail flux transport, *Geophys. Res. Lett.*, **28** (19), 3813–3816, doi:10.1029/2001GL013334.

Shabansky, V. P. (1971), Some processes in the magnetosphere, *Space Sci. Rev.*, **12** (3), 299–418.

Shprits, Y.Y., et al. (2013), Unusual stable trapping of the ultrarelativistic electrons in the Van Allen radiation belts, *Nat. phys.*, **9**, 699–703, doi:10.1038/nphys2760.

Sibeck, D. G., R.W. McEntire, A. T. Y. Lui, R. E. Lopez, and S. M. Krimigis (1987), Magnetic field drift shell splitting—cause of unusual dayside particle pitch angle distributions during storms and substorms, *J. Geophys. Res.*, **92**, 13485–13497, doi:10.1029/JA092iA12p13485.

Silin, I., I. R. Mann, R. D. Sydora, D. Summers, and R. L. Mace (2011), Warm plasma effects on electromagnetic ion cyclotron wave MeV electron interactions in the magnetosphere, *J. Geophys. Res.*, **116**, A05215, doi:10.1029/2010JA016398.

Silin, I., I. R. Mann, R. D. Sydora, D. Summers, and R. L. Mace (2011), Warm plasma effects on electromagnetic ion cyclotron wave MeV electron interactions in the magnetosphere, *J. Geophys. Res.*, **116**, A05215, doi:10.1029/2010JA016398.

Søraas F., Lundblad J. A., Maltseva N. F., Troitskaya V., and Selivanov V. (1980), A comparison between simultaneous IPDP ground-based observations and observations of energetic protons obtained by satellites, *Planet. Space Sci.*, **28** (4), 387–405.

Summers, D., R. M. Thorne, and F. Xiao (1998), Relativistic theory of wave-particle resonant diffusion with application to electron acceleration in the magnetosphere, *J. Geophys. Res.*, **103** (A9), 20487–20500, doi:10.1029/98JA01740.

Summers, D., and R. M. Thorne (2003), Relativistic electron pitch-angle scattering by electromagnetic ion cyclotron waves during geomagnetic storms, *J. Geophys. Res.*, **108**, 1143, doi:10.1029/2002JA009489.

Summers, D., B. Ni, and N. P. Meredith (2007), Timescales for radiation belt electron acceleration and loss due to resonant wave-particle interactions: 1. Theory, *J. Geophys. Res.*, **112**, A04206, doi:10.1029/2006JA011801.

Thorne, R. M. and C. F. Kennel (1971), Relativistic electron precipitation during magnetic storm main phase, *J. Geophys. Res.*, **76**, 4446–4453.

Thorne, R. M., R. B. Horne, V. K. Jordanova, J. Bortnik, and S. Glauert (2013), Interaction of EMIC waves with thermal plasma and radiation belt particles, in *Magnetospheric ULF Waves: Synthesis and New Directions* (eds. K. Takahashi, P. J. Chi, R. E. Denton, and R. L. Lysak), American Geophysical Union, Washington, D. C. doi: 10.1029/169GM14

Turner, D. L., V. Angelopoulos, W. Li, M. D. Hartinger, M. Usanova, I. R. Mann, J. Bortnik, and Y. Shprits (2013), On the storm-time evolution of relativistic electron phase space density in Earth's outer radiation belt, *J. Geophys. Res. Space Physics*, **118**, 2196–2212, doi:10.1002/jgra.50151.

Ukhorskiy, A. Y., Y. Y. Shprits, B. J. Anderson, K. Takahashi, and R. M. Thorne (2010), Rapid scattering of radiation belt electrons by storm-time EMIC waves, *Geophys. Res. Lett.*, **37**, L09101, doi:10.1029/2010GL042906.

Usanova, M. E., I. R. Mann, I. J. Rae, Z. C. Kale, V. Angelopoulos, J. W. Bonnell, K.-H. Glassmeier, H. U. Auster, and H. J. Singer (2008), Multipoint observations of magnetospheric compression-related EMIC Pc1 waves by THEMIS and CARISMA, *Geophys. Res. Lett.*, **35**, L17S25, doi:10.1029/2008GL034458.

Usanova, M. E., et al. (2010a), Conjugate ground and multisatellite observations of compression-related EMIC Pc1 waves and associated proton precipitation, *J. Geophys. Res.*, **115**, A07208, doi:10.1029/2009JA014935.

Usanova, M. E. (2010b). *Electromagnetic Ion Cyclotron Waves in a Compressed Dayside Magnetosphere*. Ph.D. Thesis. University of Alberta: Canada.

Usanova, M. E., I. R. Mann, J. Bortnik, L. Shao, and V. Angelopoulos (2012), THEMIS observations of electromagnetic ion cyclotron wave occurrence: dependence on AE, SYMH, and solar wind dynamic pressure, *J. Geophys. Res.*, **117**, A10218, doi:10.1029/2012JA018049.

Usanova, M. E., F. Darrouzet, I. R. Mann, and J. Bortnik (2013), Statistical analysis of EMIC waves in plasmaspheric plumes from Cluster observations, *J. Geophys. Res. Space Physics*, **118**, 4946–4951, doi:10.1002/jgra.50464.

Usanova, M. E., et al. (2014), Effect of EMIC waves on relativistic and ultrarelativistic electron populations: ground-based and Van Allen Probes observations, *Geophys. Res. Lett.*, **41**, doi:10.1002/2013GL059024.

Wolf, R. A., Y. Wan, X. Xing, J.-C. Zhang, and S. Sazykin (2009), Entropy and plasma sheet transport, *J. Geophys. Res.*, **114**, A00D05, doi:10.1029/2009JA014044.

Woodroffe, J. R. and R. L. Lysak (2012), Ultra-low frequency wave coupling in the ionospheric Alfvén resonator: characteristics and implications for the interpretation of ground magnetic fields, *J. Geophys. Res.*, **117**, A03223, doi:10.1029/2011JA017057.

Yahnin, A.G. and T.A. Yahnina (2007), Energetic proton precipitation related to ion-cyclotron waves, *J. Atm. Terr. Phys.*, **69**, 1690–1706.

Yahnin, A.G., T.A. Yahnina, H. Frey, V. Pierrard (2013), Sub-oval proton aurora spots: Mapping relatively to the plasmapause, *J. Atm. Terr. Phys*, **99**, 61–66.

Yuan, Z., X. Deng, X. Lin, Y. Pang, M. Zhou, P. M. E. Décréau, J. G. Trotignon, E. Lucek, H. U. Frey, and J. Wang (2010), Link between EMIC waves in a plasmaspheric plume and a detached sub-auroral proton arc with observations of Cluster and IMAGE satellites, *Geophys. Res. Lett.*, **37**, L07108, doi:10.1029/2010GL042711.

11

Multi-dimensional Analysis of Whistler-mode Waves in the Radiation Belt Region

O. Santolík

■ Abstract

Whistler-mode waves in the Earth's inner magnetosphere are believed to transfer energy between different parts of the electron phase space distribution and thus to play a significant role not only in the acceleration of radiation belt electrons to relativistic energies, but also in the losses of electrons from the radiation belts by their precipitation into the high-latitude ionosphere. Accurate description of propagation characteristics and time-frequency structure of whistler-mode waves is therefore important for successful modeling of the dynamics of the Van Allen radiation belts. We describe different techniques for determination of wave modes, propagation directions, and time-frequency structure of electromagnetic waves based on multi-component measurements of the magnetic and electric field fluctuations. We show comprehensive examples of application of these techniques to the analysis of spacecraft observations of whistler-mode chorus. Multi-component data from the Cluster and Van Allen Probes missions are used as an input for the different analysis methods whose results are compared and discussed.

Waves, Particles, and Storms in Geospace. First Edition. Georgios Balasis, Ioannis A. Daglis, Ian R. Mann.

11.1 Introduction

Electromagnetic whistler-mode waves, especially chorus and plasmaspheric hiss, are believed to play a role in the dynamics of the particle populations of the Van Allen radiation belts at different scales [e.g., Tao et al., 2013, and references therein]. For example, whistler-mode chorus waves can control the process of local acceleration of electrons in the outer belt [Miyoshi et al., 2013; Thorne et al., 2013; and references therein]. Plasmaspheric hiss has been also considered to be important for the dynamical behavior of the radiation belts, specifically in connection with the slot region between the inner and outer belts.

These waves are believed to contribute to the transfer of energy between different populations of electrons in the inner magnetosphere via wave particle interactions. Properties of these processes should be therefore included in the models of the dynamics of the Van Allen radiation belts. Description of propagation characteristics, on which all these interactions depend, is an important initial part of this task.

In this chapter we describe and discuss different techniques for determination of wave polarization and propagation directions. These analysis methods are based on *in situ* multi-component measurements of the magnetic and electric field fluctuations.

11.2 Analysis methods

The multi-component measurements of the wave magnetic and electric fields allow us to determine, for example, the average Poynting flux. Supposing the presence of a single plane wave, the direction of the wave vector can be determined. For a more complex wave field we can estimate a continuous distribution of wave energy with respect to the wave-vector direction (wave distribution function). These results are useful for the localization of sources of observed emissions.

11.2.1 Plane wave techniques

Considering a single plane wave at a frequency f and with a wave vector $\boldsymbol{k}$, we obtain the magnetic field $\boldsymbol{\mathcal{B}}$ as a function of time t and position $\boldsymbol{r}$,

$$\boldsymbol{\mathcal{B}}(t, \boldsymbol{r}) = \boldsymbol{\mathcal{B}}_0 + \mathrm{Re}\{\boldsymbol{B}(f, \boldsymbol{k}) \exp[\mathrm{i}(2\pi f t - \boldsymbol{k} \cdot \boldsymbol{r})]\}, \tag{11.1}$$

where $\boldsymbol{\mathcal{B}}_0$ is the ambient magnetic field which is varying at time scales much larger than $1/f$, $\mathrm{Re}\{c\}$ denotes the real part of a complex number c, $\mathrm{i} = \sqrt{-1}$, and $\boldsymbol{B}$ is the vector of complex magnetic field spectral amplitudes.

Under this assumption the third Maxwell's equation (the Faraday's law) becomes

$$\boldsymbol{k} \times \boldsymbol{E} = 2\pi f \boldsymbol{B}. \tag{11.2}$$

A consequence of (11.2) is that $\boldsymbol{B}$ is always perpendicular to both the wave vector $\boldsymbol{k}$ and the vector of electric field spectral amplitudes $\boldsymbol{E}$. We can therefore write

$$\mathrm{Re}\{\boldsymbol{B}\} \cdot \boldsymbol{k} = 0, \quad \mathrm{Im}\{\boldsymbol{B}\} \cdot \boldsymbol{k} = 0. \tag{11.3}$$

With the experimentally measured multi-component signals of the wave magnetic field in a predefined Cartesian coordinate system

$$\boldsymbol{\mathcal{B}} = (\mathcal{B}_1, \mathcal{B}_2, \mathcal{B}_3) \tag{11.4}$$

we can use spectral analysis methods to estimate, at a given frequency f, the corresponding complex components of a single realization of the vector of magnetic field spectral amplitudes $\boldsymbol{B}(f)$. The spectral analysis methods can be based, for example, on the windowed fast Fourier transform or wavelet analysis, both described, e.g., by Press et al. [1992]. A Hermitian magnetic spectral matrix can then be formed,

$$S_{ij}(f) = \langle B_i(f) B_j^*(f) \rangle, \quad i,j = 1 \ldots 3, \tag{11.5}$$

where c^* is a complex conjugate of c and $\langle c \rangle$ means an ensemble average of all possible realizations of a complex variable c. In practice, the ensemble average can be estimated by taking a time average of successive estimates of the spectral matrix from a measured time series. This averaging is necessary to decrease random experimental errors induced by the spectral analysis. The diagonal elements of the spectral matrix then carry information on the power spectral density of the separate magnetic field components while the off-diagonal elements additionally reflect mutual phases and coherences between the respective components.

The spectral matrices can be then used as input data for a frequency resolved multidimensional analysis of the measured magnetic field. Assuming the exact knowledge of spectral matrices and a perfect validity of the plane wave approximation, the set of two equations (11.3) can be trivially extended to a set of six mutually dependent homogeneous equations,

$$\sum_{j=1}^{3} \mathrm{Re}\left\{S_{ij}(f)\right\} k_j = 0, \sum_{j=1}^{3} \mathrm{Im}\left\{S_{ij}(f)\right\} k_j = 0, \quad i = 1 \ldots 3. \tag{11.6}$$

In a practical situation, the spectral matrix can be only approximately estimated from the measurements, with necessary random experimental errors on its components. Therefore, equations (11.6) become generally independent but still closely related. It is also clear that, solving this homogeneous set of equations for $\boldsymbol{k}$, we also solve them for $a\boldsymbol{k}$, where a is any real number. We are therefore unable to obtain the modulus of $\boldsymbol{k}$ and we can only obtain its direction with a 180° ambiguity, not resolving $\boldsymbol{k}$ and $-\boldsymbol{k}$. In this situation we can define one component of $\boldsymbol{k}$ as a fixed nonzero number and obtain an overdetermined set of six equations for two remaining unknown components of $\boldsymbol{k}$.

Selecting any two independent equations from (11.6) would give us a unique solution. For example, the method of Means [1972] is based on imaginary parts of three cross-spectra and the procedure is equivalent to solving any two of the last three equations in (11.6). Methods relying on other subsets of equations selected from (11.6) have been also designed [e.g., McPherron et al., 1972; Arthur et al., 1976; Samson and Olson, 1980]. However, for different subsets we do not generally obtain the same result from a given experimental data set. A method which takes into account all equations (11.6) and finds an optimized solution in the least-squares sense has been developed [Santolík et al., 2003a] using the singular value decomposition (SVD) technique,

$$\begin{pmatrix} \mathrm{Re}\{S\} \\ \mathrm{Im}\{S\} \end{pmatrix} = A = U\,W\,V^{\mathrm{T}} = U \begin{pmatrix} {}^1w & 0 & 0 \\ 0 & {}^2w & 0 \\ 0 & 0 & {}^3w \end{pmatrix} \begin{pmatrix} {}^1V_1 & {}^1V_2 & {}^1V_3 \\ {}^2V_1 & {}^2V_2 & {}^2V_3 \\ {}^3V_1 & {}^3V_2 & {}^3V_3 \end{pmatrix}, \tag{11.7}$$

where A is a real matrix 6×3 obtained by superposition of a real part of the spectral matrix $\mathrm{Re}\{S\}$ over its imaginary part $\mathrm{Im}\{S\}$. U is a real matrix 6×3 with orthonormal columns (which is not used in the subsequent analysis). W is a diagonal matrix containing three non-negative singular values ${}^1w, {}^2w, {}^3w$ which depend on lengths of axes of the magnetic field polarization ellipsoid. Directions of these axes can be obtained from the corresponding rows of a real 3×3 matrix V^{T} which is composed of three mutually orthogonal unit vectors. The least-squares solution of (11.6) is then given by the row of V^{T} which corresponds to the minimum singular value

$${}^kw = \min_j \left({}^jw\right). \tag{11.8}$$

Note that in the ideal case of a plane polarized wave without any experimental errors on the spectral matrix estimates we would obtain ${}^kw = 0$ but the least-squares solution can be found also for nonzero ${}^kw$. This solution can be described by two spherical angles

$$\theta = \tan^{-1}\left(\frac{\sqrt{{}^kV_1^2 + {}^kV_2^2}}{{}^kV_3}\right), \qquad \phi = \tan^{-1}\frac{{}^kV_2}{{}^kV_1}. \tag{11.9}$$

These results are expressed in a Cartesian coordinate system of the original waveform data (11.4).

In the inner magnetosphere of the Earth it is usual to work in Cartesian coordinates connected to the geocentric radius vector $\boldsymbol{r}$ pointing outward from the center of the Earth, and to the principal axis of symmetry of the plasma medium where the waves propagate, i.e., to the direction of the ambient stationary magnetic field $\boldsymbol{B}_0$. Both these directions are contained in the plane of the local magnetic meridian. The angle θ then gives us an estimate of inclination of the wave vector $\boldsymbol{k}$ from $\boldsymbol{B}_0$. The angle ϕ can be

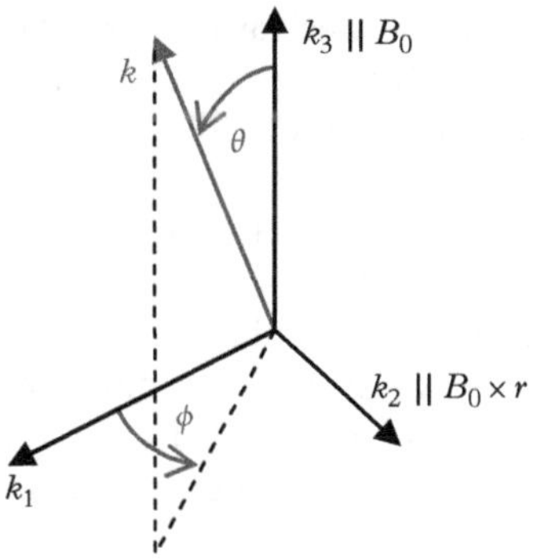

Figure 11.1 *Coordinate system linked to the ambient stationary magnetic field* $\mathbf{B}_0$ *and to the plane of the local magnetic meridian;* $\boldsymbol{r}$ *points outward from the centre of the Earth.*

defined as the azimuth of $\boldsymbol{k}$ measured around $\mathbf{B}_0$ from the plane of the local magnetic meridian: $\phi = 0°$ corresponds to the outward direction and $\phi = 90°$ points eastward (Figure 11.1). With the 180° ambiguity the direction can be only expressed in a single hemisphere, for example, $0° \leq \theta \leq 90°, -180 < \phi \leq 180°$.

The SVD technique (11.7) also allows us to characterize the polarization properties of the measured magnetic field. First of all, we can evaluate how far the experimental spectral matrix is from the ideal case of a matrix corresponding to a plane wave without any experimental errors. The planarity of polarization F of the wave magnetic field is based on the ratio of the length of the smallest axis of the polarization ellipsoid to the length of its largest axis,

$$F = 1 - \sqrt{\frac{{}^k w}{{}^p w}}, {}^p w = \max_j \left({}^j w\right), \tag{11.10}$$

where ${}^k w$ comes from equation (11.8). We obtain $F = 0$ if the polarization is completely isotropic, without any preferred direction, while $F = 1$ for the ideal plane wave case with ${}^k w = 0$. If the phases between the magnetic field components are random, the three singular values directly give us squares of the lengths of the three orthogonal axes of the random polarization ellipsoid. Results obtained from a model of a mixture of random-phase plane waves propagating in different directions [Santolík et al., 2003a] show that the planarity of polarization can be successfully used to measure the spread of modeled wave vectors, and therefore to test the validity of the plane wave assumption.

If this assumption is perfectly valid, the magnetic field does not fluctuate in the direction parallel to $\boldsymbol{k}$. The remaining magnetic field components in the plane perpendicular to $\boldsymbol{k}$ are phase locked. They form a polarization ellipse which can be characterized by the ratio of the lengths of its axes and by the sense of polarization. These properties are reflected by the ellipticity E_B of the wave magnetic field,

$$E_B = s \frac{{}^m w}{{}^p w}, \tag{11.11}$$

where ${}^{p}w$ is the largest singular value (equation 10.10), ${}^{m}w$ is the second largest singular value, and $s = -1$ when the polarization sense is left-handed with respect to $\boldsymbol{B}_0$ (sense of the ion cyclotron motion), and $s = 1$ for the right-handed polarization (sense of the electron cyclotron motion).

The resulting E_B values are between −1 and 1, while the absolute value $|E_B|$ gives the ratio of the two axes of the polarization ellipse. For example, $E_B = -1$ for the left-handed circular polarization, and $E_B = 1$ for the right-handed circular polarization. For linear polarization, two singular values are zero and only ${}^{p}w$ is non-zero. Therefore, we obtain $E_B = 0$. In such a case, the set (11.6) degenerates into a single equation, and, consequently, possible wave vector directions are contained in a plane perpendicular to the vector $({}^{p}V_1, {}^{p}V_2, {}^{p}V_3)$.

Note also that the sum of the three singular values is equal to the trace of the spectral matrix which is invariant of the coordinate system and can be used as a measure of the total power,

$$S_\Sigma = \sum_{j=1}^{3} {}^{j}w = \sum_{j=1}^{3} S_{jj}. \tag{11.12}$$

The SVD technique can also be used with combined measurements of the magnetic field and electric field components. A least-squares solution to an overdetermined set of 36 equations derived from (11.2) can be based on components of an electromagnetic spectral matrix 6 × 6 which is obtained from simultaneous measurements of three magnetic field components and three electric field components [Santolík et al., 2003a]. This method allows us to remove the 180° ambiguity and to distinguish between two antiparallel wave vector directions, obtaining the results in the full interval of spherical angles: $0° \leq \theta \leq 180°$, $-180° < \phi \leq 180°$. To verify the validity of the plane-wave assumption in this case, the electromagnetic planarity parameter has been derived to evaluate a measure of closeness of the experimental spectral matrix to equation (11.2). This parameter has been used, for example, by Santolík et al. [2004] to estimate the parallel dimension of the source of chorus emissions from multi-component and multi-point measurements of the Cluster spacecraft.

11.2.2 Wave distribution function

The wave distribution function (WDF) is defined as a continuous distribution of the wave energy density with respect to the wave vector direction [Storey, 1999]. This concept is particularly useful for all cases when the assumption of the presence of a single plane wave fails. A theoretical basis for modeling of a spectral matrix based on a known WDF (the WDF direct problem) has been defined by Storey and Lefeuvre [1974]. They considered a continuous distribution of random elementary plane waves (11.1) with respect to the direction of the wave vector $\boldsymbol{k}$ which is defined by the spherical angles θ and ϕ (see Figure 11.1). Supposing further that these elementary plane waves have random phases (no mutual coherence) and that they all occupy a narrow bandwidth around a frequency f, the spectral matrix (11.5) can be modeled as

$$S_{ij}(f) = \sum_n \iint a_{nij}(f,\theta,\phi)\; G_n(f,\theta,\phi) \sin\theta d\phi d\theta, \quad i,j = 1\ldots 3 \tag{11.13}$$

where G_n is the WDF of the nth simultaneously propagating wave mode. For example, $n = 1\ldots 2$ for the cold plasma model of the L-O and R-X free space modes. For the cold plasma model of the whistler mode, we only have $n = 1$. The integration is carried out over the full solid angle of wave-normal directions. The integration kernels a_{nij} are obtained from a theoretical model of magnetic field spectral amplitude $\boldsymbol{B_n}(f,\theta,\phi)$ for a given wave mode n,

$$a_{nij} = \frac{B_{ni}(f,\theta,\phi)\, B^*_{nj}(f,\theta,\phi)}{u_n(f,\theta,\phi)}, \quad i,j = 1\ldots 3 \tag{11.14}$$

where $u_n(f,\theta,\phi)$ is the energy density of an elementary plane wave. If the receiver (antenna on a spacecraft) has a non-negligible relative velocity with respect to the plasma medium in which the waves propagate then f can be replaced by the Doppler-shifted frequency. The models of spectral amplitudes $\boldsymbol{B_n}(f,\theta,\phi)$ and the energy density $u_n(f,\theta,\phi)$ need to take into account the physical properties of the medium. This requires the knowledge of a theoretical solution to the wave dispersion relation for given f, θ, ϕ, and n. This solution gives us the wave number $\|\boldsymbol{k}\|$ and, subsequently, the electric field spectral amplitude $\boldsymbol{E_n}(f,\theta,\phi)$. Equation (11.2) then gives us the magnetic field spectral amplitudes $\boldsymbol{B_n}(f,\theta,\phi)$. The theory of the WDF direct problem has been developed by Storey and Lefeuvre [1979, 1980], and revisited by Storey [1998] for a modeled spectral matrix consisting not only of the three magnetic field components but also for a spectral matrix 6×6 which also includes the electric field components. They used the cold plasma approximation to describe the properties of the medium.

The model described by equation (11.13) can be used to develop methods for estimation of the WDF from spacecraft measurements. This WDF inverse problem consists of reconstruction of $G_n(f,\theta,\phi)$ from (11.13) based on theoretical calculations of integration kernels $a_{nij}(f,\theta,\phi)$, using an experimentally obtained spectral matrix. The cold plasma theory has been used by Lefeuvre and Delannoy [1979], and Delannoy and Lefeuvre [1986] to develop maximum entropy methods for estimation of the WDF. Oscarsson and Rönnmark [1989] further developed these methods using a hot plasma model with a slightly different definition of the WDF and without the explicit dependence of the WDF on the wave frequency. Santolík and Parrot [1996] introduced the hot plasma theory for the frequency-dependent WDF, taking into account the Doppler effect. Santolík and Parrot [2000] compared different techniques for resolution of the WDF inverse problem based on the minimization of a least-squares merit function. They also placed the WDF results in the context of the plane wave estimates. The comparison of the obtained WDF with the results of the plane wave techniques has been shown to be useful in order to verify the results of the WDF inversions, to estimate their significance, and to help with their interpretation.

11.2.3 Instantaneous amplitude, phase, frequency, and wave vector direction

Electromagnetic waves propagating in the plasma medium can sometimes manifest rapid variations of their amplitudes and frequencies. This can also be the case of whistler-mode waves in the radiation belts and, especially, the whistler-mode chorus emissions [e.g., Cornilleau-Wehrlin et al., 1976; Santolík et al., 2003b]. In such a case, the above-mentioned spectral methods are often unable to reflect these rapid changes of the wave parameters because of their need to use sufficiently long time intervals of original waveforms for calculations of the averaged spectral matrices. In order to achieve a good time and frequency resolution of analysis of chorus elements, Santolík et al. [2003b] divided the measured waveform into short subintervals. In each of these subintervals a nonlinear least-squares procedure has been used to fit the parameters of a model. This model has assumed the presence of a harmonic signal with a single frequency and with a polynomial variation of its amplitude. Estimates of frequencies and amplitudes in a sequence of the consecutive subintervals revealed a sub-packet structure of chorus elements.

A similar but less computationally demanding technique has been introduced by Berthomier [1975], Berthomier and Cornilleau-Wehrlin [1975], and Cornilleau-Wehrlin et al. [1976], and later also used by Santolík et al. [2004] and Santolík et al. [2014a]. This technique is based on the analytic representation of the measured signal which is constructed in the frequency domain by suppressing the negative frequency components. In the time domain this is equivalent to adding an imaginary part to the real signal,

$$ {}^{a}\mathcal{B}_j(t) = \mathcal{B}_j(t) + \mathrm{i}\,\mathrm{H}\left\{\mathcal{B}_j(t)\right\} = A_j(t)\exp[\mathrm{i}\varphi_j(t)], \tag{11.15} $$

where $\mathrm{H}\left\{\mathcal{B}_j(t)\right\}$ denotes the Hilbert transform of the real signal $\mathcal{B}_j(t)$. The instantaneous amplitude $A_j(t)$ is then obtained as a modulus of this complex analytic signal ${}^{a}\mathcal{B}_j(t)$, and the instantaneous frequency $f_j(t)$ is obtained as the time derivative of the phase $\varphi_j(t)$ of the analytical signal,

$$ f_j(t) = \frac{\mathrm{d}\,\varphi_j(t)}{\mathrm{d}t}. \tag{11.16} $$

Note that a single instantaneous frequency is obtained for every given time. The method is therefore useful only for narrow-band signals when the presence of a single frequency at each given time can be successfully assumed. The frequency can vary with time but interpretation of the results becomes difficult if the wave power occupies a large frequency band or if it is dispersed over several narrow-band frequency intervals. With multi-dimensional measurements described by equation (11.4), the extension to the complex plane and calculations of the instantaneous amplitude, phase, and frequency can be separately done for each Cartesian component $\mathcal{B}_j(t), j = 1\ldots3$. Instantaneous amplitudes and/or phases are generally different for the three different components, reflecting the polarization properties of the observed waves. If, however, the three signals are mutually related, and all the instantaneous frequencies are found as equal, the observed instantaneous parameters can be interpreted as properties of a plane wave described by equation (1). The instantaneous spectral matrix can be formed,

$$S_{ij}(t) = \left\langle {}^a\mathcal{B}_i(t)\,{}^a\mathcal{B}_j(t)^* \right\rangle = \left\langle A_i(t)A_j(t)\exp\left\{\mathrm{i}\left[\varphi_i(t) - \varphi_j(t)\right]\right\}\right\rangle \quad i,j = 1\ldots 3, \tag{11.17}$$

where c^* is a complex conjugate of c and $\langle c\rangle$ again denotes an ensemble average of all possible realizations of a complex variable c. In practice, the ensemble average can be estimated by a moving average. The instantaneous spectral matrix can then be used as an input to the analysis procedures described in Section 11.2.1. For example, the wave vector direction can be estimated from the SVD of the instantaneous spectral matrix (11.17) according to equations (11.7–11.9).

11.3 Examples of results

The plane-wave techniques described in Section 11.2.1 often allow us to obtain a straightforward interpretation of results. The resulting wave vector angles and polarization parameters can be represented in time-frequency plots in a form similar to power spectrograms. The methods are computationally inexpensive and therefore suitable for a systematic analysis or implementation onboard the spacecraft. On the other hand, the wave distribution function technique described in Section 11.2.2 requires a more elaborate analysis with a number of free parameters which need to be adjusted to fit the specifics of each particular event. The same generally holds true for the techniques to determine the instantaneous amplitudes, phases, frequencies, and wave vector directions (Section 11.2.3). Examples given below show the application of all these methods to the analysis of whistler-mode waves in the radiation belts.

11.3.1 Plane wave techniques and wave distribution function methods for onboard-analyzed spectral data from the Cluster spacecraft

The first set of examples is related to the enormous data set which has been collected by the multi-spacecraft Cluster mission. Four Cluster spacecraft are still in operation since their launch in 2000. Onboard all four spacecraft, the STAFF-SA instruments [Cornilleau-Wehrlin et al., 2003] perform a multi-dimensional spectral analysis based on the signals detected by three magnetic search coil antennas and two double-sphere electric antennas. The resulting data set consists of compressed spectral matrices (equation 11.5, with an extension for the electric field signals) in 27 predefined frequency bands between 8 Hz and 4 kHz. The time resolution is 1 second for the power spectra and 4 seconds for phases and coherences in the normal telemetry mode.

These instruments on all four spacecraft recorded intense chorus during the night-side passage through the outer radiation belt region on January 20, 2004.

Figure 11.2 shows the data recorded on Cluster 2; data from other spacecraft are similar. The initial part of the electric field power spectrogram in Figure 11.2a represents the single-component high-resolution measurements obtained by the Wide-Band Data (WBD) instrument [Gurnett et al., 2001] that are available before 1938 UT. After that time we use the STAFF-SA data that can be recognized in Figure 11.2a by a limited frequency range below 4 kHz. Whistler-mode chorus is seen as an intense wave emission, initially rising in frequency from 2 to 4 kHz, as the spacecraft approaches the geomagnetic equator from the south. In the vicinity of the magnetic equatorial plane, the upper band of chorus appears above one half of the electron cyclotron frequency. The magnetic field spectrogram (according to equation 11.12) from the STAFF-SA instrument is shown in Figure 11.2b. Figure 11.2c shows the angle θ between the wave vector and the static magnetic field, as it is defined in equation (11.9), but here calculated by the singular value decomposition (SVD) method using the STAFF-SA measurements of both electric and magnetic field fluctuations mentioned at the end of Section 11.2.1 [Santolík et al., 2003a, their equation (22)]. Figure 11.2c shows that the wave vector angles are highly deviated from directions parallel or antiparallel to the magnetic field. θ values are larger than 90°at negative geomagnetic latitudes (shown as λ_m at the bottom of Figure 11.2). θ values become smaller than 90° when the spacecraft moves to positive latitudes on the northern side of the geomagnetic dipole equator.

This propagation pattern is confirmed in Figure 11.2d which shows the parallel component of the Poynting flux normalized by its standard deviation, as it is obtained from statistical uncertainties of the spectral analysis [Santolík et al., 2001, their equation (8)]. The majority of values are either below 3 or above 3. This indicates a significant antiparallel or parallel component of the Poynting vector on the southern or northern side of the geomagnetic equator, respectively. The Poynting vector has a positive divergence close to the dipole equatorial plane. The central position of the chorus source region can therefore be located also close to the equatorial plane. This position then coincides with the minimum of the magnetic field strength along any given field line, according to the dipole model. The geomagnetic field at a radial distance of 4 R_E, however, can be distorted with respect to this model, especially on the night side [see, e.g., Santolík et al., 2005, Figure 8]. Figure 11.2e shows results of experimental determination of the field strength minimum along the magnetic field line. Our multi-spacecraft method [Santolík et al., 2009] uses magnetic field measurements by fluxgate magnetometers [Balogh et al., 2001] onboard the four Cluster spacecraft, to calculate the spatial derivative of the magnetic field strength along the magnetic field direction,

$$\partial_s \|\mathcal{B}_0\| = \nabla \|\mathcal{B}_0\| \cdot \frac{\mathcal{B}_0}{\|\mathcal{B}_0\|}. \tag{11.18}$$

The field strength minimum along the magnetic field line (the true magnetic equator) is reached at 1934:26 UT.

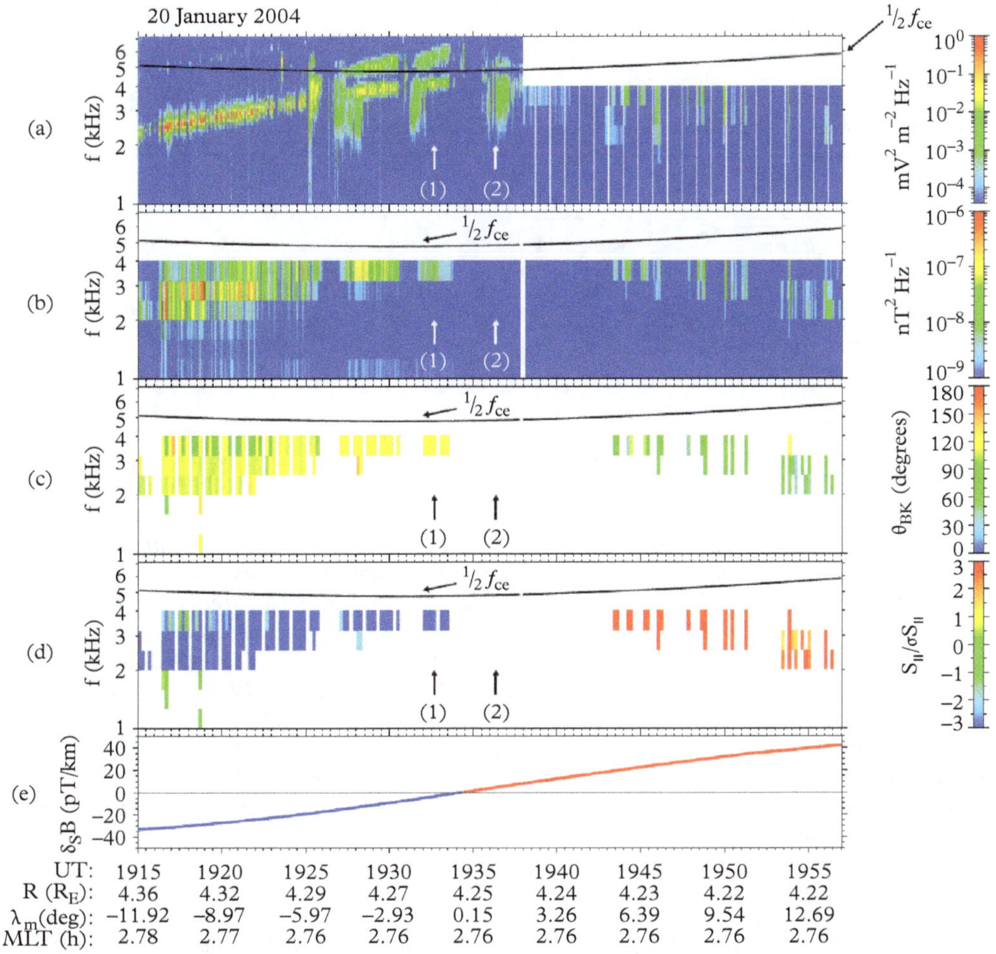

Figure 11.2 *Measurements of Cluster 2 on January 20, 2004. (a) Power spectrogram of the electric field fluctuations from the WBD measurements (before 1938 UT) and STAFF-SA measurements (after 1938 UT). In the STAFF-SA data the frequency band is limited below 4 kHz and the vertical white stripes correspond to regular data gaps. (b) Power spectrogram of the magnetic field fluctuations from the STAFF-SA measurements. (c) Angle between the wave vector and* $\mathcal{B}_0$. *(d) Parallel component of the Poynting vector normalized by its standard deviation. (e) Spatial gradient of* $\|\mathcal{B}_0\|$ *projected on the local field line. The solid black lines in Figures 11.2a–d show one half of the electron cyclotron frequency from local measurements of the static magnetic field [Balogh et al., 2001]. Vertical arrows show times of selected events 1 and 2; event 1 is analyzed in detail in Figure 11.3. Position of Cluster 2 is given at the bottom.* R*, radial distance in Earth's radii;* λ_m*, magnetic dipole latitude in degrees; MLT, magnetic local time in hours.*

(From Santolík et al., 2009.)

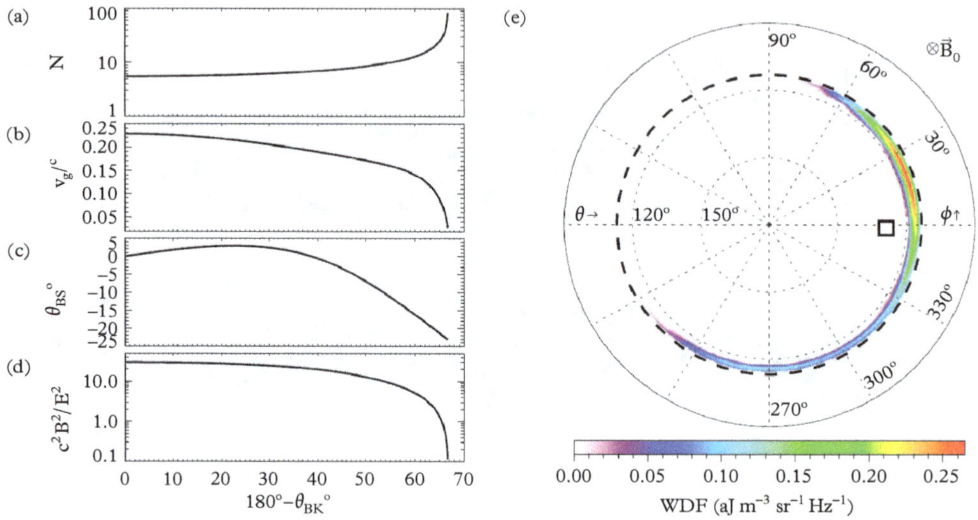

Figure 11.3 *(a–d) Results of the theoretical analysis of a whistler-mode wave at a frequency of 3.587 kHz. Cold plasma model is used and results are given as a function of the angle θ between the wave vector and $\boldsymbol{B}_0$. (a) The refractive index, (b) the group speed normalized by the speed of light, (c) angle θ_S between the group velocity and $\boldsymbol{B}_0$, and (d) ratio of the magnetic field power versus the electric field power. (e) Wave distribution function (WDF) $G_n(f, \theta, \phi)$ from equation (11.13), estimated using observations on board Cluster 4 in the frequency range 3.17–4 kHz between 1932:39.603 and 1932:43.443 on January 20, 2004 (during event 1 from Figure 11.2). WDF is plotted in a polar plot as a function of the polar angle θ between the wave vector and $\boldsymbol{B}_0$ and the angle azimuth ϕ of the wave vector, with $\phi = 0$ in the plane of the local magnetic meridian, outward from the Earth. The whistler-mode resonance angle θ_R is shown by a dashed line. Results of the plane wave analysis from Figure 11.2 are indicated by a square symbol.*
(From Santolík et al., 2009.)

Figure 11.3 shows an example of the estimation of the WDF, as it is described in Section 11.2.2. The analysis is based on Cluster 4 measurements recorded close to the true magnetic equator, well inside the chorus source region (during event 1 from Figure 11.2). For the WDF direct problem, we have used the complete cold plasma theory [Stix, 1992], assuming a magnetized hydrogen plasma. We use a central frequency $f = 3587$ Hz and the measured electron cyclotron frequency $f_{ce} = 9657$ Hz. The electron plasma frequency is estimated as $f_p = 25$ kHz, based on measurements of the WHISPER active sounder instrument [Décréau et al., 2001] onboard the Cluster spacecraft in this region. For these parameters, the cold plasma theory gives the whistler-mode resonance angle $\theta_R = 66.8°$ or $113.2°$. Selected results of the cold plasma theory for whistler-mode waves are plotted in Figures 11.3a–d as a function of θ in the interval between 0 and θ_R. Figure 11.3a shows the refractive index increasing from a value of 5.4 at $\theta = 0$ and sharply diverging at $\theta = \theta_R$. The group velocity in Figure 11.3b is decreasing close to the resonance angle. The Gendrin angle can be found from Figure 11.3c as $\theta = \theta_G > 0$ for which the angle θ_S (between the group velocity and $\mathcal{B}_0$) is zero.

Figure 11.3d shows that the relative power of magnetic field fluctuations decreases close to the resonance angle. Figure 11.3e shows the result of the estimation of the WDF. The procedure consists of fitting a multiparameter model to the measured spectral matrix of the three magnetic field components, as it is described by equation (11.13). We have used the model of discrete regions [Santolík and Parrot, 2000]. The WDF estimation procedure is based on Cluster 4 measurements during event 1 from Figure 11.2. Since we know from the previous analysis that the waves propagate southward, the results are represented in a θ interval between 90° and 180°. The estimated WDF shows that the wave energy density has a very broad peak in the azimuth angle ϕ. This peak is centered near the plane of the local magnetic meridian, in the direction outward from the Earth. On the other hand, the peak is relatively narrow as a function of the θ angle. In the previously described results of the plane wave analysis by the electromagnetic SVD method (shown by a square symbol in Figure 11.3e), most of the wave energy is found between θ_G and θ_R. The WDF maximum is still found more than 3° inward from the resonance cone. Different plane wave methods [McPherron et al., 1972; Means, 1972; Samson, 1973; Samson and Olson, 1980] or the magnetic SVD method (equation 11.9) give slightly different results, between 115.6° and 131.6° (not shown). The reason for this spread is that the assumption of the presence of a single plane wave is not entirely satisfied and each of the methods results in a different "average" direction.

11.3.2 Example of waveform measurements from the Van Allen Probes EMFISIS instrument: plane wave techniques and instantaneous wave parameters

The EMFISIS instrument suite is carried by two Van Allen Probes spacecraft (launched in 2012) to measure the background magnetic fields by a tri-axial fluxgate magnetometer and wave magnetic field fluctuations by a tri-axial search coil magnetometer. A set of wave electric and magnetic field measurements is processed by the EMFISIS Waves instrument which also uses external signals from the Electric Fields and Waves (EFW) experiment. Six inputs from sensors making 3D measurements of both the electric and the magnetic field are detected by six identical waveform receivers in a frequency range from 10 Hz to 12 kHz. The waveforms are digitized at a sampling rate of 35 kHz with a 16-bit instantaneous dynamic range [see Kletzing et al., 2013 for a complete description]. The results shown in Figure 11.4 are based on selected 5.968 s snapshots of 3D magnetic field waveforms which have been captured in the EMFISIS continuous waveform burst mode.

Figure 11.4 shows an example of a simple spectral analysis of one snapshot of 3D magnetic field waveforms recorded onboard Van Allen Probe A on November 14, 2012, after 1410:52.680 UT. The observations were made in the outer Van Allen belt at a radial distance of 5.3 Earth radii, in the equatorial source region of chorus at a magnetic dipole latitude of –2.4°, and in the morning sector at a magnetic local time of 6.9 hours. 1024-point Fast Fourier Transforms with 70% overlapping of the source waveforms in

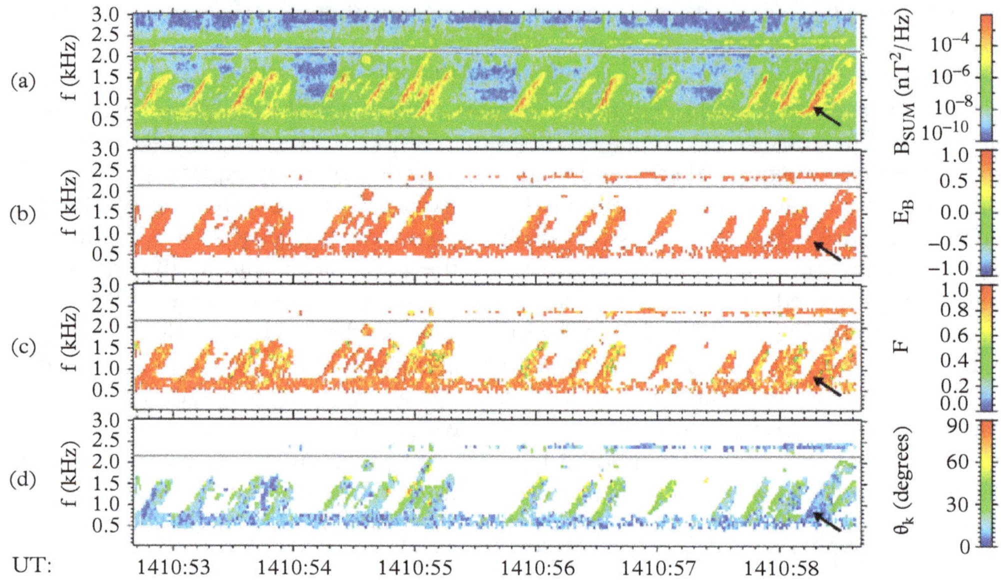

Figure 11.4 *Analysis of three-component magnetic field waveforms recorded by the EMFISIS Waves instrument onboard Van Allen Probe A on November 14, 2012. Frequency-time plots of (a) the sum of the power spectral densities of three orthogonal magnetic components according to equation (11.12), (b) the ellipticity of the magnetic field polarization with a sign corresponding to the sense of polarization—equation (11.11), (c) the planarity of the magnetic field polarization according to equation (11.10), and (d) the angle between the wave vector and the background magnetic field according to equation (11.9). A color scale is given on the right-hand side of each plot. Values in Figures 11.4b–d are plotted for power spectral densities above 10^{-6} nT^2 Hz^{-1}. Black arrows point to the time and frequency of the peak of a selected chorus element analyzed in Figure 11.5. One half of the local electron cyclotron frequency is given by a gray line in each plot. Time is given in UT at the bottom.* (From Santolík et al., 2014a.)

the time domain are used to obtain complex spectra of each component and to form averaged spectral matrices (according to equation 11.5) with a time resolution of 18 ms and with a frequency resolution of 68 Hz. The traces of the spectral matrices (equation 11.12) are shown in Figure 11.4a. We can notice a sequence of rising-frequency whistler-mode chorus elements in the lower band below one half of the electron cyclotron frequency (i.e., below $f_{ce}/2 \sim 2.16$ kHz, in this case). The ellipticity and planarity, and the angle θ between the wave vector and $\boldsymbol{B}_0$ (Figure. 11.4b, c, and d, respectively) have been obtained from the singular value decomposition of the magnetic spectral matrices [Santolík et al., 2003a] according to the above discussed equations (11.11), (11.10), and (11.9). The results show nearly circularly, right-hand polarized waves, with the magnetic-field polarization confined close to a single plane. The estimated normal direction to this plane shows that the wave vectors are often quasi-parallel to the background field at θ below 30° but sometimes they can be found also at higher angles around 60°.

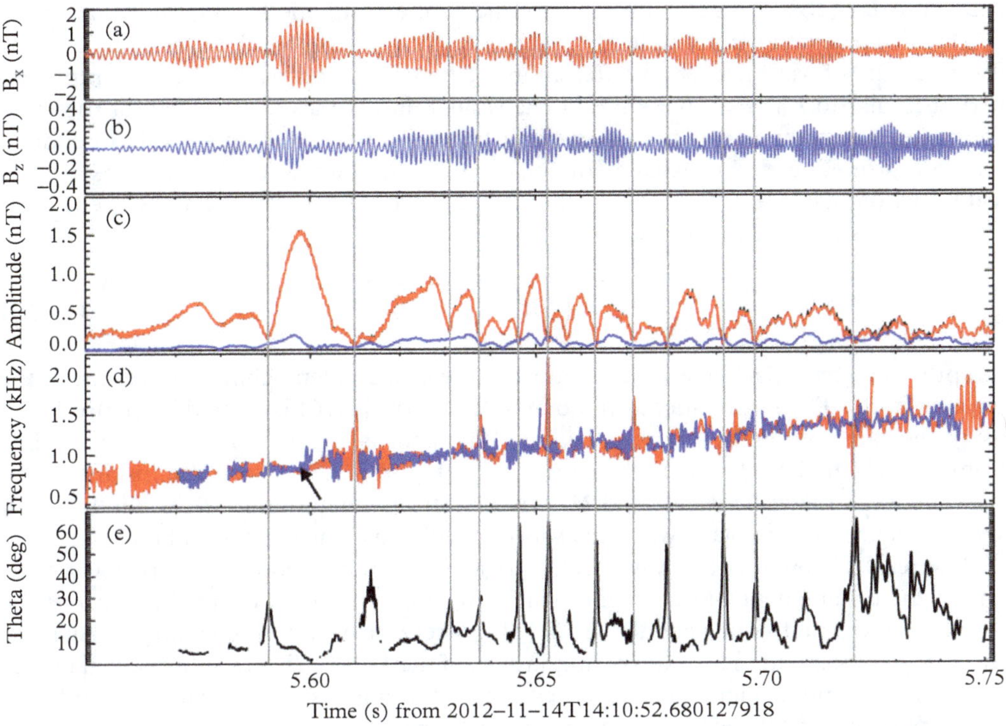

Figure 11.5 *Detailed analysis of a single chorus element marked by an arrow in Figure 11.4 and observed on November 14, 2012, after 1410:52.680. (a) Measured waveform of a magnetic field component B_X perpendicular to $\mathcal{B}_0$. (b) Waveform of the component B_Z parallel to $\mathcal{B}_0$. (c) Instantaneous amplitudes of these two waveforms according to equation (11.15), a red line shows A_X, and a blue line shows A_Z; a black line which is often hidden behind the A_X results corresponds to instantaneous total magnetic field B. (d) Instantaneous frequency according to equation (11.16) with the same color coding as in Figure 11.5a and b; the results are plotted only when the instantaneous amplitude of the corresponding component is >50 pT; a black arrow points to the same time and frequency of the maximum amplitude as in Figure 11.4. (e) Instantaneous angle between the wave vector and $\mathcal{B}_0$ obtained from equation (11.9) used for the instantaneous spectral matrix according to equation (11.17); the results are plotted only when the instantaneous amplitudes of all the magnetic field components are >50 pT. Vertical gray lines should guide the eye through all the panels by showing selected minima of A_X.*
(From Santolík et al., 2014a.)

A black arrow plotted before the end of the time interval in Figure 11.4 indicates the time and frequency of the peak of a strong chorus element which is shown in detail in Figure 11.5 using the high-resolution waveform measurements. The measured waveforms are calibrated using a frequency-dependent transfer function of the sensor-receiver chain, and transformed from the spinning spacecraft system to a system which is fixed and linked to onboard measured $\mathcal{B}_0$. Figure 11.5a and b show that the component parallel to $\mathcal{B}_0$ (B_Z) is much smaller than one of the components perpendicular to $\mathcal{B}_0$ (B_X) which reaches an amplitude of 1.5 nT (1% of the background magnetic field). The fine structure of wave packets is different in the two components. The other component perpendicular to $\mathcal{B}_0$ (B_Y—not shown) has similar amplitudes as B_X. Each of the three components was then separately pass-band filtered between 0.4 and 3 kHz to contain only the pronounced chorus elements below $f_{ce}/2$. Amplitude of the complex analytical signal (equation 11.15) is plotted in Figure 11.5c for B_Z (blue) and B_X (red). The instantaneous total magnetic field,

$$B = \sqrt{B_X^2 + B_Y^2 + B_Z^2}, \tag{11.19}$$

is represented by a black line which is most of the time hidden behind the red line of the B_X amplitude. The instantaneous frequency (equation 11.16) is plotted in Figure 11.5d using the same red and blue color coding. The chorus element is rising in frequency from 600 Hz to 1.5 kHz in both components. Figure 11.5e shows the instantaneous angle between the wave vector and $\mathcal{B}_0$ obtained from equation (11.9), while the input spectral matrix for the plane wave analysis is calculated using equation (11.17).

The results show that the wave vector angle is generally below 30° at the local maxima of the instantaneous total magnetic field, with an exception of the final part of the chorus element (at frequencies above 1.2 kHz) where we observe higher angles. All these variations correspond to the ratio of amplitudes of the B_Z and B_X waveforms seen in Figures 11.5a–b, and in Figure 11.5c. The wave vector angle is small for low B_Z and high B_X amplitudes, and it starts to grow when the B_X amplitudes decrease.

11.4 Conclusions

We have described several techniques which are useful for analysis of whistler-mode waves in the radiation belts: (1) determination of the wave vector direction, ellipticity, and planarity under the assumption of the presence of a single plane wave, using the singular value decomposition method; (2) determination of the distribution of the energy density with respect to the wave vector direction, using the wave distribution function methods; (3) determination of the instantaneous amplitude, phase, frequency, and wave vector angle, using the method of analytic signal. All these techniques are based on multi-component measurements of the magnetic and electric field fluctuations.

We have shown several examples of analysis of whistler-mode chorus using all these techniques, using data from the Cluster and Van Allen Probes spacecraft missions [Santolík et al., 2009, 2014a]. The examples were directly linked to the description of the analysis techniques. Other recent examples of results based on these methods include statistics of magnetosonic and whistler-mode waves obtained from systematic measurements of the Cluster mission [Hrbackova et al. 2015, Santolík et al., 2014b], and average parameters of whistler-mode chorus based on measurements of the THEMIS spacecraft [Taubenschuss et al., 2014]. The wave propagation and polarization parameters resulting from these techniques can be also downloaded from public archives and used as if they were direct measurements [Li et al., 2015] which illustrates the importance of proper description of the actual choice of analysis techniques for original research.

Acknowledgments

This work has been supported by EU through the FP7-Space grant agreement no. 284520 for the MAARBLE collaborative research project, by the Praemium Academiae award, and by the LH14010 grant.

REFERENCES

Arthur, C. W., R. L. McPherron, J. D. Means (1976), A comparative study of three techniques for using the spectral matrix in the wave analysis, *Radio Sci.*, **11**, 833–845.

Balogh, A., et al. (2001), The Cluster magnetic field investigation: Overview of in-flight performance and initial results, *Ann. Geophys.*, **19**, 1207–1217.

Berthomier, C. (1975), A signal analysis method, *Annales de Geophysique*, **31**, 239–251.

Berthomier, C., N. Cornilleau-Wehrlin (1975), Application of the concept of analytic signals to the determination of the amplitude and instantaneous frequency of a signal, *Annales des Telecommunications*, **30**, 224–230.

Cornilleau-Wehrlin, N., J. Etcheto, R. K. Burton (1976), Detailed analysis of magnetospheric ELF chorus—preliminary results, *J. Atm. Terr. Phys.*, **38**, 1201–1210.

Cornilleau-Wehrlin, N., et al. (2003), First results obtained by the Cluster STAFF experiment, *Ann. Geophys.*, **21**, 437–456.

Décréau, P. M. E., et al. (2001), Early results from the Whisper instrument on Cluster: An overview, *Ann. Geophys.*, **19**, 1241–1258.

Delannoy, C., F. Lefeuvre (1986), MAXENTWDF—A computer program for the maximum entropy estimation of a wave distribution function, *Comp. Phys. Comm.*, **40**, 389–419.

Gurnett, D. A., et al. (2001), First results from the Cluster wideband plasma wave investigation, *Ann. Geophys.*, **19**, 1259–1272.

Hrbackova, Z., O. Santolík, F. Nemec, E. Macusova, N. Cornilleau-Wehrlin (2015), Systematic analysis of occurrence of equatorial noise emissions using 10 years of data from the Cluster mission, *J. Geophys. Res.: Space Physics*, **120**, 1007–1021, doi:10.1002/2014JA020268.

Kletzing, C. A., et al. (2013), The Electric and Magnetic Field Instrument Suite and Integrated Science (EMFISIS) on RBSP, *Space Sci. Rev.*, **179**, 127–181, DOI 10.1007/s11214-013-9993-6.

Lefeuvre, F., C. Delannoy (1979), Analysis of random electromagnetic wave field by a maximum entropy method, *Ann. Telecommun*, **34**, 204–213.

Li, H., et al. (2015), A statistical study on the whistler waves behind dipolarization fronts, *J. Geophys. Res.: Space Physics*, **120**, 1086–1095, doi:10.1002/2014JA020474.

McPherron, R. L., C. T. Russel, P. J. Coleman, Jr. (1972), Fluctuating magnetic fields in the magnetosphere, 2, ULF waves, *Space Sci. Rev.*, **13**, 411–454.

Means, J. D. (1972), Use of the three-dimensional covariance matrix in analyzing the polarization properties of plane waves, *J. Geophys. Res.*, 77, 5551–5559.

Miyoshi, Y., R. Kataoka, Y. Kasahara, A. Kumamoto, T. Nagai, M. F. Thomsen (2013), High-speed solar wind with southward interplanetary magnetic field causes relativistic electron flux enhancement of the outer radiation belt via enhanced condition of whistler waves, *Geophys. Res. Lett.*, **40** (17), 4520–4525, doi:10.1002/grl.50916.

Oscarsson, T., K. Rönnmark (1989), Reconstruction of wave distribution functions in warm plasmas, *J. Geophys. Res.*, **94**, 2417–2428.

Press, W. H., B. P. Flannery, S. A. Teukolsky, W. T. Vetterling (1992), *Numerical Recipes*, Cambridge Univ. Press, New York.

Samson, J. C. (1973), Descriptions of the polarization states of vector processes: Applications to ULF magnetic fields, *Geophys. J. R. Astron. Soc.*, **34**, 403–419.

Samson, J. C., J. V. Olson (1980), Some comments on the descriptions of the polarisation states of waves, *Geophys. J. R. Astron. Soc.*, **61**, 115–129.

Santolík, O., M. Parrot (1996), The wave distribution function in a hot magnetospheric plasma: The direct problem, *J. Geophys. Res.*, **101** (A5), 10639–10651.

Santolík, O., M. Parrot (2000) Application of wave distribution function methods to an ELF hiss event at high latitudes, *J. Geophys. Res.*, **105**, 18885–18894.

Santolík, O., F. Lefeuvre, M. Parrot, J. L. Rauch (2001), Complete wave-vector directions of electromagnetic emissions: Application to INTERBALL-2 measurements in the nightside auroral zone, *J. Geophys. Res.*, **106**, 13,191–13, 201.

Santolík, O., M. Parrot, F. Lefeuvre (2003a), Singular value decomposition methods for wave propagation analysis, *Radio. Sci.*, **38** (1), 1010, doi:10.1029/2000RS002523.

Santolík, O., D. A. Gurnett, J. S. Pickett, M. Parrot, N. Cornilleau-Wehrlin (2003b), Spatio-temporal structure of storm-time chorus, *J. Geophys. Res.*, **108** (A7), 1278, doi:10.1029/2002JA009791.

Santolík, O., D. A. Gurnett, J. S. Pickett, M. Parrot, N. Cornilleau-Wehrlin (2004), A microscopic and nanoscopic view of storm-time chorus on March 31, 2001, *Geophys. Res. Lett.*, **31**, L02801, doi:10.1029/2003GL018757.

Santolík, O., D. A. Gurnett, J. S. Pickett, M. Parrot, N. Cornilleau-Wehrlin (2005), Central position of the source region of storm-time chorus, *Planet. Space Sci.*, **53**, 299–305, doi:10.1016/j.pss.2004.09.056.

Santolík, O., D. A. Gurnett, J. S. Pickett, J. Chum, N. Cornilleau-Wehrlin (2009), Oblique propagation of whistler mode waves in the chorus source region, *J. Geophys. Res.*, **114**, A00F03, doi:10.1029/2009JA014586.

Santolík, O., C. A. Kletzing, W. S. Kurth, G. B. Hospodarsky, S. R. Bounds (2014a), Fine structure of large-amplitude chorus wave packets, *Geophys. Res. Lett.*, **41**, 293–299, doi:10.1002/2013GL058889.

Santolík, O., E. Macúsová, I. Kolmasová, N. Cornilleau-Wehrlin, Y. de Conchy (2014b), Propagation of lower-band whistler-mode waves in the outer Van Allen belt: Systematic analysis of 11 years of multi-component data from the Cluster spacecraft, *Geophys. Res. Lett.*, **41**, 2729–2737, doi:10.1002/ 2014GL059815.

Stix, T. H. (1992), *Waves in Plasmas*, Am. Inst. of Phys., New York.

Storey, L. R. O., F. Lefeuvre (1974), Theory for the interpretation of measurements of a random electromagnetic wave field in space, in Rycroft, M. J., Reasenberg, R. D. (eds.), *Space Research XIV*. Akademie-Verlag, Berlin, pp. 381–386.

Storey, L. R. O., F. Lefeuvre (1979), The analysis of 6-component measurement of a random electromagnetic wave field in a magnetoplasma, 1, The direct problem. *Geophys. J. R. Astron. Soc.*, **56**, 255–270.

Storey, L. R. O., F. Lefeuvre (1980), The analysis of 6-component measurement of a random electromagnetic wave field in a magnetoplasma, 2, The integration kernels. *Geophys. J. R. Astron. Soc.*, **62**, 173–194.

Storey, L. R. O. (1998), Revision of the basic equations of wave distribution function analysis. *Ann. Geophys.*, **16**, 651–653.

Storey, L. R. O. (1999), The measurement of wave distribution functions, in Stuchly, M. A. (ed.), *Modern Radio Science 1999*. Oxford University Press, Oxford, pp. 249–271.

Tao, X., J. Bortnik, J. M. Albert, R. M. Thorne, W. Li (2013), The importance of amplitude modulation in nonlinear interactions between electrons and large amplitude whistler waves, *J. Atmos. Solar-Terrestrial Phys.*, **99**, 67–72.

Taubenschuss, U., et al. (2014), Wave normal angles of whistler mode chorus rising and falling tones, *J. Geophys. Res.: Space Physics*, **119**, doi:10.1002/2014JA020575.

Thorne, R. M., et al. (2013), Rapid local acceleration of relativistic radiation-belt electrons by magnetospheric chorus, *Nature*, **504**, 411–414, doi:10.1038/nature12889.

12

Extreme Variability of Relativistic Electrons in Earth's Outer Radiation Belt: An Overview and Recent Revelations

D. L. Turner and V. Angelopoulos

■ Abstract

This paper provides a brief history and general overview of observations of the extreme variability of Earth's outer radiation belt electrons. We compare and contrast observations and theory of the driving mechanisms responsible for drastic and rapid enhancements and depletions of relativistic electron intensity from the pre-CRRES (i.e., before 1990) and modern eras. The current understanding on dominant source, loss, and transport mechanisms responsible for extreme variations is presented. Particular emphasis is put on events such as rapid outer belt enhancements, sudden injections from interplanetary shock impacts, energetic electron injections from the plasma sheet, flux dropouts, and complex outer belt features like bifurcated drift orbits and double outer belts. Example cases of each type of event are provided from multipoint observations during the THEMIS and Van Allen Probes eras. We finish with a brief discussion on a few of the many outstanding questions and potentially interesting topics of future research.

Waves, Particles, and Storms in Geospace. First Edition. Georgios Balasis, Ioannis A. Daglis, Ian R. Mann.

12.1 Introduction

Earth's radiation belts were the first major scientific discovery of the space age. Based on observations from several of the world's first human-built satellites, Van Allen and Frank [1959] and Vernov and Chudukov [1960] independently reported on the existence of very energetic particles trapped in the geomagnetic field in near-Earth space. The particle intensities exhibited a two-zone structure, with two local maxima around L ~1.5 and L ~4. The L-parameter [McIlwain, 1961] is defined for any magnetic field line as the radial distance from the center of the Earth to the intersection of the field line with the magnetic equatorial plane; it is useful for radiation belt physics since it can approximately map observations at any latitude to a radial distance in Earth radii (R_E) at equatorial latitudes. Due to the nature of charged particle motion in Earth's dipole-like magnetic field, trapped energetic particles (i.e., those that are relatively immune to effects from the corotational or convective electric fields) undergo three characteristic motions: gyration around field lines, bounce between mirror points along field lines, and azimuthal drift around Earth. Due to these motions the zones of ionizing radiation are toroidal in shape, circumscribing the planet, which led to their being referred to as radiation belts. Each of the three characteristic motions of radiation belt particles can be related to an adiabatic invariant, μ, K, and Φ (or L^*, the invariant L-parameter). For detailed discussions of the motion of radiation belt particles and the adiabatic invariants, see Northrop and Teller [1960], Roederer [1970], Schulz and Lanzerotti [1974], and more recently Green and Kivelson [2004] and Ukhorskiy and Sitnov [2012].

As spacecraft launched throughout the 1960s provided more and more observational evidence, scientists developed a better understanding of radiation belt climatology [e.g., Van Allen, 1969; Vernov et al., 1969]. The inner belt consisting of both protons and electrons is relatively stable compared to the outer belt, consisting of electrons ranging in energy from hundreds of keV to several MeV, which exhibits extreme variability in intensity over a range of timescales. It was recognized early that magnetically disturbed periods, and in particular geomagnetic storms, result in drastic variations of the intensity of outer radiation belt electrons. Sudden inward shifts of the entire outer belt distribution are observed immediately following many storm sudden commencements [e.g., Vernov, 1968]. During storm main phase, the intensity of >150 keV electrons decreases abruptly [e.g., McIlwain, 1966a], but the intensity of electrons at lower energy (E > 40 keV) does not always reveal a similar decrease [e.g., Craven, 1966]. However, during the recovery phase of storms (i.e., after the minimum in the Dst index), rapid enhancements of outer belt electron intensities were often reported. The enhancement timescale is delayed based on electron energy: for E > 40 keV electrons, timescales of <3 hours were reported by Craven [1966], while for E > 150 keV, the enhancement time was longer, around ~12–36 hours [Vernov et al., 1966]. Drastic electron enhancements during recovery phases were also observed at higher energies, >240 keV to >1 MeV [e.g., Williams et al., 1968] and even >5 MeV [McIlwain, 1966b].

Russell and Thorne [1970] also showed how the "slot" region between the inner and outer electron belts could be partially filled during periods of enhanced geomagnetic activity.

Various theories were developed to explain the observations of outer radiation belt dynamics. Vernov [1968] described the inward displacement of the belt at the beginning of many storms as rapid radial diffusion, an irreversible process. Depletions during the main phase, referred to as flux dropouts [see review by Turner et al., 2012a], were explained by Dessler and Karplus [1961] as resulting from adiabatic radial transport of trapped energetic particles due to the strong variations in the global magnetic field during geomagnetic storms. Enhancements of the outer belt during geomagnetically active periods were initially thought to result from enhanced inward radial diffusion from a source of electrons in the plasma sheet, though as we'll see later, this isn't the full story. Radial diffusion theory in Earth's radiation belts, along with the formulation for momentum and pitch angle diffusion, was presented in great detail by Schulz and Lanzerotti [1974].

Wave-particle interactions are critical to quasilinear diffusion theory, as these provide the random or resonant energy kicks to the particles, breaking one or more of the adiabatic invariants and enabling diffusion of the particle distributions. Quasilinear theory proved very successful as a model to explain the radiation belts. For example, Lyons and Thorne [1973] explained a steady state of the electron radiation belts using a combination of inward radial diffusion from a source population at $L = 5.5$ and losses due to Coulomb scattering at very low L-shells and pitch angle diffusion throughout the plasmasphere. In the model, inward radial diffusion from the boundary condition at $L = 5.5$ provided a source of electrons, while the losses ultimately resulted in the formation of the slot region. This picture of the electron radiation belts—in which a source of electrons at high L-shells was accelerated by conservation of μ and K as they diffused radially inward to regions of higher magnetic field strength—was widely accepted. In this picture, storm-time depletions were dominated by adiabatic and entirely reversible variations due to variations in the global magnetic field, and true losses resulted primarily from pitch angle diffusion into the atmospheric loss cone. The strong variations observed in the belt during geomagnetically active periods were thought to simply be a result of strong variations in the diffusion coefficients due to enhanced wave activity.

This picture of the radiation belts remained *mostly* unchallenged [for some exceptions see: Baker et al., 1989; Fujimoto et al., 1990] until the early 1990s, after the launch of the Combined Release and Radiation Effects Satellite (CRRES) and Solar, Anomalous, and Magnetospheric Particle Explorer (SAMPEX) missions. CRRES was launched on July 25, 1990, into a geosynchronous transfer orbit (period ~10 hours) with an inclination of 18°. This orbit, combined with its suite of particle, fields, and waves instrumentation, made CRRES ideal for studying Earth's radiation belts [e.g., Meredith et al., 2003]. SAMPEX was launched on July 3, 1992, into a high inclination low-Earth orbit; it carried a suite of energetic particle telescopes, which allowed for observations of trapped and precipitating electrons from the outer radiation belt.

Observations from these spacecraft [e.g., Blake et al., 1992; Baker et al., 1994; Li et al., 1997] and the subsequently launched Polar spacecraft [Blake et al., 1995; Selesnick and Blake, 1997] quickly demonstrated that the extreme variations in the outer radiation belt involved more complex processes than were fully appreciated previously.

In the post-CRRES era, the newfound appreciation for the complexity of outer radiation belt dynamics has led us to where we are today. Understanding outer belt variability is becoming increasingly important as human society becomes more dependent on space-based assets, to which relativistic electrons in the outer belt pose a threat. With existing missions such as NASA's THEMIS and NOAA's GOES and POES plus the recently launched NASA Van Allen Probes, and NSF's FIREBIRD and Colorado Student Space Weather Experiment (CSSWE) CubeSats, we now have a sizeable fleet of spacecraft available to study Earth's outer belt electrons. This paper is meant to provide a primarily observational overview of our current understanding of the drivers of extreme variations in the intensity of relativistic electrons in Earth's outer radiation belt. The paper is structured as follows: The next section discusses key observations and theory since the launch of CRRES and leading up to the current era (i.e., up to 2007). The third section presents some example cases from the current era, in which (with the advent of NASA's five-satellite THEMIS mission) multi-spacecraft analysis from a near-equatorial vantage point, together with increasingly well-instrumented NOAA geosynchronous and polar orbiting satellites, has provided an unprecedented understanding of the underlying mechanisms responsible for drastic and sudden enhancements, depletions, and transport of outer belt electrons. Note that here we do not focus specifically on storm-time variations, though many of the cases of extreme variability occurred during geomagnetic storms. For a detailed overview of the behavior of the radiation belts and ring current during geomagnetic storms, please see Reeves and Daglis [this volume]. We finish with a section outlining several of the many outstanding questions and topics of future research concerning outer belt variability followed by a brief conclusion and summary.

12.2 Physical processes that can drive extreme outer belt variability

It has long been appreciated that the intensities of relativistic electrons in Earth's outer radiation belt fluctuate by orders of magnitude over a range of timescales, from minutes to decades [e.g., see reviews by Li and Temerin, 2001; Friedel et al., 2002; and Millan and Baker, 2012]. Outer belt variations are ultimately driven by fluctuations in various solar wind parameters, but that relationship is complex and relies on many internal processes within the magnetosphere, each of which is also related to activity in the solar wind [e.g., Hudson et al., 2008]. Figure 12.1 shows an example of this variability in observations from NASA's THEMIS mission [Angelopoulos, 2008]. Each THEMIS spacecraft passes through the entire outer radiation belt twice per day very close to the

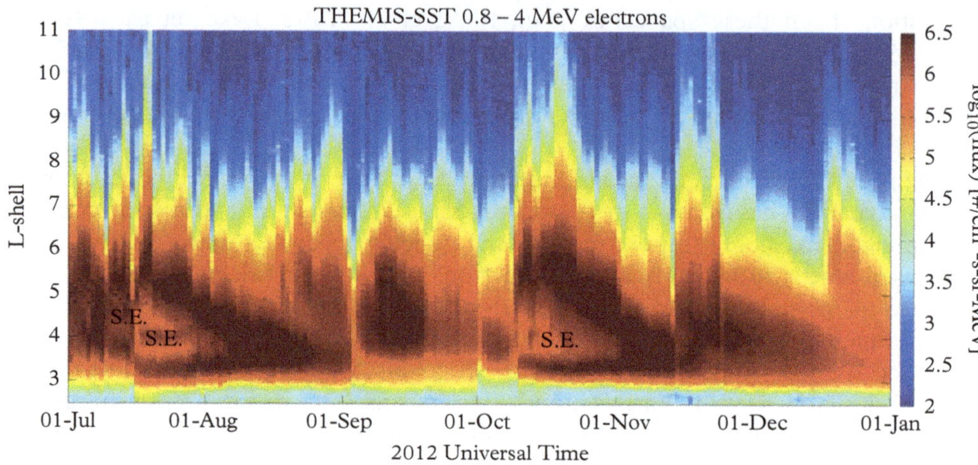

Figure 12.1 *Relativistic electron fluxes from the highest channel (0.8 to ~4 MeV) on the THEMIS Solid State Telescope (SST) instruments binned and averaged by time (daily) and L-shell ($\Delta L = 0.05$). Data are combined from all three THEMIS spacecraft (THA, THD, and THE) from July 1 to December 31, 2012. Log_{10} of flux is shown in color. The instruments sometimes suffer from saturation in the heart of the outer belt during periods in which the electron fluxes are very enhanced; such regions on the plot are marked with S.E. for "saturation effect."*

magnetic equator. The Solid State Telescopes (SSTs) on board each THEMIS satellite provide energy and pitch angle resolved fluxes of energetic ions and electrons from ~30 keV to several MeV [e.g., Turner et al., 2012b]. Figure 12.1 shows daily averaged, 0.8–4 MeV electron fluxes compiled from the SST instruments on the three THEMIS spacecraft still remaining in Earth orbit (THA, THD, and THE; THB and THC became the ARTEMIS mission in 2011) from July 1, 2012, to December 3, 2012. This period showcases how the intensity within the belt, as well as at its outer boundary can vary by more than two orders of magnitude from day to day. Rapid enhancements are visible on days like July 16, October 9 (both of which resulted in the SST instruments saturating in the heart of the outer belt), and November 14, while dropout events can be clearly seen on days like September 3 and October 1. Several of the example events that we discuss in the next section occurred during this period, as did the launch of the Van Allen Probes mission (formerly Radiation Belt Storm Probes [Mauk et al., 2012]). In this section, we present the background theory and supporting evidence of the various source, loss, and transport processes that have been identified as important to driving some of the most extreme cases of outer belt variability.

12.2.1 Sources

Understanding source processes is fundamental to understanding rapid outer belt enhancements. Reeves et al. [1998] and Li et al. [1998] examined an enhancement of outer radiation belt electrons that occurred on January 10–11, 1997. During this event, spacecraft at geosynchronous orbit (GEO) observed the fluxes of relativistic electrons increasing by more than two orders of magnitude in only around 12 hours. The enhancement was triggered by the impact of an interplanetary coronal mass ejection (ICME). The ICME in question included a period of strong, steady southward B_Z in the interplanetary magnetic field (IMF); during this period, there was strong substorm activity including rapid fluctuations of the energetic particles observed at GEO. Based on the observations, Reeves et al. [1998] concluded that the source of >2 MeV electrons was located inside of GEO during the enhancement event. Li et al. [1998] concluded that the source was the result of enhanced substorm injections introducing a seed population that was subsequently energized by the electromagnetic pulse attending the passage of the ICME shock. Both explanations contradicted the old picture of acceleration by inward radial diffusion from a source in the plasma sheet.

Around the same time, a different theoretical framework was developed to explain the source of relativistic electrons in the heart of the outer radiation belt. Following conceptual work on the topic by Temerin et al. [1994], Summers et al. [1998], and Horne and Thorne [1998] presented a theory of local acceleration of electrons from a seed population of from tens to hundreds of keV to relativistic energies via wave-particle interactions with whistler-mode chorus waves. In this framework, electrons interact with chorus waves [e.g., Li et al., 2009 and references therein] via Doppler-shifted cyclotron resonance, resulting in energy diffusion. Provided many such interactions as electrons drift around the Earth, individual electrons could be stochastically accelerated up to relativistic energies. Horne et al. [2005] showed that acceleration timescales for seed electrons (tens to a few hundreds keV) to be accelerated to ~MeV energies via interactions with chorus waves should be around 1–2 days. Such a scenario is depicted in the conceptual schematics in Figure 12.2.

One key to distinguishing between source mechanisms was to convert electron flux observations to phase space density (PSD) as a function of the adiabatic invariants (μ, K, and L*), since such a conversion effectively removes purely adiabatic variations from the data. By examining the radial distributions (i.e., in L*) of electron PSD for fixed values of μ (analogous to energy) and K (analogous to pitch angle), it was possible to look for telltale signatures of the source location of outer radiation belt electrons. This is summarized on the right-most plots in Figure 12.2a: electrons with a source in Earth's plasma sheet should have a maximum in the PSD radial distributions in the plasma sheet itself (i.e., at high L-shells), whereas electrons that are locally accelerated within the heart of the outer belt should have radial profiles that are peaked, with the peak being collocated with the waves that are causing the acceleration. For a more detailed discussion on this, see Green and Kivelson [2004] and/or Reeves et al. [2013].

As observations of the outer radiation belt improved, much evidence in favor of an internal source of relativistic electrons has appeared in the literature. Meredith et al. [2003] showed statistical results from CRRES that outer belt enhancement events occur just outside of the plasmapause during periods of prolonged substorm activity and enhanced

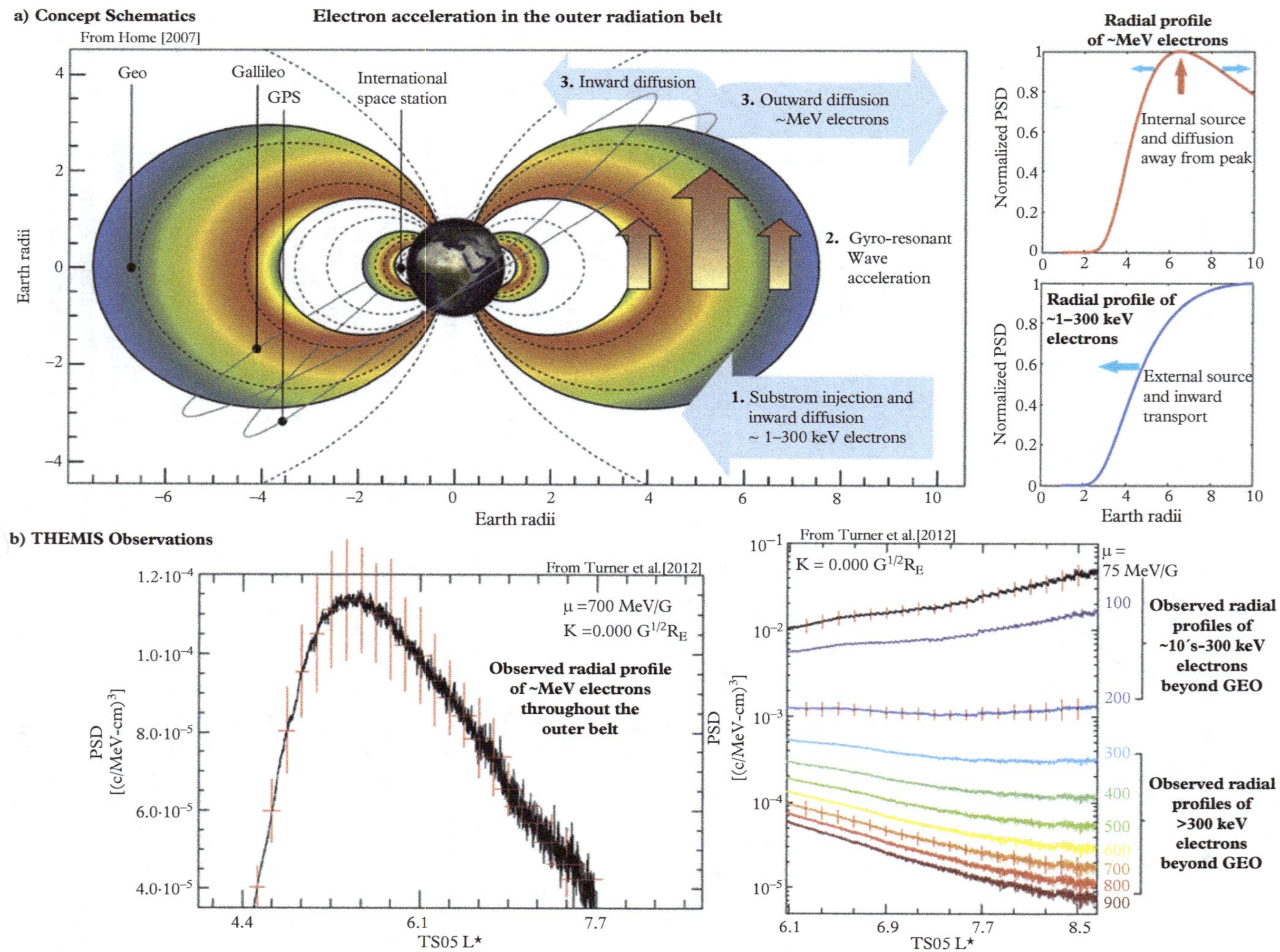
a) Concept Schematics
Electron acceleration in the outer radiation belt
From Home [2007]
Geo
Gallileo
GPS
International space station
3. Inward diffusion
3. Outward diffusion ~MeV electrons
2. Gyro-resonant Wave acceleration
1. Substrom injection and inward diffusion ~ 1–300 keV electrons
Earth radii
Radial profile of ~MeV electrons
Internal source and diffusion away from peak
Radial profile of ~1–300 keV electrons
External source and inward transport
Normalized PSD
b) THEMIS Observations
From Turner et al.[2012]
μ =700 MeV/G
K =0.000 G^1/2 R_E
Observed radial profile of ~MeV electrons throughout the outer belt
PSD [(c/MeV-cm)^3]
TS05 L*
K = 0.000 G^1/2 R_E
μ = 75 MeV/G
100
200
300
400
500
600
700
800
900
Observed radial profiles of ~10's-300 keV electrons beyond GEO
Observed radial profiles of >300 keV electrons beyond GEO

Figure 12.2 *Comparing the conceptual scenario of energy-dependent sources of radiation belt electrons to observations from THEMIS. (a) Schematic from Horne [2007] (top left) showing how ~1 to 300 keV electrons can be injected due to substorm activity (or enhanced convection) and inward radial diffusion. These electrons serve as the source population of chorus waves and the seed population that can be accelerated to >1 MeV by gyro-resonant wave-particle interactions with the chorus. This local acceleration results in a peak in the phase space density (PSD) profile around the radial distance (L-shell) of the wave-particle interactions, and—provided the third invariant is broken by wave activity—the PSD can diffuse away from this peak. Schematics at the top right show the PSD radial distributions in the equatorial plane for the scenario described above. The lower energy, ~1–300 keV electrons (bottom right of the three plots in (a)) have a source population, and thus a maxima in PSD, in the near-Earth plasma sheet, while the relativistic electrons (top right of the three plots in (a)) have a local source population within the heart of the outer belt. (b) Both plots from Turner et al. [2012b]: THEMIS observations are consistent with the above scenario. The left plot in (b) shows the PSD versus L* of relativistic, equatorially mirroring electrons, which reveals a peak at around L* ~5.5. The right plot in (b) shows the PSD versus L* over a broader range in μ (and energy) as the THEMIS spacecraft move beyond geosynchronous orbit into the near Earth plasma sheet. The outward gradients are positive for lower-energy electrons, indicating a source in the plasma sheet, but negative for relativistic electrons, indicating a source inside of GEO. Turner et al. [2012b] showed that these μ-dependent radial distributions are typical for outer belt electrons.*

chorus waves. Observational studies examining PSD of outer belt electrons at fixed values of μ and K have mostly indicated that, for relativistic electrons above ~500 keV in the heart of the outer belt, the PSD profiles in L* are most often peaked [e.g., Green and Kivelson, 2004; Chen et al., 2007; Turner et al., 2010, 2012b; Kim et al., 2010; Shprits et al., 2012]. In fact, the results presented by Turner et al. [2012b] proved remarkably consistent with the theory of local acceleration via wave-particle interactions with chorus waves. These results are summarized in Figure 12.2b, which shows THEMIS observations of electron PSD for fixed invariants. Positive radial gradients beyond GEO for electrons with energies of ~80 to ~300 keV energy in the heart of the belt indicate a source in the plasma sheet, while negative gradients beyond GEO and peaked distributions for relativistic electrons indicate a source in the inner magnetosphere, within the heart of the outer belt. Modeling efforts [e.g., Shprits et al., 2006a; Su et al., 2011; see also review by Shprits et al., 2008b] have proven successful at capturing rapid enhancements (i.e., ~2 orders of magnitude or more increase in only ~1 day) via local acceleration of electrons by whistler-mode chorus, but observationally the question remained as to whether this mechanism was able to rapidly enhance the entire outer belt of relativistic electrons in only ~1/2 day. In the next section, we examine some example cases during the THEMIS and Van Allen Probes eras that can be used to address this question.

Kataoka and Miyoshi [2008] showed that during solar cycle 23, the strongest enhancements of MeV electrons at GEO were associated with rarefaction periods of very low solar wind density and high, but declining, solar wind speed during the end of geomagnetically active periods. They concluded that such enhancements are the result of both enhanced source mechanisms *and* reduced loss mechanisms. This underlines the fact that ultimately outer belt variations result from sources dominating over losses or vice versa, and next, we also outline our current understanding of mechanisms that can result in sudden losses of outer belt electrons.

12.2.2 Losses

Being responsible for extreme depletions of the outer radiation belt, losses of relativistic electrons also play a critical role in outer belt variability. Ultimately, loss of outer belt electrons occurs due to: escape via the inner boundary (i.e., precipitation into Earth's atmosphere), escape via the outer boundary (i.e., through the magnetopause), and deceleration, which can possibly occur due to nonlinear wave-particle interactions [e.g., Bortnik et al., 2008] or outward radial transport (e.g., outward diffusion in L^*) [see review on losses by Millan and Thorne, 2007]. Pitch angle diffusion and rapid scattering of outer belt electrons into the atmospheric loss cone can occur due to wave-particle interactions with various plasma waves, such as plasmaspheric hiss [e.g., Thorne et al., 2013a], chorus [e.g., Shprits et al., 2007; Tao et al., 2012], and electromagnetic ion cyclotron (EMIC) waves [e.g., Albert, 2003; Ukhorskiy et al., 2010]. The diffusion coefficients, and thus effective loss rates, for different waves interacting with radiation belt electrons are dependent on electron energy, pitch angle, and location [e.g., Shprits et al., 2008b]. In contrast, losses through the magnetopause are relatively independent of electron energy, though there is some dependence on pitch angle due to drift-shell splitting [e.g., Kim et al., 2008]. Magnetopause shadowing is a term describing the sudden loss of electrons due to sudden inward motion of the magnetopause, which effectively empties previously closed drift shells of electrons within one drift period [e.g., Desorgher et al., 2000; Kim et al., 2010]. Finally, non-adiabatic outward radial transport, such as radial diffusion [e.g., see review by Shprits et al., 2008a], results in electrons becoming irreversibly decelerated as they move to regions of weaker magnetic field strength and longer field lines while conserving μ and K.

Flux dropouts are a good example of extreme variability of the outer radiation belt electrons. Since Li et al. [1997] and Kim and Chan [1997], it has been appreciated that adiabatic effects (i.e., the "Dst effect") alone cannot explain all of the flux variation observed during many radiation belt flux dropout events, and several of the loss mechanisms introduced above have been used to explain the sudden losses of electrons throughout the outer belt during dropouts. The two leading theories are loss to the atmosphere due to wave-particle interactions with EMIC waves [e.g., Thorne et al., 2006] and loss due to magnetopause shadowing and subsequent enhanced outward radial transport [e.g., Shprits et al., 2006b]. There has been observational evidence in favor of both mechanisms: Green et al. [2004], O'Brien et al. [2004], Selesnick [2006], Borovsky and Denton [2009], and Ukhorskiy et al. [2010] all concluded that atmospheric losses could explain dropouts; Horne et al. [2009], Loto'aniu et al. [2010], Morley et al. [2010], Matsumura et al. [2011], Meredith et al. [2011], Shprits et al. [2012], and Turner et al. [2012c] all showed results in favor of magnetopause shadowing and outward radial transport; meanwhile Bortnik et al. [2006] concluded that losses to the outer boundary dominated at higher L-shells (>~5) but atmospheric losses dominated at lower L-shells. For full details plus additional references, see the review on outer belt dropouts by Turner et al. [2012a], which concluded that *most*, but not all, dropouts are more consistent with the theory of magnetopause shadowing and outward radial transport.

When the magnetopause is suddenly compressed by an enhancement of solar wind dynamic pressure, electrons on previously closed drift shells may suddenly escape through the magnetopause within only one drift period. This results in a very distinct radial distribution of electrons in the outer radiation belt: a very sharp gradient remains in the distribution at the drift shell that maps to the innermost magnetopause distance. Provided sufficient wave activity to invalidate the third adiabatic invariant and allow for particle transport across drift shells (i.e., in L*), a cascade or spillover effect will occur, in which the many more electrons at lower L-shells that are transported to higher L-shells are not fully replaced by the very few electrons at higher L-shells that get transported to lower L-shells. This process is compounded by radial diffusion being much more effective at higher L-shells [e.g., Brautigam and Albert, 2000; Selesnick and Blake, 2000] and how electrons lose energy as they move outward in the geomagnetic field. The wave activity required for the radial transport may be manifested as enhanced ULF wave activity resulting from the variations in the solar wind that caused the magnetopause motion [e.g., Turner et al., 2012c].

An example of the effect of magnetopause shadowing and outward radial transport on a distribution of electrons in the outer belt is illustrated with the simulation results shown in Figure 12.3d. This simulation solved the one-dimensional radial diffusion equation [e.g., Shprits et al., 2008a] for a dropout event that occurred on September 30, 2012. The initial condition was based on THEMIS observations of electron PSD at μ = 750 MeV/G and K = 0 $G^{1/2}R_E$ distributed in L* throughout the belt prior to the dropout. The inner boundary, and L* = 1 was held at zero PSD, while the outer boundary condition changed in time to simulate the effects of magnetopause shadowing. As seen in Figure 12.3a, there were two sudden enhancements of solar wind dynamic pressure shortly after 11:00 UT and 23:00 UT on September 30. These resulted in sudden inward motion of the magnetopause and a reduction in the last closed drift shell (i.e., L^*_{max} in Figure 12.3c). To simulate the effect of magnetopause shadowing, the PSD at all drift shells affected by the two magnetospheric compressions were reduced to 1% of the peak value from the initial condition, which is again consistent with observations from THEMIS during the dropout. These losses were added at the beginning of the simulation and then 12 hours after, just as was observed. To compare with the THEMIS observations of the pre-to-post-dropout distributions (discussed further in the next section), the simulation was run for 19 hours using an L*-dependent diffusion coefficient (D_{LL}) based on that used by Selesnick and Blake [2000], which is representative of active conditions (see Figure 12.3d for the equation used here). This simulation shows the effect of the two magnetopause shadowing events on the outer belt. Note that the loss of PSD extends into lower L* than the innermost extent of the magnetopause shadowing

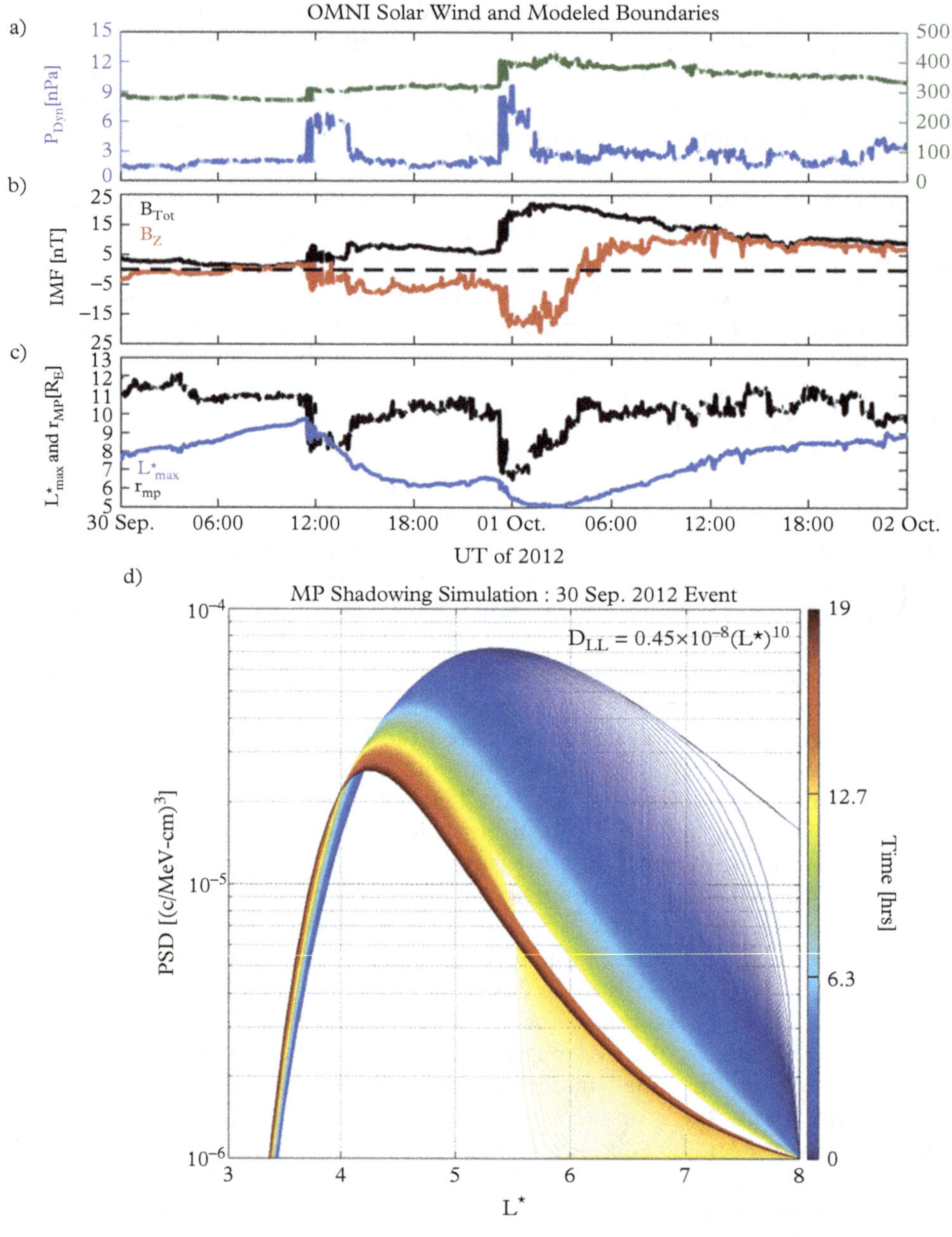

Figure 12.3 *(a), (b), and (c) From Turner et al. [2013b] OMNI solar wind dynamic pressure and speed (a), IMF strength and* Z_{GSE} *component (b), and the magnetopause standoff distance and last closed drift shell,* L^*_{max} *(c) all from the September 30, 2012, dropout period. (d) One-dimensional radial diffusion simulation results of the PSD evolution as observed by THEMIS during the dropout. See text for details.*

itself (i.e., L* ~8 and ~5.5 for the two compressions, respectively); this is because of the rapid outward radial diffusion that occurs across the sharp radial gradients in the PSD radial distribution remaining after each instance of shadowing. The simulation shows how the peak in PSD moves down in amplitude and in to lower L*, with losses at all L* >~4, in less than a day. The loss in PSD is strongest at higher L*. Also, note that at lower L*, there is actually an increase in PSD due to inward radial diffusion. Finally, this mechanism should act relatively independent of particle energy and species. These are all key signatures of this mechanism, which we will discuss further in the next section.

Magnetopause shadowing can explain ~88% of outer belt dropout events [e.g., Shprits et al., 2012]. Wave-particle interactions resulting in losses to the atmosphere may still play an important role in those other 12% of cases. Indeed, relativistic electron precipitation into the atmosphere is observed [e.g., Millan et al., 2010], and estimates of the loss rate as a percent of the total trapped population [e.g., Tu et al., 2010] are improving rapidly with increased data availability from LEO spacecraft [Li et al., 2013]. Furthermore, losses into the atmosphere likely play just as important a role as weak sources during prolonged depletion periods of the outer belt electrons, such as occurred through much of 2009 (more details on this period in the next section).

12.2.3 Transport

Radial transport processes can also result in extreme variations of the outer radiation belt electrons. In the previous section, we described how rapid outward radial transport could play a role in outer belt dropout events. Rapid transport can also lead to sudden flux enhancements of radiation belt electrons. Li et al. [1998] presented observations of injections of electrons in the outer radiation belt due to two different processes: substorms and magnetospheric impacts from interplanetary shocks. Starting with the latter, when an interplanetary shock impacts the magnetosphere, a compressional fast mode wave is launched near the subsolar point and propagates tailward. Using a simple wave model based on CRRES observations from the shock impact on March 24, 1991, Li et al. [1993] performed test-particle simulations of radiation belt electrons to examine how they would react to the shock impact. Their simulations showed how electrons are transported radially inward when they interact with the fast magnetosonic wave. The results successfully reproduced the CRRES observations of multi-MeV electrons that were injected into L < 3 following the shock impact, as was initially reported by Blake et al. [1992]. This injection formed a new radiation belt in the slot region in only around a minute. The Halloween storms of 2003 [e.g., Baker et al., 2004; Shprits et al., 2006a] also included the sudden formation of a new radiation belt within the slot region due to an injection from the impact of a CME shock in the solar wind. Once injected to such low L-shells, the multi-MeV electrons can remain for over a year after, as seen in Figure 12.4 (from Figure 1 of Baker et al. [2004]). Shock injections transport particles radially inward throughout the outer belt, which can be used to determine the PSD radial gradient just prior to the injection [Li et al., 2003a; Turner and Li, 2008; Kim et al., 2010; Turner et al., 2010], and since the particles conserve μ and K, they are also energized by the injection. Thus, shock injections are an acceleration mechanism that can result in very sudden enhancements of outer radiation belt electrons.

Energetic particles can also be suddenly injected into the inner magnetosphere during active periods related to dipolarization events and substorms [e.g., Birn et al., 1997, 1998]. Li et al. [1997, 2003b] successfully simulated energetic (tens of keV up to ~300 keV) electron injection signatures and drift echoes at GEO using a model consisting of a transient azimuthal (i.e., cross-tail) electric field signature associated with an increase in B_z, both near-Earth features of substorm dipolarization. Gabrielse et al. [2012] used typical, earthward-traveling, azimuthally localized electric field features of dipolarization fronts [e.g., Runov et al., 2011; Liu et al. 2013] and a guiding center test-particle model constrained by THEMIS multipoint observations to model energetic particle injections. With that model, they were able to explain many of the features typically observed during particle injections not only at but also beyond GEO. For example, they showed how flux depletions observed at some energies during injections can result from return flows at the edges of the fast flow channels that make up the earthward injection.

Concerning extreme cases, Ingraham et al. [2001] reported an unusual case in which an energetic electron injection observed at GEO included relativistic electrons with energy exceeding 1 MeV. This upper energy threshold was much higher than what is typically observed (i.e., a few hundred keV). The injection occurred in the recovery phase of the March 24, 1991, storm during a period of intense substorm activity. These observations showed a clear increase in the upper energy threshold of injected electrons as the ambient population of relativistic electrons also increased in the rapid and drastic outer belt enhancement event that occurred during the recovery phase of that storm. The Ingraham et al. [2001] results provide additional evidence of the importance of substorm-injected electrons during outer belt enhancement events.

Radial transport may also be relatively chaotic at high L-shells on the dayside of the magnetosphere. Due to the compression of the dayside magnetic field, the location of minimum magnetic field strength along field lines at higher L-shells (typically beyond GEO) actually moves away from the magnetic equator to higher latitudes. When trapped radiation belt electrons encounter such regions, their drift shells bifurcate, and some of them, depending on their pitch angle and bounce phase, start bouncing around the minimum-B location either above or below the equator. Such bifurcated drift orbits are often referred to as Shabansky orbits after Shabansky [1971]. Ukhorskiy et al. [2011] conducted test particle simulations to study the effects of these bifurcated drift orbits on trapped electrons. They found that some electrons undergo bifurcated drift orbits and get scattered randomly in pitch angle and L-shell. Most of those electrons remained quasi-trapped in the bifurcation region for a long time. Due to the random mixing associated with the chaotic motion, some electrons were accelerated by a net inward radial diffusion, but some electrons were also lost through the magnetopause or into the atmospheric loss cone. McCollough et al. [2012] examined EMIC wave growth due to the effects of ions undergoing bifurcated drift orbits. They found that EMIC waves

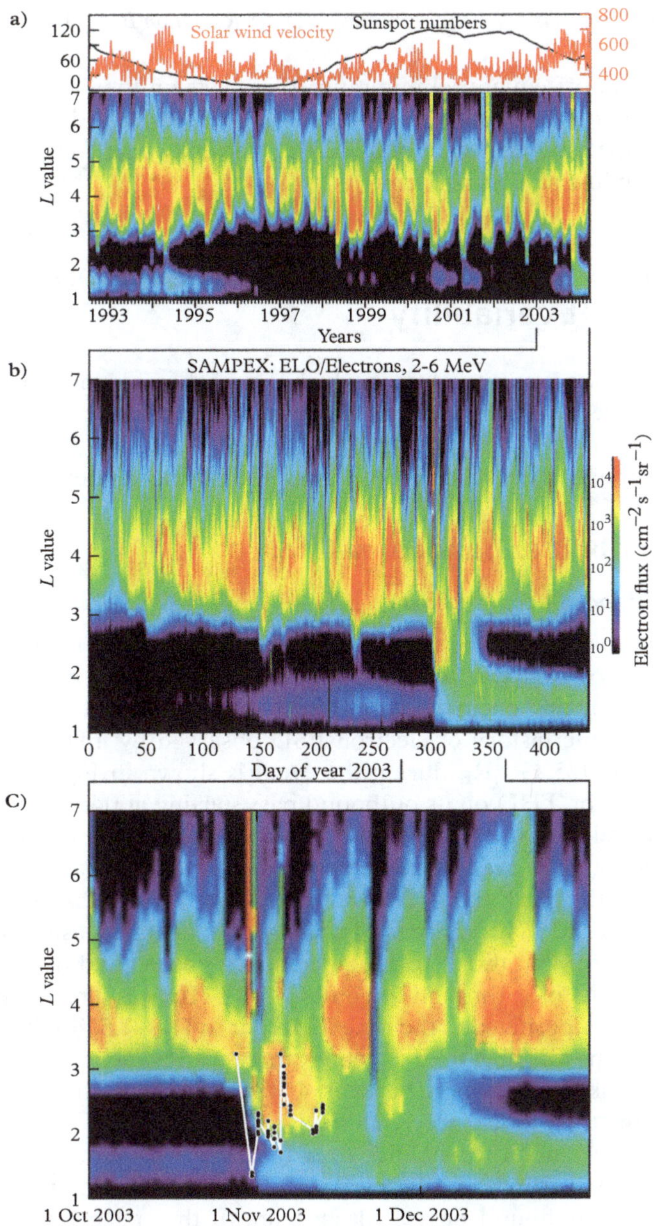

Figure 12.4 *From Baker et al. [2004]: SAMPEX observations of 2–6 MeV electron flux during the entire mission (top-most color plot), 2003 (middle) and around the Halloween 2003 storms (bottom). Flux is binned and averaged in time and L-shell and shown in color. The top plot shows averaged solar wind speed (red) and sunspot number (black) over the course of the SAMPEX mission.*

may be generated in off-equatorial minimum-B pockets on the dayside, which may explain recent observations of bidirectional EMIC waves observed at high L-shells and mid-latitudes on the dayside [Allen et al., 2013] and may also contribute to additional relativistic electron loss at high L-shells.

12.3 Recent multipoint observational examples of extreme outer belt variability

In this section, we discuss recent cases in which the mechanisms outlined in the previous section played a role in driving extreme variations of relativistic electrons in the outer belt. Each case occurred after the launch of NASA's THEMIS mission in 2007, thus allowing for multipoint observations throughout the magnetosphere.

12.3.1 Outer belt enhancements

12.3.1.1 Rapid local acceleration

We examine two cases from the recent literature in which the intensity of relativistic electrons enhanced by over two orders of magnitude throughout much of the outer belt in less than a day. The first occurred on February 4, 2011, and was presented by Turner et al. [2013]. The time history of electron PSD observed by THEMIS for μ = 1000 MeV/G and K < 0.025 $G^{1/2}R_E$ during the event is shown in Figure 12.5. The initial distribution, shown for THD on its outbound pass starting at 04:30 UT on February 5, revealed a weak, peaked distribution leftover after the main phase dropout of this storm. The next distribution, observed by THE on its outbound pass starting only 30 minutes later, revealed how this peak has grown in time. Furthermore, taking advantage of the multipoint nature of the THEMIS mission, the simultaneous measurements from these two spacecraft confirm that the distribution is indeed peaked and not just time varying. When THE started its pass through the outer belt, THD was already at higher L* and the PSD gradient was positive; by the time THE got to L* ~4.5, THD was at higher L* and the PSD gradient was negative. Together, these observations confirm that the PSD distribution was peaked somewhere between 4 < L* < 4.5 and that this peak grew over time, a telltale signature of local acceleration.

The outer belt continued to enhance over the next 24 hours, as can be seen from the subsequent PSD distributions observed by THD and THE in Figure 12.5. It should be noted that on the inbound passes (dashed lines), the spacecraft were inbound on the dayside, and due to the extreme compression of the magnetopause, L* was undefined by the Tsyganenko and Sitnov [2005] model at L* > ~5.75 on February 5 and >~6.5 on February 6. Since L* is undefined, we did not plot the PSD at higher radial distances for these distributions in that figure. However, the PSD was still calculated for fixed μ and K at these higher L-shells, which revealed that the PSD distributions were indeed peaked as shown at these times. Thus, over the course of only ~24 hours from February 5 to 6, the PSD of relativistic electrons increased by over two orders of

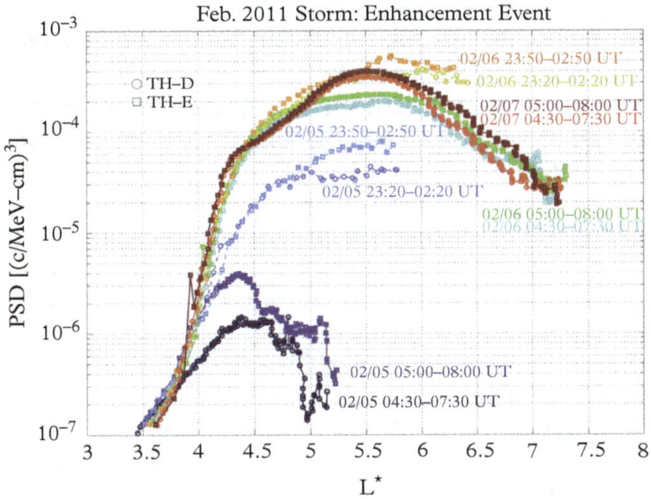

Figure 12.5 *Phase space density distributions from the February 2011 storm. Results from THEMIS-D and -E are shown with circles and squares, respectively. Distributions are color coded by time from February 5 to February 7. The UT ranges shown correspond to the periods that the spacecraft spent in the outer belt for each distribution.*

magnitude at L* > 5 throughout the belt. The time history of the distributions from THEMIS reveals that the PSD was enhanced as growing peaks that started first around L* ~4.5 and moved to higher L* throughout the event. This is strong evidence of local acceleration. Furthermore, Turner et al. [2013] showed THEMIS observations of whistler-mode chorus during the main and recovery phases of this storm; the chorus wave amplitudes were strongest first at lower L-shells during the main and early recovery phases (i.e., ~4 < L* < ~6 on February 5) and then moved to higher L-shells and covered a broader range over the next day (i.e., ~4.5 < L* < ~7 on February 6). Chorus activity was not observed on February 7, which coincides with when the peak in PSD stopped growing. Turner et al. [2013] concluded that this event provided strong evidence of rapid enhancement of relativistic outer belt electrons due to local acceleration from wave-particle interactions between seed electrons at tens to hundreds of keV and whistler-mode chorus.

Another very rapid enhancement of relativistic electrons in the outer radiation belt occurred during the October 8–9, 2012, storm. During that event, the fluxes of multi-MeV electrons in the heart of the outer belt increased by over two orders of magnitude in only ~12 hours. Reeves et al. [2013] examined PSD distributions for μ = 3433 MeV/G and K = 0.11 $G^{1/2}R_E$ through the outer radiation belt from the Van Allen Probes. They found the PSD of these multi-MeV (i.e., "ultrarelativistic") electrons evolved as growing peaks centered on L* ~4.3 during the enhancement. Furthermore, they examined PSD derived from THEMIS-SST data in the near-Earth plasma sheet, which revealed negative gradients and thus an insufficient source of these relativistic electrons in the plasma sheet throughout the event. Thorne et al. [2013b] were able to reproduce the enhancement of electrons ranging from hundreds of keV to several MeV using two-dimensional, energy and pitch angle diffusion driven by wave-particle interactions with whistler-mode chorus waves. They modeled the global chorus wave environment using a combination of Van Allen Probes and POES observations. Their results showed that the relativistic electrons could be locally accelerated from a seed population of tens to hundreds of keV electrons by chorus waves, resulting in the more than two orders of magnitude enhancement in only ~12 hours, consistent with observations.

Both events described above revealed growing peaks in PSD for relativistic electrons coincident with enhanced chorus wave activity. Those events occurred during the early recovery phase of geomagnetic storms, consistent with other recent results showing that chorus acceleration should be particularly efficient during the early recovery phase of storms [e.g., Artemyev et al., 2013; Shprits et al., 2013]. However, rapid relativistic electron enhancements need not necessarily be associated with geomagnetic storms. For example, Meredith et al. [2002] presented an outer belt enhancement observed by CRRES during a period of strong substorm activity, and Schiller et al. [2014] presented a case observed by Van Allen Probes and THEMIS in which electrons up to ~1 MeV in the outer belt enhanced by over 2.5 orders of magnitude during a 24-hour period in the absence of a storm (minimum Dst was only ~ −30 nT during a full week around the event). The latter event, however, was also associated with sustained, enhanced substorm activity based on the AL index, further evidence suggesting that enhanced and repeated substorm activity may be an important driver of radiation belt relativistic electron acceleration both during and outside of storms.

12.3.1.2 Shock injections

On St. Patrick's Day (March 17) 2013, an ICME shock impacted Earth's magnetosphere. Despite a gap in the OMNI dataset at the shock impact time, the ARTEMIS spacecraft [Angelopoulos, 2011] were in lunar orbit upstream of Earth's bow shock and observed the ICME and its shock. IMF and solar wind density and velocity observed by ARTEMIS-P2 are shown in Figure 12.6. These observations reveal that the shock hit the spacecraft, in lunar orbit, at around 05:54:30 UT. The Moon was located at ~60 R_E off of the dusk flank at the time. Compared to previously mentioned shocks, such as those associated with the March 1991, and Halloween 2003 events, the St. Patrick's Day shock was relatively weak, with the solar wind density spiking to $< 20\,cm^{-3}$ and speed only jumping to ~650 km/s; the Halloween 2003 events involved a jump in velocity up to approximately 2000 km/s [Li et al., 2009]. Regardless, at the time this paper was written, the St. Patrick's Day event was one of the clearest and strongest interplanetary shocks to impact the magnetosphere during the current solar maximum.

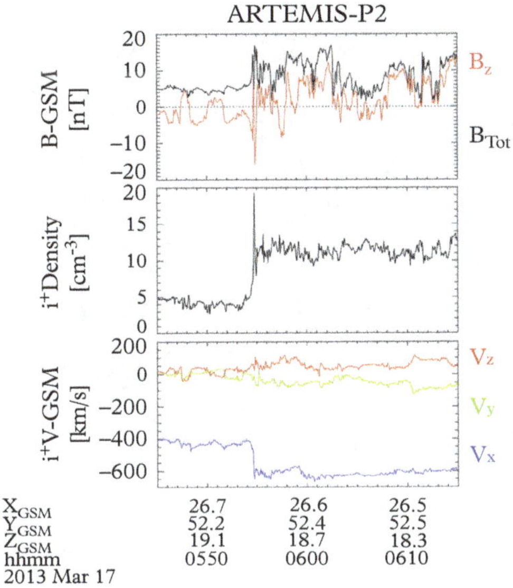

Figure 12.6 *ARTEMIS-P2 observations of the St. Patrick's Day 2013 interplanetary shock at 1 AU. The top plot shows IMF strength (B_{Tot} in black) and Z_{GSM} (B_Z in red). The middle plot shows ion density, and the bottom plot shows ion velocity in GSM coordinates. The spacecraft location is shown along with corresponding UT from March 17, 2013, on the X-axis.*

Using several of the available spacecraft in the magnetosphere, we next examine the effect this shock impact had on energetic electrons in the inner magnetosphere. Figure 12.7 shows THEMIS and GOES magnetic field and energetic electron (tens to hundreds of keV) observations during the same 30-minute period as shown in Figure 12.6. The shock impacted the magnetosphere at ~06:00 UT. At the time of the impact, THA was just outside of GEO and inbound in the noon MLT sector, while THD was also just outside of GEO but was outbound in the dawn MLT sector. THE was inside of 5 R_E at the time, also in the dawn MLT sector, and is not shown here. GOES-13 and -15 were on the night side at around 01:00 and 21:00 MLT, respectively. From Figure 12.7, the shock impact is clearly visible in the magnetic field observations from all four spacecraft. Information of the impact initially moves through the magnetosphere at the fast magnetosonic speed, which only should take a few tens of seconds to traverse the entire inner magnetosphere from the subsolar point to the start of the magnetotail. At 1-minute resolution, the GOES observations cannot resolve the delay time in the arrival of the magnetosonic wave due to its propagation through the system. However, the THEMIS observations clearly reveal this propagation effect: THA observed the compression in the magnetic field first at ~05:59:55 UT, while THE first observed it 18 seconds later at ~06:00:13 UT.

As expected based on previous studies of the effects of shock impacts on radiation belt electrons, all four spacecraft observed clear responses in the tens to hundreds of keV electrons around the time of the impact. Interestingly though, they also revealed some key differences between the different spacecraft. On the dayside near noon, where inward radial transport of electrons in response to the shock impact is most likely [Li et al., 2003a; Turner and Li, 2008], THA observed a slight increase in the flux of electrons at $\leq\sim$200 keV; above this energy, however, THA observed a decrease in

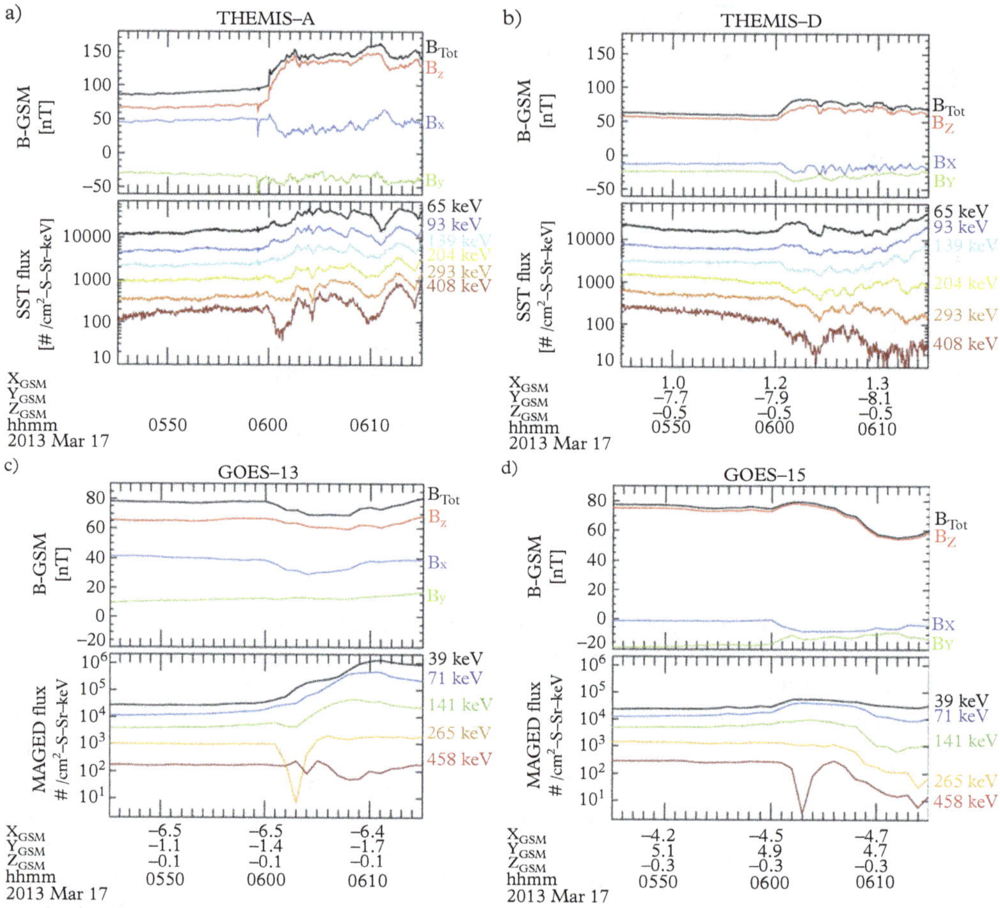

Figure 12.7 *THEMIS and GOES observations from within the magnetosphere during the St. Patrick's Day interplanetary shock impact. For each of the four spacecraft (THEMIS-A and –D and GOES-13 and -15, as labeled above each set of plots), B_{GSM} is shown in the top plot with strength in black and XYZ coordinates in blue, green, and red respectively. Energetic electron fluxes are shown from the SST instruments for THEMIS and the MAGED instruments from GOES with the corresponding energies for each channel shown listed on the right beside each line of data. For all four spacecraft, their locations in GSM coordinates are listed on the X-axis. All four show the same period of time: 05:45–06:15 UT on March 17, 2013.*

fluxes. THD observed similar features, though the transition energy was somewhere between 93 and 139 keV. These energy-dependent responses are typical of shock injections, and they indicate a μ-dependent gradient in the radial distributions of radiation belt electrons [Turner and Li, 2008; Kim et al., 2010; Turner et al., 2010]. The typical transition between positive and negative gradients from previous studies is ~200

MeV/G (corresponding to ~200 keV at GEO) [Turner et al., 2010, 2012b; see also Figure 12.2b], consistent with the THEMIS results for this event. The response is significantly different at GOES on the night side. While again, the temporal resolution at GOES is too slow to resolve the initial impact propagation delay, we note that the field strength actually decreased at GOES-13 and increased (dipolarized) at GOES-15, while the energetic electrons at both spacecraft exhibited an injection at tens of keV energies at ~06:03 UT and a depletion at one or two channels at hundreds of keV energy. These signatures are consistent with substorm activity triggered by the impact. Localized dipolarization is a hallmark of dipolarizing flux bundles, while the seemingly bizarre flux changes (some increase, some decrease) observed at both GOES spacecraft is similar to events reported by Gabrielse et al. [2012] and explained as due to the spacecraft location in relation to the injection flow channel. Since both spacecraft observed a depletion of higher-energy electrons and there was only a weak dipolarization at GOES-15, neither spacecraft intersected the earthward flow channel.

The St. Patrick's Day event ultimately resulted in an enhancement of the outer radiation belt based on Van Allen Probes data (not shown here) [Baker et al., 2014], with signatures of local acceleration that are remarkably consistent with the scenario in Figure 12.2 [Boyd et al., 2014]. As observed by THEMIS, this event revealed the classic energetic particle signatures associated with injections from an interplanetary shock impact. Furthermore, the GOES observations revealed interesting differences in the response on the night side, as expected based on previous cases [e.g., Li et al., 2003a]. GOES may also have observed one of the first energetic particle injections associated with substorm activity following the impact of the interplanetary shock; as discussed in the previous subsection, injections like this may have played a critical role in the overall enhancement of the belt during the recovery phase of this storm.

12.3.1.3 Energetic particle injections from the plasma sheet

Energetic particle injections from the plasma sheet provide both the source population of several to tens of keV electrons that generate chorus waves and the seed population of tens to hundreds of keV electrons that can be accelerated to relativistic energies (e.g., see Figure 12.2). After a prolonged period of sustained substorm activity, injections of ~MeV electrons at GEO have also been observed [Ingraham et al., 2001], indicating a sufficient source of relativistic electrons at L-shells beyond GEO for that event. For these reasons, it is critically important to understand the nature and characteristics of energetic particle injections.

On July 3, 2012, a fortuitous alignment of the ARTEMIS, THEMIS, Geotail, and GOES spacecraft allowed Angelopoulos et al. [2013] to study in detail the nature of tail reconnection and the results both tailward and Earthward of the reconnection site. The multipoint observations revealed that information of the reconnection travels away from the reconnection site via reconnection fronts of recently reconnected magnetic flux and plasma. These reconnection fronts exhibit significant structure on the electron inertial scales in depth, perpendicular to the local field, which can be responsible for the majority of energy conversion, from magnetic energy stored in the stretched magnetotail to plasma heating, during substorm activity. On the Earthward side of the reconnection

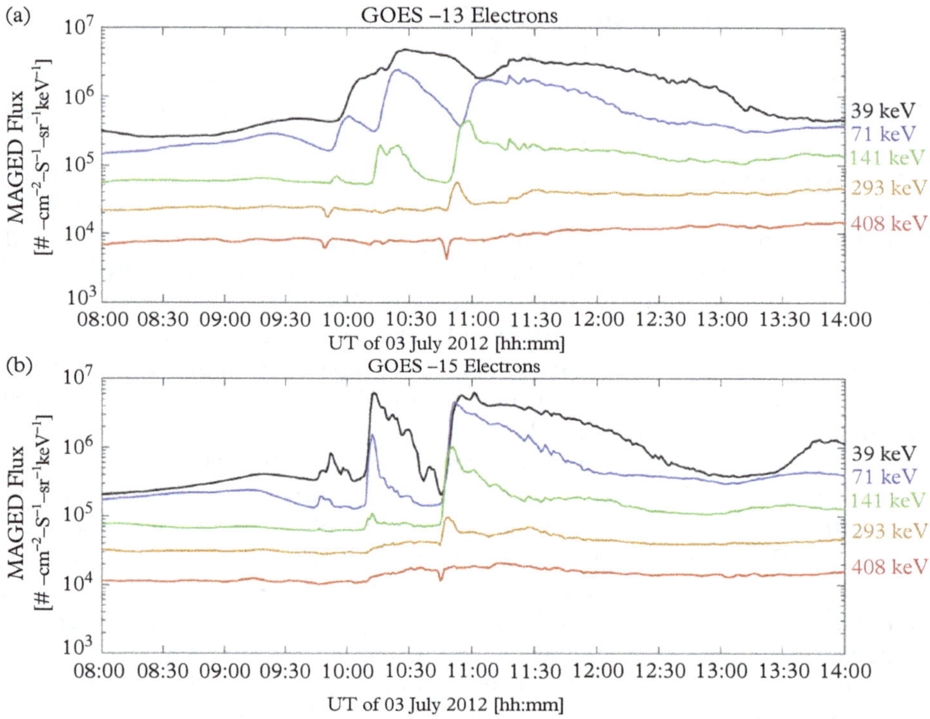

Figure 12.8 *Energetic electron injections observed by GOES-13 (a) and -15 (b) from July 3, 2012. For each satellite, electron flux data from the MAGED instruments are shown from the instruments' five energy channels as listed on the right beside each data line.*

region, these fronts are what have been referred to previously as dipolarization fronts [e.g., Runov et al., 2011]. It is the fields and flow properties around these fronts that are responsible for the rapid inward transport of energetic particles that can manifest themselves as energetic particle injections observed in the inner magnetosphere. For example, Figure 12.8 shows GOES-13 and -15 observations from July 3, 2012, during the arrival of reconnection fronts at the inner magnetosphere from the reconnection site, located 23R_E down the tail. During this period, there were three energetic electron injections, observed at GOES-15 around ~09:45, ~10:09, and ~10:45 UT. These injections correspond directly to reconnection fronts observed earlier in the plasma sheet by THEMIS and ARTEMIS [Angelopoulos et al., 2013]. At GOES-15, at post-midnight, the injections were dispersionless, whereas at GOES-13, near 06:00 MLT the injections were dispersed. The increased dispersion at GOES-13 is simply a result of the energy-dependent drift time as electrons move eastward around the system between GOES-15 and -13.

One interesting feature about this case was how the upper energy threshold on the injected electrons at GEO increased from the first to the third injection. For the first injection, GOES-15 (-13) observed an upper energy limit of 71 keV (141 keV), while for the second and third injections, this upper limit increased to 141 keV (141 keV) and 293 keV (293 keV), respectively. This indicates that there was some heating or acceleration of hundreds of keV electrons in the plasma sheet between the first and last injections, presumably due to the arrival of additional reconnection fronts; the electromagnetic waves associated with the fronts operate on a seed population of progressively increasing energy and flux. Also of interest are the decreases in flux observed by GOES-13 for 293 and 408 keV during the first injection and by both GOES for 408 keV during the last injection. These decreases may be either due to an insufficient source of these electrons at higher L-shells [e.g., Turner et al., 2012b] or entrapment of high-energy electron drift paths in the return flow of injection flow channels [e.g., Gabrielse et al., 2012]. Interestingly, there was an enhancement of >0.8 MeV electrons in the radiation belt on the same day as these injections (e.g., Figure 12.1). The July 3, 2012, series of energetic particle injections from the plasma sheet provides additional evidence of how sustained substorm activity can result in an increase of the upper energy threshold of injected electrons. This may provide an explanation for the ~MeV injections observed by Ingraham et al. [2001] and presents an interesting topic for future work (see details later).

12.3.2 Outer belt depletions

12.3.2.1 Dropout events

Flux dropout events represent one of the most drastic types of depletions that can occur in the outer radiation belt. Here, we showcase two dropout events presented in the recent literature, which highlight important clues as to the dominant loss mechanisms active during dropouts. On January 6, 2011, a dropout resulted in the sudden loss of relativistic electron flux throughout the outer belt above L ~4. Turner et al. [2012c] examined this event using two GOES spacecraft at GEO, the three THEMIS spacecraft, which observed the dropout throughout the belt along the magnetic equator, and six POES spacecraft, which are spread out over a broad range of MLT at low-Earth orbit (LEO). The dropout was clear in the observations from all eleven available spacecraft. Each POES satellite provides simultaneous measurements of trapped electrons near the edge of the loss cone and atmospheric electrons within the atmospheric loss cone. Taking advantage of this, Turner et al. [2012c] presented evidence that the loss of electrons to the atmosphere during the dropout was insufficient to explain the loss observed throughout the outer belt. Based on THEMIS and GOES data used to calculate PSD for fixed values of the three adiabatic invariants, it was clear that the dropout involved true losses from the system and wasn't simply a result of the "Dst effect." Additional key evidence included: the dropout started immediately after the magnetopause was compressed by the high pressure solar wind in a stream-interaction region; and the global ULF wave activity, which should facilitate enhanced radial transport of outer belt electrons, was enhanced by several orders of magnitude over a broad range of L-shells throughout the outer belt following the impact of the stream-interface region. Based on all of these, Turner et al. [2012c] concluded that the losses must have resulted from magnetopause shadowing and subsequent outward transport.

Since Van Allen Probes launched, the radiation belt science community has been afforded an unprecedented level of temporal, spatial, and energy resolution for observations of relativistic electrons in the outer radiation belt. Taking advantage of the addition of these two new spacecraft to the existing fleet capable of observing the outer belt electrons, Turner et al. [2014] studied a dropout that occurred on September 30, 2012, to fully quantify the energy, pitch angle, and L* ranges and timescales of electrons affected by the dropout. Like the January 2011 event, the September 30, 2012, dropout started after a sudden compression of the magnetosphere associated with an ICME sheath (Figure 12.3a and b). Also, POES observations revealed insufficient loss of relativistic electrons into the atmosphere throughout the majority of the pre-existing belt ($L > 4$). THEMIS and Van Allen Probes electron fluxes were converted to PSD distributions in L* for fixed values of μ and K. The results from THEMIS are shown in Figure 12.9a. Consistent with the Van Allen results, the pre-dropout PSD distribution was broadly peaked, with the peak location around L* of ~5.5. Immediately following the first sudden compression of the magnetosphere, a sharp gradient in the PSD distribution was observed around $6 < L^* < 6.5$, after which, the loss moved to lower L-shells over the next ~19 hours until the distribution shown in red from 06:30 UT in Figure 12.9a was observed by THEMIS. Also of interest, after the second sudden compression of the magnetosphere (see Figure 12.3c), another sharp gradient was observed in the PSD distributions from THEMIS and Van Allen at around $L^* \sim 5.5$.

The observed losses during this dropout were consistent with expectations from magnetopause shadowing and subsequent outward transport, as shown in the data-model comparisons in Figure 12.9b. Figure 12.9b compares the percent change in PSD for $\mu = 750$ MeV/G, $K \sim 0\ G^{1/2}R_E$ electrons as observed by THEMIS (difference between the September 30, 05:30 and October 1, 06:30 pre- and post-dropout curves in Fig. 12.9a, normalized to the pre-dropout values) and also simulated using the simple radial diffusion model shown in Figure 12.3d (note too that Figures 12.9a and 12.3d are on the same scales). The comparison shows that the simple model captures the nature of the dropout remarkably well, including both the loss up to > 90% of the pre-dropout PSD at $L^* > \sim 4.5$ and the enhancement of PSD due to inward diffusion at $L^* < \sim 4.5$. It is gratifying that such a simplified "toy model" can so accurately simulate the behavior of the outer belt; we interpret the agreement as strong evidence in support of magnetopause shadowing and subsequent outward radial transport being the dominant loss mechanism during this dropout event. Using test-particle simulations in global magnetohydrodynamic fields simulated using the solar wind conditions for this event, Hudson et al. [2014] drew the same conclusions. For this dropout and two others in September and October 2012, they found magnetopause shadowing could explain the dropout throughout the majority of the outer belt with additional losses to the atmosphere occurring at lower L-shells.

The two Van Allen Probes provided a significant amount of additional evidence in support of the loss being dominated at $L^* > \sim 4$ by magnetopause shadowing and outward transport as well as evidence of additional losses from a different mechanism at lower L*. Turner et al. [2014] showed that the dropout was effective for electrons over a broad range in energy, from only a few hundreds of keV to multi-MeV and over the full

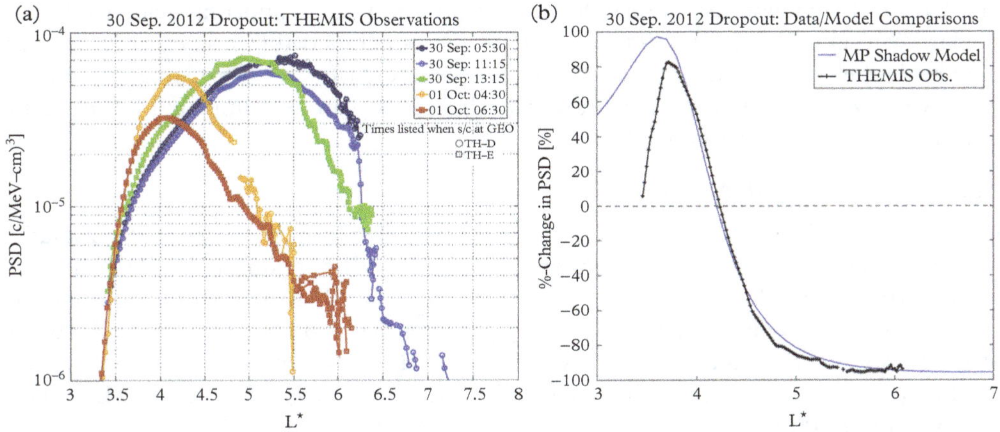

Figure 12.9 *(a) THEMIS observations of the PSD L*-distributions during the September 30, 2012, dropout event. The format is the same as shown in Figure 12.5, except the times listed correspond to when the spacecraft (THEMIS-D or -E) were at geosynchronous orbit. (b) Comparing the percent difference from the THEMIS observations shown in (a) to the results from the simple model shown in Figure 12.3d. For the observations (black line with markers), the pre-dropout distribution was taken from the September 30, 05:30 UT curve and the final from the October 1, 06:30 UT curve. The observed dropout started shortly after 11:00 UT, and the simulation was run for 19 hours with the results from the start and stop times shown here with the solid blue line. Note the dashed line at zero denotes the boundary between increased PSD (positive percentage difference) and decreased PSD (negative percentage difference).*

range of equatorial pitch angles. From PSD analysis, the dropout included actual loss of more than 90% of electrons throughout most of the belt ($L^* > \sim 4$), and the loss was effective over a period of only a few hours ($< \sim 1/2$ day). Losses started first at higher L* and moved in to lower L* over time, and the losses were ultimately greatest at higher L-shells. Furthermore, Van Allen Probes revealed that there were similar losses observed in energetic (hundreds of keV to >1 MeV) ring current protons, which also exhibited some of the same characteristics as the electrons (i.e., sharp cutoffs in the radial distributions following sudden inward magnetopause motion, loss starting at high L-shells and moving in over time, and stronger losses at higher L-shells). Turner et al. [2014] concluded that the only loss mechanism theory that can explain simultaneously all of these features is magnetopause shadowing and subsequent outward radial transport. However, Van Allen Probes revealed notably different results for $L^* < 4$. There, additional losses of predominantly very high-energy (i.e., multi-MeV), high-K (i.e., mirroring at high magnetic latitudes) electrons were also observed. The energy and pitch angle dependence of the loss explains why it was not observed in the THEMIS results shown in Figure 12.9a (also consistent with Van Allen Probes results, not shown here). Turner et al. [2014] speculated that these losses may have resulted due to atmospheric losses from wave-particle interactions with EMIC waves outside of the plasmapause. If so, this event also provided: (1) evidence of how EMIC waves can result in rapid, energy- and pitch angle-dependent loss of relativistic electrons; and (2) supporting evidence for the scenario presented in Bortnik et al. [2006], who concluded that loss during dropouts may be L-shell dependent. Loss at higher L-shells (e.g., $L^* > \sim 4$ or 5) can be dominated by outward transport, while at lower L-shells, loss can be dominated by scattering into the atmospheric loss cone.

12.3.2.2 Prolonged depletions

During the previous solar minimum, the relativistic electrons in Earth's outer radiation belt essentially vanished or were at very low intensity throughout all of 2009 and into early 2010. During this period, the solar wind speed and IMF strength were unusually low, with the 27-day averaged speed barely going above 400 km/s and the IMF staying around only 4 nT [Kataoka and Miyoshi, 2010]. Kataoka and Miyoshi [2010] studied this period of prolonged outer belt depletion using GOES data from GEO. They compared the 2009 solar minimum period to that from 1996 to 1997, which also had below-average solar wind levels but did not display the extremely weak outer radiation belt. They concluded that the extremely weak IMF in the recent solar minimum, when combined with the slow solar wind, resulted in lower storm and substorm activity, and was the primary factor in suppressing outer belt source processes and the exceedingly low levels of relativistic electrons there.

Using THEMIS data, Lee et al. [2013] also studied the prolonged outer belt depletion during the last solar minimum. They reported how the outer belt essentially disappeared during several intervals throughout the period, one of which lasted up to two months before a weak form of the belt reformed. They highlighted the importance of outer belt losses during the prolonged depletion period, showing that during the periods where the belt disappeared, the plasmasphere had expanded to high L-shells (beyond GEO). This expansion of the plasmasphere impacted the outer belt in two ways: (1) the supply of tens to hundreds of keV electrons from the plasma sheet due to energetic particle injections and enhanced convection (e.g., Figure 12.2) was greatly diminished, affecting outer belt source processes; and (2) loss processes due to pitch angle scattering from interactions with hiss waves within the highly inflated plasmasphere [e.g., Lyons and Thorne, 1973] were effective throughout the vast majority of the outer belt. Thus, the prolonged depletion of the outer belt around 2009 provided an excellent opportunity to understand the extreme effects that can result from longer-term competition between source and loss processes of the outer belt electrons.

12.3.3 Complex outer belt structures: remnant belts

We next review one more type of extreme variation of the outer belt electrons, since it likely results due to a dynamic interplay of several of the loss and source processes discussed in this paper. Baker et al. [2013] and Turner et al. [2013] independently reported two different cases of a bifurcation of the outer radiation belt, in which the belt exhibited two distinct peaks in intensity of relativistic electrons. Typically, the outer radiation belt has a single peak in intensity, somewhere between $4 < L < 6$. However, shortly after the launch of the Van Allen Probes, the spacecraft observed a dropout on September 3, 2012, which left in its wake a remnant belt of several MeV electrons between

~3 < L < ~3.5. However, over the next several days, another belt started forming at higher L (L > ~4) resulting in two distinct peaks in intensity separated by a second slot region [Baker et al., 2013]. The double outer belt structure persisted for over 25 days at multi-MeV energies, until another dropout occurred on September 30, 2012 (see details of that dropout above). Examining the formation and decay of the innermost of the two belts, Thorne et al. [2013a] showed that the belt formed over a broader range of energy than was observable with the Van Allen Probes REPT instruments [Baker et al., 2012], down to at least 300 keV as observed by THEMIS. However, the innermost belt exhibited an energy-dependent decay, which Thorne et al. [2013] showed was due to energy-dependent losses due to interactions with plasmaspheric hiss. This explained the THEMIS observations and why the lower-energy MagEIS instruments [Blake et al., 2013] did not observe the double outer belt structure when those instruments were turned on later in September. Baker et al. [2013] and Thorne et al. [2013a] demonstrated the importance of the plasmasphere in the formation of such a complex outer belt structure.

Using THEMIS observations, Turner et al. [2013] studied a separate instance of a double outer belt structure that occurred on February 2, 2011. Like the September 2013 event, this double outer belt structure formed after an outer belt dropout. However, Turner et al. [2013] examined relativistic electron PSD for fixed values of the first and second adiabatic invariants, which provided further insight into the nature of the formation of the double outer belt. Figure 12.10 showcases their results. Before the

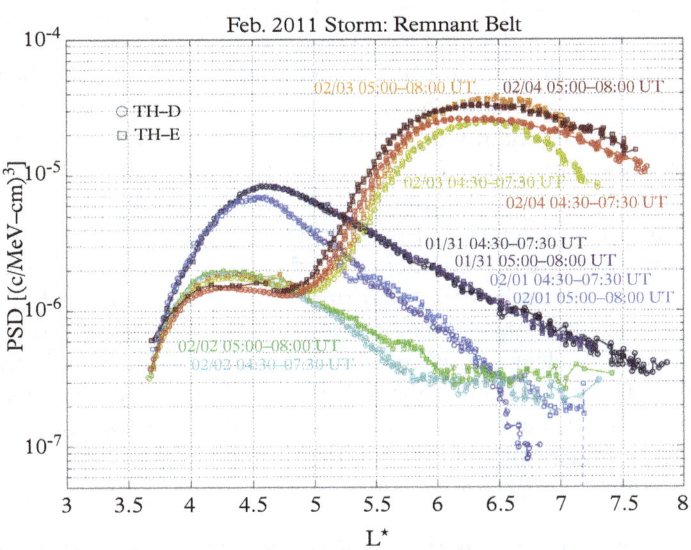

Figure 12.10 *Phase space density distributions from the February 2011 remnant belt event. The format is the same as that used for Figure 12.5 with the corresponding times shown on the plot. This period reveals how a double outer belt structure could form out of the remnant belt after a dropout.*

event, the outer belt PSD distribution in L* for electrons with μ = 1000 MeV/G and K < 0.025 $G^{1/2}R_E$ was peaked around L* ~4.5, as shown with the dark blue curve from early in the day on January 31, 2011. Following a sudden increase in the solar wind dynamic pressure and magnetopause compression on January 31, there was an outer radiation belt dropout. During the dropout, the peak in PSD decreased and moved closer to L* ~4 and the losses started at higher L* and moved to lower L* over time, as seen from the blue and cyan curves in Figure 12.10. As we discussed earlier in this paper, these signatures are typical of dropouts dominated by losses from magnetopause shadowing and outward radial transport. The remnant belt after this dropout is shown in the green curve from February 2, 2011, on Figure 12.10. However, within less than 24 hours, a new outer belt formed at higher L*, evidently independent of the remnant belt. This resulted in a double peaked PSD distribution shown in the yellow, orange, and red curves. The two peaks were even more pronounced in flux observations. Based on estimates of the plasmapause from this period, this new outer belt formed just outside of the plasmapause, which had expanded out to L* > 4.5 on February 2. The slow decay in the remnant belt from February 2 to 4 is also consistent with slow losses due to plasmaspheric hiss, as was shown for the September 2012 case by Thorne et al. [2013]. Turner et al. [2013] also showed that this new outer belt formed over the same period and at the same L* range in which the peak chorus wave activity was observed. This double outer belt structure was eradicated by another dropout event during the February 4, 2011, storm event discussed in Section 12.3.1.1.

These two events showcase how the various source, loss, and transport processes discussed here, combined with the nature of the plasmasphere, can conspire to result in the formation of two outer radiation belts. The dropout at the beginning of the formation is necessary to produce the remnant belt of electrons, which will later become the inner of the two outer radiation belts. If the plasmapause then expands rapidly outward over the next day or so, this remnant belt is essentially shielded by the plasmasphere, within which ULF activity is damped so inward radial diffusion from the new outer belt that forms outside the plasmapause cannot rapidly engulf the remnant belt's peak. Within the plasmasphere, those electrons in the remnant belt are primarily affected by slow, energy-dependent decay due to interactions with plasmaspheric hiss. The plasmapause also serves as a critical boundary for the formation of the new belt at higher L*. Chorus waves outside of the plasmapause can accelerate a seed population of freshly injected tens to hundreds of keV electrons up to relativistic energies, resulting in the formation of the new belt. This formation scenario was confirmed by the three-dimensional modeling effort presented in Shprits et al. [2013], who successfully simulated the evolution of the September 2012 "storage ring" event.

12.4 Outstanding questions and topics for future work

In the last two decades, the radiation belt community has conducted an overhaul in the theoretical framework of the drivers of extreme variations of relativistic electrons in Earth's outer radiation belt. However, many important questions of increasing specificity regarding the processes at work remain. Thanks to the unprecedented level and quality of observations by missions like NASA's Van Allen Probes and THEMIS the field has the potential of addressing these questions and usher a qualitative change in our understanding of outer radiation belt dynamics. Next, we highlight just a few of the many outstanding questions and topics that may prove important to further understanding extreme variations of Earth's outer belt electrons.

12.4.1 Concerning sources

We have stressed here the importance of energetic particle injections and local acceleration to enhancements of the outer radiation belt. One major outstanding question concerning the role of energetic electron injections is: What is the upper energy threshold of energetic electron injections, and how does this change with various activity indicators? As Ingraham et al. [2001] showed, direct injections of ~MeV electrons may be possible following periods of sustained substorm activity. Also, Figure 12.8 shows GOES observations of how the upper energy threshold of injected electrons increases during subsequent injections. It is important to understand the processes active in the magnetotail responsible for this acceleration of plasma sheet electrons and the extent to which this acceleration can be effective. Since it is injected electrons that form the source and seed populations for local acceleration by wave-particle interactions within the heart of the outer belt (e.g., Figure 12.2), a higher upper energy threshold on the injected electrons may play a critical role in enabling that local acceleration to be more effective for reaching higher energy levels over shorter timescales.

Another question that may prove to be important to sudden enhancements, or depletions, of radiation belt electrons is: How important is the role of nonlinear wave-particle interactions between whistler-mode waves and energetic electrons for acceleration and/or loss? Large amplitude whistler waves have been observed in Earth's outer radiation belts [e.g., Wilson et al., 2011]. Such waves invalidate several of the assumptions fundamental to quasilinear diffusion theory, and with test-particle simulations, researchers have only started to study how interactions with such nonlinear waves can affect outer belt electrons. Albert [2002] found that phase-bunching during wave-particle interactions can lead to deceleration, while phase-trapping leads to very rapid acceleration of a small number of the test particles. Bortnik et al. [2008] found similar results, though added that the type of interaction may be dependent on the latitude (and thus wave normal angle) at which it takes place. Tao et al. [2012] examined the effects of amplitude modulation in electron interactions with chorus waves. They found that amplitude modulation—as is observed in chorus wave packets—has a significant effect on phase-bunching and phase-trapping during the interactions. These results demonstrate the complexity of wave-particle interactions and the necessity for better understanding their role in accelerating and scattering outer belt electrons.

12.4.2 Concerning losses

Many questions still remain on the nature of losses of electrons through the magnetopause. For example, is electron loss through the magnetopause like an open drain that always operates or is it dependent on magnetic connection to the magnetosheath? This is a complex question, and it has significant implications for loss of electrons from the outer radiation belt since the relativistic population typically has a negative PSD gradient beyond GEO and radial diffusion is faster at higher L-shells (as discussed throughout this chapter). If the magnetopause was just a constant sink for electrons, then it should play a major role in controlling the population at lower L-shells: in particular it could limit the PSD gradient and the maximum height of the PSD peak at lower L-shells. This limit would also depend on the radial diffusion coefficient due to magnetospheric wave activity. However, the magnetopause is an ion scale layer, and when low-latitude reconnection is not occurring on the dayside (i.e., periods of northward IMF), the equatorial magnetopause is a tangential discontinuity with no magnetic connection to the magnetosheath. So, is it possible for outer belt electrons to escape through the magnetopause under northward IMF conditions? The answer to this might be dependent on the electron energy; if the electron gyro-radius is large enough (i.e., the electron is energetic enough) to be comparable to the thickness of the magnetopause, then they might escape. This should prove to be an interesting topic for future work.

Shprits et al. [2012] studied dropouts during the CRRES period and showed that 22 of 25 dropouts (88%) were associated with enhancements in solar wind dynamic pressure just prior to the event. The remaining three cases occurred under relatively steady solar wind dynamic pressure. This raises the question: What causes outer belt dropouts that are not related to magnetopause shadowing events? EMIC waves may play a dominant role in those other events, and they may also contribute to the magnetopause shadowing events as well, as was indicated by Bortnik et al. [2006] and Turner et al. [2014]. Therefore, it would be important to study how many dropout events include losses due to EMIC waves and over what ranges of L-shells, pitch angles, and energies are those losses effective. The latter is a question that will require multipoint observations, combining simultaneous measurements from equatorial, high-altitude and polar, LEO satellites to address.

12.4.3 Concerning transport

Claudepierre et al. [2013] showed evidence of resonant interactions between ~60 keV drifting electrons and poloidal ULF waves. Theoretically, drift resonance between electrons and ULF waves could lead to rapid acceleration of outer belt electrons [e.g., Elkington et al., 2003]. However, observational evidence of such a mechanism has proven elusive, so the question remains as to the importance of drift resonance in relativistic electron acceleration in the outer belt. New observations from the Van Allen Probes, which provide an unprecedented level of energy resolution, should help address this question.

Significant effects from transport can also occur around locations where the global magnetic field varies from its typical, dipole-like geometry, and there are many outstanding questions concerning such scenarios. For example, the test-particle simulations by Ukhorskiy et al. [2011] demonstrated that electrons may experience significant transport in L and sudden scattering in pitch angle when they undergo bifurcated drift orbits, yet there has not been a thorough observational analysis of the effects of such orbits on outer belt electrons drifting at high L-shells. THEMIS provides a near-ideal dataset for such an analysis, since it provides multipoint, pitch-angle resolved observations of energetic electrons near the magnetic equator from high L-shells on the dayside. The ring current and partial ring current also play a role in altering the drift trajectories of outer belt electrons [e.g., Ukhorskiy et al., 2006], which is another topic with several interesting areas for future research using multipoint observations.

12.5 Conclusions

This chapter focused on extreme variations of the outer radiation belt electrons from a primarily observational standpoint. We have provided a brief history of the topic as well as an overview of the current understanding concerning processes that result in the drastic enhancements and depletions that are observed so often in relativistic electron intensity in the outer belt. Furthermore, we have presented some recent example cases of sudden enhancements, injections from interplanetary shock impacts, energetic electron injections associated with dipolarization fronts, flux dropout events, prolonged depletions, and double outer belt structures that can form out of the remnant belt population after dropouts. These examples showcase how, with the availability of an unprecedented number of spacecraft observing the outer belt electrons simultaneously, we are currently in a "golden era" for radiation belt physics. Multipoint observations from missions like NASA's THEMIS and Van Allen Probes have already provided strong evidence addressing previous outstanding questions such as the dominant acceleration mechanism for relativistic electrons or the dominant loss mechanism during dropouts. However, many outstanding questions remain as indicated in the previous section.

Despite our current level of understanding, the ability to predict how the outer belt will respond to given solar wind inputs for individual events has proven exceedingly challenging. Currently, radiation belt models are able to reasonably predict the longer-term climatology or reproduce individual events after their outcomes were already observed. The challenge of predicting the extreme variability of the outer belt is an outcome of the belt's very nature: outer belt variations are the result of a complex interplay between multiple source, loss, and transport processes. The picture is further complicated by the often-random individual outcomes of each of those processes provided similar input conditions. In many ways the system is analogous to terrestrial weather: a complex set of often coupled driving processes results in chaotic responses of the system. However, terrestrial weather has proven that some level of predictability is possible. That predictability is provided by coupling complex global models which capture or otherwise parameterize all of the underlying physics, with a global network of observatories which provide the models with input data in near real time. Therefore, we must continue pushing our model capabilities while simultaneously increasing the number of observation points throughout the system.

Acknowledgments

D. L. Turner and V. Angelopoulos are thankful for funding support from NASA (THEMIS mission: contract NAS5-02099; and a grant: NNX12AJ55G) and the Monitoring, Analyzing, and Assessing the Radiation Belt Loss and Energization (MAARBLE) project funded under the European Commission's (EC) FP7 framework (note that this work reflects the authors' views, and the EC is not liable for any use that may be made of the information contained herein). We would also like to thank the THEMIS and Van Allen Probes teams, NOAA's NGDC and NASA's CDAWeb online databases, the SpacePy (http://spacepy.lanl.gov/) development team, and the THEMIS Data Analysis Software (TDAS: http://themis.ssl.berkeley.edu/software.shtml) development team.

REFERENCES

Albert, J. M. (2002), Nonlinear interaction of outer zone electrons with VLF waves, *Geophys. Res. Lett.*, **29**, 8, doi:10.1029/2001GL013941.

Albert, J. M. (2003), Evaluation of quasi-linear diffusion coefficients for EMIC waves in a multispecies plasma, *J. Geophys. Res.*, **108** (A6), 1249, doi:10.1029/2002JA009792.

Allen, R. C., et al. (2013), Multiple bidirectional EMIC waves observed by Cluster at middle magnetic latitudes in the dayside magnetosphere, *J. Geophys. Res. Space Physics*, 118, doi:10.1002/jgra.50600.

Angelopoulos, V. (2008), The THEMIS Mission, *Space Sci. Rev.*, **141**, 5–34, doi:10.1007/s11214-008-9336-1.

Angelopoulos, V. (2011), The ARTEMIS Mission, *Space Sci. Rev.*, **165**, 3–25, doi:10.1007/s11214-010-9687-2.

Angelopoulos, V., et al. (2013), Electromagnetic energy conversion at reconnection fronts, *Science*, **341**, 1478, doi:10.1126/science.1236992.

Artemyev, A. V., O. V. Agapitov, D. Mourenas, V. Krasnoselskikh, and L. M. Zelenyi (2013), Storm-induced energization of radiation belt electrons: Effect of wave obliquity, *Geophys. Res. Lett.*, **40**, doi:10.1002/grl.50837.

Baker, D. N., J. B. Blake, L. B. Callis, R. D. Belian, and T. E. Cayton (1989), Relativistic electrons near geostationary orbit: Evidence for internal magnetospheric acceleration, *Geophys. Res. Lett.*, **16**, 559.

Baker, D. N., et al. (1994), Relativistic electron acceleration and decay time scales in the inner and outer radiation belts: SAMPEX, *Geophys. Res. Lett.*, **21**, 6, 409–412.

Baker, D. N., S. G. Kanekal, X. Li, S. P. Monk, J. Goldstein, and J. L. Burch (2004), An extreme distortion of the Van Allen belt arising from the 'Hallowe'en' solar storm in 2003, *Nature*, **432**, 878–881.

Baker, D. N., et al. (2012), The Relativistic Electron-Proton Telescope (REPT) instrument on board the Radiation Belt Storm Probes (RBSP) spacecraft: Characterization of Earth's radiation belt high-energy particle populations, *Space Sci. Rev.*, **179**, 337–381, doi:10.1007/s11214-012-9950-9.

Baker, D. N., et al. (2013), A long-lived relativistic electron storage ring embedded in Earth's outer Van Allen Belt, *Science*, **340**, doi:10.1126/science.1233518.

Baker, D. N., et al. (2014), Gradual diffusion and punctuated phase space density enhancements of highly relativistic electrons: Van Allen Probes observations, *Geophys. Rev. Lett.*, **41**, doi:10.1002/2013GL058942.

Birn, J., M. F. Thomsen, J. E. Borovsky, G. D. Reeves, D. J. M. R. D. Belian, and M. Hesse (1997), Substorm ion injections: Geosynchronous observations and test particle orbits in three-dimensional dynamic MHD fields, *J. Geophys. Res.*, **102**, 2325.

Birn, J., et al. (1998), Substorm electron injections: Geosynchronous observations and test particle simulations, *J. Geophys. Res.*, **103**, 9235.

Blake, J. B., W. A. Kolasinski, R. W. Fillius, and E. G. Mullen (1992), Injection of electrons and protons with energies of tens of MeV into L < 3 on March 24, 1991, *Geophys. Res. Lett.*, **19**, 8, 821–824.

Blake, J. B., et al. (1995), CEPPAD: Comprehensive energetic particle and pitch angle distribution experiment on POLAR, *Space Sc. Rev.*, **71**, 531–562.

Blake, J. B., et al. (2013), The Magnetic Electron Ion Spectrometer (MagEIS) instruments aboard the Radiation Belt Storm Probes (RBSP) spacecraft, *Space Sci. Rev.*, **179** (1–4), 383–421, doi:10.1007/s11214-013-9991-8.

Borovsky, J. E. and M. Denton (2009), Relativistic-electron dropouts and recovery: A superposed epoch study of the magnetosphere and solar wind, *J. Geophys. Res.*, **114**, A02201, doi:10.1029/2008JA013128.

Bortnik, J., et al. (2006), Observation of two distinct, rapid loss mechanisms during the November 20, 2003, radiation belt dropout event, *J. Geophys. Res.*, **111**, A12216, doi:10.1029/2006JA011802.

Bortnik, J., R. M. Thorne, and U. S. Inan (2008), Nonlinear interaction of energetic electrons with large amplitude chorus, *Geophys. Res. Lett.*, **35**, L21102, doi:10.1029/2008GL035500.

Boyd, A. J., et al. (2014), Quantifying the radiation belt seed population in the March 17, 2013, electron acceleration event, *Geophys. Res. Lett.*, **41**, doi:10.1002/2014GL059626.

Brautigam, D. H. and J. M. Albert (2000), Radial diffusion analysis of outer radiation belt electrons during the October 9, 1990, magnetic storm, *J. Geophys. Res.*, **105** (A1), 291–309.

Claudepierre, S. G., et al. (2013), Van Allen Probes observation of localized drift resonance between poloidal mode ultra-low frequency waves and 60 keV electrons, *Geophys. Res. Lett.*, **40**, doi:10.1002/grl.50901.

Chen, Y., G. D. Reeves, and R. H. W. Friedel (2007), The energization of relativistic electrons in the outer Van Allen radiation belt, *Nat. Phys.*, doi:10.1038/nphys655.

Craven, J. D. (1966), Temporal variations of electron intensities at low altitudes in the outer radiation zone as observed with satellite Injun 3, *J. Geophys. Res.*, **71**, 5643–5663.

Desorgher, L., P. Bühler, A. Zehnder, and E. O. Flückiger (2000), Simulation of the outer radiation belt electron flux decrease during the March 26, 1995, magnetic storm, *J. Geophys. Res.*, **105** (A9), 21,211–21,223, doi:10.1029/2000JA900060.

Dessler, A. J. and R. Karplus (1961), Some effects of diamagnetic ring currents on Van Allen radiation, *J. Geophys. Res.*, **66**, 8, 2289–2295.

Elkington, S. R., M. K. Hudson, and A. A. Chan (2003), Resonant acceleration and diffusion of outer zone electrons in an asymmetric geomagnetic field, *J. Geophys. Res.*, **108** (A3), 1116, doi:10.1029/2001JA009202.

Friedel, R. H. W., G. D. Reeves, and T. Obara (2002), Relativistic electron dynamics in the inner magnetosphere – a review, *J. Atmos. Sol.-Terr. Phys.*, **64**, 265–682.

Fujimoto, M. and A. Nishida (1990), Energization and anisotropization of energetic electrons in the Earth's radiation belt by the recirculation process, *J. Geophys Res.*, **95**, 4265.

Gabrielse, C., V. Angelopoulos, A. Runov, and D. L. Turner (2012), The effects of transient, localized electric fields on equatorial electron acceleration and transport toward the inner magnetosphere, *J. Geophys. Res.*, **117**, A10213, doi:10.1029/2012JA017873.

Green, J. C. and M. G. Kivelson (2004), Relativistic electrons in the outer radiation belt: Differentiating between acceleration mechanisms, *J. Geophys. Res.*, **109** (A3), A03213, doi:10.1029/2003JA010153.

Green, J. C., T. G. Onsager, T. P. O'Brien, and D. N. Baker (2004), Testing loss mechanisms capable of rapidly depleting relativistic electron flux in the Earth's outer radiation belt, *J. Geophys. Res.*, **109**, A12211, doi:10.1029/2004JA010579.

Horne, R. B. and R. M. Thorne (1998), Potential waves for relativistic electron scattering and stochastic acceleration during magnetic storms, *Geophys. Res. Lett.*, **25** (15), 3011–3014.

Horne, R. B., et al. (2005), Timescale for radiation belt electron acceleration by whistler mode chorus waves, *J. Geophys. Res.*, **110**, A03225, doi:10.1029/2004JA010811.

Horne, R. B. (2007), Acceleration of Killer Electrons, *Nat. Phys.*, **3**, 590–591.

Horne, R. B., M. M. Lam, and J. C. Green (2009), Energetic electron precipitation from the outer radiation belt during geomagnetic storms, *Geophys. Res. Lett.*, **36**, L19104, doi:10.1029/2009GL040236.

Hudson, M. K., B. T. Kress, H.-R. Mueller, J. A. Zastrow, and J. B. Blake (2008), Relationship of the Van Allen radiation belts to solar wind drivers, *J. Atmos. Sol.-Terr. Phys.*, **70**, 708–729, doi:10.1016/j.jastp.2007.11.003.

Hudson, M. K., et al. (2014), Simulated magnetopause losses and Van Allen Probe flux dropouts, *Geophys. Res. Lett.*, **41**, doi:10.1002/2014GL059222.

Ingraham, J. C., et al. (2001), Substorm injection of relativistic electrons to geosynchronous orbit during the great magnetic storm of March 24, 1991, *J. Geophys. Res.*, **106** (A11), 25,759–25,776.

Kataoka, R. and Y. Miyoshi (2008), Average profiles of the solar wind and outer radiation belt during the extreme flux enhancement of relativistic electrons at geosynchronous orbit, *Ann. Geophys.*, **26**, 1335–1339.

Kataoka, R. and Y. Miyoshi (2010), Why are relativistic electrons persistently quiet at geosynchronous orbit in 2009? *Space Weather*, **8**, S08002, doi:10.1029/2010SW000571.

Kim, H.-J. and A. A. Chan (1997), Fully adiabatic changes in storm time relativistic electron fluxes, *J. Geophys. Res.*, **102** (A10), 22107–22116.

Kim, H.-J., E. Zesta, K.-C. Kim, Y. Shprits, Y. Shi, and L. R. Lyons (2010), Estimation of radial gradients of phase space density from POLAR observations during a quiet period prior to a sudden solar wind dynamic pressure enhancement, *J. Geophys. Res.*, **115**, A12249, doi:10.1029/2010JA015722.

Kim, K. C., et al. (2008), Numerical calculations of relativistic electron drift loss effect, *J. Geophys. Res.*, **113**, A09212, doi:10.1029/2007JA013011.

Kim, K. C., D.-Y. Lee, H.-J. Kim, E. S. Lee, and C. R. Choi (2010), Numerical estimates of drift loss and Dst effect for outer radiation belt relativistic electrons with arbitrary pitch angle, *J. Geophys. Res.*, **115**, A03208, doi:10.1029/2009JA014523.

Lee, D.-Y., et al. (2013), Long-term loss and re-formation of the outer radiation belt, *J. Geophys. Res. Space Physics*, **118**, doi:10.1002/jgra.50357.

Li, X., I. Roth, M. Temerin, J. R. Wygant, M. K. Hudson, and J. B. Blake (1993), Simulation of the prompt energization and transport of radiation belt particles during the March 24, 1991, SSC, *Geophys. Res. Lett.*, **20**, 2423–2426, doi:10.1029/93GL02701.

Li, X. et al. (1997), Multisatellite observations of the outer zone electron variation during the November 3–4, 1993, magnetic storm, *J. Geophys. Res.*, **102** (A7), 14123–14140.

Li, X., et al. (1998), Energetic electron injections into the inner magnetosphere during the Jan. 10-11, 1997 magnetic storm, *Geophys. Res. Lett.*, **25**, 2561–2564.

Li, X. and M. A. Temerin (2001), The Electron Radiation Belt, Space Science Reviews, **95** (1/2), 569.

Li, X., et al. (2003a), Energetic particle injections in the inner magnetosphere as a response to an interplanetary shock, *J. Atmos. Sol.-Terr. Phys.*, **65**, 233–244.

Li, X., T. E. Sarris, D. N. Baker, W. K. Peterson, and H. J. Singer (2003b), Simulation of energetic particle injections associated with a substorm on August 27, 2001, *Geophys. Res. Lett.* , **30** (1), 1004, doi:10.1029/2002GL015967.

Li, X., et al. (2009), Modeling the deep penetration of outer belt electrons during the "Halloween" magnetic storm in 2003, *Space Weather*, 7, S02004, doi:10.1029/2008SW000418.

Li, X., et al. (2013), First results from CSSWE CubeSat: Characteristics of relativistic electrons in the near-earth environment during the October 2012 magnetic storms, *J. Geophys. Res. Space Physics*, **118**, doi:10.1002/2013JA019342.

Li, W. et al. (2009), Global distribution of whistler-mode chorus waves observed on the THEMIS spacecraft, *Geophys. Res. Lett.*, **36**, L09104, doi:10.1029/2009GL037595.

Liu, J., V. Angelopoulos, A. Runov, and X.-Z. Zhou (2013), On the current sheets surrounding dipolarizing flux bundles in the magnetotail: The case for wedgelets, *J. Geophys. Res. Space Physics*, **118**, doi: 10.1002/jgra.50092.

Loto'aniu, T. M., et al. (2010), Relativistic electron loss due to ultralow frequency waves and enhanced outward radial diffusion, *J. Geophys. Res.*, **115**, A12245, doi:10.1029/2010JA015755.

Lyons, L. R. and R. M. Thorne (1973), Equilibrium structure of radiation belt electrons, *J. Geophys. Res.*, **78** (13), 2142–2149.

Matsumura, C., Y. Miyoshi, K. Seki, S. Saito, V. Angelopoulos, and J. Koller (2011), Outer radiation belt boundary location relative to the magnetopause: Implications for magnetopause shadowing, *J. Geophys. Res.*, **116**, A06212, doi:10.1029/2011JA016575.

Mauk, B. H., N. J. Fox, S. G. Kanekal, R. L. Kessel, D. G. Sibeck, and A. Ukhorskiy (2012), Science objectives and rationale for the Radiation Belt Storm Probes mission, *Space Sci. Rev.*, doi:10.1007/s11214-012-9908-y.

McCollough, J. P., S. R. Elkington, and D. N. Baker (2012), The role of Shabansky orbits in compression-related electromagnetic ion cyclotron wave growth, *J. Geophys. Res.*, **117**, A01208, doi:10.1029/2011JA016948.

McIlwain, C. E. (1961), Coordinates for mapping the distribution of magnetically trapped particles, *J. Geophys. Res.*, **66**, 3681.

McIlwain, C. E. (1966a), Ring current effects on trapped particles, *J. Geophys. Res.*, **71** (15), 3623–3628.

McIlwain, C. E. (1966b), Processes acting upon outer zone electrons, *Preprint UCSD-SP-66-5*, Presented at Solar Terrestrial Physics Symposium, Belgrade, 1966.

Meredith, N. P., M. Cain, R. B. Horne, R. M. Thorne, D. Summers, and R. R. Anderson (2003), Evidence for chorus-driven electron acceleration to relativistic energies from a survey of geomagnetically disturbed periods, *J. Geophys. Res.*, **108** (A6), 1248, doi:10.1029/2002JA009764.

Meredith, N. P., R. B. Horne, M. M. Lam, M. Denton, J. E. Borovsky, and J. C. Green (2011), Energetic electron precipitation during high-speed solar wind stream driven storms, *J. Geophys. Res.*, **116**, A05223, doi:10.1029/2010JA016293.

Millan, R. M., K. B. Yando, J. C. Green, and A. Y. Ukhorskiy (2010), Spatial distribution of relativistic electron precipitation during a radiation belt depletion event, *Geophys. Res. Lett.*, **37**, L20103, doi:10.1029/2010GL044919.

Millan, R. M. and R. M. Thorne (2007), Review of radiation belt relativistic electron losses, *J. Atmos. Sol.-Terr. Phys.*, **69**, doi:10.1016/j.jastp.2006.06.019.

Millan, R. M. and D. N. Baker (2012), Acceleration of particles to high energies in Earth's radiation belts, *Space Sci. Rev.*, **173**, 103–131, doi:10.1007/s11214-012-9941-x.

Morley, S. K., et al. (2010), Dropouts of the outer electron radiation belt in response to solar wind stream interfaces: Global Positioning System observations, *Proc. Roy. Soc. A,* Vol. **466**, 3329–3350, doi:10.1098/rspa.2010.0078.

Northrop, T. G. and E. Teller (1960), Stability of the adiabatic motion of charged particles in the Earth's field, *Phys. Rev.*, **117**, 215–225.

O'Brien, T. P., M. D. Looper, and J. B. Blake (2004), Quantification of relativistic electron microburst losses during the GEM storms, *Geophys. Res. Lett.*, **31**, L04802, doi:10.1029/2003GL018621.

Reeves, G. D., et al. (1998), The relativistic electron response at geosynchronous orbit during the January 1997 magnetic storm, *J. Geophys. Res.*, **103** (A8), 17559–17570.

Reeves, G. D., et al. (2013), Electron acceleration in the heart of the Van Allen radiation belts, *Science*, **341**, 991, doi:10.1126/science.1237743.

Roederer, J. G. (1970), Dynamics of Geomagnetically Trapped Radiation, Springer-Verlag, New York.

Runov, A., V. Angelopoulos, X.-Z. Zhou, X.-J. Zhang, S. Li, F. Plaschke, and J. Bonnell (2011), A THEMIS multicase study of depolarization fronts in the magnetotail plasma sheet, *J. Geophys. Res.*, **116**, A05216, doi:10.1029/2010JA016316.

Russell, C. T. and R. M. Thorne (1970), On the structure of the inner magnetosphere, *Cosmic Electrodynamics*, **1**, 67.

Schiller, Q., X. Li, L. Blum, W. Tu, D. L. Turner, and J. B. Blake (2014), A non-storm time enhancement of relativistic electrons in the outer radiation belt, *Geophys. Rev. Lett.*, **41**, 1–6, doi:10.1002/2013GL058485.

Schulz, M. and L. J. Lanzerotti (1974), Particle Diffusion in the Radiation Belts, Springer-Verlag, New York.

Selesnick, R. S. and J. B. Blake (1997), Dynamics of the outer radiation belt, *Geophys., Res. Lett.*, **24**, 11, 1347–1350.

Selesnick, R. S. and J. B. Blake (2000), On the source location of radiation belt relativistic electrons, *J. Geophys. Res.*, **105** (A2), 2607–2624.

Selesnick, R. S. (2006), Source and loss rates of radiation belt relativistic electrons during magnetic storms, *J. Geophys. Res.*, **111**, A04210, doi:10.1029/2005JA011473.

Shabansky, V. P. (1971), Some processes in magnetosphere, *Space Sci. Rev.*, **12** (3), 299.

Shprits, Y. Y., et al. (2006a), Acceleration mechanism responsible for the formation of the new radiation belt during the 2003 Halloween solar storm, *Geophys. Res. Lett.* **33**, L05104, doi:10.1029/2005GL024256.

Shprits, Y. Y., et al. (2006b), Radial diffusion driven by losses at magnetopause, *J. Geophys. Res.*, **111**, A11214, doi:10.1029/2006JA011657.

Shprits, Y. Y., N. P. Meredith, and R. M. Thorne (2007), Parameterization of radiation belt electron loss timescales due to interactions with chorus waves, *Geophys. Res. Lett.*, **34**, L11110.

Shprits, Y. Y., S. R. Elkington, N. P. Meredith, and D. Subbotin (2008a), Review of modeling losses and sources of relativistic electrons in the outer radiation belt I: Radial transport, *J. Atmos. Sol.-Terr. Phys.*, **70**, doi:10.1016/j.jastp.2008.06.008.

Shprits, Y. Y., D. Subbotin, N. P. Meredith, and S. R. Elkington (2008b), Review of modeling of losses and sources of relativistic electrons in the outer radiation belt II: Local acceleration and loss, *J. Atmos. Sol.-Terr. Phys.*, **70**, doi:10.1016/j.jastp.2008.06.014.

Shprits, Y. Y., M. Daae, and B. Ni (2012), Statistical analysis of phase space density buildups and dropouts, *J. Geophys. Res.*, **117**, A01219, doi:10.1029/2011JA016939.

Shprits, Y. Y., et al. (2013), Unusual stable trapping of ultrarelativistic electrons in the Van Allen radiation belts, *Nat. Phys.*, **9**, doi:10.1038/NPHYS2760.

Su, Z., F. Xiao, H. Zheng, and S. Wang (2011), Radiation belt electron dynamics driven by adiabatic transport, radial diffusion, and wave-particle interactions, *J. Geophys. Res.*, **116**, A04205, doi:10.1029/2010JA016228.

Summers, D., R. M. Thorne, and F. Xiao (1998), Relativistic theory of wave-particle resonant diffusion with application to electron acceleration in the magnetosphere, *J. Geophys. Res.*, **103** (A9), 20487–20500.

Tao, X., J. Bortnik, R. M. Thorne, J. M. Albert, and W. Li (2012), Effects of amplitude modulation on nonlinear interactions between electrons and chorus waves, *Geophys. Res. Lett.*, **39**, L06102, doi:10.1029/2012GL051202.

Temerin, M., I. Roth, M. K. Hudson, J. R. Wygant (1994), New paradigm for the transport and energization of radiation belt particles (abstract), *Eos. Trans. AGU*, **75**, Fall Meet. Suppl., 538, 1994.

Thorne, R. M., R. B. Horne, V. K. Jordanova, J. Bortnik, and S. Glauer (2006), Interaction of EMIC waves with thermal plasma and radiation belt particles, in *Magnetospheric ULF Waves: Synthesis and New Directions* (eds. K Takahashi, P. J. Chi, R. E. Denton, and R. L. Lysak), AGU Monograph Series, Washington D. C., doi:10.1029/169GM14.

Thorne, R. M., et al. (2013a), Evolution and slow decay of an unusual narrow ring of relativistic electrons near L $\sim$ 3.2 following the September 2012 magnetic storm, *Geophys. Res. Lett.*, **40**, doi:10.1002/grl.50627.

Thorne et al. (2013b), Rapid acceleration of relativistic radiation belt electrons by magnetospheric chorus, *Nature*, **504**, doi:10.1038/nature12889.

Tsyganenko, N. A. and M. I. Sitnov (2005), Modeling the dynamics of the inner magnetosphere during strong geomagnetic storms, *J. Geophys. Res.*, **110**, A03208, doi:10.1029/2004JA010798.

Tu, W., R. S. Selesnick, X. Li, and M. D. Looper (2010), Quantification of the precipitation loss of radiation belt electrons observed by SAMPEX, *J. Geophys. Res.*, **115**, A07210, doi:10.1029/2009JA014949.

Turner, D. L. and X. Li (2008), Radial gradients of phase space density of the outer radiation belt electrons prior to sudden solar wind pressure enhancements, *Geophys. Res. Lett.*, **35**, L18101, doi:10.1029/2008GL034866.

Turner, D. L., X. Li, G. D. Reeves, and H. J. Singer (2010), On phase space density radial gradients of Earth's outer-belt electrons prior to sudden solar wind pressure enhancements: Results from distinctive events and a superposed epoch analysis, *J. Geophys. Res.*, **115**, A01205, doi:10.1029/2009JA014423.

Turner, D. L., S. Morley, Y. Miyoshi, B. Ni, and C.-L. Huang (2012a), Outer radiation belt flux dropouts: Current understanding and unresolved questions, in *Dynamics of Earth's Radiation Belts and Inner Magnetosphere* (eds. D. Summers, I. R. Mann, D. N. Baker, and M. Schulz), pp. 195–212, AGU Geophysical Monograph Series, Vol. 199, Washington, D.C., doi:10.1029/2012GM001310.

Turner, D. L., V. Angelopoulos, Y. Shprits, A. Kellerman, P. Cruce, and D. Larson (2012b), Radial distributions of equatorial phase space density for outer radiation belt electrons, *Geophys. Res. Lett.*, **39**, L09101, doi:10.1029/2012GL051722.

Turner, D. L., Y. Y. Shprits, M. Hartinger, and V. Angelopoulos (2012c), Explaining sudden losses of relativistic electrons during geomagnetic storms, *Nat. Phys.*, **8**, doi:10.1038/NPHYS2185.

Turner, D. L., V. Angelopoulos, W. Li, M. D. Hartinger, M. Usanova, I. R. Mann, J. Bortnik, and Y. Shprits (2013), On the storm-time evolution of relativistic electron phase space density in Earth's outer radiation belt, *J. Geophys. Res. Space Physics*, **118**, 2196–2212, doi:10.1002/jgra.50151.

Turner, D. L., et al. (2014), On the cause and extent of outer radiation belt losses during the September 30, 2012, dropout event, *J. Geophys. Res.*, **119**, doi:10.1002/2013JA019446.

Ukhorskiy, A. Y., B. J. Anderson, P. C. Brandt, and N. A. Tsyganenko (2006), Storm time evolution of the outer radiation belt: Transport and losses, *J. Geophys. Res.*, **111**, A11S03, doi:10.1029/2006JA011690.

Ukhorskiy, A. Y., Y. Y. Shprits, B. J. Anderson, K. Takahashi, and R. M. Thorne (2010), Rapid scattering of radiation belt electrons by storm-time EMIC waves, *Geophys. Res. Lett.*, **37**, L09101, doi:10.1029/2010GL042906.

Ukhorskiy, A. Y., M. I. Sitnov, R. M. Millan, and B. T. Kress (2011), The role of drift orbit bifurcations in energization and loss of electrons in the outer radiation belt, *J. Geophys. Res.*, **116**, A09208, doi:10.1029/2011JA016623.

Ukhorskiy, A. Y. and M. I. Sitnov (2012), Dynamics of radiation belt particles, *Space Sci. Rev.*, **179** (1–4), 545–578, doi:10.1007/s11214-012-9938-5.

Van Allen, J. A. and L. A. Frank (1959), Radiation around the Earth to a radial distance of 107,400 km, *Nature*, **183**, 430–434.

Van Allen, J. A. (1969), Charged particles in the magnetosphere, *Rev. Geophys.*, 7 (1, 2), 233–255.

Vernov, S. N. and A. E. Chudukov (1960), Investigation of radiation in outer space, Moscow Cosmic Ray Conference, in Syrovatsky, S.I. (Ed.), *Proceedings of the Moscow Cosmic Ray Conference*, Vol. III, International Union of Pure and Applied Physics, Moscow, p. 19.

Vernov, S. N., V. M. Driatsky, S. N. Kuznetsov, Yu. I. Logachev, E. N. Sosnovets, and V. G. Stolpovsky (1966), Behavior of radiation belts and anomalous absorption of cosmic radiation in the auroral zone during magnetic storms in February 1964, *Geomagn. Aeronomy*, **6**, 1–6.

Vernov, S. N. (1968), The earth's radiation belts, *Ann. IQSY*, **4**.

Vernov, S. N., E. V. Gorchakov, S. N. Kuznetsov, Yu. I. Logachev, E. N. Sosnovets, and V. G. Stolpovsky (1969), Particle fluxes in the outer geomagnetic field, *Rev. Geophys.*, 7 (1, 2), 257–280.

Williams, D. J., J. F. Arens, and L. J. Lanzerotti (1968), Observations of trapped electrons at low and high altitudes, *J. Geophys. Res.*, 73, 5673–5696.

Wilson, L. B., III, et al. (2011), The properties of large amplitude whistler mode waves in the magnetosphere: Propagation and relationship with geomagnetic activity, *Geophys. Res. Lett.*, **38**, L17107, doi:10.1029/2011GL048671.

13

Flux Enhancement of Relativistic Electrons Associated with Substorms

Y. Miyoshi, R. Kataoka, and Y. Ebihara

■ Abstract

The outer belt flux enhancements occur in association with high-speed coronal hole streams. We show that the southward interplanetary magnetic field (IMF) within a coronal hole stream controls the flux enhancement of the relativistic electrons. Moreover, the amplitudes of the southward IMF of Alfvénic fluctuations in a coronal hole stream are controlled by the Russell–McPherron effect and continuous substorms can be observed during such periods. Therefore, the relativistic electron flux variations are also related to the Russell–McPherron effect and prolonged substorm activity may be of great importance for these variations. During periods of prolonged substorm activity, continuous hot electron injections from the plasma sheet occur, which is free energy source that can generate whistler mode waves. In fact, continuous whistler mode wave activity is observed during the periods and these waves produce relativistic electrons by the cross-energy couplings between thermal plasma, hot electrons, and relativistic electrons via wave–particle interactions. The southward IMF controls the flux enhancement of the outer belt electrons via wave–particle interactions and prolonged substorm activities are essential to cause the accelerations of relativistic electrons.

Waves, Particles, and Storms in Geospace. First Edition. Georgios Balasis, Ioannis A. Daglis, Ian R. Mann.

13.1 Introduction

Relativistic electron flux of the outer belt increases during magnetic storms, especially during the storm recovery phase. An earlier statistical study indicated that the flux increases relative to the pre-storm level in 50% of storms, it recovers to the pre-storm level in 25% of storms, and it decreases after the storm main phase in 25% of storms [Reeves et al., 2003].

Later studies demonstrated that the flux variations during a storm depend on the large-scale solar wind structures [Miyoshi and Kataoka, 2005; Kataoka and Miyoshi, 2006]. The intense southward interplanetary magnetic field (IMF) of coronal mass ejections (CMEs) and corotating interaction regions (CIRs) cause severe magnetic storms. However, the largest storms as classified by the Dst index are CME-driven storms. Furthermore, flux variations of the outer belt ($\sim L > 4$) during the recovery phase of storms are different between CME-driven storms and CIR-driven storms: for example, flux enhancements of the outer belt during CIR-driven storms are generally faster and stronger than that those during CME-driven storms.

During the recovery phase of CIR-driven storms, in which the largest flux enhancement of relativistic electrons of the outer belt is observed, the high-speed coronal hole streams arrive at the Earth. Alfvénic fluctuations are typically observed in the high-speed coronal hole streams [Tsurutani et al., 2006], and the southward IMF within the coronal hole streams leads to continuous substorm activity or so-called High-Intensity Long-Duration Continuous AE (auroral electrojet) Activity (HILDCAAs) [Tsurutani and Gonzalez, 1987]. It has been presumed that the relativistic electron flux enhancement is associated with the continuous substorm activities. In this chapter, we show the importance of continuous substorm activity for large flux enhancements of the outer belt electron fluxes and briefly discuss recent results regarding the associated mechanisms. First, we demonstrate the IMF Bz dependence of the outer belt flux enhancement by the superposed epoch analysis associated with the large-scale structure of the solar wind. Then, we investigate the physical process behind the IMF Bz dependence. We suggest that the IMF Bz produces the enhanced condition of the whistler mode chorus waves in the inner magnetosphere and resultant acceleration of relativistic accelerations takes place. Finally, we discuss the causal relationship between the IMF Bz dependence and the flux enhancement and we show the importance of cross-energy coupling process via wave–particle interactions.

13.2 Solar wind parameter dependence

Several studies have shown that the solar wind speed is a primary parameter that causes flux enhancements of the outer belt [e.g., Paulikas and Blake, 1979; Baker et al., 1986; O'Brien et al., 2001; Reeves et al., 2011]. However, the outer belt electrons do not always increase greatly when high-speed streams arrive at the Earth [Kim et al., 2006]; hence, there must be parameters other than solar wind speed that contribute to the flux enhancement.

Besides the solar wind speed, there are studies that have demonstrated the importance of a southward IMF in the production of large flux enhancements. Blake et al. [1997] showed that high-speed solar wind streams can have a strong effect on the energetic electron population associated with the southward IMF. Iles et al. [2002] showed that flux enhancements of the outer belt occur in conjunction with the southward IMF in the high-speed solar wind streams during the storm recovery phase.

From these IMF Bz dependences, it is expected that substorms are important for the large flux enhancements of the outer belt. In fact, Meredith et al. [2002] suggested that the gradual acceleration of electrons during storms is effective when there are periods of prolonged substorm activity during the storm recovery phase.

In order to clarify the importance of substorms for large flux enhancements of the outer belt, it is essential to study the IMF Bz effect for large flux enhancements of the outer belt statistically. The geoefficiency of the high-speed coronal hole streams is strongly controlled by the Russell–McPherron effect [Russell and McPherron, 1973]. According to the Russell–McPherron effect, geoefficiency is enhanced during the time intervals of so-called spring-toward fall-away (STFA) because the IMF polarities of toward and away have a large projection component in the southward IMF around the spring and fall equinox, respectively. The southward component of the IMF in a high-speed stream is always significantly suppressed or enhanced because of the STFA rule, since the amplitudes of IMF Bz from both the Russell–McPherron effect and Alfvénic fluctuations are comparable amplitudes.

The coronal hole stream events can be classified into two groups by considering the STFA rule as follows: the IMF sector after the stream interface is toward in spring (February, March, April, and May) or away in fall (August, September, October, and November) (Miyoshi and Kataoka [2008a]; Miyoshi et al. [2013]). The southward IMF Bz high-speed streams are denoted as SBz-HSS hereafter, whereas the northward IMF Bz high-speed streams are denoted as NBz-HSS. As a result of the above categorization, SBz-HSS and NBz-HSS have the same solar wind structures for different IMF Bz offset after the stream interface. Considering these points, we made a database of the stream interface crossing (Miyoshi and Kataoka, 2008a, b; Miyoshi et al., 2013). The details for the event selection are described by Miyoshi and Kataoka [2008a]. Using this database, we have done several analyses to identify the importance of the IMF Bz for the flux enhancements.

Figure 13.1 shows the superposed epoch analysis for solar wind profiles for SBz-HSS (light) and NBz-HSS (dark). The stream interface crossing corresponds to the zero time. From top to bottom, magnetic field strength B, IMF Bz (geocentric solar magnetospheric (GSM) coordinates), fluctuation of IMF Bz, solar wind speed V, density N, proton temperature T, the dynamic pressure P_{dyn}, the AE index, the cross polar cap potential, and the Kp and Dst indices. The typical variations of CIR and subsequent high-speed coronal hole streams can be seen here. The solar wind speed and temperature increase and the solar wind density drops following the high-speed coronal hole streams [e.g., Gosling et al., 1978; Richardson et al., 2006]. Following the STFA rule, the average IMF Bz is systematically shifted southward after $t = 0$ in the SBz-HSS, while the IMF Bz is shifted northward in the NBz-HSS. Other solar wind parameters are essentially similar between the SBz-HSS and NBz-HSS, so that investigations on the pure IMF Bz effect in high-speed streams are possible.

Different geomagnetic activity can be observed during SBz-HSS and NBz-HSS events. Continuous AE and Kp activity is present during SBz-HSS events, and the prolonged AE activity corresponds to the HILDCAA periods. During these periods, enhanced convection is also expected. Differences in evolutions of the polar cap potential can be seen between the two groups. The polar cap potential increases to 40 kV at the stream interface arrival, and then the potential returns to modest activity levels during the SBz-HSS period. On the contrary, the polar cap potential does not significantly change during the NBz-HSS events. Differences in the Dst index are also seen. During the SBz-HSS, the recovery of the Dst index is slow, thus indicating that there are continuous hot ions inputs from the plasma sheet.

Using the database of the stream interface crossing, Figure 13.2 shows the superposed >2 MeV electron flux for the SBz-HSS and NBz-HSS as observed at geosynchronous earth orbit (GEO) [Miyoshi and Kataoka, 2008a]. The data are obtained by the GOES satellites at GEO. One half day before the stream interface arrival, the electron flux begins to decrease in both SBz-HSS and NBz-HSS. In the SBz-HSS events, the flux recovers to the pre-CIR level at t = 1.0 day and then increases above the pre-CIR level (Figure 13.2a). In contrast, the flux continues to decrease over the first day and then gradually recovers to the pre-CIR level at t = 2.0 days during the NBz-HSS. After t = 3.0 days, the flux is over the pre-CIR level, but the amount of the flux enhancement is relatively weak.

The results clearly indicate that the flux enhancement in the outer radiation belt is significantly controlled by the IMF Bz through the Russell–McPherron effect. As shown in Figure 13.1, solar wind parameters except for IMF Bz are essentially the same between SBz-HSS and NBz-HSS, so we conclude that the IMF Bz is an important parameter to control flux enhancements of the outer belt. McPherron et al. [2009] also pointed out an importance of the Russell-McPherron effect for the large flux enhancement of the outer belt.

Figure 13.3 shows the solar wind parameter dependence of the flux enhancement of the outer belt as a function of the IMF Bz, i.e., by considering the different solar wind speed. Since the STFA rule controls the average IMF Bz for a few days after the stream interface, the averaged IMF Bz for 3 days after the stream interface can be used for the analysis. Figure 13.3a shows the coronal hole stream events where the solar wind speed is faster than 500 km/s, and Figure 13.3b shows the coronal hole stream events where the solar wind speed is slower than 500 km/s. These data demonstrate that the flux enhancement in the outer belt depends on the southward IMF. When the solar wind speed is high, the dependences on the IMF can be seen more clearly. The solar wind speed dependence is consistent with the results of previous studies [Paulikas and Blake, 1979; Baker et al., 1986; Weigel et al., 2003], and our results additionally show that the solar wind speed is not a sufficient condition for the large flux enhancement, i.e., the largest flux enhancement was observed in high-speed coronal hole streams with a southward IMF offset of a few nT. Li et al. [2011] pointed out that there is a strong

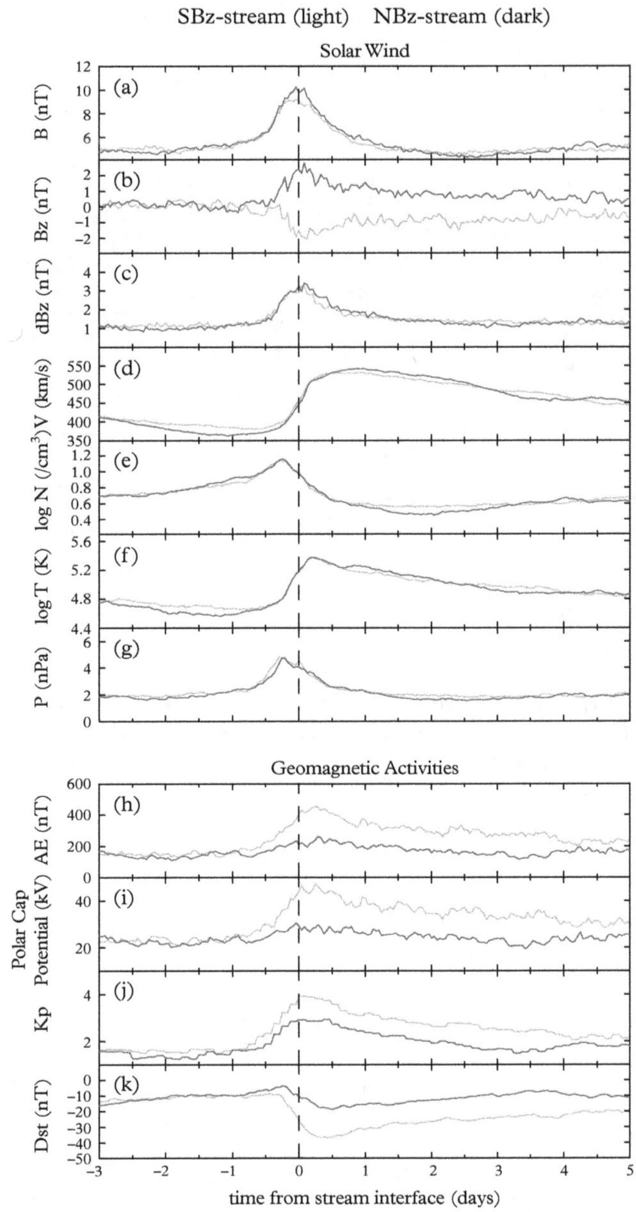

Figure 13.1 *Superposed epoch analysis for solar wind parameters and geomagnetic activities. SBz-HSS and NBz-HSS events are shown in light and dark curves, respectively. From top to bottom, magnetic field strength B, IMF Bz (geocentric solar magnetospheric (GSM) coordinates), fluctuation of IMF Bz, solar wind speed V, density N, proton temperature T, the dynamic pressure Pdyn, the AE index, the cross polar cap potential derived from the polar cap (PC) index [Troshichev et al., 1996], and the Kp and Dst indices.*
(After Miyoshi et al., 2013.)

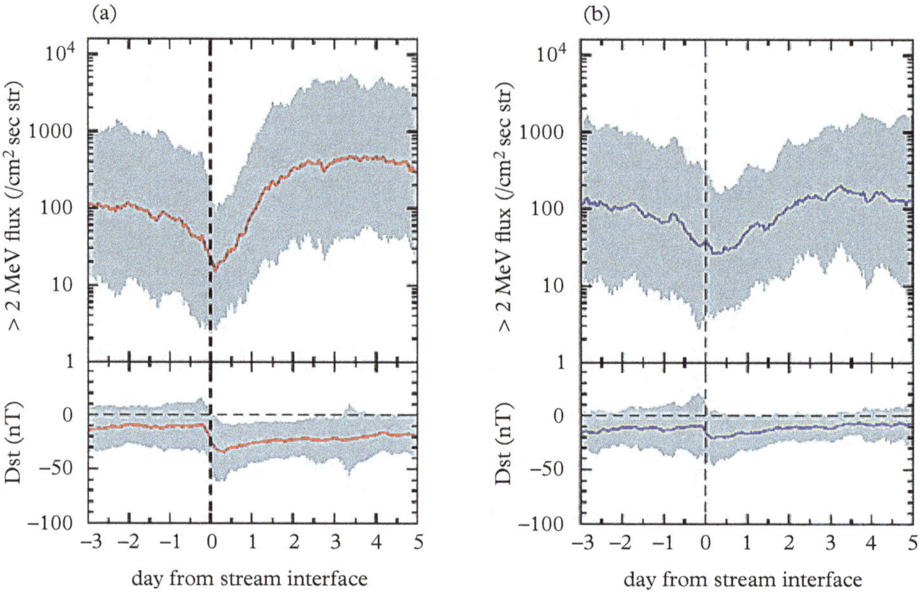

Figure 13.2 *Superposed epoch analysis for the flux of >2.0 MeV electrons as observed by GOES for (a) SBz and (b) NBz events. The bottom panels indicate the Dst index.*
(After Miyoshi and Kataoka, 2008a.)

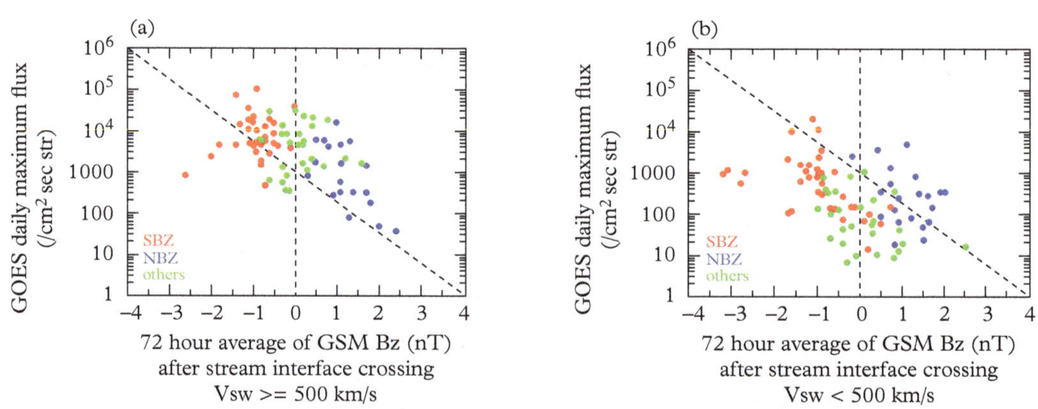

Figure 13.3 *Scatter diagram of the daily maximum flux at GEO at t = 3 days after the stream interface crossing. Horizontal axis is the 3-day average of IMF Bz after the stream interface: (a) fast and (b) slow stream events. Red and blue circles correspond to SBz-HSS and NBz-HSS, respectively, and green circles show the other high-speed coronal hole stream events.*
(After Miyoshi and Kataoka, 2008a.)

connection to the southward IMF for large flux enhancements of the outer belt and suggested that the observed correlation between the relativistic electron flux and solar wind speed might be explained by the southward IMF associated with fast solar wind.

While the southward IMF embedded in the high-speed coronal hole stream is too small to cause the storm main phase, it is a suitable condition for continuous substorm activities, i.e., HILDCAAs. It is expected that the acceleration mechanisms of relativistic electrons are activated during the continuous substorms in the SBz-HSS events. In the next section, we discuss the possible physical processes behind the IMF Bz dependencies.

13.3 Physics behind the solar wind parameter dependence

In regard to the possible acceleration processes, different acceleration mechanisms have been proposed such as radial transportation by geomagnetic pulsations in the Pc5 ultra low frequency (ULF) range and cyclotron resonance with whistler mode waves in the very low frequency (VLF) range [e.g., Shprits et al., 2008; Ebihara and Miyoshi, 2011]. The solar wind parameter dependence as shown in the previous section should be consistent with the solar wind parameter dependence on the acceleration mechanisms. That is, if we find the same parameter dependences for key parameters of an acceleration mechanism as the relativistic electron flux variations, the acceleration mechanism should be a primary process to cause the flux enhancement.

The drift resonance caused by the Pc5 ULF waves drives the radial transport of relativistic electrons; specifically, the conservation of the first and second adiabatic invariants causes the acceleration of electrons when electrons move toward the Earth [e.g., Elkington et al., 1999]. Liu et al. [1999] and Summers and Ma [2000] have proposed non-adiabatic accelerations by Pc5 ULF waves. Since the excitation of Pc5 ULF waves is primarily correlated with solar wind speed [Mathie and Mann, 2000; Mann et al., 2012], the positive correlation between the solar wind speed and the relativistic electron flux enhancement is consistent with the dependence. However, the solar wind speed is the only sufficient condition for the flux enhancement, and the flux enhancement largely depends on the IMF Bz, so that another mechanism should also be effective for the flux enhancement of the outer belt.

The cyclotron resonance with whistler mode waves is another important acceleration mechanism [Summers et al., 1998; Miyoshi et al., 2003; Horne et al., 2005]. The cyclotron resonance with whistler mode waves occurs inside the outer belt, and consequently, a peak of the radial profile of the phase space density represents important evidence. Reeves et al. [2013] identified the evolution of the phase space density peak inside the outer belt for the October 2012, storm event from Van Allen Probes observation data [Mauk et al., 2012], and this provides definitive evidence for the internal acceleration. Thorne et al. [2013] showed that the observed whistler mode waves caused the flux enhancement of the outer belt electrons during this event.

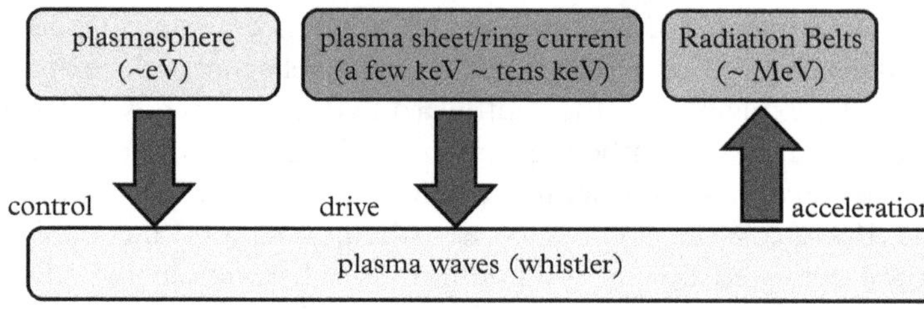

Figure 13.4 *Schematic figure for the cross-energy coupling via wave–particle interactions.*

Acceleration caused by the whistler mode waves represents a coupled process among different energy plasma/particle populations, i.e., a cross-energy coupling process [Miyoshi et al., 2012]. Figure 13.4 is a schematic diagram that illustrates the relativistic electron acceleration process caused by cross-energy coupling via cyclotron resonance. There are several key parameters that control this process. Hot electrons that come from the plasma sheet into the inner magnetosphere represent a free energy source that generates whistler mode waves. That is, the adiabatic transport naturally produces the temperature anisotropy that provides free energy for the generation of whistler mode waves. The generation of whistler wave activity depends on the ratio of the plasma frequency and the gyro-frequency, and the variation of the thermal plasma density controls the efficiency of wave generation as well as the acceleration.

If this cross-energy coupling process is a primary process for the acceleration, STFA dependence for these key parameters should be observed, i.e., the southward IMF should clearly control the hot electron enhancement, subsequent wave growth, and accelerations of relativistic electrons.

First, we discuss the seed populations. Figure 13.5 shows the superposed magnetic local time (MLT) diagram of hot electron density and temperature at GEO [Miyoshi et al., 2013]. The data are provided by the Los Alamos National Laboratory Magnetospheric Plasma Analyzer (LANL/MPA) [Denton et al., 2006]. Figure 13.5a and b show electron density and temperature for SBz-HSS, while Figure 13.5c and d show that for NBz-HSS. While the electron temperature increases after the stream interface in both Figure 13.5b and d [Denton et al., 2006], Figure 13.5b shows that the electron temperature becomes high in the SBz-HSS in which HILDCAAs occur. The density also increases in association with the stream interface crossing, while the densities are not different between the SBz-HSS and the NBz-HSS as shown in Figure 13.5a and c. Since these populations represent free energy sources to generate whistler mode waves in the inner magnetosphere, differences in the plasma sheet electron temperature will lead to a different evolution of whistler mode wave excitation [MacDonald et al., 2008].

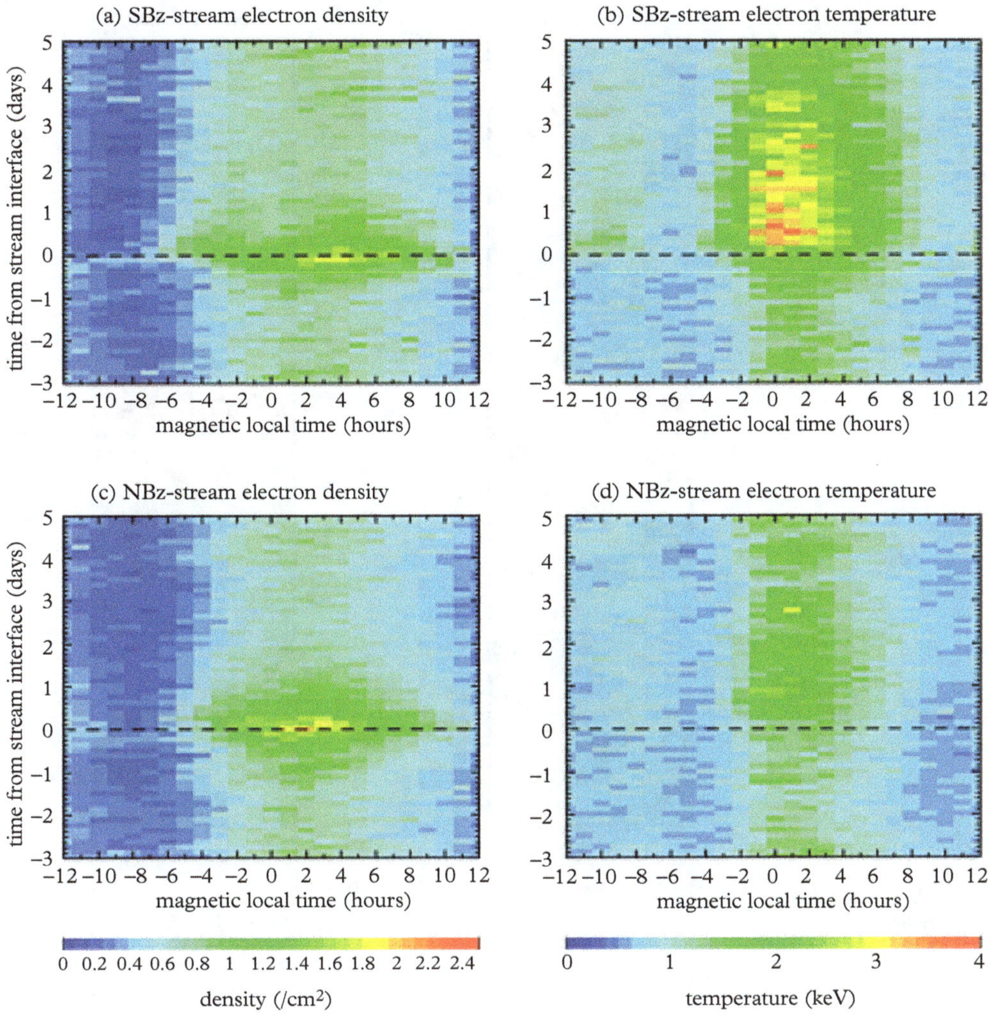

Figure 13.5 *Superposed epoch analysis for the density and temperature of hot electrons at GEO. (a) SBz-HSS and (b) NBz-HSS events.*
(After Miyoshi et al., 2013.)

Similar differences can be seen in other key parameters. Figure 13.6 shows the superposed epoch analysis for 30–100 keV electrons in the inner magnetosphere, thermal plasma density, whistler mode chorus waves, and relativistic electron fluxes in the inner magnetosphere [Miyoshi et al., 2013].

During the SBz-HSS events, the thermal plasma density decreases at $L > 4$, thus indicating a shrinkage of plasmapause. Hot electrons enter into the low-density region from the plasma sheet. Hot electrons act as a free energy source for the generation of whistler mode waves; conditions wherein the ratio of the plasma frequency to cyclotron

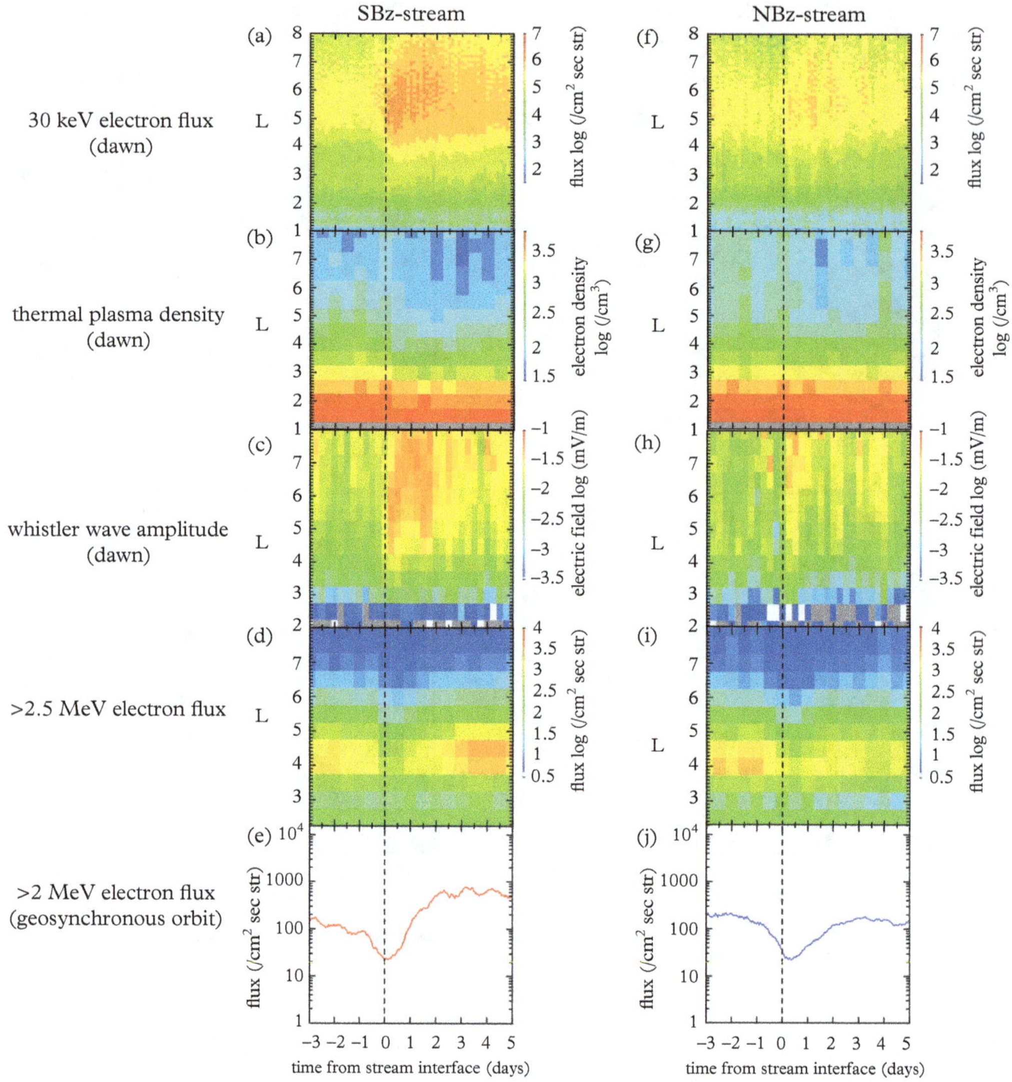

Figure 13.6 *Superposed epoch analysis for hot electron (30–100 keV) flux at the dawnside (a), (f), thermal plasma density at the dawnside (b), (g), whistler wave activities at the dawn side (c), (h), and >2.5 MeV electron flux (d), (i) and >2 MeV electron flux at GEO (e), (j). (left) SBz-HSS events and (right) NBz-HSS events.*
(After Miyoshi et al., 2013.)

frequency is small are preferable for the growth of whistler mode waves. In the SBz-HSS, such preferable conditions are achieved in the inner magnetosphere for a few days. The amplitude of whistler mode waves is enhanced at $L > 4$ in this region, and wave activities are accompanied by hot electron enhancement during the SBz-HSS events.

The inner magnetosphere environment in the NBz-HSS events is different from that in the SBz-HSS events. In the NBz-HSS events, the thermal plasma density does not change very much; that is, the shrinkage of the plasmapause is not remarkable. Although transient enhancements of hot electrons are observed during the NBz-HSS events, the average flux does not increase. These differences in the behavior of thermal plasma and hot electrons lead to a different evolution of whistler mode wave activity. Although the generation of whistler mode waves is enhanced just after the stream interface crossing during NBz-HSS events, the average amplitude of whistler mode waves in the inner magnetosphere in the NBz-HSS events is weaker than that in SBz-HSS events by more than one order of magnitude.

Li et al. [2012] have done similar statistical analyses on whistler mode waves and seed electrons as a free energy source. The evolution of the chorus wave activity and the source electron population was found to be highly dependent on the preceding IMF Bz orientation, i.e., continuous substorm activities. Lyons et al. [2005] showed that enhanced convection leads to enhanced waves that accelerate the relativistic electrons, which is consistent with the findings of this study.

A case study for two CIR-storms shows a clear causal relationship between the solar wind parameter dependences and the large flux enhancement of the outer belt [Miyoshi et al., 2007]. Here, we discuss typical CIR-driven storms in November 1993. Figure 13.7 shows the solar wind and IMF variations in November 1993, obtained from the Geotail satellite. The solar wind speed in the former high-speed stream increased up to 800 km/s, while the speed in the latter high-speed stream increased up to 600 km/s.

The IMF sector polarity in the coronal hole streams was different during these two events. The two storms were driven by CIRs, and magnetic field fluctuations were observed in both coronal hole streams. Figure 13.7(f) shows the azimuthal angle of the IMF, which is indicative of the IMF sector polarity. In the former high-speed stream, the IMF sector polarity was away, so that the southern IMF offset was significant because of the Russell–McPherron effect. Conversely, the IMF sector polarity was toward in the latter high-speed stream and northward offset was observed.

As we have already seen in the statistical analysis (Figure 13.6), the IMF sector polarity controls the continuous substorm activities during the period of the high-speed coronal hole streams by the STFA rule. The maximum AE index in the former streams and the latter streams were about 2300 nT and 1500 nT, respectively. The recovery phase of the former storm lasted for about one week and was associated with prolonged and enhanced AE activities, i.e., HILDCAAs, while the intense AE activity was seen only around the main phase and the early recovery phase for the latter storm.

Significant differences between these two streams are found in the key parameters for the cyclotron resonance as shown in Figure 13.8 [Miyoshi et al., 2007]. Figure 13.8a shows the variation in relativistic electrons (>2 MeV) at GEO. The large flux enhancements can be seen in association with the arrival of the coronal hole streams, and then they gradually increased and remained at a high level. However, the flux did not show any significant enhancements above pre-storm levels in the latter storms.

Figure 13.8b shows the sub-relativistic electrons at GEO, which are considered to be the seed population for the production of relativistic electrons. The flux enhancements

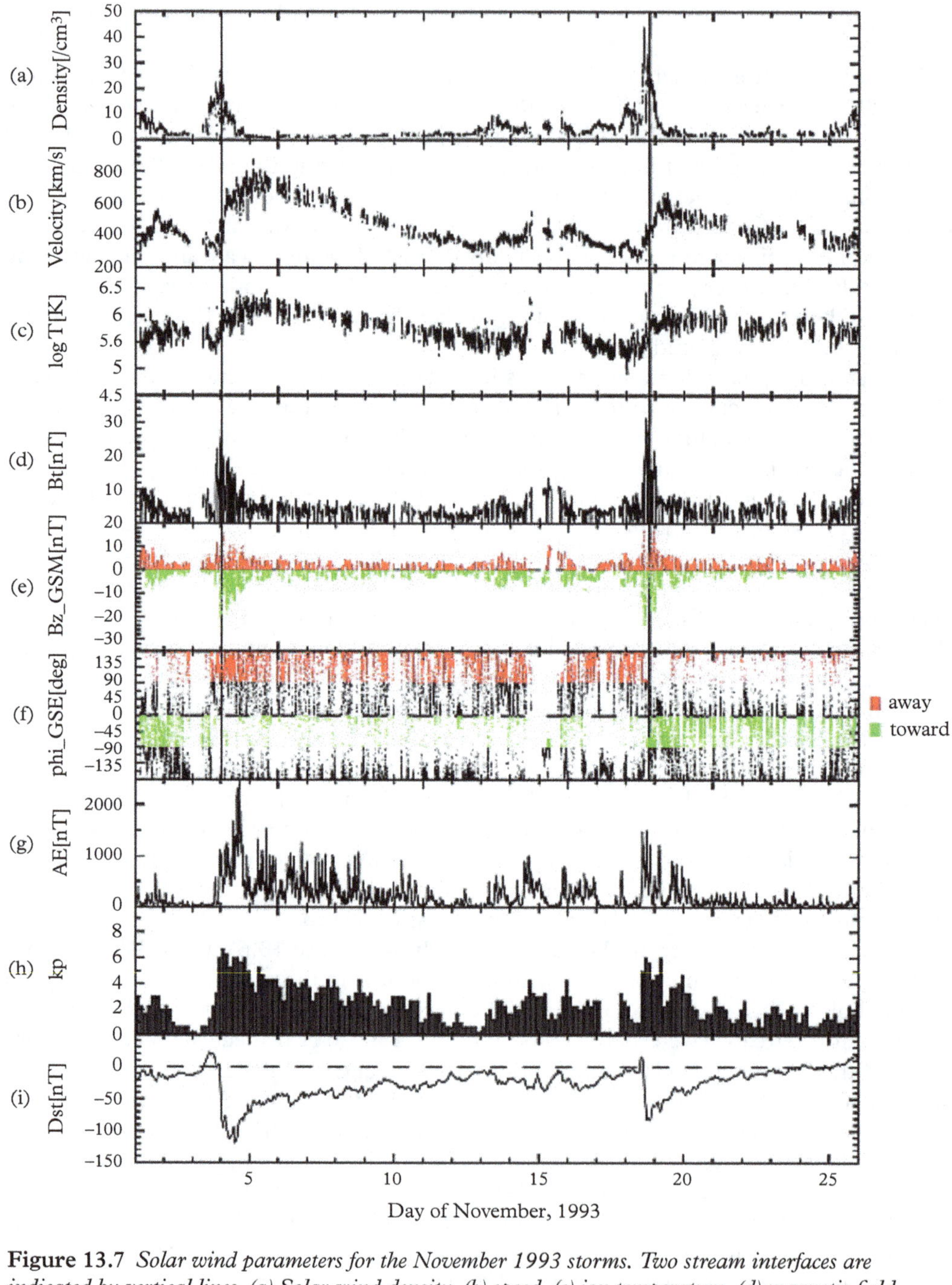

Figure 13.7 *Solar wind parameters for the November 1993 storms. Two stream interfaces are indicated by vertical lines. (a) Solar wind density, (b) speed, (c) ion temperature, (d) magnetic field strength, and (e) IMF Bz in geocentric solar magnetospheric (GSM) coordinates. The red (green) color indicates the positive (negative) IMF Bz. (f) Azimuth angle of the IMF in geocentric solar ecliptic (GSE) coordinates. The red (green) color indicates the away (toward) sector polarity. (g) AE, (h) Kp, and (i) Dst indices.*

(After Miyoshi et al., 2007.)

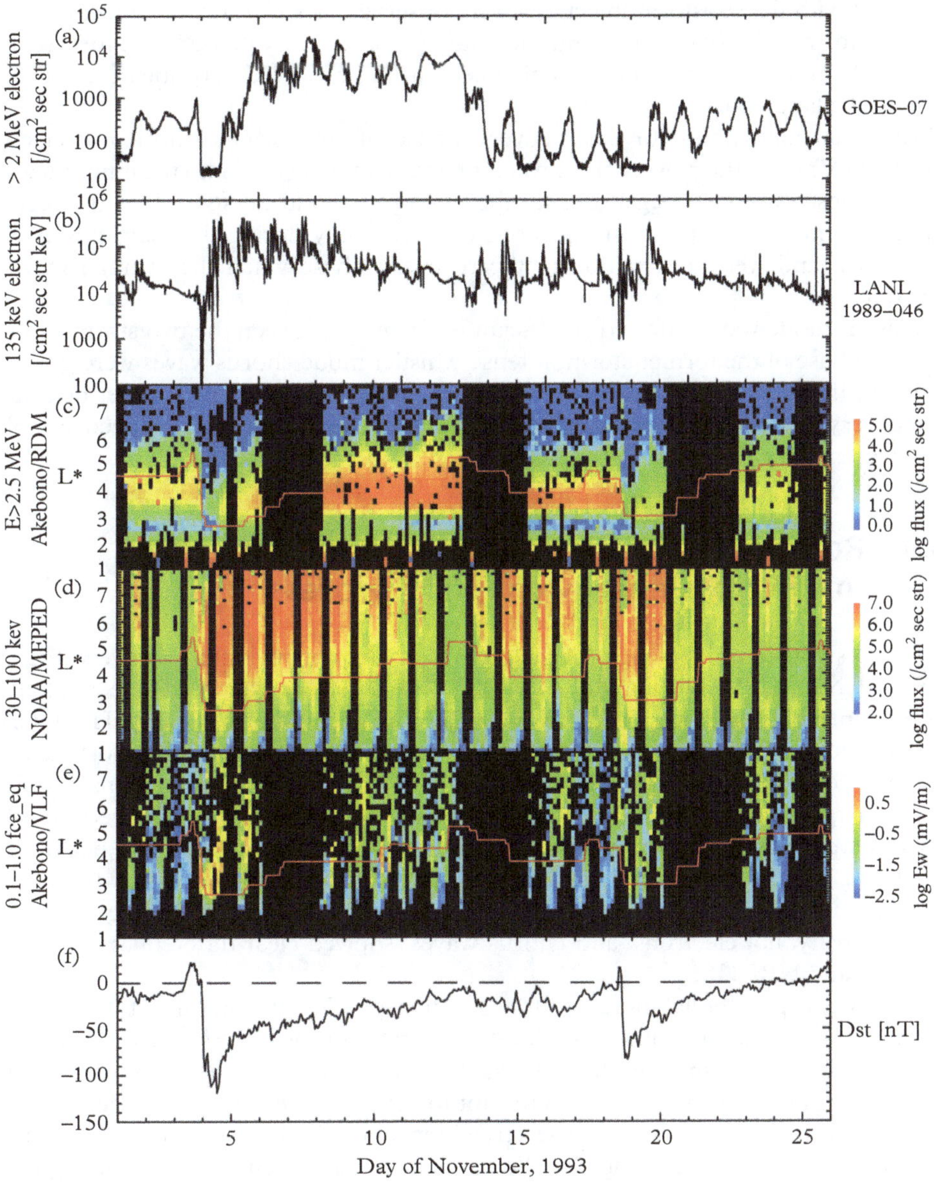

Figure 13.8 *(a) Electron flux (>2.0 MeV) at the geosynchronous orbit. (b) The >135 keV electron flux at the geosynchronous orbit. (c) The >2.5 MeV electron flux. (d) The 30–100 keV electron flux at the dawnside. (e) Whistler mode waves at the dawnside. The red line shows the empirical plasmapause location. (f) The Dst index.*
(After Miyoshi et al., 2007.)

associated with the substorm injections were observed as spiky flux enhancements during both storms. The injections continued during the recovery phase in the former storm with a higher level of flux, while flux enhancement was observed only during the recovery phase in the latter storm.

From the main phase to the recovery phase of the former storm, hot electrons (30–100 keV) from the plasma sheet were intermittently injected into the inner magnetosphere. These injections began during the main phase and persisted into the recovery phase. In contrast, hot electron injections occurred only during the main phase of the latter storm and were absent during the recovery phase when the average IMF was northward.

Whistler mode waves showed significant differences between the two storms. During the main phase of the former storm, intense whistler mode chorus waves were generated outside of the plamapause, and then waves were detected in the entire outer belt during the recovery phase. However, chorus wave activities were low during the recovery phase for the latter storm.

13.4 Role of continuous substorm activities in the cross-energy coupling process

We found significant differences of the outer belt flux enhancements between the SBz-HSS and NBz-HSS. During the SBz-HSS, the interplanetary Alfvénic fluctuations caused continuous substorm activity and enhanced convection, so-called HILDCAAs. The injections of the plasma sheet particles occurred in the night-side portion of the plasma sheet during HILDCAAs [Sørras et al., 2004]. These populations should be a free energy source to generate whistler mode chorus waves. Prolonged substorm activities were observed in the SBz-HSS, while such substorms did not occur in the NBz-HSS. Besides the outer belt flux variations, all key parameters for the electron accelerations associated with whistler wave particle interactions, including the thermal plasma density, hot electrons, and chorus waves, showed clear differences between the SBz-HSS and NBz-HSS.

The polar cap potential increased to a few tens of kilovolts after the stream interface crossing and maintained a moderate level of activity during the SBz-HSS. These polar cap potentials are too small to drive the storm main phase, but continuous transport of electrons from the plasma sheet is important to drive whistler mode wave generations.

A substorm cycle consists of the stretching and dipolarization of magnetic field lines in the tail region. During the dipolarization, strong dawn-dusk electric field is observed [Roux et al., 1991; Maynard et al., 1996; Sergeev et al., 1998; Ohtani et al., 2007; Nishimura et al., 2008], which could transport electrons inward. According to a global magneto hydrodynamics (MHD) simulation, the dipolarization does not proceed smoothly because of imbalance between the plasma pressure force and the Lorentz force [Ebihara and Tanaka, 2013]. As a consequence of the imbalance, the electric field shows an oscillation with a period of a few minutes near geosynchronous orbit. The oscillation of the electric field results in "multiple" injections of electrons into the inner magnetosphere [Ebihara and Tanaka, 2013]. Such "multiple" injections are observed by satellites [Baker et al., 1982, Sergeev et al., 1998] together with oscillation of the electric field. Substorms may also have a direct influence on electrons with energies greater than 100 keV that is a seed population for subsequent accelerations by wave–particle interactions. Near geosynchronous orbit, such energetic electrons may be decelerated due to betatron acceleration during the stretching (substorm growth phase), whereas they may be accelerated by the electric field and/or lost due to magnetopause shadowing during the dipolarization (substorm expansion phase) [Ebihara and Tanaka, 2013]. Due to both continuous injections associated with substorms and enhanced convections, the plasma sheet electrons were continuously supplied to the inner magnetosphere, which caused the whistler mode wave intensification as shown in Figure 13.6.

During the enhanced convection in the SBz-HSS, the thermal plasma density decreased at $L > 4$, i.e., a shrinkage of the plasmapause was detected. Decreases of the thermal plasma density will cause large growth rates in the whistler mode waves [Li et al., 2010], so that the preferable situation for wave growth is achieved in the SBz-HSS. However, the thermal plasma density did not change significantly due to the NBz-HSS; i.e., the shrinkage of the plasmapause was not remarkable.

Meredith et al. [2011] performed superposed epoch analysis during CIR-driven storms and found that the flux of trapped and precipitating electrons (>30 keV) increases immediately following storm onset and remains elevated during the passage of the HSSs. The increased precipitating fluxes are probably due to the increased equatorial electron flux and enhanced scattering by chorus waves, whose wave amplitudes also remain elevated after the storm onset.

Figure 13.9 illustrates how substorms are important for the large flux enhancement via wave–particle interactions. Continuous substorms and enhanced convections have two roles in the production of the flux enhancements of relativistic electrons. Continuous hot electron enhancements provide a free energy source to generate whistler mode waves [Omura et al., 2009]. Moreover, the plasmapause moves earthward due to the enhanced convection, which provides suitable conditions to generate whistler mode waves through the formation of the region for the small ratio of the plasma frequency (fp) and the electron cyclotron frequency (fce). Chorus waves are generated through the non-linear wave–particle interactions after the linear growth in the small fp/fce region, i.e., outside the plasmapause. Continuous chorus and wave–particle interactions cause the flux enhancement of relativistic electrons in the outer belt. These situations are naturally realized during the period of the SBz-HSS. Continuous Alfvénic fluctuations with southward offset embedded in the high-speed streams cause the continuous substorm activities and the enhanced convections in the inner magnetosphere. Absence of continuous substorms, however, does not produce hot electron enhancement in the inner magnetosphere as well as subsequent whistler mode wave generation, and hence, no enhancement of the relativistic electrons is observed under such conditions of the NBz-HSS.

Relativistic electron flux of the outer belt tends to increase largely during the recovery phase of CIR-driven storms rather than during CME-driven storms [Miyoshi and Kataoka, 2005; Kataoka and Miyoshi, 2006]. As shown in this chapter, the

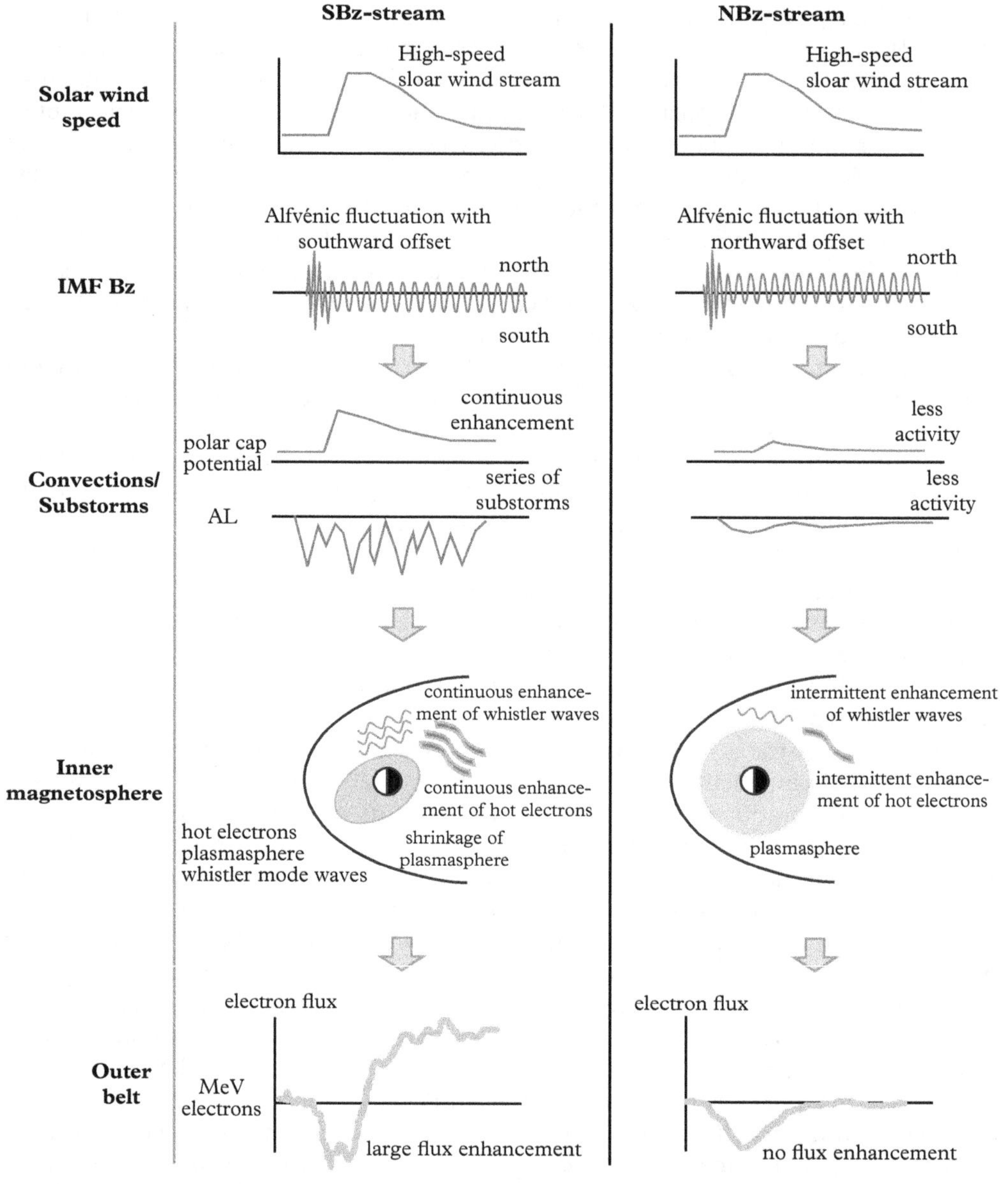

Figure 13.9 *Schematic diagram of solar wind–radiation belt coupling. Prolonged substorm activities driven by the intermittent IMF B z with its average being negative cause the flux enhancement in the outer belt via wave–particle interactions. All key parameters for the accelerations have clear differences between SBz-HSS and NBz-HSS, which is consistent with the concept that wave–particle interactions are a primary process for electron accelerations.*

high-speed coronal hole streams as well as the SBz cause large flux enhancements in the outer belt through continuous substorm activities. The IMF Bz tends to turn northward continuously for a longer time interval than usual during the recovery phase of CME-driven storms, and it is expected that the differences in the evolution of the outer belt during CME-driven storms and CIR-driven storms may also be a result of the differences in the evolution of the IMF Bz during the recovery phase of these storms.

It is interesting to note that these IMF Bz dependences can be seen in the long-term variations of the outer belt. Kataoka and Miyoshi [2010] reported that the outer belt electron flux was unusually small in 2009. They compared several solar wind parameters between 2009 and 1997, and they found that the solar wind speed was very slow in 2009 but a comparably low level as of 1997. The IMF was quite different between 1997 and 2009, so that weak IMF Bz, i.e., the weak substorm activity, may cause unusual low flux of the outer belt electrons in 2009.

13.5 Summary

In this chapter, we discussed a possible role of prolonged substorm activities in the production of large flux enhancements of outer belt electrons via wave–particle interactions. These situations are naturally observed as HILDCAAs driven by continuous Alfvénic fluctuations of IMF embedded in the high-speed coronal hole streams. The Russell–McPherron effect controls the amplitude of the southward IMF by the STFA rule. New satellite observations such as Van Allen Probes [Mauk et al., 2012] and ERG [Miyoshi et al., 2012] will shed lights on details of cross-energy couplings that cause large flux enhancement of relativistic electrons via wave–particle interactions.

By considering certain key characteristics, it is possible to forecast which coronal hole streams will cause HILDCAAs and large flux enhancements in the outer belt. Specifically, information on the IMF sector polarity of forthcoming coronal hole streams as well as the season can be used to determine whether the STFA rule works in an appropriate manner. Such probabilistic forecasts have been proposed [Miyoshi and Kataoka, 2008b] for the outer belt electron flux, and now space weather forecasts are becoming more feasible as our understanding on the causal relationship between solar wind-radiation belts coupling is improved.

Acknowledgments

This study was supported by Grants-in-Aid for Scientific Research (grant numbers 23340146, 23224011, 15H05815, 15H03732) from the Japan Society for the Promotion of Science (JSPS). This work was also supported by the JSPS Program for Advancing Strategic International Networks to Accelerate the Circulation of Talented Researchers under grant number G2602.

REFERENCES

Baker, D. N., et al. (1982), Observation and modeling of energetic particles at synchronous orbit on July 29, 1977, *J. Geophys. Res.*, **87**, 5917–5932.

Baker, D. N., J. B. Blake, R. W. Klebesadel, and P. R. Higbie (1986), Highly relativistic electrons in the Earth's outer magnetosphere: 1. Lifetimes and temporal history 1979–1984, *J. Geophys. Res.*, **91**, 4265.

Blake, J. B., D. N. Baker, N. Turner, K. W. Olgilvie, and R. P. Lepping (1997), Correlation of changes in the outer-zone relativistic-electron population with upstream solar wind and magnetic field measurements, *Geophys. Res. Lett.*, **24**, 927.

Denton, M. H., et al. (2006), Geomagnetic storms driven by ICME- and CIR-dominated solar wind, *J. Geophys. Res.*, **111**, A07S07, doi:10.1029/2005JA011436.

Ebihara, Y., and Y. Miyoshi (2011), Dynamic inner magnetosphere: A tutorial and recent advances, in *The dynamic magnetosphere*, IAGA Special Sopron Book Series, vol. 3. (ed. W. Liu and M. Fujimoto), 145–187, Springer, Dordrecht, The Netherlands, doi:10.1007/978-94-007-0501-2_9.

Ebihara, Y., and T. Tanaka (2013), Fundamental properties of substorm time energetic electrons in the inner magnetosphere, *J. Geophys. Res. Space Physics*, **118**, 1589–1603, doi:10.1002/jgra.50115.

Elkington, S. R., M. K. Hudson, and A. A. Chan (1999), Acceleration of relativistic electrons via drift-resonant interaction with toroidal-mode Pc-5 ULF oscillations, *Geophys. Res. Lett.*, **26**, 3273–3276.

Gosling, J. T., J. R. Asbridge, S. J. Bame, and W. C. Feldman (1978), Solar wind stream interface, *J. Geophys. Res.*, **83**, 1401–1412.

Horne, R. B., R. M. Thorne, S. A. Glauert, J. M. Albert, N. P. Meredith, and R. R. Anderson (2005), Timescale for radiation belt electron acceleration by whistler mode chorus waves, *J. Geophys. Res.*, **110**, A03, 225, doi:10.1029/2004JA010811.

Iles, R. H. A., A. N. Fazakerley, A. D. Johnstone, N. P. Meredith, and P. Buhler (2002), The relativistic electron response in the outer radiation belt during magnetic storms, *Ann. Geophys.*, **20**, 957–965.

Kataoka, R., and Y. Miyoshi (2006), Flux enhancement of radiation belt electrons during geomagnetic storms driven by coronal mass ejections and corotating interaction regions, *Space Weather*, **4**, S09004, doi:10.1029/2005SW000211.

Kataoka, R., and Y. Miyoshi (2010), Why are relativistic electrons persistently quiet at geosynchronous orbit in 2009? *Space Weather*, **8**, S08002, doi:10.1029/2010SW000571.

Kim, H.-J., K. C. Kim, D. Lee, and G. Rostoker (2006), Origin of geosynchronous relativistic electron events, *J. Geophys. Res.*, **111**, A03208, doi:10.1029/2005JA011469.

Li, W., R. M. Thorne, Y. Nishimura, J. Bortnik, V. Angelopoulos, J. P. McFadden, D. E. Larson, J. W. Bonnell, O. Le Contel, A. Roux, and U.Auster (2010), THEMIS analysis of observed equatorial electron distributions responsible for the chorus excitation, *J. Geophys. Res.*, **115**, A00F11, doi:10.1029/2009JA014845.

Li, X., M. Temerin, D. N. Baker, and G. D. Reeves (2011), Behavior of MeV electrons at geosynchronous orbit during last two solar cycles, *J. Geophys. Res.*, **116**, A11207, doi:10.1029/2011JA016934.

Li, W., R. Thorne, J. Bortnik, R. McPherron, Y. Nishimura, V. Angelopoulos, and I. G. Richardson (2012), Evolution of chorus waves and their source electrons during storms driven by corotating interaction regions, *J. Geophys. Res.*, **117**, A08209, doi:10.1029/2012JA017797.

Liu, W. W., G. Rostoker, and D. N. Baker (1999), Internal acceleration process of relativistic electrons by large-amplitude ULF pulsations, *J. Geophys. Res.*, **104**, 17391–17407.

Lyons, L. R., D. –Y. Lee, R. M. Thorne, R. B. Horne, and A. J. Smith (2005), Solar wind-magnetosphere coupling leading to relativistic electron energization during high-speed streams, *J. Geophys. Res.*, **110**, A11202, doi:10.1029/2005JA011254.

MacDonald, E. A., M. H. Denton, M. F. Thomsen, and S. P. Gary (2008), Superposed epoch analysis of a whistler instability criterion at geosynchronous orbit during geomagnetic storms, *J. Atm. Solar-Terr. Phys.*, **70**, 1789–1796.

Mann, I. R, K. R. Murphy, L. G. Ozeke, I. J. Rae, D. K. Milling, A. Kale, and F. Honary (2012), The role of Ultralow Frequency Waves in radiation belt dynamics, in *Dynamics of the Earth's radiation belts and inner magnetosphere*, Geophys. Monograph Ser., 199 (ed. D. Summers et al.), doi:10.1029/2012GM001349.

Mathie, R. A., and I. R. Mann (2000), A correlation between extended intervals of ULF wave power and storm-time geosynchronous relativistic electron flux enhancements, *Geophys. Res. Lett.*, **27**, 3261–3264, doi:10.1029/2000GL003822.

Mauk, B. H., N. J. Fox, S. G. Kanekal, R. L. Kessel, D. G. Sibeck, and A. Ukhorskiy (2012), Science objectives and rationale for the radiation belt storm probes mission, Space Sci. Rev., 179, 3–27, doi:10.1007/s11214-012-9908-y.

Maynard, N. C., W. J. Burke, E. M. Basinska, G. M. Erickson, W. J. Hughes, H. J. Singer, A. G. Yahnin, D. A. Hardy, and F. S. Mozer (1996), Dynamics of the inner magnetosphere near times of substorm onsets, *J. Geophys. Res.*, **101** (A4), 7705–7736, doi:10.1029/95JA03856.

McPherron, R. L., D. N. Baker, and N. U. Crooker (2009), Role of the Russell-McPherron effect in the acceleration of relativistic electrons, *J. Atm. Solar-Terr. Phys.*, **71**, 1032–1044, doi:10.1016/j.jastp.2008.11.002.

Meredith, N. P., R. B. Horne, and R. R. Anderson (2002), Substorm dependence of chorus amplitude: Implications for the acceleration of electrons to relativistic energies, *J. Geophys. Res.*, **106**, 13163–13178.

Meredith, N. P., R. B. Horne, M. M. Lam, M. H. Denton, J. E. Borovsky, and J. C. Green (2011), Energetic electron precipitation during high-speed solar wind stream driven storms, *J. Geophys. Res.*, **116**, A05223, doi:10.1029/2010JA016293.

Miyoshi, Y., and R. Kataoka (2005), Ring current ions and radiation belt electrons during geomagnetic storms driven by coronal mass ejections and corotating interaction regions, *Geophys. Res. Lett.*, **32**, L211105, doi:10.1029/2005GL024590.

Miyoshi, Y., and R. Kataoka (2008a), Flux enhancement of the outer radiation belt electrons after the arrival of stream interaction region, *J. Geophys. Res.*, **113**, A03S09, doi:10.1029/2007JA012506.

Miyoshi, Y., and R. Kataoka (2008b), Probabilistic space weather forecast of the relativistic electron flux enhancement at geosynchronous orbit, *J. Atm. Solar-Terr. Phys.*, **70**, 475–481.

Miyoshi, Y., A. Morioka, T. Obara, H. Misawa, T. Nagai, and Y. Kasahara (2003), Rebuilding process of the outer radiation belt during the November 3, 1993, magnetic storm NOAA and EXOS-D observations, *J. Geophys. Res.*, **108**, 1004, doi:10.1029/2001JA007542.

Miyoshi, Y., A. Morioka, R. Kataoka, Y. Kasahara, and T. Mukai (2007), Evolution of the outer radiation belt during the November 1993, storms driven by corotating interaction region, *J. Geophys. Res.*, **112**, A05210, doi:10.1029/2006JA012148.

Miyoshi, Y. et al. (2012), The energization and radiation in geospace (ERG) project, in *Dynamics of the Earth's radiation belts and inner magnetosphere*, Geophys Monograph Series, 199, 103–116, doi:10.1029/2012GM001304.

Miyoshi, Y., et al. (2013), High-speed solar wind with southward interplanetary magnetic field causes relativistic electron flux enhancement of the outer radiation belt via enhanced condition of whistler waves, *Geophys. Res. Lett.*, **40**, 1–6, doi:10.1002/grl.50916.

Nishimura, Y., et al. (2008), Large-amplitude wave electric field in the inner magnetosphere during substorms, *J. Geophys. Res.*, **113** A7, A07202.

O'Brien, T. P., et al. (2001), Which magnetic storms produce relativistic electrons at geosynchronous orbit? *J. Geophys. Res.*, **106**, 15533–15544.

Ohtani, S., et al. (2007), Cluster observations in the inner magnetosphere during the April 18, 2002, sawtooth event: Dipolarization and injection at r = 4.6 RE, *J. Geophys. Res.*, **112**, A08213, doi:10.1029/2007JA012357.

Omura, Y., M. Hikishima, Y. Katoh, D. Summers, and S. Yagitani (2009), Nonlinear mechanisms of lower-band an upper-band VLF chorus emissions in the magnetosphere, *J. Geophys. Res.*, **114**, A07217, doi:10.1029/2009JA014206.

Paulikas, G., and J. B. Blake (1979), Effects of the solar wind on magnetospheric dynamics: Energetic electrons at the geosynchronous orbit, in *Quantitative modeling of magnetospheric processes*, Geophys. Monograph Ser., 21, edited by W. Olsen, p21, AGU, Washington D. C.

Reeves, G. D., K. L. McAdams, R. H. W. Friedel, and T. P. O'Brien (2003), Acceleration and loss of relativistic electrons during geomagnetic storms, *Geophys. Res. Lett.*, **30**, 10, doi:10.1029/2002GL016513.

Reeves, G. D., et al. (2011), On the relationship between relativistic electron flux and solar wind velocity: Paulikas and Blake revisited, *J. Geophys. Res.*, **116**, A022213, doi:10.1029/2010JA015735.

Reeves, G. D., et al. (2013), Electron acceleration in the heart of the Van Allen Radiation Belts, *Science*, **30**, 341, 6149, 991–994, doi:10.1126/science.1237743.

Richardson, I.G., et al. (2006), Major geomagnetic storms (Dst<-100 nT) generated by corotating interaction regions, *J. Geophys. Res.*, **111**, A07S09, doi:10.1029/2005JA011476.

Roux, A., et al. (1991), Plasma sheet instability related to the westward traveling surge, *J. Geophys. Res.*, **96** (A10), 17697–17714, doi:10.1029/91JA01106.

Russell, C. T., and R. L. McPherron (1973), Semiannual variation of geomagnetic activity, *J. Geophys. Res.*, **78**, 92–108.

Sergeev, V., et al. (1998), Event study of deep energetic particle injections during substorm, *J. Geophys. Res.*, **103** (A5), 9217–9234.

Shprits, Y. Y., S. R. Elkington, N. P. Meredith, and D. A. Subbotin (2008), Review of modeling of losses and sources of relativistic electrons in the outer radiation belt 2: Local acceleration and loss, *J. Atm. Solar Terr. Phys.*, **70**, 1694–1713, doi:10.1016/j.jastp.2008.0.014.

Søraas, F., K. Asrsnes, K. Oksavik, M. I. Sandanger, D. S. Evans, and M. S. Greer (2004), Evidence for particle injection as the cause of Dst reduction during HILDCAA events, *J. Atm. Solar-Terr. Phys.*, **66**, 177.

Summers, D., R. M. Thorne, and F. Xiao (1998), Relativsitic theory of wave-particle resonant diffusion with application to electron acceleration in the magnetosphere, *J. Geophys. Res.*, **103**, 20487–20500.

Summers, D., and C. Ma (2000), Rapid acceleration of electrons in the magnetosphere by fast-mode MHD waves, *J. Geophys. Res.*, **105**, 15887–15895.

Thorne, R. M., et al. (2013), Rapid local acceleration of relativistic radiation-belt electrons by magnetospheric chorus, *Nature*, **504**, doi:10.1038/nature12889.

Troshichev, O., H. Hayakawa, A. Matsuoka, T. Mukai, and K. Tsuruda (1996), Cross polar cap diameter and voltage as a function of PC index and interplanetary quantities, *J. Geophys. Res.*, **101**, 13429–13435, doi:10.1029/5JA03672.

Tsurutani, B. T., and W. D. Gonzalez (1987), The cause of high intensity long-duration continuous AE activity (HILDCAAs): Interplanetary Alfvén wave trains, *Planet. Space Sci.*, **35**, 405.

Tsurutani, B. T., et al. (2006), Corotating solar wind streams and recurrent geomagnetic activity: A review, *J. Geophys. Res.*, **111**, A07S01, doi:10.1029/2005JA011273.

Weigel, R. S., A. J. Klimas, and D. Vassiliadis (2003), Precursor analysis and prediction of large-amplitude relativistic electron fluxes, *Space Weather*, **1** (3), 1014, doi:10.1029/2003SW000023.

14

Linkages Between the Radiation Belts, Polar Atmosphere and Climate: Electron Precipitation Through Wave Particle Interactions

M. A. Clilverd, C. J. Rodger, M. E. Andersson, A. Seppälä, and P. T. Verronen

■ Abstract

In this chapter we report on the processes by which plasma waves occurring in geospace drives electron precipitation, generating excess ionization at altitudes of 40-100 km, thereby affecting the polar atmosphere and ultimately coupling to surface climate. Excess ionization is created through wave-particle interaction processes in the outer radiation belt which can scatter energetic electrons into the atmosphere. The processes define the electron energies involved, the location of the wave-particle interaction regions, and the strength of the pitch-angle diffusion taking place. The ionization produced in the upper atmosphere results in the generation of odd nitrogen and odd hydrogen species, which can catalytically destroy ozone. Through the coupling of chemical and dynamical changes in the middle and upper atmosphere, changes in atmospheric dynamics can

Waves, Particles, and Storms in Geospace. First Edition. Georgios Balasis, Ioannis A. Daglis, Ian R. Mann.

propagate down into the troposphere. There they have been observed to modulate the large scale wintertime temperature patterns in the med- to high-latitude northern and southern hemispheres.

14.1 Introduction

The space around the Earth is filled with fast moving particles trapped in our magnetic field. These particles are principally electrons and protons, forming two "belts" separated by a "slot" region. The belts were initially discovered (Van Allen and Frank, 1959) in 1958 by the first US satellite, Explorer I, and were named the "Van Allen radiation belts" in honour of their discoverer, James Van Allen of the University of Iowa. As most satellites spend their lives inside the radiation belts, they are directly affected by this environment. Earth-orbiting satellites can be damaged or even lost (Lam et al., 2012; Clilverd et al., 2012) due to increased high-energy electron fluxes in the Earth's radiation belts (Van Allen and Frank, 1959; Millan and Baker, 2012). In particular, the outer radiation belt (located 3.5–8 Earth radii from the Earth's centre) is highly dynamic (Thorne, 2010; Horne et al., 2005), with fluxes changing by 1000 times on timescales of hours to days (Morley et al., 2010a; Baker et al., 2004).

This high dynamism is due to large and rapid changes in the fluxes in the particles, caused by the acceleration and loss of energetic radiation belt electrons during and after geomagnetic storms (e.g., Reeves et al., 2003). One of the major routes for the loss of these electrons is into the atmosphere (Green et al., 2004; Millan and Thorne, 2007), in a process termed "precipitation". Due to the configuration of the geomagnetic field, the outer radiation belt electrons precipitate into the polar atmosphere, depositing energy into the mesosphere and upper stratosphere. This precipitation alters the ionization rate over a wide altitude range; a 20 keV electron deposits the majority of its energy at an altitude slightly below 100 km, but higher energy electrons penetrate into the atmosphere more deeply, such that an ~4 MeV electron causes the largest changes at 40 km altitude (Turunen et al., 2009), directly altering the ionospheric electron number density (and hence conductivity).

In Fig. 14.1 we show the altitudes versus ionization rate for mono-energetic electron beams with a flux of 100 electrons $cm^{-2}s^{-1}sr^{-1}$. The energy ranges span from 20 keV to 4 MeV, representing relatively high energy auroral electrons through to relativistic electron precipitation. The electron ionization rates make use of the expressions given by Rees (1989), and the individual lines represent energies of interest in terms of satellite measurements often discussed in this publication, i.e., POES and GOES electron detectors.

In this chapter we report on the processes by which plasma waves occurring in the space around the Earth can drive precipitation, generating excess ionization at altitudes of 40–100 km, affecting the polar atmosphere and coupling to surface climate. The study of these processes includes observations from the ground, balloons and satellites, as well as modelling and climate data itself. This is an example of the multi-disciplinary nature of modern geophysics.

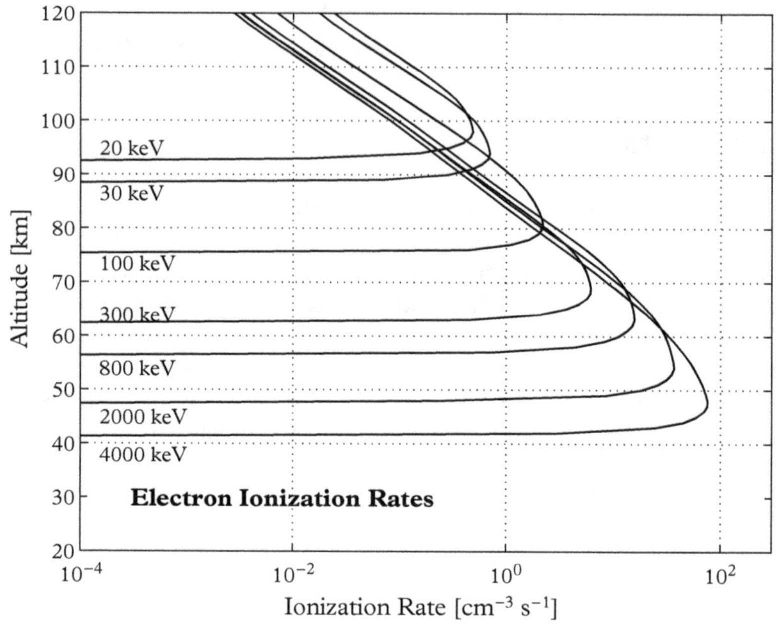

Figure 14.1 *Altitude versus ionization rate for mono-energetic electron beams with a flux of 100 electrons* $cm^{-2}s^{-1}sr^{-1}$.

14.2 Overview of coupling process

When considering the linkages between the radiation belts, polar atmosphere and climate, via electron precipitation driven by wave particle interactions, we are envisaging the flowchart shown in Fig. 14.2. The fundamental driver of the system can be thought of as the geomagnetic storm, where processes energize the radiation belt particles, either through, or as a consequence of, wave-particle interactions. What drives the geomagnetic storm has been a topic of many of the previous chapters in this book and space physics texts (e.g., Hargreaves, 1992), and thus we do not consider it further here.

As a result of the geomagnetic storm, both VLF and ULF waves are enhanced, which in turn, can enhance the transport of electrons within the outer radiation belt and also drive wave-particle interactions between energetic electrons and magnetospheric waves such as VLF chorus, plasmaspheric hiss, electromagnetic ion cyclotron waves (EMIC), and Pc5 micropulsations. These interactions are known to accelerate electrons from seed populations of ~30 keV to energies of >1 MeV while at the same time perturbing a fraction of the trapped population onto paths which are likely to interact with the Earth's atmosphere (e.g. Thorne, 2010, and references within; Omura

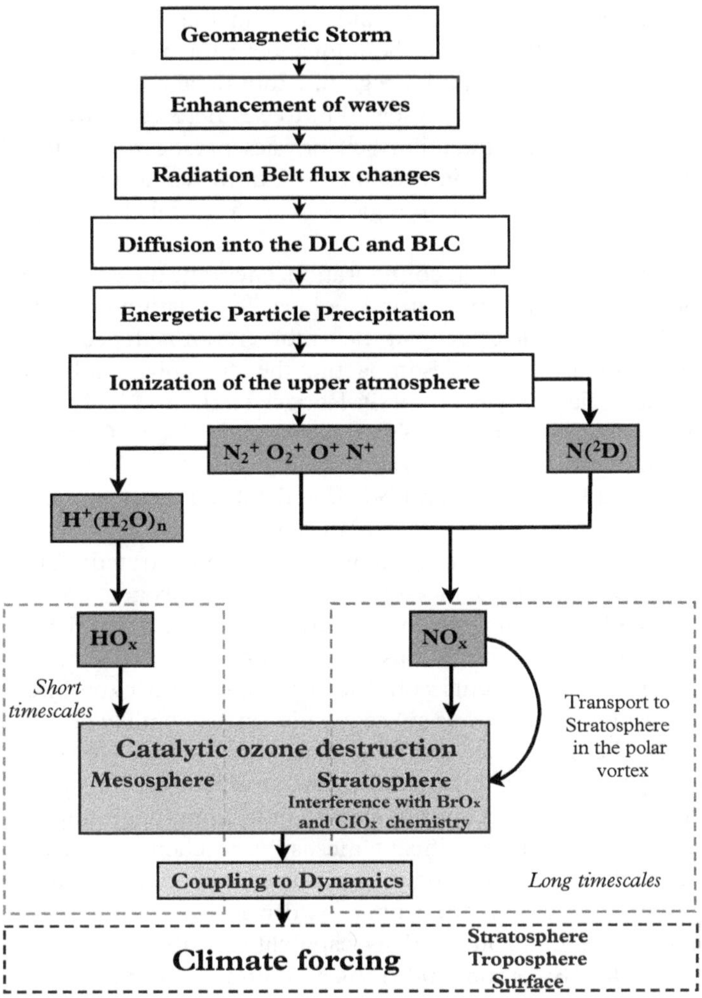

Figure 14.2 *Flowchart depicting the linkage from radiation belt to climate.*

and Zhao, 2013). Electrons which magnetically mirror at altitudes that are so low that they will hit the neutral atmosphere and be lost are said to be in the bounce-loss cone (BLC). This threshold altitude is generally set at ~100 km, although the literature include studies with thresholds ranging from 90 to 120 km. Mirroring altitudes vary with the strength of the Earth's magnetic field and thus with latitude and longitude. In and around the longitudes of the South Atlantic Magnetic Anomaly (SAMA) the mirror altitudes are at their lowest for a fixed geomagnetic latitude, and thus electrons that successfully mirror at other longitudes might still scatter in this region as they drift around the Earth. These electrons are said to be in the drift-loss cone (DLC) mirroring hundreds of kilometres above the atmosphere for most longitudes, but will be lost into the atmosphere in the SAMA-region. More detailed descriptions on the basic properties of the radiation belts may be found elsewhere (e.g. Walt, 1984; Spjeldvik and Rothwell, 1985). From a radiation belt physics perspective both BLC and DLC electrons are lost from the belts via precipitation, however, to the atmosphere the distinction is significant and can result in regions of preferential electron precipitation into the atmosphere.

Due to the Earth's magnetic field configuration, energetic particle precipitation occurs mainly in the polar auroral and sub-auroral regions, i.e., at latitudes higher than 45°. The altitudes at which these particles deposit their momentum is dependent on their energy spectrum, with lower energy particles impacting the atmosphere at higher altitudes than their more energetic relatives (Rees, 1989; Rodger et al., 2007). Precipitating charged particles produce NO_x ($NO_x = N + NO + NO_2$) and HO_x ($HO_x = H + OH + HO_2$) through ionization or dissociative ionization of N_2 and O_2 molecules, which results in the formation of N_2^+, O_2^+, N^+, O^+, and NO^+, and subsequent ion chemistry (Verronen and Lehmann, 2013). This is shown schematically in Fig. 14.2.

Modelling and experimental observations have demonstrated significant chemical alterations, particularly in the concentrations of odd nitrogen (NO_x) and odd hydrogen (HO_x), which can catalytically destroy ozone (Crutzen, 1970; Brasseur and Solomon, 2005). For some time it has been recognized that very intense energetic particle precipitation (EPP) events could lead to significant ozone destruction in the polar middle atmosphere, which was subsequently experimentally observed during solar proton events (e.g., Seppälä et al., 2004; Seppälä et al., 2006). However, there has also been growing evidence that geomagnetic storms produce high fluxes of energetic electron precipitation (e.g. Rodger et al., 2007), with modelling suggesting energetic electron precipitation can also lead to significant mesospheric chemical changes in the polar regions (Rodger et al., 2010) affecting the chemical makeup of the polar atmosphere.

The radiative balance of the atmosphere is driven by its composition, particularly through the abundance of ozone and its capability of absorbing solar UV radiation efficiently (Brasseur and Solomon, 2005). As NO_x and HO_x influence the ozone abundance at upper stratosphere and mesospheric levels, the radiative balance can be altered. Through this mechanism atmospheric wind patterns can be affected, influencing the propagation of atmospheric waves within the stratosphere and troposphere. As Fig. 14.2 shows, this step-by-step process provides a pathway to link geomagnetic storm perturbations in the radiation belts, with chemical changes in the upper atmosphere, and ultimately, with the dynamics of the lower atmosphere. In the next sections we will describe each of these links in more detail.

14.3 Waves driving precipitation

Previous chapters in this book have highlighted the types of waves that can energy exchange with outer radiation belt electrons through gyroresonant interactions.

Magnetospheric wave modes include whistler-mode chorus outside of the plasmasphere, electromagnetic ion cyclotron waves (EMIC) on or around the plasmapause, and whistler-mode hiss inside the plasmasphere. Whistler mode chorus can accelerate electrons to higher energies, while chorus, hiss and EMIC waves can all scatter electrons into the DLC and BLC through resonant pitch angle scattering. Other losses of radiation belt electrons can occur through radial diffusion and magnetopause shadowing. However, in this chapter we are particularly interested in the loss into the atmosphere of electrons as a result of pitch angle scattering. As an electron drifts in longitude it may well experience each of these wave modes within one orbit of the Earth, depending on the L-shell of the drift orbit. As a result, all of the wave modes may contribute to the overall rates of electron precipitation, with different contributions occurring for different electron energies (Horne, 2002). The schematic of the distributions of chorus, plasmaspheric hiss, and EMIC waves relative to the shape of the plasmasphere shown in Summers et al. (2007) encapsulates the picture we have in mind here.

Although the DLC is only a small fraction of the pitch angle distribution of electrons when compared with the BLC it does play an important role as far as the atmospheric impact of radiation belt electrons is concerned. When electrons are scattered towards the loss-cone from quasi-trapped pitch angles they will either diffuse deep into the BLC or remain close to the outer edge of the BLC, i.e., the DLC. In these cases the diffusion is known either as strong or weak (Kennel and Petschek, 1966). Calculations of bounce-averaged quasi-linear diffusion coefficients for chorus, hiss, and EMIC waves have shown that the scattering strength is dependent on the wave mode type, the electron energy, the amplitude and the distribution of the waves, and the plasma density distribution (Summers and Thorne, 2003; Summers et al., 2007). If strong diffusion dominates the pitch angle scattering process, driving electrons into the BLC, any precipitation will be independent of geographic longitude. However, if weak diffusion dominates, the DLC comes more into play, and a bias would be expected with geographic longitude, particularly in the southern hemisphere around the SAMA.

Key to the type and efficiency of pitch angle diffusion of any given wave mode are the amplitude and latitudinal distribution of the waves. Meredith et al. (2003) have shown the MLT and L-shell distributions of chorus for non-disturbed, moderately-disturbed, and severely-disturbed geomagnetic activity levels. Typically the regions of intense chorus waves increase in MLT and L-shell range as geomagnetic activity is enhanced. Chorus occurs in two frequency bands, spanning either side of the half electron gyrofrequency ($f_{ce}/2$). In Fig. 14.3 we show the lower band chorus intensity (0.1–0.5 f_{ce}) as a function of latitude and longitude as observed by the low attitude satellite DEMETER for quiet (Kp<4) and disturbed (Kp≥4) geomagnetic conditions. The intensity of the observed chorus waves is 2–3 orders of magnitude higher during periods of disturbed geomagnetic activity than during non-disturbed conditions. No significant variations in longitude are observed, and no hemispheric bias is present either. However, over the typical range of DEMETER observations (2<L<5) there is a noticeable increase in wave intensity with increasing geomagnetic latitude. Increasing wave intensity

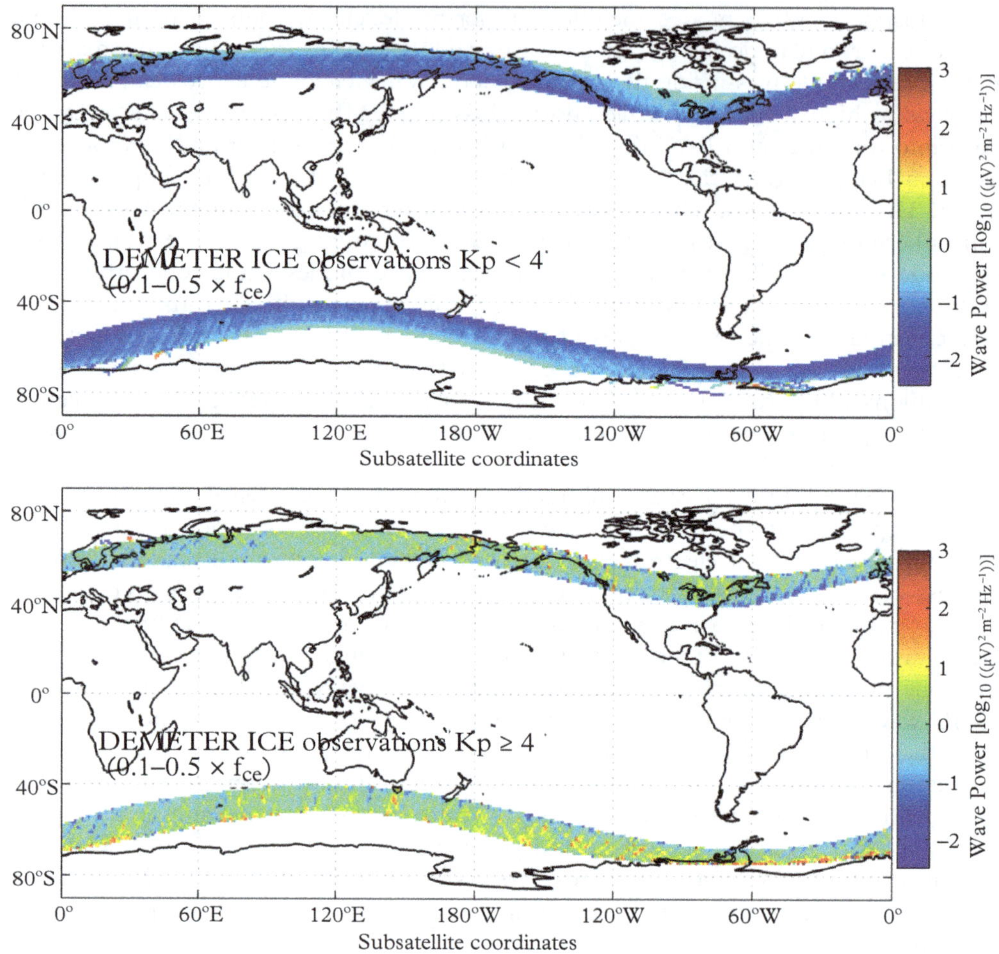

Figure 14.3 *DEMETER lower band chorus wave maps, showing variation with Kp, representing quiet and disturbed geomagnetic activity conditions.*

with geomagnetic activity is likely to change the pitch angle scattering regime from one of weak diffusion to strong diffusion, depending on the electron energy, and thus change the precipitation patterns from a South Atlantic DLC-dominated picture into a more uniformly distributed BLC-dominated picture.

14.4 Variations in precipitation

One way to study the linkage between geomagnetic activity, the dynamic nature of the outer radiation belt trapped electron fluxes, and electron precipitation into the

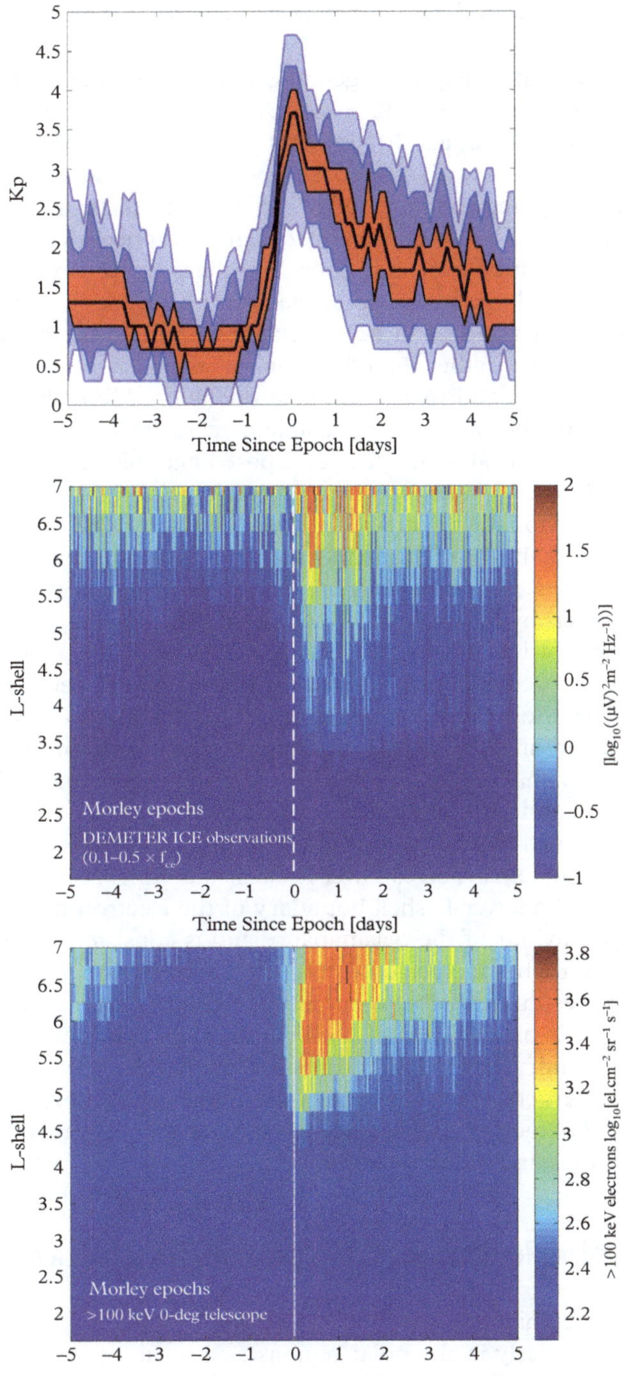

Figure 14.4 *DEMETER chorus wave and POES electron precipitation characteristics during the Morley superposed epoch geomagnetic storm events. The superposed epoch median of the quantity is given by a black line. The 95% confidence interval for the median is given by the red band. The blue bands mark the interquartile range and the 95% confidence interval about it (light blue).*

atmosphere, is through the analysis of geomagnetic storms triggered by the arrival of a high speed solar wind stream interface (SWSI) and associated dropouts in energetic electron fluxes (e.g., Morley et al., 2010b). In the Morley statistical studies, utilizing 9 GPS-borne particle detectors, a superposed epoch analysis around the arrival of 67 SWSIs showed a strong repeatable "signal" of a rapid electron flux dropout (Morley et al., 2010a) followed by a recovery with higher trapped electron fluxes. Hendry et al. (2012) used the Morley epochs to show that the dropouts in trapped flux are followed ~3 hours later by large increases of electron precipitation into the atmosphere. These precipitation events appear to be associated with the time that the trapped electron fluxes begin to recover rather than the dropouts themselves, and may signify that a simultaneous acceleration and loss process is in operation. We make use of the Morley epochs again in showing the link between geomagnetic storms, wave intensities, and electron precipitation. Figure 14.4 shows the superposed analysis of the geomagnetic index Kp, the DEMETER lower band chorus wave intensity, and the POES satellite observed >100 keV electron precipitation flux for the Morley epochs. The POES >100 keV electron precipitation fluxes show similar temporal and L-shell variations as the DEMETER wave panel. As expected the Kp index increases at epoch day 0 with a steady recovery back to pre-storm levels by about day 5. The DEMETER chorus wave intensities show an increase of about 2 orders of magnitude at epoch day 0, with a recovery lasting about 5 days. High electron precipitation fluxes are triggered by the geomagnetic storm and extend from L = 7 (the observational limit of DEMETER) to L = 4.5 at the peak of the geomagnetic disturbance, with magnitude increases of a factor of ~1.5. In this case the POES fluxes are zonally averaged, and provide no information about the relative contributions of BLC and DLC flux.

The L-shell over which the chorus waves are observed increases as a result of the geomagnetic storm, with an expansion to lower L-shells that mirrors the dynamic behaviour of the plasmapause. The lower L-shell boundary of the electron precipitation also mirrors the expected behaviour of the plasmapause and is consistent with the storm event analysed in detail by Lichtenberger et al. (2013). From Figure 14.4 we can infer that there is a strong correspondence between chorus wave intensity and electron precipitation flux, and that it is primarily associated with regions outside of the plasmasphere. Given the work of Hendry et al. (2012) it is possible that the waves are contributing to the acceleration of energetic electrons in the outer radiation belt, while at the same time enhancing the loss of a fraction of those electrons to the atmosphere, i.e., in the pumping up and subsequent draining of the radiation belts.

14.5 Atmospheric impact of electron precipitation

Energetic electron precipitation into the atmosphere results in the ionization of neutral constituents, particularly in the polar regions as a result of the configuration of the Earth's magnetic field. Strong ionization sources can enhance the ion pair production levels in the D-region above that provided by Lyman-alpha radiation and EUV (Brasseur and Solomon, 2005). Electron precipitation contributes to the production of odd nitrogen NO_x ($N + NO + NO_2$) through the dissociation of molecular nitrogen, and odd hydrogen HO_x ($H + OH + HO_2$) through chemistry associated with ion pair production, water cluster ion formation, and subsequent neutralization (Verronen and Lehmann, 2013). A schematic of this process is shown in Figure 14.2.

NO_x can have a long chemical lifetime (months) if it is produced in the dark polar wintertime when it can be transported to lower altitudes and latitudes, or if it's produced directly at stratospheric altitudes (Siskind et al., 1997). HO_x has a short chemical lifetime (hours) and so is generated by electron precipitation in situ, and only for as long as the ionization continues. Both NO_x and HO_x can destroy odd oxygen through catalytic reactions, and hence play an important role in the ozone balance of the middle atmosphere. As NO_x and HO_x have such different characteristics, we consider them separately in the following sections.

14.5.1 Odd nitrogen

When we consider NO_x that is created by energetic electron precipitation, we are primarily interested in that produced in the polar regions during the dark wintertime. In these conditions the NO_x gases can survive for long periods, and be transported significant distances, particularly horizontally by winds within the stratospheric polar vortex, and vertically by subsiding air masses at the winter pole. As a result of these transport conditions, it is difficult to separate the contribution of in situ direct formation of NO_x by electron precipitation, and indirect NO_x being transported from another region where it could have been generated at a different altitude by electron precipitation with different energy.

A simple picture would be of a wintertime column of enhanced NO_x generated by particle precipitation from 40 to 120 km, and confined to a region polewards of 60° (geographic latitude) by the stratospheric polar vortex. At the summer pole the abundance of NO_x would be lower due to photodissociation from solar radiation, and the absence of the polar vortex. This picture is complicated by the time varying nature of electron precipitation due to geomagnetic storms, the radiation belt processes that influence the type of wave modes that act to scatter electrons, and the dynamic behaviour of the stratospheric polar vortex (Solomon et al., 1999; Harvey et al., 2009).

Direct NO_x production by radiation belt electron precipitation has been observed by the sub-mm radiometer at Troll, Antarctica, during the winter of 2008 (Newnham et al., 2011). NO was observed at 70–80 km altitudes at L=4.5 following a geomagnetic storm, and was well correlated with ~300 keV electron precipitation energies observed by POES. In a superposed epoch analysis of recurrent geomagnetic storms at the same site Newnham et al. (2013) confirmed the direct production of 70–80 km NO by radiation belt electron precipitation, but also observed a second NO signature at 70–80 km ~7 days after the geomagnetic storm. The second pulse of NO could be a result of downward transport of auroral NO_x created by lower energy electron precipitation, or horizontal transport. Given the long lifetime of NO_x in wintertime conditions it is difficult to be certain of the ionization source without high quality modelling of atmospheric transport conditions.

Indirect NO_x has been observed by satellites, and its transport to lower altitudes studied extensively. Randall et al. (2005, 2007, 2009) analysed several indirect NO_x events that occurred during the northern hemisphere wintertime. Enhancements of NO_x were observed to descend from ~80 km altitude to ~30 km altitude over several months, with descent rates of 5–10 km/month. Although an overall descent of winter polar air is expected, the enhanced NO_x was particularly associated with sudden stratospheric warmings (SSW), where the stratospheric polar vortex is disrupted and then formed again, with the stratopause occurring at much higher altitudes than normal. Thus the enhanced NO_x is likely to have originated at altitudes as high as the aurora at >100 km. The electron precipitation that generates the NO_x under these circumstances is likely to be relatively soft, with energies of <30 keV, or possibly auroral energies of only 1 keV or so. However, when high energy electron precipitation events occur during an indirect NO_x descent period it will add to the abundance of NO_x at the appropriate altitude for the energy, and be transported down to the stratosphere as well (Clilverd et al., 2009).

There are significant differences in the occurrence of indirect NO_x events in the northern and southern hemisphere. In the southern hemisphere the stratospheric polar vortex forms regularly each winter and transports NO_x down into the stratosphere in most years (Randall et al., 2007). In the northern hemisphere the wintertime polar vortex is affected by the highly varying dynamical conditions and is often disrupted during the course of the winter. As a result indirect NO_x enhancements are only intermittently observed during the winter in the northern hemisphere (Seppälä et al., 2007; Randall et al., 2009).

In the stratosphere NO_x catalytically destroys ozone. The combination of direct and indirect NO_x sources makes attribution of stratospheric ozone changes more challenging, particularly when considering the influence of radiation-belt processes in preference to auroral production. However, Clilverd et al. (2009) reported the observation of enhanced direct NO_x at ~50 km generated by relativistic electron precipitation (~2 MeV), and the subsequent reduction of ozone at altitudes as low as 40 km. Randall et al. (2005) showed that large enhancements of NO_x transported downwards in the northern hemisphere polar vortex caused long lasting ozone reductions at 40 km altitude from March to at least July 2004. The NO_x was generated at mesospheric and upper stratospheric altitudes as a result of large geomagnetic storms in October and November 2003, with contributing geomagnetic activity continuing into January 2004. Vertical transport times for the NO_x to travel from ~80 km to 40 km altitudes were about 3 months, and the NO_x influence on ozone lasted for a further 4 months, indicating that NO_x effects can be long-lasting and significantly displaced from the generation region.

Rodger et al. (2010) showed that direct radiation-belt electron precipitation occurring at L = 3 during a large geomagnetic storm could increase NO_x abundance by 80%

at 60–75 km altitudes, and decrease ozone by up to 35%, with the effect lasting beyond a week. The calculations were undertaken by the Sodankylä Ion and Neutral Chemistry model (SIC), which includes a comprehensive description of the ion chemistry involved (Verronen et al., 2005). The results showed that the large initial short-term ozone depletion peaks were caused by HO_x, with a smaller, but longer lasting ozone loss arising from the NO_x enhancements. Figure 14.5 shows the change in ionization from 50–100 km due to the 70 keV–2.5 MeV electron precipitation fluxes that were calculated during a large geomagnetic storm (–130 nT) that occurred in September 2005, the resulting NOx enhancements, and consequent ozone reductions that would be produced at L = 3 in the southern hemisphere if polar winter conditions (i.e., if the storm occurred in July) were applied. The level of ozone reduction in this case is equivalent to that of a large solar proton event, lasts longer than a typical solar proton event, and has the potential to occur more frequently throughout the 11-year solar activity cycle (Rodger et al., 2010).

More recently Daae et al. (2012) reported the ground-based observation of an enhancement of direct NO_x generated in situ by electron precipitation (~300 keV) during a geomagnetic storm in 2009, with a co-incident 17–67% reduction of ozone abundance between 70–75 km. It is likely that much of the initial ozone depletion was caused by HO_x generated at the same time as the NO_x (Verronen et al., 2013), although only NO_x is able to survive long enough to descend towards the stratosphere influencing ozone abundance along the way, as was observed. Attribution of the influence of energetic electron precipitation on ozone is one of a combined role of NO_x and HO_x production during the precipitation event, with NO_x having a longer-term influence after the precipitation ceases, involving horizontal and vertical transport. The production of HO_x by energetic electron precipitation is discussed in more detail in the next section.

14.5.2 Odd Hydrogen

In polar regions energetic electron precipitation can enhance HO_x as a result of production through ionization and water cluster ion chemistry (Verronen and Lehmann, 2013). Below altitudes of 80 km there is enough H_2O for water cluster ion formation, and at these altitudes the chemical lifetime of HO_x is up to a day. Thus HO_x concentrations respond rapidly to electron precipitation events, and are not subject to the influence of significant horizontal or vertical transport once the precipitation forcing finishes. Taking advantage of this property means that HO_x can be used to monitor the regions and altitudes over which electron precipitation from the outer radiation belt occurs.

The first observation of OH production due to energetic electron precipitation was published by Verronen et al. (2011). They found a strong correlation between 100 and 300 keV loss cone electron count rates observed in the outer radiation belt by POES MEPED, and the night-time OH concentrations in the middle mesosphere at 71–78 km altitude observed by the Microwave Limb Sounder (MLS) on board the Aura satellite. At 55–65° magnetic latitude (L = 3.0–5.6) increases in electron precipitation by

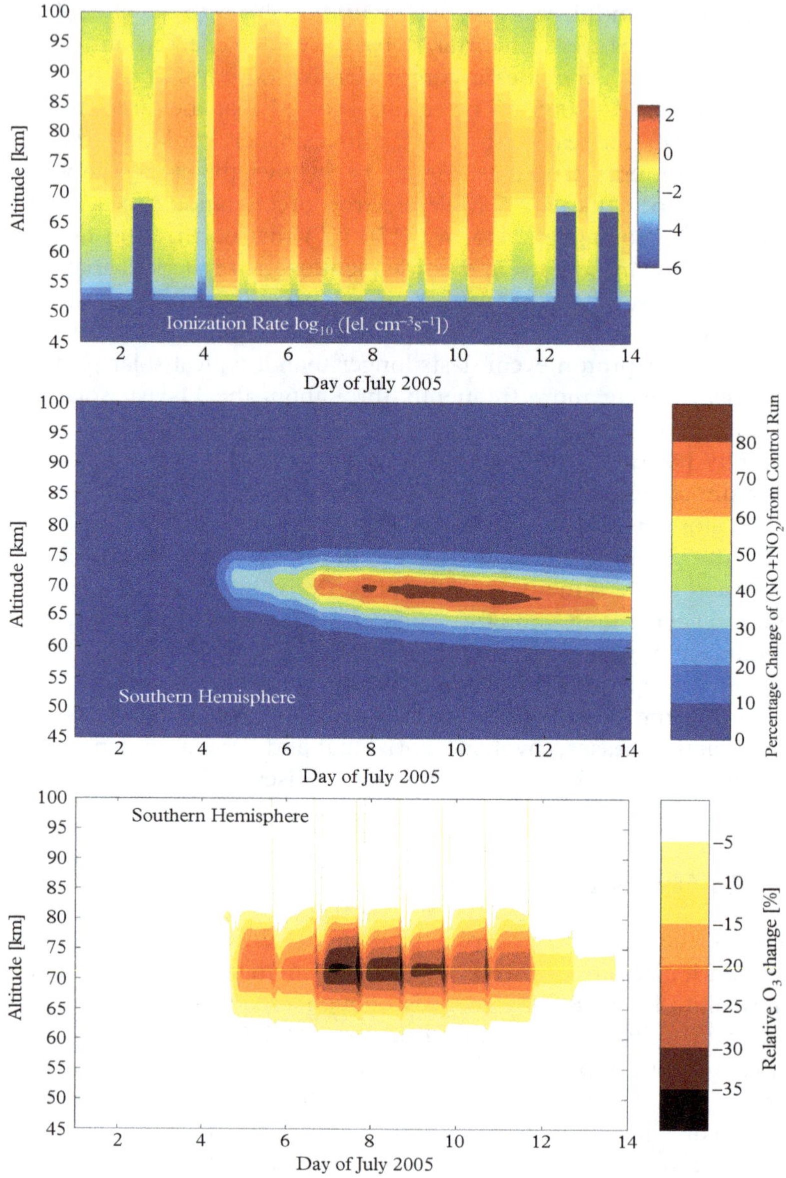

Figure 14.5 *Modelling the impact of radiation-belt electron precipitation between 45 and 100 km during an electron precipitation event at L = 3 in wintertime in the southern hemisphere. Top panel: ionization levels, middle panel: NO_x abundance, and bottom panel: O_3 destruction.*

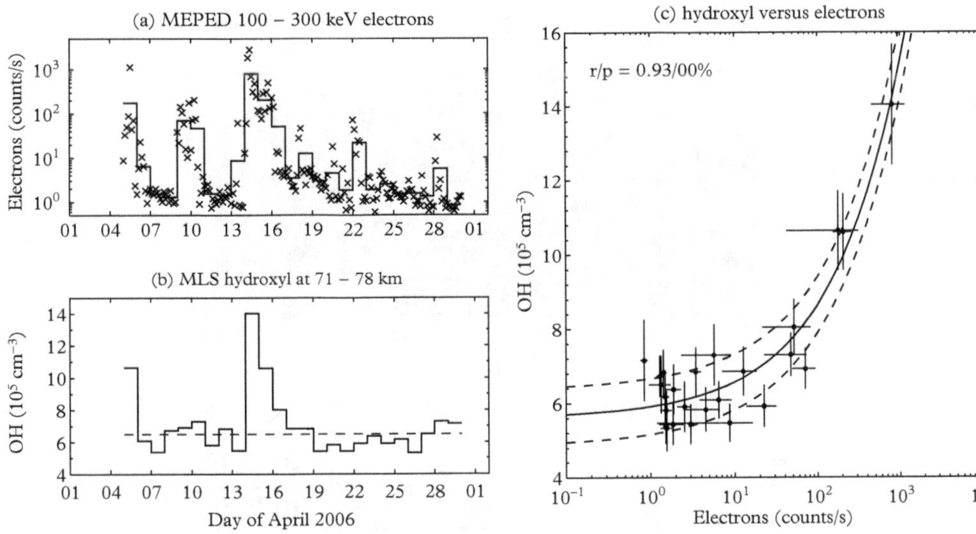

Figure 14.6 *OH concentration at 71–78 km versus electron count rate during April 2006. (a) 3-hour (X) and daily (solid line) mean electron count rates measured by MEPED. (b) Daily mean OH concentrations measured by MLS. (c) Daily mean OH versus daily mean electrons (dots). Horizontal and vertical lines mark the standard error of the mean. The solid and dashed lines mark a line fit and its estimated standard deviation. r is the correlation coefficient and p is the related random chance probability.*

two orders of magnitude were accompanied by increases in night-time OH concentration of 100%. In Fig. 14.6 we summarize the findings made by Verronen et al. (2011) showing the POES MEPED and MLS data from April 2006. A large geomagnetic storm occurred on 13 April which was followed by high fluxes of 100–300 keV electron precipitation into the atmosphere. High concentrations of OH were observed at 70–78 km altitudes, which as Figure 14.6 shows, is consistent with in situ ionization from 100–300 keV electron precipitation. The relationship between the two parameters is given in the panel on the right, with a linear fit (solid line) and standard deviations (dashed lines) shown. In this monthly case, over 85% (r^2) of the OH daily variation could be explained by changes in radiation belt electron precipitation.

Andersson et al. (2012) extended the study period to 2004–2009, also comparing POES and MLS nighttime data on a month by month basis. Over the 55–65° magnetic latitude range (L = 3.0–5.6) high correlation between POES electrons and MLS OH was found when high fluxes of 100–300 keV precipitating electrons were observed, while 34% of the 65 months analysed showed clear correlation. The highest correlations were found at 75 km altitudes, although the influence on OH production by electron precipitation was detected as low as 52 km, equivalent to fluxes of ~3 MeV electrons. Similar results were found in both hemispheres, and clear effects on OH production

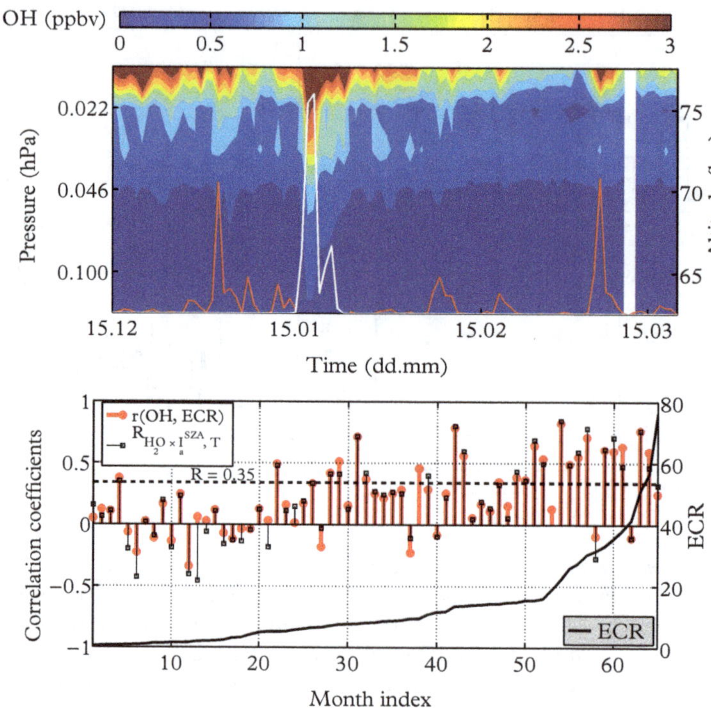

Figure 14.7 *Left panel: Daily mean OH mixing ratio (ppbv) at geomagnetic latitudes 55–65° shown for altitudes 63–78 km for the period 15 December 2004 to 15 March 2005. Daily mean electron count rate (ECR) is marked with a red line and daily mean proton fluxes are marked as white line. Right panel: Correlation r (red bar) and partial correlation (black line) between OH and ECR at 55–65° shown for altitude 75 km (0.022 hPa) sorted in ascending order by monthly mean ECR.*

were found over the 55–72° magnetic latitude range (L = 3.0–10). Thus, radiation belt forcing of chemical change in the upper atmosphere (50–80 km) was confirmed and shown to be a significant influence on the overall variability of HO_x at mid-high latitudes in both hemispheres. Figure 14.7 provides an example of the analysis undertaken by Andersson et al. (2012) showing 3 months of MLS OH observations (left panel) and the month by month correlation analysis, with the months sorted by electron count rate (ECR) on the 100–300 keV range.

The analysis of MLS OH observations has also been used to investigate the impact of weak and strong scattering processes that take place in the outer radiation belt. Andersson et al. (2013) showed that during periods of high POES 100–300 keV electron count rate more OH was produced in the southern hemisphere than the northern hemisphere, and that the majority in the south occurred in the region of the Weddell

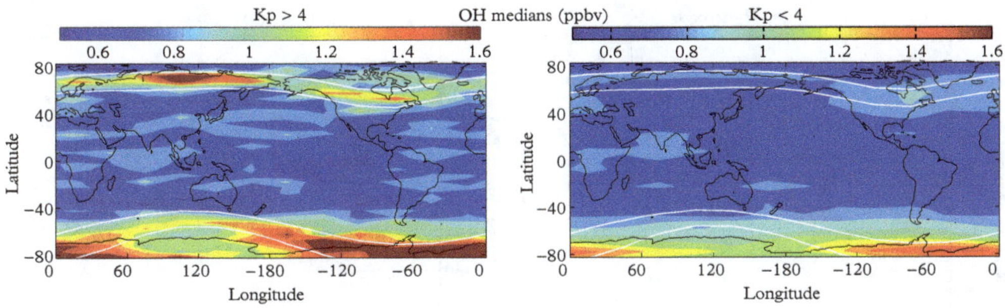

Figure 14.8 *Maps of OH volume mixing ratio for high (Kp>4) and low (Kp<4) geomagnetic activity levels. L = 3 and L = 10 contours are shown.*

Sea poleward of the SAMA (–120° to 60°E). As discussed earlier in this chapter, the bias towards the Weddell Sea region is indicative of the dominance of weak diffusion processes preferentially pushing electrons into the DLC. Figure 14.8 shows the geographic variation of MLS nighttime OH concentration at 75 km, with the left panel representing disturbed geomagnetic conditions (Kp>4) and the right panel representing quiet geomagnetic conditions (Kp<4). L = 3 and L = 10 contours are plotted for both hemispheres as white lines, representing the footprint of the outer radiation belt super-imposed on the atmosphere. From Fig. 14.8 we can take two things: 1) during quiet geomagnetic conditions there is still some OH generation in the Weddell Sea region suggesting a background drizzle of electron precipitation from the outer radiation belt; 2) during high geomagnetic conditions the OH is observed over an almost continuous range of longitudes (particularly when both hemispheres are combined) and this suggests that strong diffusion into the BLC is taking place.

Once the HO_x has been generated by energetic electron precipitation it has the ability to catalytically destroy ozone. Verronen et al. (2013) used the daily mean electron energy flux spectrum, determined from POES and DEMETER electron precipitation detectors, to constrain the SIC model. The ionization rate enhancements produced by the model were able to reproduce the day-to-day variability of OH and ozone observed by MLS during four large precipitation events. Figure 14.9 shows the modelling results from SIC during an example geomagnetic storm in January 2005, at 74 km in the northern hemisphere. Overall, the model predicted OH increases at 60–80 km, reaching several hundred percent at 70–80 km during peak electron precipitation forcing. Increases in OH were followed by ozone depletion, of several tens of percent. The magnitude of modeled changes were not only similar to those observed by MLS, but were large enough to be comparable to the effects of individual solar proton events.

14.6 Linkages to polar surface climate

Changes in the abundance of ozone in the stratosphere and mesosphere (30–100 km) can lead to changes in the radiative balance of the atmosphere. In this section we describe the analysis that has lead to an increased understanding of the way in which radiation

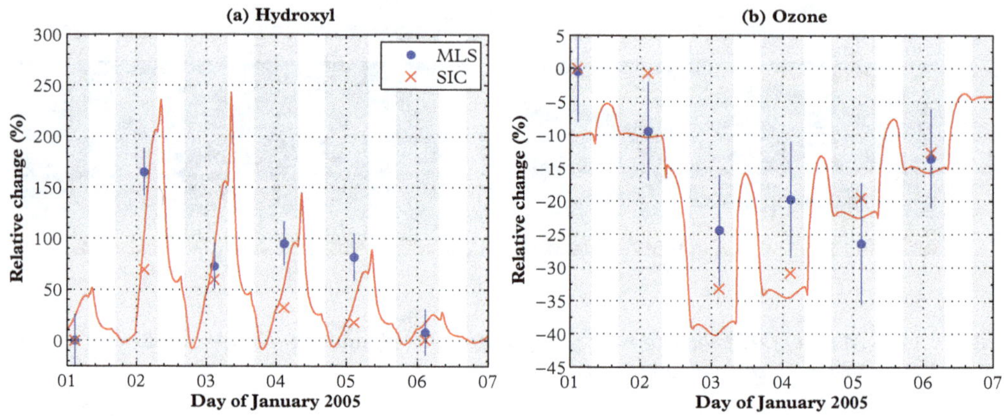

Figure 14.9 *Comparison between modelled and observed relative change of OH (a) and ozone (b) at 74 km for January NH. Red Line: SIC data showing 100 × (EPR/CTR −1), where EPR and CTR are gas concentrations from the electron and control runs, respectively. Red X marks: Same as Red Line, except that CTR is replaced by 1st-day result from the electron precipitation run, and shown only at the local solar time of MLS observations. Blue Circles: MLS data showing the change with respect to the observations on the day before the electron forcing peak. Grey shading marks the local times with solar zenith angle >100°, i.e. nighttime.*

belt processes not only link to the polar atmosphere, but could ultimately couple to polar climate.

14.6.1 Influence on polar climate

The influence of NO_x and HO_x chemistry on polar surface temperatures was first suggested by modelling results presented by Rozanov et al. (2005). The chemistry–climate model, UIUC CCM, was run with and without a NO_x source, generated by energetic particle precipitation between the upper stratosphere and lower-thermosphere. The resultant polar surface temperature changes were strongest during the winter months, and showed complex positive and negative temperature patterns which were different in each hemisphere. The conclusions of the study were that NO_x generated at high altitude by energetic electron precipitation was able to influence surface temperatures through the coupling of chemical and dynamical changes in the middle and upper atmosphere. This prediction was confirmed by Seppälä et al. (2009) who showed that the polar surface temperature differences found in ERA-40 re-analysis data, grouped into two subsets consisting of high and low geomagnetic activity, were consistent with Rozanov et al. (2005). The temperature ranges involved were ±5 K, with some regions experiencing warmer winters when geomagnetic activity was high, and some experiencing cooler winters. Seppälä et al. (2009) also confirmed these regional findings with winter temperature analysis from individual meteorological sites. Figure 14.10 shows the wintertime northern and southern hemisphere surface temperature differences for high geomagnetic activity (i.e., high Ap)–low Ap. In the northern hemisphere the figure shows warming

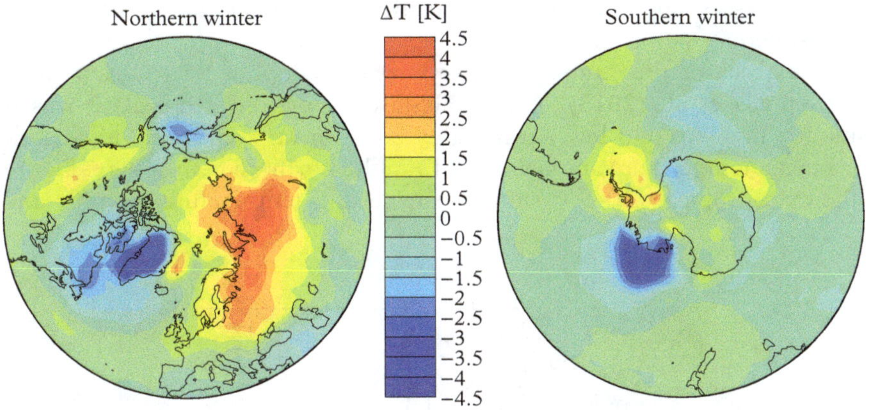

Figure 14.10 *Wintertime northern and southern hemisphere surface temperature differences for high geomagnetic activity (i.e., high A_p)–low A_p.*

over Russia and Canada, and cooling over Greenland and the Bering Sea. In the southern hemisphere the figure shows warming over the Antarctic Peninsula, and cooling over the Bellingshausen Sea. Other regions show less structured variations.

These findings were confirmed by Baumgaertner et al. (2011) in a comprehensive analysis using the ECHAM5/MESSy Atmospheric Chemistry (EMAC) model. ECHAM5/MESSy includes submodels describing tropospheric and middle atmosphere processes and their interaction with oceans, land and human influences (Jöckel et al., 2006). Baumgaertner et al. (2011) concluded that similar surface temperature patterns were found and were related to NO_x production due to increased geomagnetic activity and resultant energetic electron precipitation. The NO_x production leads to ozone depletion in the stratosphere which causes changes in the radiative budget and the mean meridional circulation. In the presence of sunlight ozone absorbs solar short-wave radiation resulting in heating of the atmosphere. An opposing effect is present in the absence of sunlight, such as during the dark polar winter, when ozone absorbs long-wave radiation from the surface, with the different wavelength range leading to a cooling effect (Langematz et al., 2003). Ozone changes can therefore potentially lead to temperatures changes even during the polar night influencing the strength of the polar vortex (Langematz et al., 2003). Associated positive stratospheric Northern Annular Mode (NAM) anomalies (and Southern Annular Mode SAM anomalies) propagate into the troposphere, where typical positive NAM/SAM surface pressure and temperature patterns occur, as were found by Seppälä et al. (2009). NAM and SAM are large scale patterns of climate variability arising from internal variation in atmospheric dynamics and they are recognized as the most important climate variability patterns at mid- and high latitudes, e.g. in horizontal and vertical wind and temperature fields. Enhanced geomagnetic activity and NO_x production appear to trigger positive NAM/SAM phases at the surface which are constrained to wintertime in the polar regions because of radiation belt source processes, and the complex interaction of chemistry and dynamical coupling in the atmosphere.

Further analysis of the stratosphere-troposphere coupling mechanism was undertaken by Seppälä et al. (2013). A re-analysis dataset from 1958–2012 was used to determine the high Ap and low Ap monthly zonal mean zonal wind anomalies (ΔU) and zonal mean temperature anomalies (ΔT) in the northern hemisphere. Figure 14.11 shows the extra-tropical (20–90° latitudes) wind and temperature monthly anomalies

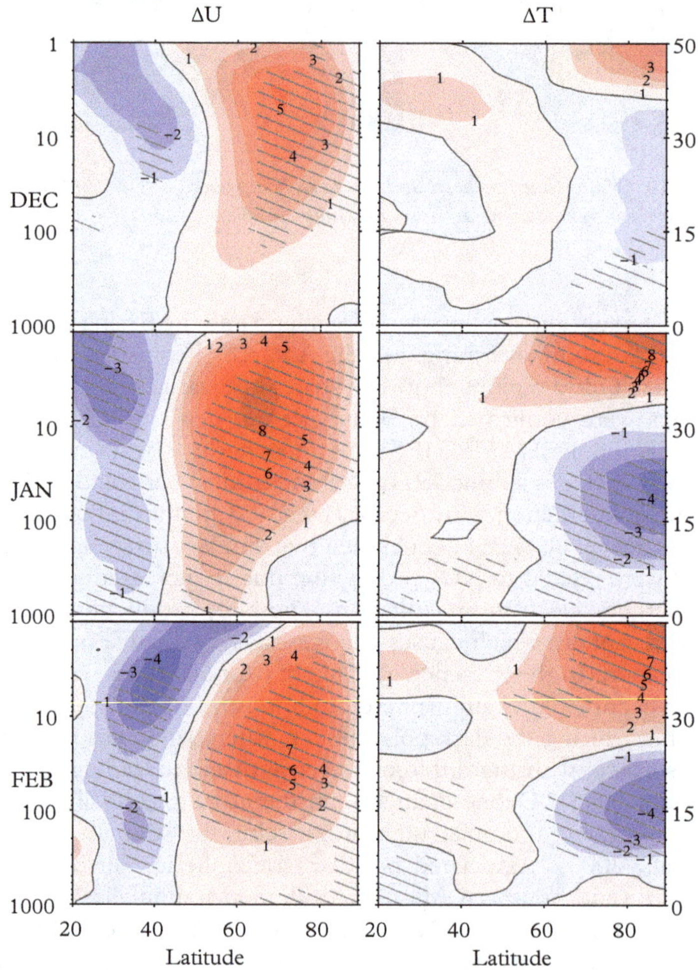

Figure 14.11 *Monthly zonal mean zonal wind anomalies (ΔU) and zonal mean temperature anomalies (ΔT) in the northern hemisphere for high A_p conditions. The atmospheric pressure levels in hPa are shown on the left and corresponding altitudes in km on the right. The shading corresponds to ≥90 % statistical significance levels.*

for December, January, and February for high Ap conditions. A region of strengthening zonal wind anomalies (more eastward) and warmer temperatures appear in the upper stratosphere and are observed to slowly descend in altitude at high latitudes. As a result, a stronger polar night jet persists longer than normal, with anomalously high polar winds across the stratosphere, i.e., a stronger polar vortex, resulting in positive NAM anomalies.

14.6.2 Possible significance to regional weather variability

Regional climate variability of the extra-tropical Northern Hemisphere (NH) is largely controlled by the NAM in the stratosphere, and its counter-part in the troposphere, the North Atlantic Oscillation (NAO). Under certain conditions, the low frequency effect of the NAM spans from the stratosphere into the troposphere (Baldwin et al., 2003). Near the surface and over the North Atlantic region, the NAO and the NAM are almost indistinguishable from each other. The NAO is a large-scale mode (i.e., pattern) of climate variability that has major impact on the weather and climate either side of the North Atlantic, especially north-western Europe. Downward coupling of stratospheric circulation changes to the surface accounts for the majority of change in regional surface climate over Europe and North America (Scaife et al., 2005). When the NAO/NAM is positive, there is a net displacement of Arctic air towards the subtropics near the Azores. This strengthens westerly winds in the North Atlantic resulting in warmer moist air flowing towards the adjacent regions, and milder maritime winters. In a negative NAM/NAO air displaces towards the Arctic, bringing weaker westerlies over the North Atlantic, and colder than normal maritime winters.

Stronger and more significant correlations have been found to exist between geomagnetic activity and the NAO than with the 11-yr solar UV cycle (Bochníček and Hejda, 2005), and solar wind dynamic pressure has been shown to be correlated with the winter NAM (Lu et al., 2008). To study the relationship between geomagnetic forcing and the NAO during the northern hemisphere winter, Li et al. (2011) analysed the 150-year time series of the geomagnetic aa index. Over the whole 150 year period studied only about 8% of the variance in wintertime NAO is explained by the wintertime aa index. However, the aa–NAO relationship becomes clearer and more robust statistically when only the data from the declining phase of even-numbered 11-year solar cycles are included. This may be related to the dominant polarity of the inter-planetary magnetic field with respect to the ecliptic plane, as energy shifts between the Sun's toroidal and poloidal magnetic fields. In this case 34% of the variance of the winter NAO can be explained by the geomagnetic aa index. Thus, in the northern hemisphere, geomagnetic forcing of NAO weather patterns is strongly associated with the declining phase of the 11-year solar cycle. This phase of the cycle is characterized by recurrent geomagnetic activity which drives large changes in outer radiation belt acceleration and loss processes, involving storms which are of long duration, and produce high fluxes of relativistic electrons (Borovsky and Denton, 2006). All these observations are consistent with the idea that radiation belt electron precipitation fluxes play a role in atmospheric chemistry and dynamics, ultimately linking to surface climate variability.

In the southern hemisphere the equivalent to the NAM is the southern annular mode (SAM). The coupling of radiation belt electron precipitation into the upper atmosphere has many similarities with the northern hemisphere, although key processes show differences, such as the increased stability of the wintertime polar vortex over Antarctica. Seppälä et al. (2009) found clear indications of geomagnetic activity impact on polar surface temperatures, with patterns that are consistent with the modulation of the SAM. Other work has shown the influence of the SAM on rainfall precipitation over New Zealand, and thus indicates the potential of geomagnetic activity influence on southern hemisphere regional weather variability (Renwick and Thompson, 2006).

As outlined throughout this chapter, wave-particle interaction processes in the outer radiation belt scatter energetic electrons into the atmosphere creating excess ionization. The processes influence the locations and altitudes of the chemical changes, through defining the electron energies involved, the location of the wave-particle interaction regions relative to the plasmapause, and the strength of the pitch-angle diffusion taking place. The ionization produced in the upper atmosphere results in the generation of NO_x and HO_x species, which can catalytically destroy ozone. Resultant changes in atmospheric dynamics propagate down into the troposphere and modulate the wintertime weather patterns in the northern and southern hemispheres. In recent years the linkage between the radiation belts, energetic electron loss processes, the atmosphere, and regional climate variability has been investigated with increasing understanding. However, details about the exact coupling mechanism, and its impact on climate, still need to be fully determined.

Acknowledgments

M. A. Clilverd was supported by the NERC project NE/J008125/1. C. J. Rodger was supported by the New Zealand Marsden Fund. M. E. Andersson and P. T. Verronen were supported by the Academy of Finland (Projects #136225 and #140888). A. Seppälä was supported by the Academy of Finland (Projects #258165 and #265005).

REFERENCES

Andersson, M. E., Verronen, P. T., Rodger, C. J., Clilverd, M. A., and Wang, S. (2013). *Atmos. Chem. Phys. Discuss.*, **13**(7), 19895–19919.

Andersson, M. E., Verronen, P. T., Wang, S., Rodger, C. J., Clilverd, M. A., and Carson, B. R. (2012). *J. Geophys. Res.*, **117**, D09304.

Baker, D. N., Kanekal, S. G., Li, X., Monk, S. P., Goldstein, J., and Burch, J. L. (2004). *Nature*, **432**, 878–881.

Baldwin, M. P., Stephenson, D. B., Thompson, D. W. J., Dunkerton, T. J., Charlton, A. J., and O'Neill, A. (2003). *Science*, **301**, 636–640.

Baumgaertner, A. J. G., Seppälä, A., Joeckel, P., and Clilverd, M. A. (2011). *Atmos. Chem. Phys.*, **11**(9), 4521–4531.

Bochníček, J. and Hejda, P. (2005). *J. Atmos. Sol. Terr. Phys.*, **67**(1–2), 17–32.

Borovsky, J. E. and Denton, M. H. (2006). *J. Geophys. Res.*, **111**, A07S08.

Brasseur, G. P. and Solomon, S. (2005). *Aeronomy of the Middle Atmosphere* (3rd revised and enlarged edn). Springer, Dordrecht.

Clilverd, M. A., et al. (2012). *J. Geophys. Res.*, **117**, A12213.

Clilverd, M. A., Seppälä, A., Rodger, C. J., Mlynczak, M. G., and Kozyra, J. U. (2009). *J. Geophys. Res.*, **114**, A04305.

Crutzen, P. J. (1970). *Q. J. R. Meteorol. Soc.*, **96**, 320–325.

Daae, M., Espy, P. J., Nesse Tyssøy, H., Newnham, D. A., Stadsnes, J., and Søraas, F. (2012). *Geophys. Res. Lett.*, **39**, L21811.

Green, J. C., Onsager, T. G., O'Brien, T. P., and Baker, D. N. (2004). *J. Geophys. Res.*, **109**, A12211.

Hargreaves, J. K. (1992). *The solar-terrestrial environment.* Cambridge Atmospheric and Space Science Series. Cambridge University Press, Cambridge, UK.

Harvey, V. L., Randall, C. E., and Hitchman, M. H. (2009). *J. Geophys. Res.*, **114**, D22105.

Hendry, A. T., Rodger, C. J., Clilverd, M. A., Thomson, N. R., Morley, S. K., and Raita, T. (2012). *Geophys. Monogr. Ser.*, **199**, 213–223.

Horne, R. B. (2002). In *The contribution of wave-particle interactions to electron loss and acceleration in the Earth's radiation belts during geomagnetic storms* (ed. W. R. Stone), pp. 801–828.

Horne, R. B., et al. (2005). *Nature*, **437**, 227–230.

Jöckel, P., et al. (2006). *Atmos. Chem. Phys.*, **6**, 5067–5104.

Kennel, C. F. and Petschek, H. E. (1966). *J. Geophys. Res.: Space Physics (1978–2012)*, **71**(1), 1–28.

Lam, H. L., Boteler, D. H., Burlton, B., and Evans, J. (2012). *Space Weather*, **10**, S10003.

Langematz, U., Kunze, M., Krüger, K., Labitzke, K., and Roff, G. L. (2003). *J. Geophys. Res.*, **108**, 4027.

Li, Y., Lu, H., Jarvis, M. J., Clilverd, M. A., and Bates, B. (2011). *J. Geophys. Res.*, **116**, 16109.

Lichtenberger, J., et al. (2013). *J. Space Weather Space Clim.*, **3**, A23.

Lu, H., Jarvis, M. J., and Hibbins, R. E. (2008). *J. Geophys. Res.*, **113**(D23).

Meredith, N. P., Horne, R. B., Thorne, R. M., and Anderson, R. R. (2003). *Geophys. Res. Lett.*, **30**(16), 1871.

Millan, R. M. and Baker, D. N. (2012). *Space Sci. Rev.*, **173**, 103–131.

Millan, R. M. and Thorne, R. M. (2007). *J. Atmos. Sol. Terr. Phys.*, **69**(3), 362–377.

Morley, S. K., Friedel, R. H. W., Cayton, T. E., and Noveroske, E. (2010a). *Geophys. Res. Lett.*, **37**, L06102.

Morley, S. K., et al. (2010b). *Proceedings of the Royal Society A: Mathematical, Physical and Engineering Science*, **466**, 3329–3350.

Newnham, D. A., et al. (2011). *Geophys. Res. Lett.*, **38**, L20104.

Newnham, D. A., et al. (2013). *J. Geophys. Res.*, **118**, 7874–7885.

Omura, Y. and Zhao, Q. (2013). *J. Geophys. Res.*, **118**(8), 5008–5020.

Randall, C. E. et al. (2005). *Geophys. Res. Lett.*, **32**, L05802.

Randall, C. E., et al. (2007). *J. Geophys. Res.*, **112**, D08308.

Randall, C. E., et al. (2009). *Geophys. Res. Lett.*, **36**, L18811.

Rees, M. H. (1989). *Physics and chemistry of the upper atmosphere.* Cambridge atmospheric and space science series. Cambridge University Press, Cambridge, UK.

Reeves, G. D., McAdams, K. L., Friedel, R. H. W., and O'Brien, T. P. (2003). *Geophys. Res. Lett.*, **30**, 1529.

Renwick, J. and Thompson, D. (2006). *Water Atmos*, **14**(2).

Rodger, C. J., et al. (2010). *J. Geophys. Res.*, **115**, A11320.

Rodger, C. J., et al. (2007). *J. Geophys. Res.*, **112**, 11307.
Rozanov, E., Callis, L. B., Schlesinger, M., Yang, F., Andronova, N., and Zubov, V. (2005). *Geophys. Res. Lett.*, **32**, L14811.
Scaife, A. A., Knight, J. R., Vallis, G. K., and Folland, C. K. (2005). *Geophys. Res. Lett.*, **32**(18), L18715.
Seppälä, A., Lu, H., Clilverd, M. A., and Rodger, C. J. (2013). *J. Geophys. Res.*, **118**(5), 2169–2183.
Seppälä, A., Randall, C. E., Clilverd, M. A., Rozanov, E., and Rodger, C. J. (2009). *J. Geophys. Res.*, **114**, A10312.
Seppälä, A., et al. (2007). *Geophys. Res. Lett.*, **34**(12), L12810.
Seppälä, A., et al. (2004). *Geophys. Res. Lett.*, **31**(19), L19107.
Seppälä, A., et al. (2006). *Geophys. Res. Lett.*, **33**(7), L07804.
Simpson, I. R., Blackburn, M., and Haigh, J. D. (2009). *J. Atmos. Sci.*, **66**, 1347.
Siskind, D. E., Bacmeister, J. T., Summers, M. E., and Russell, J. M. III (1997). *J. Geophys. Res.*, **102**(D3), 3527–3545.
Solomon, S. C., Barth, C. A., and Bailey, S. M. (1999). *Geophys. Res. Lett.*, **26**(9), 1259–1262.
Spjeldvik, W. N. and Rothwell, P. L. (1985). In *Handbook of Geophysics and the Space Environment* (ed. A. S. Jursa). Air Force Geophysics Laboratory, Springfield, Virginia.
Summers, D., Ni, B., and Meredith, N. P. (2007). *J. Geophys. Res.*, **112**(A4), A04207.
Summers, D. and Thorne, R. M. (2003). *Journal of Geophysical Research: Space Physics (1978–2012)*, **108**(A4), 1143.
Thorne, R. M. (2010). *Geophys. Res. Lett.*, **37**(2), 22107.
Turunen, E., Verronen, P. T., Seppälä, A., Rodger, C. J., Clilverd, M. A., Tamminen, J., Enell, C.-F., and Ulich, Th. (2009). *J. Atmos. Sol. Terr. Phys.*, **71**, 1176–1189.
Van Allen, J. A. and Frank, L. A. (1959). *Nature*, **183**, 430–434.
Verronen, P. T., Andersson, M. E., Rodger, C. J., Clilverd, M. A., Wang, S., and Turunen, E. (2013). *J. Geophys. Res.*, **118**, 11419–11428.
Verronen, P. T. and Lehmann, R. (2013). *Ann. Geophys*, **31**(5), 909–956.
Verronen, P. T., Rodger, C. J., Clilverd, M. A., and Wang, S. (2011). *J. Geophys. Res.*, **116**, D07307.
Verronen, P.T., et al. (2005). *J. Geophys. Res.*, **110**, A09S32.
Walt, M. (1984). *Introduction to Geomagnetically Trapped Radiation*. Cambridge University Press, Cambridge, UK.

15

Energetic Particles and Waves in the Outer Planet Radiation Belts

N. Krupp, E. Roussos, C. Paranicas, A. Sicard, G. Hospodarsky, and Y. Shprits

■ Abstract

The radiation belts of the outer planets are among the harshest environments in our solar system. In extremely strong internal closed magnetic field configurations energetic particles up to several hundred MeV energies are trapped and bounce back and forth along the magnetic field lines emitting waves in a whole variety of frequencies. Charged particles in strong magnetic fields gyrate about the magnetic field line, bounce along it, and drift perpendicular to it simultaneously. Large ground-based or Earth-orbit-based telescopes as well as remote and in-situ measurements aboard spacecraft have been used to characterize those intense radiation environments of the gas giants in our solar system. Pitch angle distributions, energy spectra and composition measurements help to better understand the processes acting and help to identify the particle sources and acceleration mechanisms as well as the sources of waves in the huge magnetospheres. This chapter gives a summary of the present knowledge of the radiation belts of Jupiter and Saturn.

15.1 Introduction

The radiation belts of the outer planets are among the harshest environments in our solar system. In extremely strong internal closed magnetic field configurations energetic

Waves, Particles, and Storms in Geospace. First Edition. Georgios Balasis, Ioannis A. Daglis, Ian R. Mann.

Table 15.1 *Spacecraft missions to Jupiter and Saturn*

Target planet	Spacecraft	Date	flyby/orbiter
Jupiter	Pioneer 10	1973	flyby
	Pioneer 11	1974	flyby
	Voyager 1	1979	flyby
	Voyager 2	1979	flyby
	Ulysses	1992	flyby
	Galileo	1995–2003	orbiter
	Cassini	2001	flyby
	New Horizons	2007	flyby
	JUNO	2016–2017	orbiter
	JUICE	2030–2033	orbiter
Saturn	Pioneer 11	1979	flyby
	Voyager 1	1980	flyby
	Voyager 2	1981	flyby
	Cassini	2004–2017	orbiter

particles up to several hundred MeV energies are trapped and bounce back and forth along the magnetic field lines emitting waves in a whole variety of frequencies. Jupiter's and Saturn's magnetospheres are rotation dominated. Charged particle drift paths close around the whole planet to substantial planetary distances, unlike in the case of Earth (Chen, 1970). The combination of a strong internal magnet and quasi-stable trapping allows the fluxes of energetic ions and electrons to become very large. In this chapter the available in-situ measurements of Jupiter's and Saturn's radiation belts of instruments onboard flyby and orbiting missions are reviewed as well as current modelling approaches. Table 15.1 summarizes the spacecraft mission dates to Jupiter and Saturn.

Before the review of the current understanding about particles and waves in the radiation belts of the Jupiter and Saturn, and because some of the moons play an important role in those huge plasma laboratories, a short overview of the Jovian and the Saturnian magnetosphere will be given.

A sketch of Jupiter's magnetosphere is shown in Fig. 15.1. Basically Jupiter's magnetosphere can be subdivided in three regions, namely the inner magnetosphere including the radiation belts, the middle magnetosphere with an equatorial plasma sheet, and the outer magnetosphere including the huge magnetotail region of the system. The major plasma source of the Jovian magnetosphere is the moon Io orbiting Jupiter at 5.9 Jovian radii R_J ($1\ R_J = 71492$ km) providing more than a ton/sec of oxygen and sulfur ions. Minor other sources are the ionosphere of the planet, the solar wind, and galactic cosmic rays as well as other moons. The whole system is powered by the fast rotation of the planet. Due to the centrifugal forces the charged particles moving around and along the magnetic field lines are radially driven outwards creating an equatorial magnetodisk and associated plasma disk. Since the rotation axis and the magnetic dipole axis of Jupiter are tilted to each other by 9.7 degrees the magnetosphere including the magnetodisk wobble up and down with respect to the rotational equator of the system. Plasma and

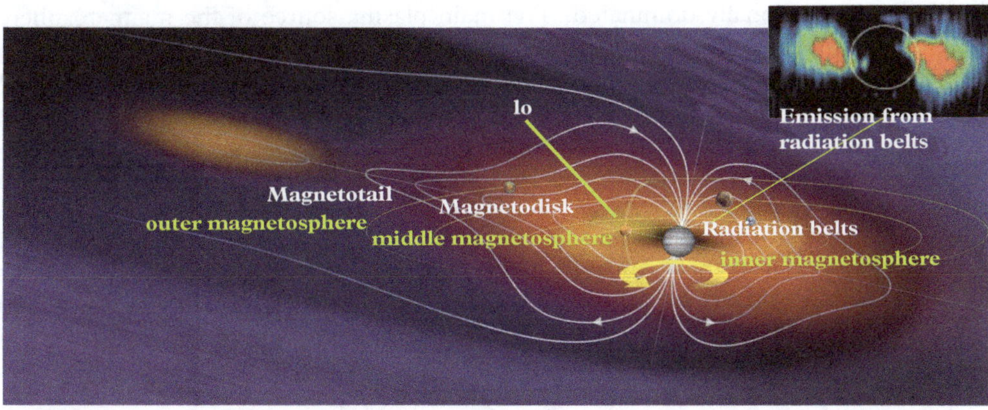

Figure 15.1 *Key regions of the Jovian magnetosphere.*

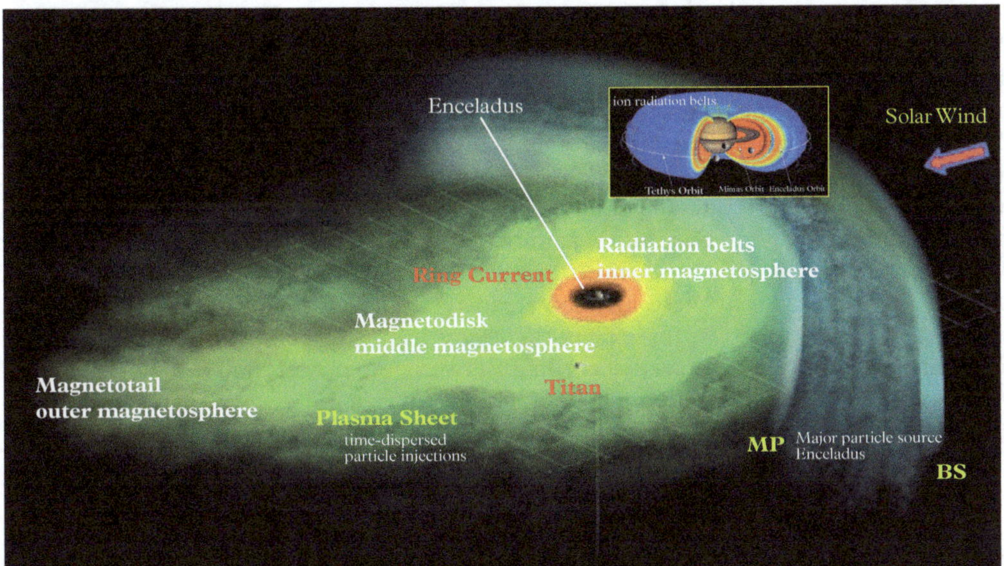

Figure 15.2 *Key regions of the Saturnian magnetosphere.*

energetic charged particles in the Jovian magnetosphere are nearly corotating up to a certain distance from the planet and sub-corotating beyond. All the details of the ongoing physical processes in the Jovian magnetosphere and the high variability can be found in the literature (Bagenal et al. (2009)) and (Dessler, 1983) and will not be explained here again.

Similarly to Fig. 15.1 a schematics of the Saturnian system is shown in Fig. 15.2. For Saturn also inner, middle and an outer magnetosphere are distinguished. The Kronian system is also rotationally dominated. The main plasma source of the magnetosphere is the moon Enceladus, at 3.98 Saturnian radii R_S (1 $R_S = 60268$ km) very close to the planet. Enceladus provide several hundred kg/s of water group ions into the magnetosphere and the centrifugal forces acting the same as described for Jupiter building an equatorial magnetodisk. Different from the Jupiter system are mainly two things. First, rotation and magnetic dipole axes are parallel to each other, and secondly, Saturn has more icy moons in the inner magnetosphere. Those moons (except Enceladus) act mainly as particle absorbers for magnetospheric particles along their L-shells around Saturn leaving a gap in the particle distributions (see insert in Fig. 15.2). For Jupiter this process is less effective because of the tilt between dipole and rotation axes and due to completely different diffusion rates at Jupiter filling up those gaps in the distribution functions faster than in the case of Saturn. Again for the details of those processes the reader is referred to the literature (Dougherty et al., 2009).

15.2 Charged particles

When a charged particle is trapped in a magnetic field, i.e. subjected to the Lorentz force, its trajectory can be decomposed into three elementary distinct motions: the gyration around the magnetic field line, the bounce between two mirror points along the magnetic field line, and the azimuthal drift around the planet. The combination of these three motions defines a surface, the drift shell, characterized by the McIlwain parameter, L (equatorial distance from the planet in a dipole magnetic field), and the equatorial pitch angle of the particle α_{eq}. Each of these motions is associated with adiabatic invariants: the "relativistic" magnetic moment M, the second invariant $\mathcal{J}$ and the magnetic flux Φ. A detailed description of these invariants can be found in (Roederer, 1970). Close to the planet, it is well known that the particles are trapped by the strong magnetic field of Jupiter and are well organized according to two parameters L and B/B_{eq} (Divine and Garrett, 1983), where B is the magnetic field magnitude at the spacecraft location and B_{eq} is the field magnitude at the magnetic equator along the same field line.

15.2.1 Jupiter

Jupiter's magnetic field and its radiation belts have been discovered by the detection of decametric radio emissions and synchrotron radiation (Burke and Franklin, 1955). Figure 15.3 shows the measurements taken onboard the Cassini spacecraft during the Jupiter flyby with a peak emission near 1.4 $R_{\mathcal{J}}$. A detailed description can be found in (Bolton et al., 2002; de Pater and Dunn, 2003). In-situ particles and fields measurements during the first flyby missions of Pioneer 10 and 11 went as far in as 1.6 $R_{\mathcal{J}}$ confirmed the existence of a trapped charged particle population in radiation belts. The first orbiting mission Galileo was dedicated to explore the Galilean moons. Therefore and due to the harsh environment only a few orbits around the planet went inside Europa's L-shell at 9.4 $R_{\mathcal{J}}$. Galileo's first and the few last orbits passed through the Io torus before the spacecraft crashed into Jupiter's atmosphere. Figure 15.4 shows as an example the measured

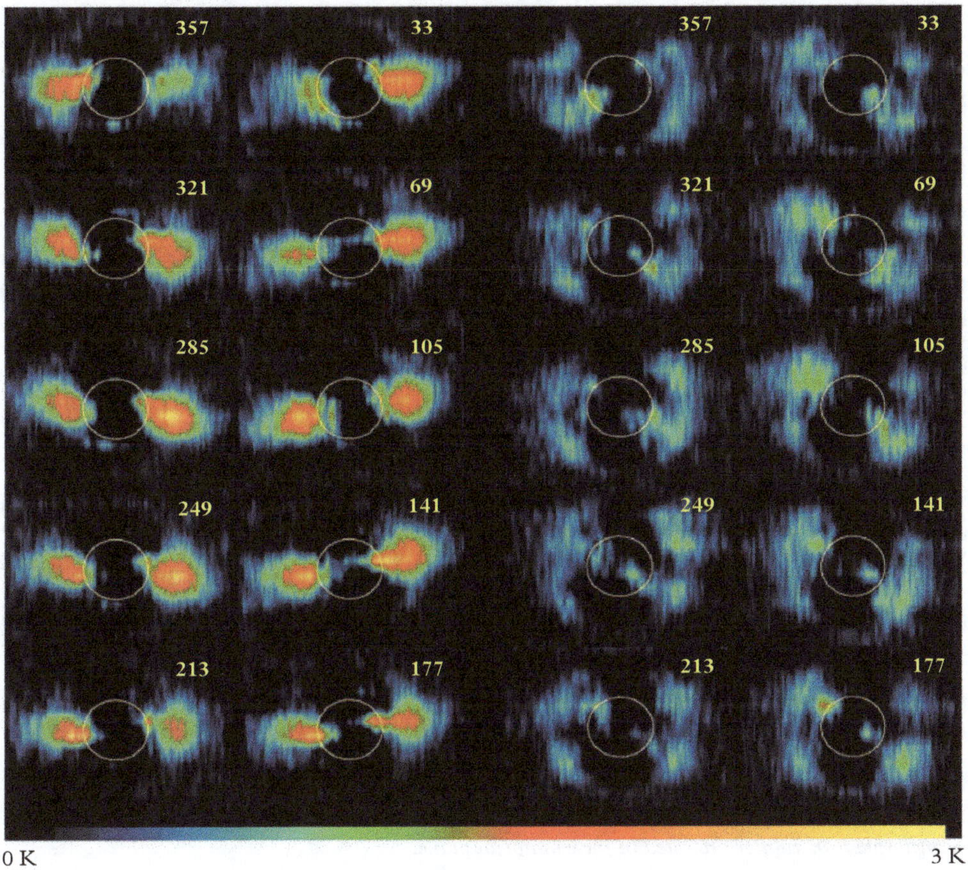

Figure 15.3 *Colour maps (0–3 Kelvin) of Jupiter's synchrotron emission (distribution of ultra-relativistic electrons (~ 50 MeV)) in Jupiter's inner radiation belts taken by the Cassini spacecraft during the Jupiter flyby end of 2000/beginning of 2001. The numbers at each plot mark the longitude in the SIII coordinate system. Left: horizontal linear polarization; right: vertical linear polarization.*
(Bolton et al., 2002).

count rate of electrons (1.5–10.5 MeV) inside L = 10 in the Jovian magnetosphere. The innermost orbit is the last Galileo orbit J35 before the spacecraft stopped sending data.

At Jupiter, not only protons but also heavy ions in the keV/nucleon energy range are important as well. Mauk et al. (2004) have created intensity-distance profiles for all the important energetic ions at Jupiter (see Figure 15.5). The number of protons with energies above 50 keV is higher than the heavier ions but the particle pressures in the inner magnetosphere are dominated by oxygen and sulfur (whose ultimate source is Io). As expected for a radiation belt structure, the intensities increase toward the planet with a peak near Europa's orbital distance. Inward of that distance, a decrease in overall

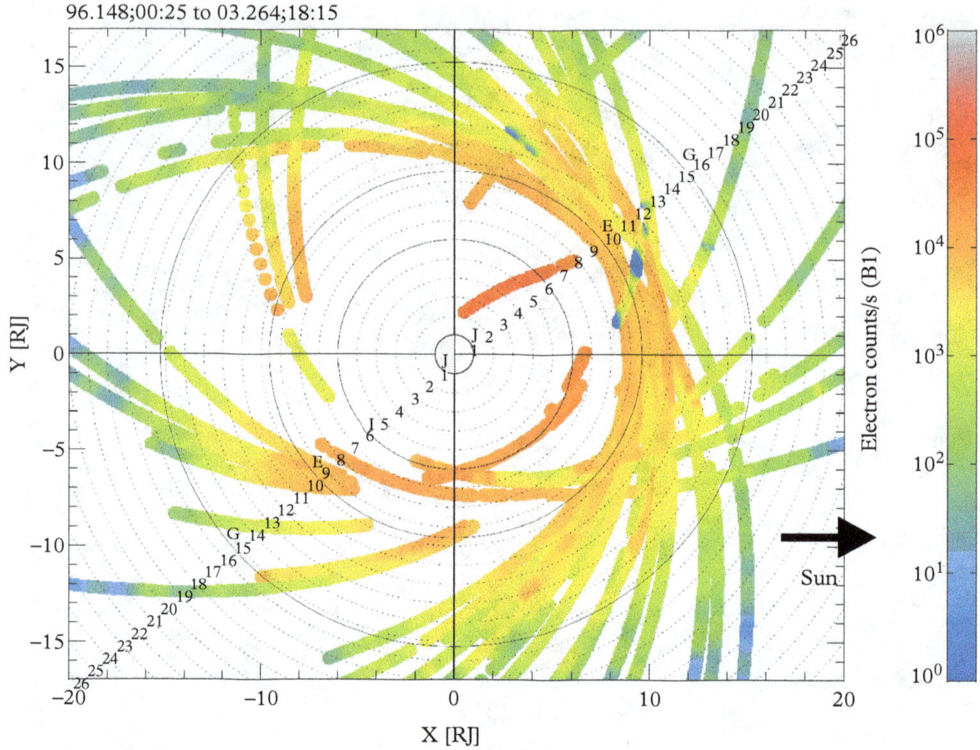

Figure 15.4 *Color-coded count rate of electrons (1.5–10.5 MeV) as measured between 1995 and 2003 by the Energetic Particles Detector EPD onboard the Galileo spacecraft in the inner magnetosphere of Jupiter projected in the equatorial plane.*

intensity of energetic ions, likely due to losses in charge-exchange collisions with the neutral gas, is measured. Smyth and Marconi (2006) have modelled the distributions of major neutral species in this region and shown that there are strong peaks in neutral density between the orbital distances of Io and Europa (see, for example, their Figure 9). Other representations of the energetic ions and electrons can be found in the literature. For instance, energy spectra for energetic Jovian ions and electrons (sometimes with comparisons to Garrett's models) can be found in Cooper et al. (2001), Paranicas et al. (2001), Paranicas et al. (2003), Mauk et al. (2004), to name a few.

In addition the Galileo Probe provided the opportunity to measure electrons, protons, He, C, and S in the 3-530 MeV/nucleon range even closer to the planet. The particle measurements during the probe's descent between 1.25 and about 6 R_J is shown in Figure 15.6 (Fischer et al., 1996; Bolton et al., 2004). The fluxes increase radially inward and peak at about 2.1 R_J inside the orbit of moon Amalthea (2.54 R_J) followed by a sharp decrease and a second peak at about 1.5 R_J.

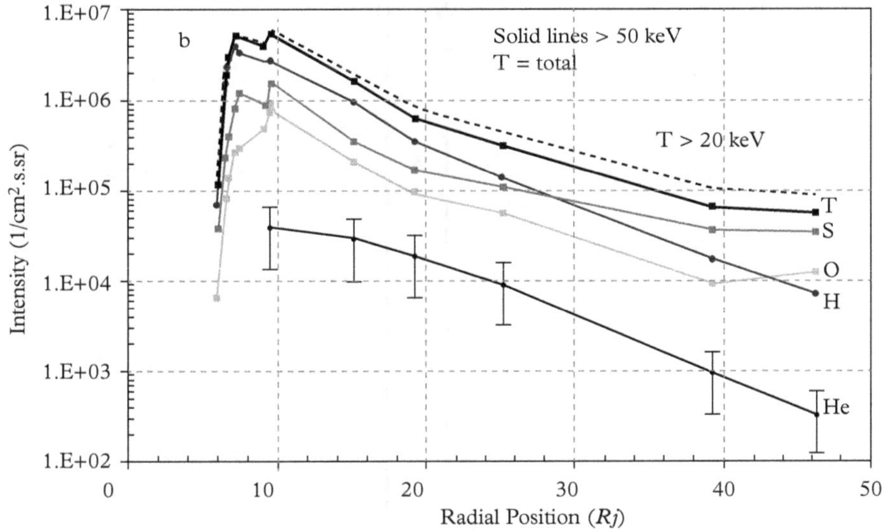

Figure 15.5 *Intensities of various energetic ions as a function of distance from Jupiter as measured by the EPD instrument onboard Galileo.*
(Mauk et al., 2004)

Pitch angle distributions Galileo particle and fields measurements provided the opportunity to study the pitch angle distribution of energetic ions and electrons in the Jovian magnetosphere. One of the major findings was that the electron pitch angle distribution changed from being bi-directional along the magnetic field direction outside of 10–17 R_J, while inside that distance a trapped pitch angle distribution was observed with maximum fluxes near 90 ° (Tomás et al., 2004a; Tomás et al., 2004b) as visible in Figure 15.7a. This location is thought to be the transition region between a more dipolar magnetic field lines and a more plasma sheet-like magnetic field or the outer edge of the Jovian radiation belts. At slightly different interpretation of the pitch angle "boundary" was given by (Mauk and Saur, 2007). The authors suggested that the boundary is related to upward auroral acceleration. Similar pitch angle distributions with a boundary at about L = 10 has been reported at Saturn and is described below in the Saturn section (Carbary et al., 2011).

Figure 15.7b shows examples of ion pitch angle distributions at four different L-shells for protons (right) and sulfur ions (left) (Mauk et al., 2004). Protons and sulfur ions show different pitch angle distributions as a function of L. The sulfur pitch angle distributions change from being trapped near 7.8 R_J to more butterfly-like (losses of particles with 90 degree pitch angles) closer in at 6.8 R_J. For protons, the opposite seems to be the case where a trapped population is observed at 6.8 R_J changing to butterfly further out. A possible explanation could be different interaction processes of heavy ions and protons with the neutral gas in the Io torus.

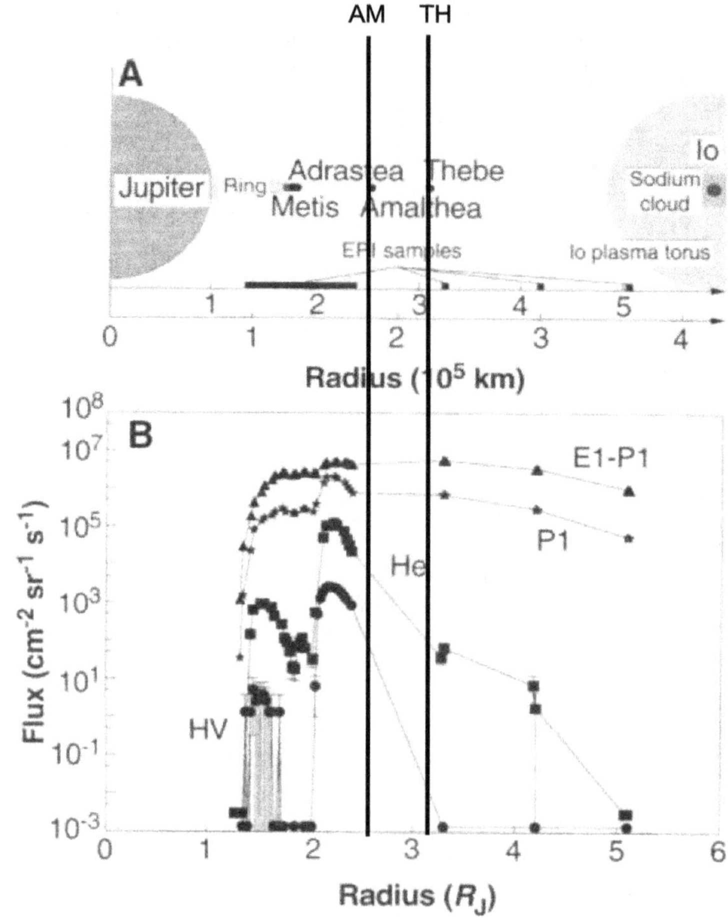

Figure 15.6 *Particle fluxes as a function of distance from Jupiter as obtained by the EPI instrument onboard the Galileo Probe. The various channels are responding to electrons 3.2–450 MeV; protons 42–131 MeV; Helium 42–530 MeV/nucleon; Carbon 75–168 MeV/nucleon; Sulfur 125–210 MeV/nucleon. Details can be found in Fischer et al. (1996).*

From synchrotron emission measurements two components in relativistic electron pitch angle distributions at small L-shells (1–3 R_J) have been concluded in Jupiter's radiation belts: an anisotropic pancake component and an isotropic component for details see Santos-Costa and Bolton (2008); Bolton et al. (2004).

Time variability and asymmetries The time variability in particles and fields data vary between the planetary rotation period (about 10 hours) to a few minutes. Variations on longer time scales cannot be easily assessed from in-situ data, as the radiation

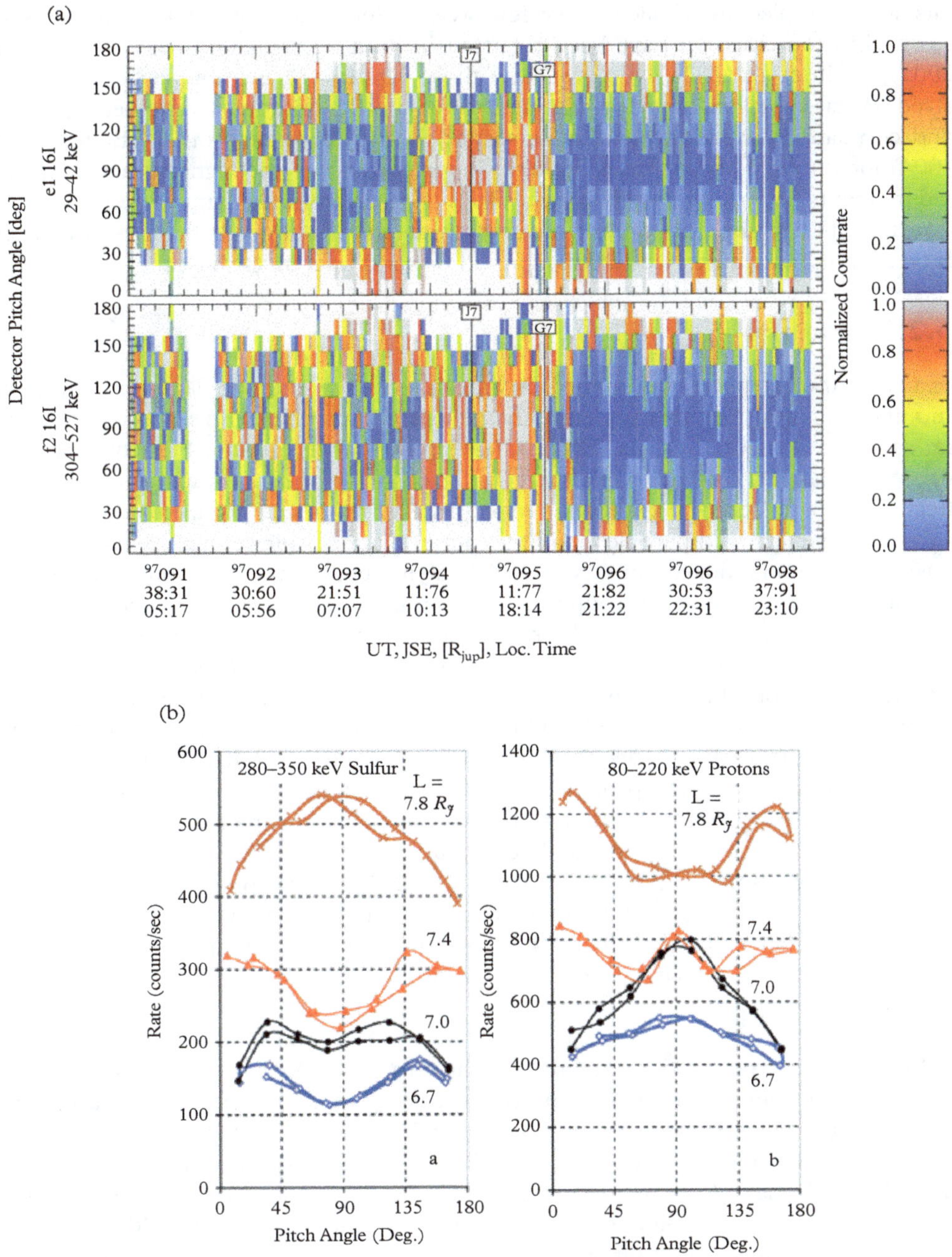

Figure 15.7 *(a) Electron pitch angle distribution as a function of distance from Jupiter (from Tomás et al. [2004a]); (b) Ion pitch angle distributions for different L-values.*
(Mauk et al., 2004).

belts were sampled by Galileo every few weeks. Short- and long-term variations in the synchrotron emission have been reported in Jupiter's radiation belts Bolton et al. (2002); Klein et al. (2001). The cause of those variations have not yet been fully resolved. Possible explanations include radial diffusion due to atmospheric heating, solar wind correlations, electron pitch angle scattering etc. As pointed out by Bolton et al. (2004) longitudinal asymmetries are introduced because of the higher order magnetic field moments and the beaming characteristics of the synchrotron emission.

Periodicities in the particle and fields parameters of about 40–60 min have been reported in the literature (see Krupp et al. (2004) and references therein).

15.2.2 Saturn

Because of the extensive neutral torus at Saturn, energetic particle fluxes are limited in a number of ways. Ions can undergo charge exchange with neutrals and exit the system as energetic neutral atoms (ENAs), leaving behind a cold ion. Electrons can lose energy by interacting with the electrons and nuclei in the gas and thus be cooled (and effectively "lost" from the energetic range). At Saturn, injections are easily identified in the energetic charged particle data in the inner magnetosphere because ambient and quasi-stably trapped fluxes can be very low. Dialynas et al. (2009) computed ion energy spectra for protons and watergroup ions between the radial distances of about 10 to 20 R_S. Paranicas et al. (2012) created ion and electron spectra at the orbital distances of several planetary satellites using mission-averaged Cassini data (see Figure 15.8). Interestingly they found a peak in the proton energy spectra, which cover also the MeV energy range, due to the CRAND (Cosmic ray albedo neutron decay) contribution. Roussos et al. (2011) has previously established a strong link to CRAND when they correlated the fluxes of energetic protons at Saturn to the solar cycle. Energetic particle pitch angle distributions at Saturn have not been well explained in the literature. Armstrong et al. (2009) found proton pitch angle distributions close to the planet, which do not follow a simple trapping shape. Carbary et al. (2011) have found very field-aligned energetic electrons outward of about $10R_S$, with trapped or pancake distributions inward of that distance (see their figure 4 and the discussion in the Jupiter section above). It is not clear that the physical processes causing these "boundaries" are similar, but possible.

L-shell Dependence The structure of Saturn's radiation belts differs significantly in electrons and ions. For each of the two species, L-shell profiles change also if we are below or above the range of few hundred keV. The most detailed study of L-shell profiles has been done by Kollmann et al. (2011) and is based on almost 8 years of MIMI/LEMMS data. Below a few hundred keV, fluxes of both electrons and ions peak around $L = 9$, in a region identified as the ring current (Sergis et al., 2007). Energy spectra in that source region rarely show any significant foreground beyond a few MeV (for electrons or ions).

Inside $L = 9$, fluxes experience a drop which is sharper for ions than for electrons. The drop in ions is caused by charge exchange, which is very efficient for both protons

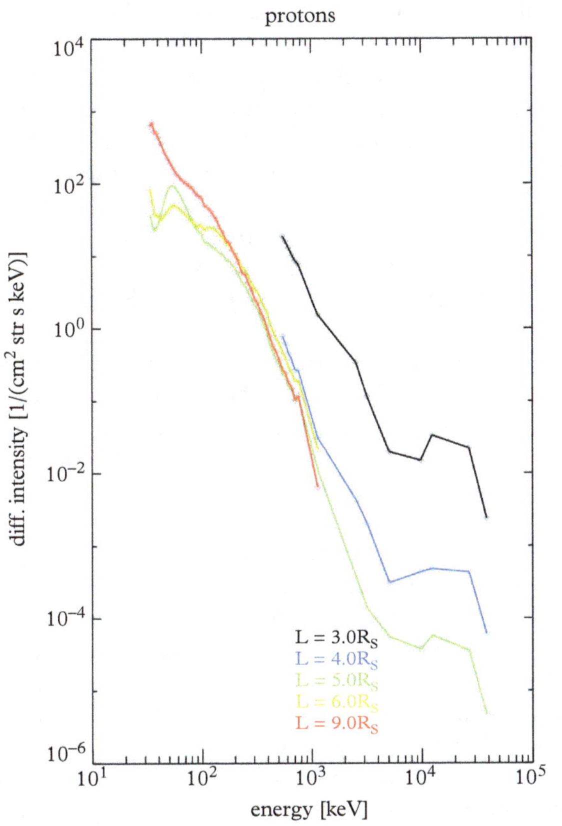

Figure 15.8 *Intensity vs energy spectra of protons in the Jovian magnetosphere measured at L-shells between 3 and 9 R_J marked by different colours.*
(Paranicas et al., 2012).

and water group ions below 100 keV (Dialynas et al., 2009). The actual driver for keV electron losses is less clear. Kollmann et al. (2011) shows that energy losses to neutrals are probably more significant than to dust, but both loss rates still comparable. Pitch angle scattering into the loss cone may also be significant, but that has not been directly demonstrated. Earlier studies suggested that moons may dominate losses at their corresponding L-shells (Randall, 1994), but in the view of Cassini data, moon absorption effects on electrons, while present locally (Carbary et al., 1983; Paranicas et al., 2005; Roussos et al., 2005; Roussos et al., 2007), could be secondary on average. Moon L-shells are not empty of electrons, except maybe above the 5–10 MeV range (Selesnick, 1993).

Inside L = 9 and down to L = 5, average keV electron and ion fluxes drop monotonically. Localized enhancements are attributed to transient injection events (Paranicas et al., 2008). Injections are considered the main process for transporting keV particles from the ring current to low L-shells. The apparent recovery of average keV particle fluxes inside L = 5 at least partly attributed to penetrating radiation that affects all detectors of Cassini's MIMI suite. Foreground of keV electrons has been measured down to the orbit of Mimas (Jones et al., 2006b; Jones et al., 2006a) but the actual level of fluxes is still questionable due to the aforementioned reason.

Beyond few hundred keV or in the MeV range, L-shell profiles of ions are relatively stable in intensity over both short and long time scales. The outer boundary beyond which ion fluxes drop to background typically coincides with the L-shell of Saturn's moon Tethys (L = 4.89). These ion belts comprise different sectors, each separated from the other by ion depleted regions that are centered on the L-shells of planet's inner icy moons (Paranicas et al., 2008; Armstrong et al., 2009; Krupp et al., 2009). Fluxes within these belts originate from nuclear collisions between Galactic Cosmic Rays and the planet's main rings and atmosphere (CRAND process) (Blake et al., 1983; Cooper, 1983; Roussos et al., 2011; Kollmann et al., 2013).

On the contrary, MeV electrons have a smooth and continuous flux increase towards low L-shells, peaking at about L = 3 (Kollmann et al., 2011; Tang and Summers, 2012), and dropping sharply only at the outer edge of the main rings. The outer boundary of those belts extends on average out to about L = 7-8 (inner edge of the ring current) (Roussos et al., 2014).

Asymmetries and time dependence Separation of Saturn's belt populations to electrons/ions and to the two broad keV/MeV energy ranges is also relevant when studying longitudinal asymmetries and time variations. Both keV ions and electrons show a clear flux or phase space density local time asymmetry pointing close to the noon to midnight direction (Carbary et al., 2009; Kollmann et al., 2011), with fluxes being typically higher around midnight. Analysis from Andriopoulou et al. (2012); Thomsen et al. (2012); Wilson et al. (2013) indicates that this is due to a convective electric field that on average points between 15 and 30 deg from the midnight direction and towards dawn. Effects of this electric field on keV energetic particle fluxes have been detected between L = 4 and L = 10. Inside L = 4, data limitations prevent any identification of relevant signatures in keV particles. Outside L = 10 proton and oxygen measurements in the keV energy range show also asymmetries with respect to the SLS longitude (with SLS we refer to the longitude system for Saturn which is locked to the period of Saturn kilometric radiation and its period has been found to vary with time described in detail at http://www-pw.physics.uiowa.edu/sls4/). The SLS sector at which ion fluxes peak coincides well with that sector of the peak ENA emissions detected from INCA, indicating the presence of a preferred SLS sector where large-scale energetic ion injections occur (Carbary et al., 2010). No obvious SLS organization of energetic electrons has been identified (Carbary et al., 2009). In terms of temporal changes of keV particles, several studies indicate a series of planetary or solar rotational periodicities in the system (Carbary and Mitchell, 2013), but apart from that no clear pattern (e.g. with respect to seasonal or solar cycle changes) has been detected.

Moving to the MeV range, ion fluxes ($L < 5$) were found to be local time symmetric (Paranicas et al., 2010a) and unaffected by the noon to midnight convective electric field mentioned earlier. This is because gradient and curvature drifts of energetic ions enforce the azimuthal corotation drift to levels that radial velocities from convection are negligible. In terms of time, variability of MeV ion belts has two components. The one is a weak modulation related to the solar cycle control of galactic cosmic ray influx to the Saturnian system, as has been measured by Cassini (Roussos et al., 2011). This affects the source rate of MeV protons associated with the CRAND process. The second concerns transient extensions of the ion belts beyond the orbit of Tethys, that may last several months. These occur after the interaction of Saturn's magnetosphere with a Solar Energetic Particle event (SEP) (Roussos et al., 2008) and surprisingly have no impact on the belts inside Tethys's orbit.

MeV electron belts are not symmetric in longitude (Paranicas et al., 2010b), but it is unknown if this asymmetry is transient or whether if it has any organization as a function of local time or SLS longitude. The fact that electron channels in the MeV range have very broad and complex responses complicates any relevant analysis. In terms of time, both the extension and the content of the MeV electron belts varies significantly. For instance, the outer boundary can take values in the range of $4 < L < 10$ (Roussos et al., 2014) (Fig. 15.9). This striking difference with respect to the MeV proton belts has its origin primarily on the nature of energetic particle absorption by Saturn's icy moons. While this is a very effective local process for energetic electrons, (Carbary et al., 1983; Paranicas et al., 2005; Roussos et al., 2005; Roussos et al., 2007), it is restricted within very narrow L-shell corridors, because electrons have a small gyroradius. Radial transport processes allow electrons to cross these corridors before they encounter a moon, meaning that the electron belt content can be modulated from dynamics in the highly variable middle magnetosphere (Krimigis et al., 2007). Roussos et al. (2014) showed the time variability of the radiation belt extension in L-shell (see Figure 15.10). The extension of the radiation belts appears to be highly variable, and this variability appears to be reflected in the count rate levels at low L-shells. The authors report also on sudden and sometimes extreme belt expansions which can be isolated or recurrent. In other cases belts shrink down to L = 4 and can stay in that state for 1–3 months. The reason of the variability could be internally or externally driven. No final conclusion could be drawn so far.

15.2.3 Modelling of charged particles in the Jovian and Kronian radiation belts

Jupiter's radiation belts models To simulate the relativistic electron and ion dynamics in the radiation belts, one must clearly identify the most important physical processes to take into account. The main physical processes are : radial diffusion due to field fluctuations, pitch angle diffusion (for electrons only) and energy friction with atmosphere, plasmasphere and rings, pitch angle friction due to synchrotron radiation (for electrons only), absorption by moons, charge exchange and nuclear interaction with atmosphere (for protons only). For helium ions, the physical processes are the same than in the case of protons. However, the charge exchange can be a source of particle as well as a loss of particle. Indeed, by charge exchange of a particle with an atmosphere a He^+ ion can become a He^{2+} ion, which leads to a loss of He^+ but a source of He^{2+}. Thus, the ion populations of different charge states are linked and depend on each other. Finally, all these physical processes are controlled by the magnetic field. It scales the energy range

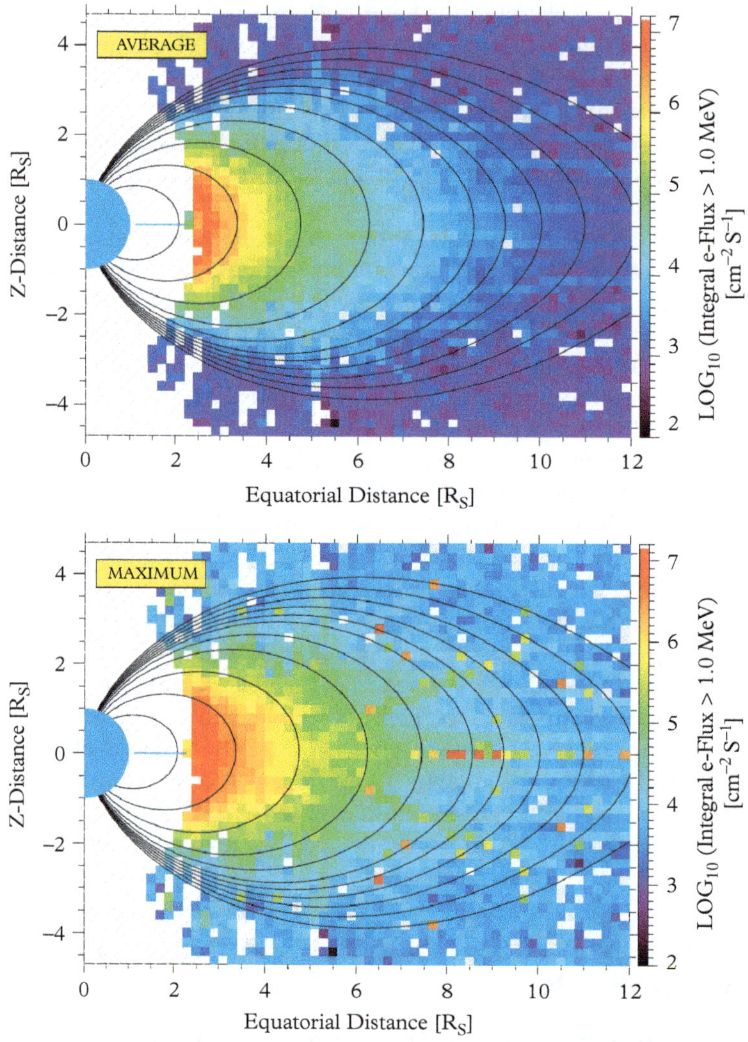

Figure 15.9 *Color-coded maps of the omnidirectional electron integral upper limit flux (> 1 MeV) in the inner Saturnian magnetosphere measured with the MIMI/LEMMS instrument onboard Cassini. The top panel shows the average case, while the bottom shows the maximum flux per bin. Field lines based on the K06 model are also drawn and symmetry across the magnetic equator has also been assumed.*
(Roussos et al., 2014).

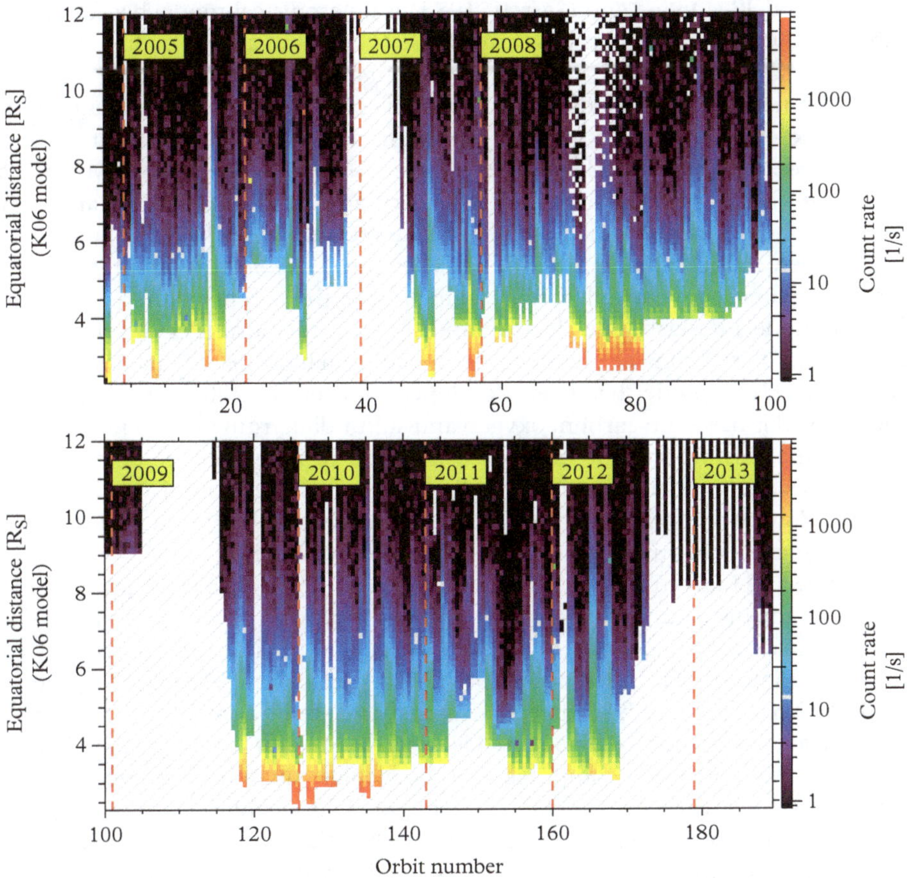

Figure 15.10 *Count rate spectrogram of a LEMMS channel sensitive to the MeV electron omnidirectional flux as a function of Cassini orbit number and equatorial field line distance. The latter is calculated in with a magnetic field model of Saturn's magnetosphere. Orbit durations vary, which is why there are more orbits e.g. in 2008 (orbit period of 1 week) than in other years (period of about 1 month). Hatched regions indicate no data in the respective bins. Several empty bins for large distances during 2008 is due to the high Cassini velocity (in terms of equatorial field line distance) and the small bin size used to create this spectrogram. The plot is adapted from Roussos et al. (2014).*

of the charged particles present in the Jovian environment and plays an important role in their spatial distribution. As a simple internal magnetic field model is no more realistic beyond Io's orbit, the use of a magnetic field model taking into account the presence of Io torus is required.

The first quantitative theoretical models of Jupiter's magnetosphere were developed by Coroniti (1974); Kennel and Coroniti (1975). But it was not before 1983 that the first quantitative model of the radiation belts of Jupiter was developed. This "Divine and Garrett model" (Divine and Garrett, 1983) is an empirical model based on Pioneer measurements with an update in the early 2000s (Garrett et al., 2005; Jun et al., 2005). This updated so called "GIRE model", is a combination of the "Divine and Garrett model" and a new electron model based on Galileo data, which have greatly enriched our knowledge of the radiation belts of Jupiter. In addition to these empirical models, a physical model of Jupiter's radiation belts, called *Salammbô* model, is developed since 1998 (Santos-Costa and Bourdarie, 2001; Sicard and Bourdarie, 2004) and is based on models of the terrestrial radiation belts. More recently Sicard-Piet et al. (2011) has developed a new model, "JOSE model", based on the *Salammbô* model in the inner part of the radiation belts and on Galileo data in the rest of the magnetosphere.

All these models describe the electron and proton radiation belts of Jupiter but it is well known that heavy ions also exist in the Jovian radiation belts. That's why a Heavy Ions Counter (HIC) model has been developed (Jun et al., 2005) (EVANS03). This empirical model is based on carbon, oxygen and sulfur data from Galileo mission. Before we describe the existing models in detail it is useful to note that the energetic proton model includes a strong bite-out at the radial distance of the moon Io and a change in character at Europa's orbital distance. Energetic protons at both Jupiter and Saturn can be removed to very low fluxes at the orbits of planetary satellites.

The Divine and Garrett model (Divine and Garrett, 1983) is a compact and quantitative model of the distribution of charged particles between 1 eV and several MeV in the Jovian magnetosphere. The model is based primarily on in-situ data returned by experiments on the Pioneer (Van Allen et al., 1974; Van Allen et al., 1975) and Voyager (Vogt et al., 1979a; Vogt et al., 1979b) spacecraft, supplemented by earth-based observations and theoretical considerations. The model is described by a set of 14 parameters at each L-shell where the model was computed and to calculate fluxes at intermediate L-shells the parameters are linearly interpolated. The range of applicability of the energetic electron model extends to the Jovian magnetopause, while that of the protons extends out to $L = 12$. The electron model includes a pitch angle dependence within $L = 16$ but is considered isotropic beyond that point. The proton model includes a pitch angle dependence within $L = 12$. Major features of the magnetic field, thermal plasma, and trapped particle distributions are modelled, such as ring and satellite absorption signatures and corotational flow within the Io plasma torus and the current disk. The model populations are assumed independent of time, longitude, and direction azimuth about the field line. The pitch angle-dependent proton and electron models were fitted using L-shell, magnetic field strength B, pitch angle, and energy E as the independent variables.

The GIRE model (Galileo Interim Electron Environment model) (Garrett et al., 2003; Jun et al., 2005) is an update of the Divine and Garrett model described above. This update is based on Galileo EPD data (Williams et al., 1992) and concerns only trapped electrons. Ten-minute averages of the EPD data have been used to form an extensive database of observations of the Jovian radiation belts between Jupiter orbit insertion (JOI) in 1995 and end of the mission 2003. In Figure 15.11 the modelled fluxes of electrons and protons in Jupiter's radiation belts are shown.

The *Salammbô*-3D model, initially developed to model the Earth's radiation belts, has been adapted to model Jupiter's radiation environment (Santos-Costa and Bourdarie, 2001; Sicard and Bourdarie, 2004). It allows the calculation of electron, proton and

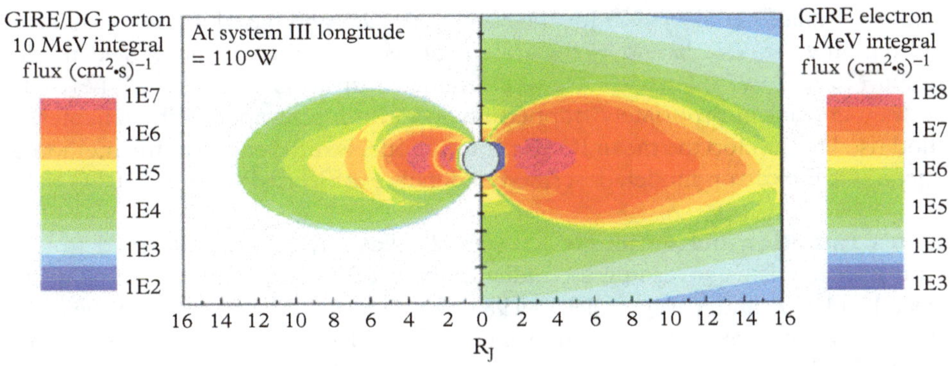

Figure 15.11 *Color-coded integral flux of > 10 MeV protons (left) and >1 MeV electrons (right) as a function of distance from the centre of Jupiter in planetary radii [R_J] as derived from the Divine-Garrett/GIRE models of the Jovian radiation belts.*
(Paranicas et al., 2009).

Helium ion distribution functions everywhere in the inner radiation belts, even for non-equatorial particles. The spatial coverage extends from the Jupiter's "surface" to L = 9.5 (close to Europa orbit) for electrons and protons and to L = 6 for Helium. The valid energy coverage extends from roughly 1 MeV to 600 MeV for electrons and from 1 MeV to 1 GeV for protons (note that the exact energy range depends on the L shell). The model makes use of the adiabatic invariant theory and solves the governing Fokker-Planck transport equation (Santos-Costa and Bourdarie, 2001) to solve this three-dimensional diffusion equation in the E, α_{eq}, L (energy, equatorial pitch angle, L) phase-space based on a finite difference method. Studying electrons, protons and ions in parallel adds constraints to the model. The local environment like the dust particles distribution, moon characteristics, etc. must be the same for all species studied reducing the uncertainties. A slightly modified *Salammbô* code has been developed principally used to study correlations between the Jovian synchrotron variability and particles injection and their transport, solar-wind conditions, solar radio fluxes, Galileo EPD data and the geometric parameter DE (Santos-Costa et al., 2008).

The HIC model (Evans et al., 2003; Jun et al., 2005) is based on the data of the Heavy Ion Counter (HIC) (Garrard et al., 1992) on board the Galileo spacecraft. The HIC model uses data from 31 of the 35 Galileo orbits of Jupiter covering radial distances from 2.5 R_J to well past 30 R_J. The HIC model covers the energy range between ~6 and ~200 MeV/nucleon from 2.8 R_J to > 30R_J, and defines three heavy ions populations: carbon, oxygen and sulfur. The sources of oxygen are the Jovian moons and the solar wind. Sulfur is mainly from the moon Io and is well represented in the inner magnetosphere. For distances greater than 15 R_J, the sulfur signal is erratic and has large error bars. Carbon is thought to be of solar origin but inside of 15 R_J statistically insignificant in the HIC energy range.

The latest model of Jovian radiation belts is the Jovian Onera Specification Electron model JOSE (Sicard-Piet et al., 2011). It provides electron and proton flux from the planet to 100 R_J. JOSE is mainly based on Galileo/EPD measurements between 1995 and 2004 (Williams et al., 1992) for the region $L > 9.5$ and based on *Salammbô* results in the region $L < 9.5$. L and B/B_{eq} are calculated with the magnetic field model of Khurana (1992; 1997). Particles are well organized up to L = 20 whereas beyond this value they are not appropriately trapped anymore, not relevant for this chapter. The method used to develop the mean JOSE proton model is different from the method used for mean JOSE electron model, described in details in (Sicard-Piet et al., 2011).

Saturn's radiation belts models Several studies have been performed to understand the mechanisms governing the radiation belts of Saturn. The first space missions Pioneer 11 (1979) and Voyager (1980–1981) led to a preliminary study of the electron radiation belts, protons and ions of Saturn (Fillius and Ip, 1980; Randall, 1994; Simpson et al., 1980; Thomsen and van Allen, 1980; Van Allen, 1983). The Cassini mission, the first orbiting spacecraft around Saturn 2004–2017, has enriched the understanding of Saturn's magnetosphere and the radiation belts dramatically (Armstrong et al., 2009; Krupp et al., 2009; Paranicas et al., 2012; Roussos et al., 2011; Santos-Costa et al., 2003; Lorenzato et al., 2012). Santos-Costa et al. (2003) developed the first physical model of the radiation belts of Saturn; Garrett et al. (2005) enriches the empirical model of Divine (1990) through the SATRAD (SATurn RADiation belts) model. Finally, Lorenzato et al. (2012) developed an adaptation of the physical model *Salammbô* to Saturn's radiation belts.

Divine (1990) used published data from several charged particle experiments aboard the Pioneer 11, Voyager 1, and Voyager 2 spacecraft during their flybys at Saturn to generate numerical models for the electron and proton radiation belts between 2.3 and 13 R_S. The model described the electron distributions at energies between 0.04 and 10 MeV and the proton distributions at energies between 0.14 and 80 MeV. This "Divine" model was intended to predict particle intensity, flux, and fluence for the Cassini mission analysis. Based on this study the SATRAD model of the Kronian radiation environment has been developed (Garrett et al., 2005). The SATRAD model consists of electron and proton particle fluxes in the B-L coordinate system determined with a magnetic field model for evaluating the model fluxes.

The *Salammbô* physical model, first developed for the Earth and then adapted for Jupiter's environment as described above in the Jupiter section, is based on Pioneer 11, Voyager 2 and Cassini particle data. It provides the electron and proton distribution functions in the inner part of the radiation belts of Saturn. Taking into account electrons and protons with energies between 10 keV and 5 MeV the model considers the interactions of the radiation belt particles with the extended neutral environment (atmospheres and gas torii), plasma, dust (ring systems), waves and particle transport and the particle sources. Similar to the Jupiter model, a finite differential method is used to solve the three-dimensional diffusion equation (Santos-Costa and Bourdarie, 2001; Lorenzato et al., 2012). The relative importance of the various physical processes shows that near the planet ($L < 2.27$), radiation belt electron interaction with ring dust is the more important physical process (no energetic electrons in this region of the magnetosphere); out from the planet ($2.27 < L < 6$), radial diffusion and local interactions of electrons with moons are the major physical processes for high energy electrons and govern Saturn's radiation belt dynamics. The results suggest that the interaction of electrons with neutral particles is not a major physical process for high energy electrons, in accordance with previous work (Kollmann et al., 2011; Santos-Costa et al., 2003), but can be important for low energy electrons ($E < 50keV$). Comparisons of the *Salammbô* results with the Cassini measured data (Cassini mean flux) show a good consistency from about 300 keV to a few MeV. This model shows also that wave-particle interactions play a significant role in the Kronian radiation belts dynamics. The results also predict a strong production of energetic neutral hydrogen atoms, but it predicts also a very weak synchrotron emission confirming the weakness of this radiation for Saturn (Van Allen and Grosskreutz, 1989).

15.3 Plasma waves

Plasma wave emissions have been detected at all of the planets that have been visited by spacecraft equipped with plasma wave instruments. Wave-particle interactions are involved in both the acceleration and loss of radiation belt particles (Kennel and Petschek, 1966; Omura and Summers, 2006; Mauk and Fox, 2010; Tang and Summers, 2012) and play a major role in radiation belt dynamics at Jupiter (Tao et al., 2011a; Horne et al., 2008) and possibly Saturn (Hospodarsky et al., 2008; Mauk et al., 2009; Menietti et al., 2013).

15.3.1 Jupiter observations

The Voyager flybys of Jupiter provided the first opportunities to directly measure plasma waves in the Jovian magnetosphere (Scarf et al., 1979b; Gurnett et al., 1979; Gurnett and Scarf, 1983; Kurth and Gurnett, 1991). The orbiting Galileo spacecraft provided additional opportunities to study both the Jovian plasma waves (Gurnett et al., 1996; Menietti et al., 2012) and the plasma waves associated with some of the Jovian moons (Gurnett et al., 1996; Gurnett et al., 1998; Kurth et al., 2001). The Cassini spacecraft also visited the Jovian system during its flyby of Jupiter in 2000, however, it did not enter the inner magnetosphere. Although many of the same plasma waves (whistler mode hiss and chorus, electron cyclotron harmonic (ECH) and upper hybrid resonance (UHR) waves that are observed at Earth are also detected at Jupiter, the majority of studies have concentrated on the strong whistler mode chorus emission (Scarf et al., 1979a; Thorne and Tsurutani, 1979; Inan et al., 1983; Thorne et al., 1997; Bolton et al., 1997; Xiao et al., 2003; Katoh and Omura, 2011) and its possible role in particle dynamics (Horne et al., 2008; Katoh et al., 2011).

Jovian chorus is detected primarily near the edge of the Io torus from L-shells of about 6 to about 12 and has many of the characteristics of chorus observed at Earth, including small scale structures with time scales less than one second and two chorus bands separated by a gap at frequencies at about half the electron cyclotron frequency f_{ce} (Coroniti et al., 1980; Coroniti et al., 1984). Peak amplitudes of chorus detected by Voyager were found to be about 0.26 mV/m (10 pT assuming parallel propagation) and that the chorus scattered electrons with energy of a few keV (Coroniti et al., 1980; Coroniti et al., 1984). Horne et al. (2008) and Menietti et al. (2008a) have performed statistical surveys of the occurrence and strength of the Jovian chorus detected by Galileo. They found that chorus typically occurred between 400 Hz and 8 kHz with peak amplitudes $\sim 10^{-10} V^2/m^2 Hz$ (~0.1 mV/m and ~3 pT). The maximum power and occurrence frequency was observed a few degrees above the magnetic equator with the amplitude and frequency decreasing rapidly as the spacecraft went to higher magnetic latitudes. Horne et al. (2008) used the wave power spectral density from the Galileo measurements to calculate quasi-linear pitch angle and energy diffusion coefficients, and solved a two-dimensional diffusion equation over the region of 5 to 20 R_J. Over time scales of ~30 days, fluxes of 1 to 6 MeV electrons were found to increase by more than an order of magnitude. These electrons are then transported towards the planet via radial diffusion and accelerated to even higher (~50 MeV) energies by betatron and Fermi processes, resulting in the intense radiation belts detected close to the planet.

Tao et al. (2011b) expanded on the work of Horne et al. (2008) by examining the effect of different density models for the injection events and also the effect of latitude dependent wave normal angles for the chorus. They found that various density differences inside the injection event compared to the density outside can significantly change the time scales of energization (both increase and decrease certain energies). The latitude dependent wave normal angle distribution has no obvious effect on 10 MeV electron fluxes, but it reduces the fluxes of 1–3 MeV electrons by about a factor of three. Thus, realistic density models of the injection events and wave normal angle of chorus waves during propagation should be taken into account in future studies to more accurately model electron flux evolution of Jovian electrons.

15.3.2 Saturn observations

The Saturnian magnetosphere has characteristics similar to both the Jovian and Earth systems, including many of the plasma wave emissions detected at Earth and Jupiter. The Voyager spacecraft during their flybys of Saturn detected a variety of plasma and radio waves (Gurnett et al., 1981; Scarf et al., 1982; Scarf et al., 1983), including whistler-mode chorus and hiss (Scarf et al., 1984), and ECH and UHR emissions (Kurth et al., 1983). The Radio and Plasma Wave Science (RPWS) instrument (Gurnett et al., 2004) on the Cassini orbital mission to Saturn has allowed these waves to be studied in much more detail. Like Jupiter, the majority of the studies of plasma waves at Saturn have concentrated on observations of whistler mode chorus. Only a few short periods of chorus were detected by the Voyager spacecraft during their flybys of Saturn (Scarf et al., 1983). The high-resolution data showed a band of chorus present below 0.5 f_{ce} (from about 0.2 to 0.4 f_{ce}). The chorus was primarily hiss-like (diffuse) with some rising structures, but the temporal variations were "unusually slow" (Scarf et al., 1983). No chorus above 0.5 f_{ce} was detected by Voyager at Saturn. Scarf et al. (1983; 1984) examined the possibility of wave-particle interaction and pitch angle scattering of the electrons in Saturn's inner magnetosphere from the detected chorus, but found that the amplitudes were too small to play a significant role. However, during the period of the Voyager flybys the Saturnian magnetosphere was not very active and Kurth and Gurnett (1991) suggested that the chorus emission could play a more significant role in particle dynamics during more active periods. The Cassini spacecraft has detected a variety of wave emissions at Saturn (Gurnett et al., 2005; Hospodarsky et al., 2008; Menietti et al., 2008b; Menietti et al., 2008c; Mauk et al., 2009; Kurth et al., 2009; Menietti et al., 2012). Figure 15.12 (Hospodarsky et al., 2012) shows typical spectrograms of the electric and magnetic field intensities of the plasma wave emissions measured by RPWS during a pass through Saturn's inner magnetosphere in 2005. The intensities are plotted as decibels (dB) above background and the white line shows f_{ce} derived from the magnetometer instrument (Dougherty et al., 2004). Whistler mode emission is observed both during the inbound and the outbound trajectory. Following the definition of Hospodarsky et al. (2008) we will call this emission chorus, although at times the emission may appear hiss-like. Electrostatic ECH and UHR waves are also present, and the Saturn Kilometric Radio (SKR) emission is observed above about 20 kHz throughout most of this period. The whistler mode chorus emissions shown in Figure 15.12 are similar to the chorus detected by the Voyager spacecraft with frequencies only below 0.5 f_{ce}. The majority of the chorus detected at Saturn has this characteristic and was called "magnetospheric" chorus by Hospodarsky et al. (2008). "Magnetospheric" chorus is detected by the RPWS instrument during most orbits of Saturn when Cassini is within about 30 degrees of the magnetic equator and primarily between L shells of about 4 to 8. The ECH emission typically occurs at multiple harmonics of f_{ce}, specifically $\sim (n + 1/2)f_{ce}$, where n is an integer. The UHR emissions can be used to estimate the electron plasma density since the UHR frequency, f_{UHR}, is given by $f_{pe}^2 + f_{ce}^2 = f_{UHR}^2$. By measuring the UHR frequency and obtaining f_{ce} from the measured magnetic field strength, the electron plasma density n_e, can be calculated from $f_{pe} = 8980 n_e^{1/2}$. Persoon et al. (2005; 2006; 2009) have developed a plasma density model for the inner magnetosphere using the UHR frequency observed by the Cassini spacecraft at Saturn.

ECH and whistler mode emissions show very different characteristics inside and outside of injection events at Saturn. Injection events are planetward injections of hot, tenuous plasma produced by the interchange instability as the colder, denser plasma from the inner magnetosphere flows outward in magnetospheres of rapidly rotating planets such as Saturn (Mitchell et al., 2009; Mauk et al., 2009 and references therein). Outside of injection events the chorus emissions are usually hiss like and are only observed at $< 0.5\ f_{ce}$ (Hospodarsky et al., 2008) and the ECH emissions are primarily found in the first harmonic band centered near 1.5 f_{ce}, with weaker, more sporadic bands at the higher harmonics (Menietti et al., 2008b). During injection events the characteristics of the ECH emissions change drastically with an increase in intensity and harmonic structure (Rymer et al., 2008; Menietti et al., 2008c; Tao et al., 2011b) (Fig. 15.13).

Chorus detected in association with local plasma injections defined as "injection event" chorus by Hospodarsky et al. (2008) also changes characteristics. For many plasma injection events, chorus emissions are detected both above and below 0.5 f_{ce}, with a gap in the emission at $\sim$ 0.5 f_{ce} and often contain structure (usually a series of

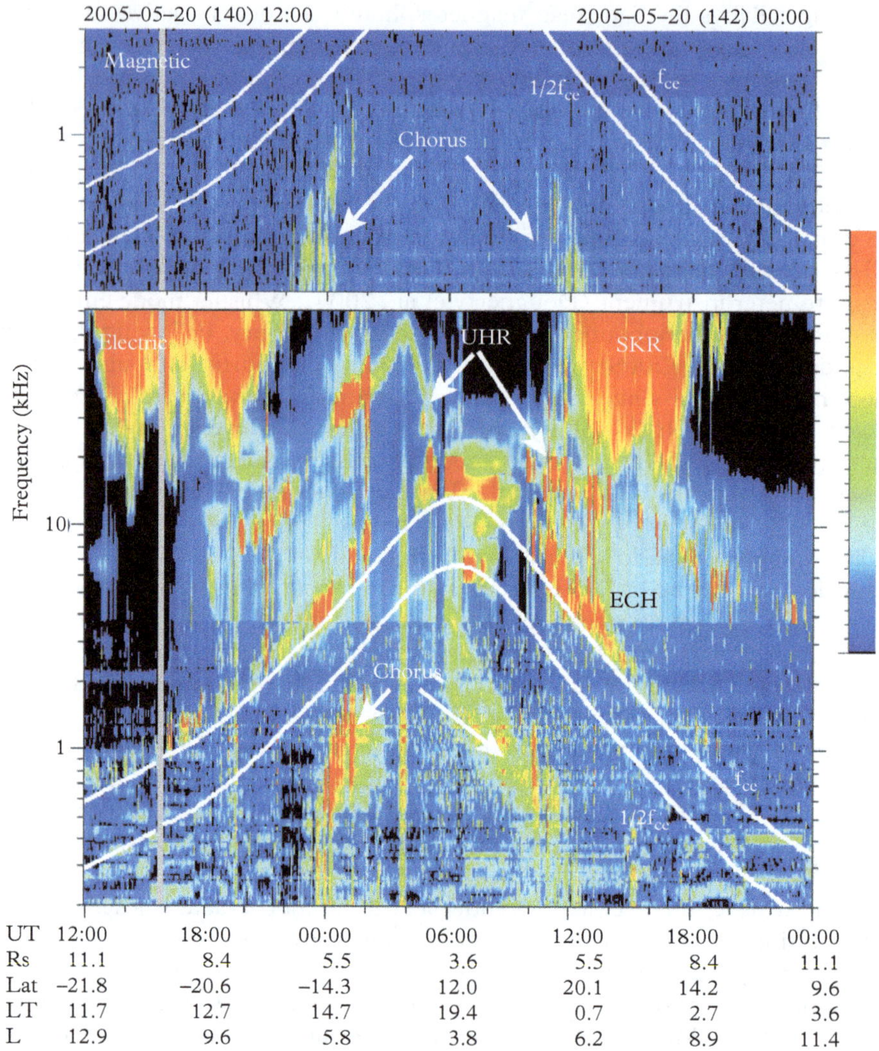

Figure 15.12 *Magnetic and electric field frequency-time spectrograms showing many of the plasma waves detected in the inner magnetosphere of Saturn.*
(Hospodarsky et al., 2012).

rising tones) at a much smaller timescale (less than a second to a few seconds) than the "magnetospheric" chorus (Hospodarsky et al., 2008). This chorus is detected for only the few minutes Cassini is in the injection event, and is not observed outside of the event. This change in the ECH emission inside of an injection event can easily be seen from about 17:35:20 to about 17:40:15 UT in Figure 15.13 (from Menietti et al. 2008c).

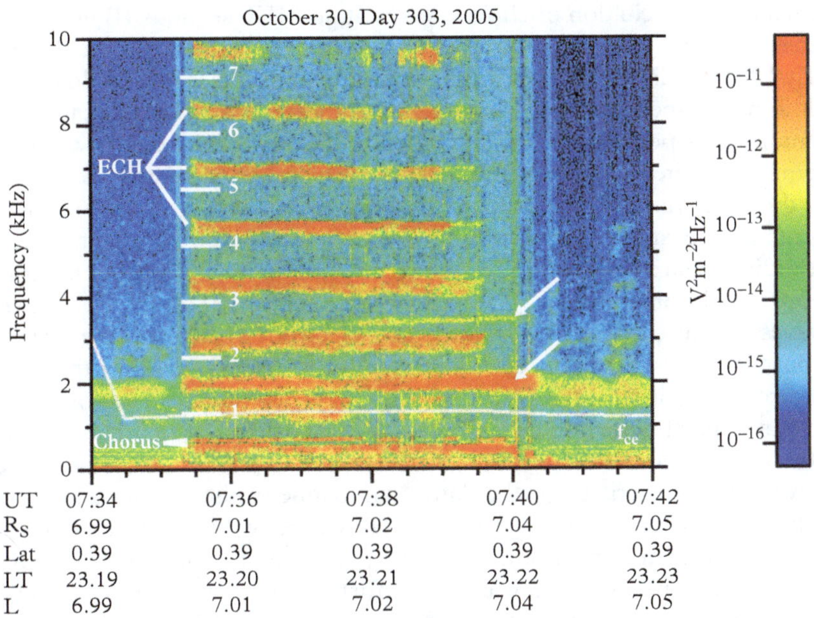

Figure 15.13 *High-resolution wideband wave electric field frequency-time spectrogram obtained during an plasma injection event at Saturn. The enhanced electrostatic cyclotron harmonic wave spectra inside the injection event is easy to see starting at about 07:35:20 UT.*
(Menietti et al., 2008c).

An initial examination of the peak wave amplitudes by Hospodarsky et al. (2008) of the chorus emissions at Saturn (both inside and outside of the injection events) found amplitudes (~1 mV/m and 0.04 nT) much smaller than the peak amplitudes that have been reported at Earth (~30 mV/m and 1 nT), but larger by than the amplitudes detected at Saturn by Voyager. The detection of these larger amplitudes suggests that chorus may be responsible for some pitch-angle and energy diffusion at Saturn although this needs to be investigated in more detail.

The final part of the Cassini mission includes a series of high-inclination orbits (very similar orbits than the JUNO spacecraft in Jovian system at the same time) that cross the equatorial region between the inner edge of the rings and the upper atmosphere (altitudes of few thousand km) in late 2016/early 2017. These orbits should provide a rich data set of fields and particles in the region of Saturn's newly discovered inner radiation belt.

15.3.3 Modelling wave-particle interaction

While magnetospheres of all three planets considered in this section are driven by different processes and have very different special scales and magnitudes of magnetic fields, all of them support excitation of chorus waves at all MLT sections (Bolton et al., 1997; Thorne et al., 1997; Rymer et al., 2008; Liu et al., 2010; Menietti et al., 2012; Spasojevic and Shprits, 2013). Chorus waves can produce loss of particles into the loss cone, which is likely dominating the loss in the outer magnetospheres. If pitch angle scattering is faster than the radial transport processes, and other loss is negligible, chorus wave scattering will determine the pitch-angle distributions of electrons. While it has been clearly established for the Earth radiation belts that chorus waves can also produce a net acceleration of electrons, the question of whether this is true in the case for Jupiter remains open, as the latitudinal distribution of waves is currently unknown. The upcoming Juno mission will provide this information. If waves are confined to 10–15 degrees of the equator, chorus waves may be capable of accelerations of electrons in the outer region and providing seed population for further acceleration to ultra-relativistic energies that are a source of synchrotron radiation, peaking around 1.4 $R_{\mathcal{J}}$. While three-dimensional simulations similar to that of Earth radiation belts (Varotsou et al., 2005; Shprits et al., 2008a; Shprits et al., 2009; Albert et al., 2009; Su et al., 2010) will be required to accurately evaluate the role of wave-particle interactions, the Saturnian magnetosphere is not a likely candidate for providing energy for relativistic electrons. How exactly the Saturnian magnetosphere sustains a strong radiation environment in the presence of strong losses due to neutrals still remains largely unknown and will require further research and detailed modeling. A detailed knowledge of wave properties and radial transport properties is required for the estimation of the relative contribution of various wave particle interactions.

Modeling of the inner portion of trapped radiation in Jupiter (Santos-Costa and Bourdarie, 2001; Santos-Costa et al., 2008) and Saturn (Santos-Costa et al., 2003) included radial diffusion, interaction with neutrals, and, for Jupiter, synchrotron radiation loss. Up until recently, the wave-particles interaction with the outer region of the planetary radiation environment had not been intensively studied and the physical processes responsible for the acceleration and loss of electrons remained not well understood. Whistler-mode chorus waves are present in the magnetospheres of the Earth (Tsurutani and Smith, 1977; Meredith et al., 2001; Meredith et al., 2002; Meredith et al., 2003; Santolík et al., 2003; Santolík et al., 2004 Bunch et al., 2011; Bunch et al., 2012, Jupiter (Gurnett and Scarf, 1983; Menietti et al., 2008a), and Saturn (Gurnett et al., 1981; Gurnett et al., 2005; Hospodarsky et al., 2008). The term "chorus" has been historically used to describe structured emissions (e.g. rising tones, falling tones in frequency with time or burst). In recent years, many studies of the Earth's magnetosphere simply referred to all whistler-mode emissions outside of the plasmasphere as chorus. Chorus waves are present at all MLT sectors for all three planets. While properties of waves (e.g. amplitudes, spectral properties, latitudinal distributions, distributions of wave-normal angles, structure of the tones) show a remarkable variability, the average amplitudes of waves are similar for the Earth and Jupiter while a factor of only approximately 3–4 lower for Saturn (Menietti et al., 2012). The effect of chorus waves on terrestrial relativistic electrons has been extensively studied during the last decade. Modelling and observations showed that for Earth's radiation belts, wave-particle interactions with whistler-mode chorus waves play a most important role for the acceleration and loss of electrons in the outer radiation belts (see review papers by

Millan and Thorne (2007); Shprits et al. (2008a; 2008b); Millan and Baker (2012) and numerous references therein). Horne et al. (2008) solved for pitch angle and energy diffusion, ignoring mixed diffusion, and showed that, similar to the Earth's radiation belts, relativistic electrons in the Jovian radiation belts can be accelerated locally by chorus waves. The free energy of chorus waves comes from lower energy electrons (Kennel and Petschek, 1966; Mauk and Fox, 2010). In the described process of excitation of waves and local acceleration, waves allow the transfer of energy from the low-energy population of electrons to the high-energy population. Horne et al. (2008) further argued that electrons accelerated by waves can provide a seed population for further acceleration to ultra-relativistic energies in the inner region of Jupiter's magnetosphere (Bolton et al., 2002; Tomás et al., 2004b). Shprits et al. (2012) confirmed the results of Horne et al. (2008) with the same 2D code but now also accounting for the mixed diffusion. They also performed the sensitivity simulations that showed that the balance between acceleration and loss depends on a number of parameters assumed for the simulations. In particular, simulations with waves extending to high latitudes showed that wave-particle interactions produced a net loss of electrons while simulations where waves were confined to only 10 degrees of the equator as was assumed by Horne et al. (2008) produced significant acceleration of multi-MeV electron sources. Since the high-latitude observations of waves are not currently available, it is still an open question whether wave particles interactions with chorus results in acceleration or loss of electrons. Woodfield et al. (2013) further showed that bandwidth of the power spectrum can significantly change the scattering rates. The plasma environment of Saturn turned out to be rather different from that of Jupiter and Earth. The ratio of plasma- to gyro-frequency is the key parameter that determines the effectiveness of the wave-particle interactions (Horne et al., 2005).

While the ratio of the plasma-frequency-to-gyro-frequency in the equatorial plane of the magnetospheres of Earth and Jupiter are relatively low (on the scale of few), the values for Saturn are estimated to be above 10 for the entire equatorial plane (Figure 15.14). The process of acceleration by chorus waves is a result of diffusion in energy. As electron energy spectrum is usually monotonically decreasing with energy (bumps in the tail can become rapidly Landau unstable) quasi-linear interactions with waves will produce flattening in the energy spectrum, which results in the hardening of the spectrum and acceleration of particles to higher energies Shprits et al. (2012) noticed that the shape of the initial spectrum of electrons plays a most important role for determining the outcome of the competition of acceleration and loss due to chorus waves. Formation of the energy spectrum upon which waves will be acting depends on other processes such as interchange instability injections that have not been yet modelled together with pitch-angle, energy, and mixed scattering by chorus waves. This shows that conclusions made in this initial study based on 2D simulations can be only interpreted as preliminary. Nevertheless, they provide insight into the complicated and delicate balance between acceleration and loss cause by these processes. Interactions with other types of waves may play a very important role for the dynamics of the planetary magnetosphere. Hiss waves that are often observed in coincidence with chorus and are often observed on Jovian and Saturnian spectrograms may play a significant role in scattering radiation belt electrons. Recent studies by Menietti et al. (2012) observed strong Z-mode waves and ECH waves that may interact with particles and may potentially produce significant acceleration and loss.

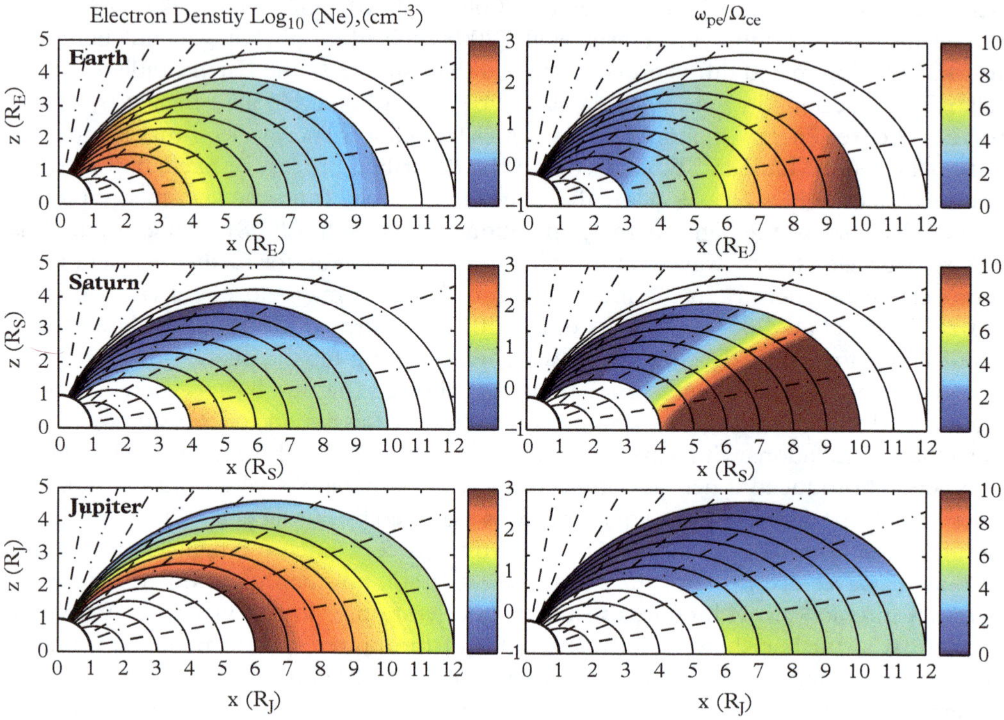

Figure 15.14 *North-south cut of the model electron density* $log_{10}(N_e)[cm^{-3}]$ *(left) and ratio of plasma to gyrofrequency* $\omega_{pe}/|\Omega_{ce}|$ *distribution (right) in the inner magnetosphere of Earth, Saturn, and Jupiter as a function of planetary radii.*
(Shprits et al., 2012).

While sources of Ultra Low Frequency (ULF) waves may be very different for the Earth and outer planets, qualitatively they play a similar role for the Earth and outer planets. Wave particle interactions with ULF waves provide random radial displacement and may result in acceleration during inward transport and loss during outward transport on Earth (Kellogg, 1959; Fälthammar, 1965; Tverskoy, 1965; Shprits and Thorne, 2004) or outer planets (Brice and McDonough, 1973; Siscoe and Summers, 1981; Kivelson et al., 1997).

15.4 Future missions to the outer planets

The future of outer planets exploration is currently limited to two missions for Jupiter. Firstly JUNO (arrival at Jupiter 2016) will investigate the polar Jovian magnetosphere and aurora. The trajectory of the spacecraft will be in favor to study the inner magnetosphere inside of about 30 R_J. Therefore JUNO will dramatically enhance our knowledge about the Jovian radiation belts. JUNO will carry a plasma wave instrument to measure chorus and other plasma waves during its trajectory at lower magnetic latitudes, providing new insight on the plasma wave properties of emissions found at the inner edge of the Jovian radiation belts. The Particle instrumentation onboard JUNO will directly measure the radiation belt populations. The second Jupiter mission is the ESA JUpiter ICy moon Explorer JUICE (arrival at Jupiter 2030) that will concentrate on the Galilean moons, especially Ganymede, and the interaction with the magnetosphere. During two flybys at the moon Europa the sophisticated field and particle instruments will provide additional plasma, energetic particles and plasma wave measurements in the region of interest at Jupiter. JUICE will significantly improve our understanding of the role wave-particle interactions play for the multi-energy and multi-scale radiation environments of the outer planets.

For all the other outer planets, however, only mission concepts exist and no new mission is currently under development.

Acknowledgments

This work is supported by the Max Planck Society and by the German Space Agency DLR under contracts 50 OH 1101, 50 QJ 90032, 50 OO 1002, 50 OO 1206, 50 QJ 1301. CP would like to acknowledge support from NASA's Outer Planets Research program.

REFERENCES

Albert, J. M., Meredith, N. P., and Horne, R. B. (2009). Three-dimensional diffusion simulation of outer radiation belt electrons during the 9 October 1990 magnetic storm. *J. Geophys. Res.*, **114**, 9214.

Andriopoulou, M., Roussos, E., Krupp, N., Paranicas, C., Thomsen, M., Krimigis, S., Dougherty, M. K., and Glassmeier, K.-H. (2012). A noon-to-midnight electric field and night-side dynamics in Saturn's inner magnetosphere, using microsignature observations. *Icarus*, **220**, 503–513.

Armstrong, T. P., Taherion, S., Manweiler, J., Krimigis, S., Paranicas, C., Mitchell, D., and Krupp, N. (2009). Energetic ions trapped in Saturn's inner magnetosphere. *Planetary and Space Science*, **57**, 1723–1731.

Blake, J. B., Hilton, H. H., and Margolis, S. H. (1983). On the injection of cosmic ray secondaries into the inner Saturnian magnetosphere. I - Protons from the CRAND process. *J. Geophys. Res.*, **88**, 803–807.

Bolton, S. J., Janssen, M., Thorne, R., Levin, S., Klein, M., Gulkis, S., Bastian, T., Sault, R., Elachi, C., Hofstadter, M., Bunker, A., Dulk, G., Gudim, E., Hamilton, G., Johnson, W. T. K., Leblanc, Y., Liepack, O., McLeod, R., Roller, J., Roth, L., and West, R. (2002). Ultra-relativistic electrons in Jupiter's radiation belts. *Nature*, **415**, 987–991.

Bolton, S. J., Thorne, R. M., Bourdarie, S., de Pater, I., and Mauk, B. (2004). *Jupiter's inner radiation belts*, pp. 671–688.

Bolton, S. J., Thorne, R. M., Gurnett, D. A., Kurth, W. S., and Williams, D. J. (1997). Enhanced whistler-mode emissions: Signatures of interchange motion in the Io torus. *Geophys. Res. Lett.*, **24**, 2123–2126.

Brice, N. M. and McDonough, T. R. (1973). Jupiter's radiation belts. *Icarus*, **18**, 206–219.

Bunch, N. L., Spasojevic, M., and Shprits, Y. Y. (2011). On the latitudinal extent of chorus emissions as observed by the Polar Plasma Wave Instrument. *J. Geophys. Res.: Space Physics*, **116**, 4204.

Bunch, N. L., Spasojevic, M., and Shprits, Y. Y. (2012). Off-equatorial chorus occurrence and wave amplitude distributions as observed by the Polar Plasma Wave Instrument. *J. Geophys. Res.: Space Physics*, **117**, 4205.

Burke, B. F. and Franklin, K. L. (1955). Observations of a Variable Radio Source Associated with the Planet Jupiter. *J. Geophys. Res.:*, **60**, 213–217.

Carbary, J. F., Hamilton, D. C., Christon, S. P., Mitchell, D. G., and Krimigis, S. M. (2010). Longitude dependences of energetic H^+ and O^+ at Saturn. *J. Geophys. Res.: Space Physics*, **115**, 7226.

Carbary, J. F., Krimigis, S. M., and Ip, W.-H. (1983). Energetic particle microsignatures of Saturn's satellites. *J. Geophys. Res.*, **88**, 8947–8958.

Carbary, J. F. and Mitchell, D. G. (2013). Periodicities in Saturn's magnetosphere. *Reviews of Geophysics*, **51**, 1–30.

Carbary, J. F., Mitchell, D. G., Krupp, N., and Krimigis, S. M. (2009). L shell distribution of energetic electrons at Saturn. *J. Geophys. Res. (Space Physics)*, **114**, 9210.

Carbary, J. F., Paranicas, C., Mitchell, D. G., Krimigis, S. M., and Krupp, N. (2011). Energetic electron spectra in saturn's plasma sheet. *J. Geophys. Res.: Space Physics*, **116**(A7), n/a–n/a.

Chen, A. J. (1970). Penetration of low-energy protons deep into the magnetosphere. *J. Geophys. Res.*, **75**, 2458.

Cooper, J. F. (1983). Nuclear cascades in Saturn's rings - Cosmic ray albedo neutron decay and origins of trapped protons in the inner magnetosphere. *J. Geophys. Res.*, **88**, 3945–3954.

Cooper, J. F., Johnson, R. E., Mauk, B. H., Garrett, H. B., and Gehrels, N. (2001). Energetic Ion and Electron Irradiation of the Icy Galilean Satellites. *Icarus*, **149**, 133–159.

Coroniti, F.Ṽ. (1974). Energetic Electrons in Jupiter's Magnetosphere. *Astrophys. J. Suppl. Ser.*, **27**, 261–281.

Coroniti, F. V., Scarf, F. L., Kennel, C. F., and Kurth, W. S. (1984). Analysis of chorus emissions at Jupiter. *J. Geophys. Res.: Space Phypics*, **89**, 3801–3820.

Coroniti, F. V., Scarf, F. L., Kennel, C. F., Kurth, W. S., and Gurnett, D. A. (1980). Detection of Jovian whistler mode chorus - Implications for the Io torus aurora. *Geophys. Res. Lett.*, **7**(1), 45–48.

de Pater, I. and Dunn, D. E. (2003). VLA observations of Jupiter's synchrotron radiation at 15 and 22 GHz. *Icarus*, **163**, 449–455.

Dessler, A. J. (ed.) (1983). *Physics of the Jovian magnetosphere.* Cambridge Univ. Press, New York.

Dialynas, K., Krimigis, S. M., Mitchell, D. G., Hamilton, D. C., Krupp, N., and Brandt, P. C. (2009). Energetic ion spectral characteristics in the Saturnian magnetosphere using Cassini/MIMI measurements. *J. Geophys. Res. (Space Physics)*, **114**, 1212.

Divine, N. (1990). Saturn energetic electron and proton spectra for Cassini. *JPL IOM 5217-90-023.*

Divine, N. and Garrett, H. B. (1983). Charged particle distributions in Jupiter's magnetosphere. *J. Geophys. Res.*, **88**, 6889–6903.

Dougherty, M. K., Esposito, L.W., and Krimigis, S.M. (2009). *Saturn from Cassini-Huygens.* Springer.

Dougherty, M. K., Kellock, S., Southwood, D. J., Balogh, A., Smith, E. J., Tsurutani, B. T., Gerlach, B., Glassmeier, K.-H., Gleim, F., Russell, C. T., Erdos, G., Neubauer, F. M., and Cowley, S. W. H. (2004). The Cassini Magnetic Field Investigation. *Space Sci. Rev.*, **114**, 331–383.

Evans, R., Garrett, H.B, Cohen, C.M.S., and Stone, E.C. (2003). Galileo heavy ion radiation model. *JPL D-24813.*

Fälthammar, C.-G. (1965). Effects of Time-Dependent Electric Fields on Geomagnetically Trapped Radiation. *J. Geophys. Res.*, **70**, 2503–2516.

Fillius, W. and Ip, W. H. (1980). Trapped radiation belts of Saturn - First look. *Science*, **207**, 425–431.

Fischer, H. M., Pehlke, E., Wibberenz, G., Lanzerotti, L. J., and Mihalov, J. D. (1996). High-Energy Charged Particles in the Innermost Jovian Magnetosphere. *Science*, **272**, 856–858.

Garrard, T. L., Gehrels, N., and Stone, E. C. (1992). The Galileo heavy element monitor. *Space Sci. Rev.*, **60**, 305–315.

Garrett, H. B., Jun, I., Ratliff, J.M., Evans, R.W., Clough, G.A., and McEntire, R.E. (2003). Galileo Interim Radiation Electron model. *JPL Publication 03-006.*

Garrett, H. B., Levin, S. M., Bolton, S. J., Evans, R. W., and Bhattacharya, B. (2005, February). A revised model of Jupiter's inner electron belts: Updating the Divine radiation model. *Geophys. Res. Lett.*, **32**, L04104.

Gurnett, D. A., et al. (2005). Radio and Plasma Wave Observations at Saturn from Cassini's Approach and First Orbit. *Science*, **307**, 1255–1259.

Gurnett, D. A., et al. (2004). The Cassini Radio and Plasma Wave Investigation. *Space Sci. Rev.*, **114**, 395–463.

Gurnett, D. A., Kurth, W. S., Roux, A., Bolton, S. J., and Kennel, C. F. (1996). Galileo Plasma Wave Observations in the lo Plasma Torus and Near lo. *Science*, **274**, 391–392.

Gurnett, D. A., Kurth, W. S., Roux, A., Bolton, S. J., Thomsen, E. A., and Groene, J. B. (1998). Galileo plasma wave observations near Europa. *Geophys. Res. Lett.*, **25**, 237–240.

Gurnett, D. A., Kurth, W. S., Roux, A., J., Bolton S., and Kennel, C. F. (1996). Evidence for a magnetosphere at Ganymede from plasma-wave observations by the Galileo spacecraft. *Nature*, **384**, 535–537.

Gurnett, D. A., Kurth, W. S., and Scarf, F. L. (1979). Plasma wave observations near Jupiter - Initial results from Voyager 2. *Science*, **206**, 987–991.

Gurnett, D. A., Kurth, W. S., and Scarf, F. L. (1981). Plasma waves near Saturn - Initial results from Voyager 1. *Science*, **212**, 235–239.

Gurnett, D. A. and Scarf, F. L. (1983). *Plasma waves in the Jovian magnetosphere*, pp. 285–316.

Horne, R.B., Thorne, R.M., Glauert, S.A., Menietti, J.D., Shprits, Y.Y., and Gurnett, D.A (2008). Gyro-resonant electron acceleration at Jupiter. *Nature Physics*.

Horne, R. B., Thorne, R. M., Glauert, S. A., Albert, J. M., Meredith, N. P., and Anderson, R. R. (2005). Timescale for radiation belt electron acceleration by whistler mode chorus waves. *J. Geophys. Res.: Space Physics*, **110**, 3225.

Hospodarsky, G. B., et al. (2008). Observations of chorus at Saturn using the Cassini Radio and Plasma Wave Science instrument. *J. Geophys Res.: Space Physics*, **113**, 12206.

Hospodarsky, G. B., et al. (2012). Plasma Wave Observations at Earth, Jupiter, and Saturn. *Washington DC American Geophysical Union Geophysical Monograph Series*, **119**, 415–430.

Inan, U. S., Helliwell, R. A., and Kurth, W. S. (1983). Terrestrial versus Jovian VLF chorus - A comparative study. *J. Geophys. Res: Space Physics*, **88**, 6171–6180.

Jones, G. H., et al., (2006a). Formation of Saturn's ring spokes by lightning-induced electron beams. *Geophys. Res. Lett.*, **33**, 21202.

Jones, G. H., et al. (2006b). Enceladus' Varying Imprint on the Magnetosphere of Saturn. *Science*, **311**, 1412–1415.

Jun, I., Garrett, H. B., and Evans, R. W. (2005). High-Energy Trapped Particle Environments at Jupiter: An Update. *IEEE Transactions on Nuclear Science*, **52**, 2281–2286.

Katoh, Y. and Omura, Y. (2011). Amplitude dependence of frequency sweep rates of whistler mode chorus emissions. *J. Geophys. Res.: Space Physics*, **116**, 7201.

Katoh, Y., et al. (2011). Whistler mode chorus enhancements in association with energetic electron signatures in the Jovian magnetosphere. *J. Geophys. Res.: Space Physics*, **116**, 2215.

Kellogg, P. J. (1959). Van Allen radiation of solar origin. *Nature*.

Kennel, C. F. and Coroniti, F. V. (1975). Is Jupiter's magnetosphere like a pulsar's or Earth's. *Space Sci. Rev.*, **17**, 857–883.

Kennel, C. F. and Petschek, H. E. (1966). Limit on Stably Trapped Proton Fluxes. *J. Geophys. Res.*, **71**(1), 1–28.

Khurana, K. K. (1992). A generalized hinged-magnetodisc model of Jupiter's nightside current sheet. *J. Geophys. Res: Space Physics*, **97**, 6269–6276.

Khurana, K. K. (1997, June). Euler potential models of Jupiter's magnetospheric field., **102**, 11295–11306.

Kivelson, M. G., Khurana, K. K., Russell, C. T., and Walker, R. J. (1997). Intermittent short-duration magnetic field anomalies in the Io torus: Evidence for plasma interchange? *Geophys. Res. Lett.*, **24**, 2127–2130.

Klein, M. J., et al. (2001). Cassini-Jupiter Microwave Observing Campaign: DSN and GAVRT Observations of Jovian Synchrotron Radio Emission. In *Planetary Radio Emissions V* (ed H. O. Rucker, M. L. Kaiser, and Y. Leblanc), p. 221.

Kollmann, P., Roussos, E., Paranicas, C., Krupp, N., and Haggerty, D. K. (2013, January). Processes forming and sustaining Saturn's proton radiation belts. *Icarus*, **222**, 323–341.

Kollmann, P., et al. (2011). Energetic particle phase space densities at Saturn: Cassini observations and interpretations. *J. Geophys. Res.*, **116**, A05222.

Krimigis, S.M., Sergis, N., Mitchell, D.G., Hamilton, D.C., and Krupp, N. (2007). A dynamic, rotating ring current around Saturn. *Nature*, **450**, 1050.

Krupp, N., et al. (2009). Energetic particles in Saturn's magnetosphere during the Cassini nominal mission (July 2004-July 2008). *Planet. Space Sci.*, **57**, 1754–1768.

Krupp, N., et al. (2004). Dynamics of the Jovian magnetosphere, in *Jupiter – The Planet, Satellites and Magnetosphere*, vol. 1 (eds F. Bagenal, T. E. Dowling and W. B. McKinnon), Cambridge University Press, Cambridge. pp. 617–638.

Kurth, W. S., et al. (2009). Auroral Processes; in *Saturn from Cassini – Huygens* (eds M. Dougherty, L.W. Esposito and S.M. Krimigis), Springer, Berlin, p. 333.

Kurth, W. S. and Gurnett, D. A. (1991). Plasma waves in planetary magnetospheres. *J. Geophys. Res.*, **96**, 18977+.

Kurth, W. S., Gurnett, D. A., Persoon, A. M., Roux, A., Bolton, S. J., and Alexander, C. J. (2001). The plasma wave environment of Europa. *Planet. Space Sci.*, **49**, 345–363.

Kurth, W. S., Gurnett, D. A., Scarf, F. L., and Barbosa, D. D. (1983). A survey of electrostatic waves in Saturn's magnetosphere. *J. Geophys. Res.*, **88**, 8959–8970.

Liu, X., Hill, T. W., Wolf, R. A., Sazykin, S., Spiro, R. W., and Wu, H. (2010). Numerical simulation of plasma transport in Saturn's inner magnetosphere using the Rice Convection Model. *J. Geophys. Res.: Space Physics*, **115**, 12254.

Lorenzato, L., Sicard, A., and Bourdarie, S. (2012). A physical model for electron radiation belts of Saturn. *J. Geophys Res.: Space Physics*, **117**, 8214.

Mauk, B. H. and Fox, N. J. (2010). Electron radiation belts of the solar system. *J. Geophys. Res.: Space Physics*, **115**, 12220.

Mauk, B. H., et al. (2009). Fundamental Plasma Processes in Saturn's Magnetosphere; in *Saturn from Cassini – Huygens* (eds M. Dougherty, L. W. Esposito ans S. M. Krimigis), Springer, Berline. p. 281.

Mauk, B. H., et al. (2004). Energetic ion characteristics and neutral gas interactions in Jupiter's magnetosphere. *J. Geophys. Res.*, **109**, 9.

Mauk, B. H. and Saur, J. (2007). Equatorial electron beams and auroral structuring at Jupiter. *J. Geophys. Res.: Space Physics*, **112**, 10221.

Menietti, J. D., Horne, R. B., Gurnett, D. A., Hospodarsky, G. B., Piker, C. W., and Groene, J. B. (2008a). A survey of Galileo plasma wave instrument observations of Jovian whistler-mode chorus. *Annales Geophysicae*, **26**, 1819–1828.

Menietti, J. D., Santolik, O., Rymer, A. M., Hospodarsky, G. B., Gurnett, D. A., and Coates, A. J. (2008b). Analysis of plasma waves observed in the inner Saturn magnetosphere. *Annales Geophysicae*, **26**, 2631–2644.

Menietti, J. D., Santolik, O., Rymer, A. M., Hospodarsky, G. B., Persoon, A. M., Gurnett, D. A., Coates, A. J., and Young, D. T. (2008c). Analysis of plasma waves observed within local plasma injections seen in Saturn's magnetosphere. *J. Geophys. Res.: Space Physics*, **113**, 5213.

Menietti, J. D., Schippers, P., Katoh, Y., Leisner, J. S., Hospodarsky, G. B., Gurnett, D. A., and Santolik, O. (2013). Saturn chorus intensity variations. *J. Geophys. Res.: Space Physics*, **118**, 5592–5602.

Menietti, J. D., Shprits, Y. Y., Horne, R. B., Woodfield, E. E., Hospodarsky, G. B., and Gurnett, D. A. (2012). Chorus, ECH, and Z mode emissions observed at Jupiter and Saturn and possible electron acceleration. *J. Geophys. Res.: Space Physics*, **117**, 12214.

Meredith, N. P., Horne, R. B., and Anderson, R. R. (2001). Substorm dependence of chorus amplitudes: Implications for the acceleration of electrons to relativistic energies. *J. Geophys. Res.*, **106**, 13165–13178.

Meredith, N. P., Horne, R. B., Iles, R. H. A., Thorne, R. M., Heynderickx, D., and Anderson, R. R. (2002). Outer zone relativistic electron acceleration associated with substorm-enhanced whistler mode chorus. *J. Geophys. Res.: Space Physics*, **107**, 1144.

Meredith, N. P., Horne, R. B., Thorne, R. M., and Anderson, R. R. (2003). Favored regions for chorus-driven electron acceleration to relativistic energies in the Earth's outer radiation belt. *J. Geophys. Res.: Space Physics*, **30**, 1871.

Millan, R. M. and Baker, D. N. (2012). Acceleration of Particles to High Energies in Earth's Radiation Belts. *Space Sci. Rev.*, **173**, 103–131.

Millan, R. M. and Thorne, R. M. (2007). Review of radiation belt relativistic electron losses. *Journal of Atmospheric and Solar-Terrestrial Physics*, **69**, 362–377.

Mitchell, D. G., Carbary, J. F., Cowley, S. W. H., Hill, T. W., and Zarka, P. (2009). The Dynamics of Saturn's Magnetosphere; in *Saturn from Cassini – Huygens* (eds M. Dougherty, L.W. Esposito and S.M. Krimigis), Springer, Berlin. p. 257.

Omura, Y. and Summers, D. (2006). Dynamics of high-energy electrons interacting with whistler mode chorus emissions in the magnetosphere. *Journal of Geophysical Research (Space Physics)*, **111**, 9222.

Paranicas, C., Carlson, R. W., and Johnson, R. E. (2001). Electron bombardment of Europa. *Geophys. Res. Lett.*, **28**, 673–676.

Paranicas, C., Cooper, J. F., Garrett, H. B., Johnson, R. E., and Sturner, S. J. (2009). Europa's Radiation Environment and Its Effects on the Surface, In *Europa* (eds R. T. Pappalardo, W. B. McKinnon, K.K. Khurana). The University of Arizona Press, Tucson in collaboration with Lunar and Planetary Institute, Houston. p. 529.

Paranicas, C., Mauk, B. H., McEntire, R. W., and Armstrong, T. P. (2003). The radiation environment near Io. *Geophys. Res. Lett.*, **30**, 1–1.

Paranicas, C., et al. (2010a). Asymmetries in Saturn's radiation belts. *J. Geophys. Res.: Space Physics*, **115**, 7216.

Paranicas, C., et al. (2008). Sources and losses of energetic protons in Saturn's magnetosphere. *Icarus*, **197**, 519–525.

Paranicas, C., et al. (2005). Evidence of Enceladus and Tethys microsignatures. *Geophys. Res. Lett.*, **32**, 20101.

Paranicas, et al. (2010b). Transport of energetic electrons into Saturn's inner magnetosphere. *J. Geophys. Res.: Space Physics*, **115**, 9214.

Paranicas, C., et al. (2012). Energetic charged particle weathering of Saturn's inner satellites. *Planet. Space Sci.*, **61**, 60–65.

Persoon, A. M., Gurnett, D. A., Kurth, W. S., and Groene, J. B. (2006). A simple scale height model of the electron density in Saturn's plasma disk. *Geophys. Res. Lett.*, **33**, 18106.

Persoon, A. M., et al. (2005). Equatorial electron density measurements in Saturn's inner magnetosphere. *Geophys. Res. Lett.*, **32**, 23105.

Persoon, et al. (2009). A diffusive equilibrium model for the plasma density in Saturn's magnetosphere. *J. Geophys. Res.*, **114**, 4211.

Randall, B. A. (1994). Energetic electrons in the magnetophere of Saturn. *Journal Geophys. Res.*, **99**, 8771–8785.

Roederer, J. G. (1970). *Dynamics of geomagnetically trapped radiation.* Springer, New York.

Roussos, E., et al. (2007). Electron microdiffusion in the Saturnian radiation belts: Cassini MIMI/LEMMS observations of energetic electron absorption by the icy moons. *J. Geophys. Res.*, **112**, 6214.

Roussos, E., et al. (2008). Discovery of a transient radiation belt at Saturn. *Geophys. Res. Lett.*, **35**, 22106.

Roussos, E., et al. (2014). The variable extension of Saturn's electron radiation belts. *Planet. Space Sci.*, **104**, 3–17.

Roussos, E., et al. (2011). Long- and short-term variability of Saturn's ionic radiation belts. *J. of Geophys. Res.: Space Physics*, **116**, A02217.

Roussos, E., et al. (2005). Low energy electron microsignatures at the orbit of Tethys: Cassini MIMI/LEMMS observations. *Geophys. Res. Lett.*, **32**, 24107.

Rymer, A.M., et al. (2008). Electron circulation in Saturn's magnetosphere. *J. Geophys. Res.*, **113**.

Santolík, O., Gurnett, D. A., Pickett, J. S., Parrot, M., and Cornilleau-Wehrlin, N. (2003, July). Spatio-temporal structure of storm-time chorus. *J. Geophys. Res.: Space Physics*, **108**, 1278.

Santolík, O., Gurnett, D. A., Pickett, J. S., Parrot, M., and Cornilleau-Wehrlin, N. (2004). A microscopic and nanoscopic view of storm-time chorus on 31 March 2001. *Geophys. Res. Lett.*, **31**, 2801.

Santos-Costa, D., Blanc, M., Maurice, S., and Bolton, S. J. (2003). Modeling the electron and proton radiation belts of Saturn. *Geophys. Res. Lett.*, **30**(20), 6–1.

Santos-Costa, D. and Bolton, S. J. (2008). Discussing the processes constraining the Jovian synchrotron radio emission's features. *Planet. Space Sci.*, **56**, 326–345.

Santos-Costa, D., Bolton, S. J., Thorne, R. M., Miyoshi, Y., and Levin, S. M. (2008). Investigating the origins of the Jovian decimetric emission's variability. *J. Geophys. Res.: Space Physics*, **113**, 1204.

Santos-Costa, D. and Bourdarie, S. A. (2001). Modeling the inner Jovian electron radiation belt including non-equatorial particles. *Planet. Space Sci.*, **49**, 303–312.

Scarf, F.L., Frank, L.A., Gurnett, D.A., Lanzerotti, L.J., Lazarus, A., and Sittler, E.C. (1984). Measurements of plasma, plasma waves, and suprathermal charged particles in Saturn's inner magnetosphere. In *Saturn*, University of Arizona Press, Tucson. pp. 318–353.

Scarf, F. L., Coroniti, F. V., Gurnett, D. A., and Kurth, W. S. (1979a). Pitch-angle diffusion by whistler mode waves near the Io plasma torus. *Geophys. Res. Lett.*, **6**, 653–656.

Scarf, F. L., Gurnett, D. A., and Kurth, W. S. (1979b). Jupiter plasma wave observations - an initial Voyager 1 overview. *Science*, **204**, 991–995.

Scarf, F. L., Gurnett, D. A., Kurth, W. S., and Poynter, R. L. (1982). Voyager 2 plasma wave observations at Saturn. *Science*, **215**, 587–594.

Scarf, F. L., Gurnett, D. A., Kurth, W. S., and Poynter, R. L. (1983). Voyager plasma wave measurements at Saturn. *J. Geophys. Res.*, **88**, 8971–8984.

Selesnick, R. S. (1993). Micro- and macro-signatures of energetic charged particles in planetary magnetospheres. *Adv. Space Res.*, **13**, 221–230.

Sergis, N., Krimigis, S. M., Mitchell, D. G., Hamilton, D. C., Krupp, N., Mauk, B. M., Roelof, E. C., and Dougherty, M. (2007). Ring current at Saturn: Energetic particle pressure in Saturn's equatorial magnetosphere measured with Cassini/MIMI. *Geophys. Res. Lett.*, **34**, 9102.

Shprits, Y. Y., Elkington, S. R., Meredith, N. P., and Subbotin, D. A. (2008a). Review of modeling of losses and sources of relativistic electrons in the outer radiation belt I: Radial transport. *Journal of Atmospheric and Solar-Terrestrial Physics (JASTP)*, **70**, 1679–1693.

Shprits, Y. Y., Menietti, J. D., Gu, X., Kim, K. C., and Horne, R. B. (2012). Gyroresonant interactions between the radiation belt electrons and whistler mode chorus waves in the radiation environments of Earth, Jupiter, and Saturn: A comparative study. *J. Geophys. Res.: Space Physics*, **117**, 11216.

Shprits, Y. Y., Subbotin, D., and Ni, B. (2009). Evolution of electron fluxes in the outer radiation belt computed with the VERB code. *J. Geophys. Res.*, **114**, 11209.

Shprits, Y. Y., Subbotin, D. A., Meredith, N. P., and Elkington, S. R. (2008b). Review of modeling of losses and sources of relativistic electrons in the outer radiation belt II: Local acceleration and loss. *Journal of Atmospheric and Solar-Terrestrial Physics*, **70**, 1694–1713.

Shprits, Y. Y. and Thorne, R. M. (2004). Time dependent radial diffusion modeling of relativistic electrons with realistic loss rates. *Geophys. Res. Lett.*, **31**, 8805.

Sicard, A. and Bourdarie, S. (2004). Physical Electron Belt Model from Jupiter's surface to the orbit of Europa. *J. Geophys. Res.: Space Physics*, **109**, 2216.

Sicard-Piet, A., Bourdarie, S., and Krupp, N. (2011). JOSE: A New Jovian Specification Environment Model. *IEEE Transactions on Nuclear Science*, **58**, 923–931.

Simpson, J. A., Bastian, T. S., Chenette, D. L., McKibben, R. B., and Pyle, K. R. (1980). The trapped radiations of Saturn and their absorption by satellites and rings. *J. Geophys. Res.*, **85**, 5731–5762.

Siscoe, G. L. and Summers, D. (1981). Centrifugally driven diffusion of Iogenic plasma. *J. Geophys. Res.*, **86**, 8471–8479.

Smyth, W. H. and Marconi, M. L. (2006). Europa's atmosphere, gas tori, and magnetospheric implications. *Icarus*, **181**, 510–526.

Spasojevic, M. and Shprits, Y. Y. (2013). Chorus functional dependencies derived from CRRES data. *Geophys. Res. Lett.*, **40**, 3793–3797.

Su, Z., Xiao, F., Zheng, H., and Wang, S. (2010). STEERB: A three-dimensional code for storm-time evolution of electron radiation belt. *J. Geophys. Res.*, **115**, 9208.

Tang, R. and Summers, D. (2012). Energetic electron fluxes at Saturn from Cassini observations. *J. Geophys. Res.: Space Physics*, **117**, 6221.

Tao, C., Badman, S. V., and Fujimoto, M. (2011a). UV and IR auroral emission model for the outer planets: Jupiter and Saturn comparison., **213**, 581–592.

Tao, X., et al. (2011b). Importance of plasma injection events for energization of relativistic electrons in the Jovian magnetosphere. *J. Geophys. Res.*, **116**, 1206.

Thomsen, M. F., et al. (2012). Saturn's inner magnetospheric convection pattern: Further evidence. *J. Geophys. Res.: Space Physics*, **117**, 9208.

Thomsen, M. F. and van Allen, J. A. (1980). Motion of trapped electrons and protons in Saturn's inner magnetosphere. *J. Geophys. Res.*, **85**, 5831–5834.

Thorne, R. M., Armstrong, T. P., Stone, S., Williams, D. J., McEntire, R. W., Bolton, S. J., Gurnett, D. A., and Kivelson, M. G. (1997). Galileo evidence for rapid interchange transport in the Io torus. *Geophys. Res. Lett.*, **24**, 2131–2134.

Thorne, R. M. and Tsurutani, B. T. (1979). Diffuse Jovian aurora influenced by plasma injection from Io. *Geophys. Res. Lett.*, **6**, 649–652.

Tomás, A., Woch, J., Krupp, N., Lagg, A., Glassmeier, K.-H., Dougherty, M. K., and Hanlon, P. G. (2004a). Changes of the energetic particles characteristics in the inner part of the Jovian magnetosphere: a topological study. *Planetary and Space Science*, **52**, 491–498.

Tomás, A. T., Woch, J., Krupp, N., Lagg, A., Glassmeier, K.-H., and Kurth, W. S. (2004b). Energetic electrons in the inner part of the Jovian magnetosphere and their relation to auroral emissions. *J. Geophys. Res.: Space Physics*, **109**, 6203.

Tsurutani, B. T. and Smith, E. J. (1977). Two types of magnetospheric ELF chorus and their substorm dependences., **82**, 5112–5128.

Tverskoy, B. A. (1965). Transport and Acceleration of Charged Particles in the Earth's Magnetosphere. *Geomagnetism and Aeronomy*, **5**, 617.

Van Allen, J. A. (1983). Absorption of energetic protons by Saturn's Ring G. *J. Geophys. Res.*, **88**, 6911–6918.

Van Allen, J. A., Baker, D. N., Randall, B. A., and Sentman, D. D. (1974). The magnetosphere of Jupiter as observed with Pioneer 10: 1. Instrument and principal findings. *J. Geophys. Res.*, **79**, 3559.

Van Allen, J. A. and Grosskreutz, C. L. (1989). Relativistic electrons in Saturn's inner magnetosphere and an estimate of their synchrotron emission. *J. Geophys. Res.*, **94**, 8731–8738.

Van Allen, J. A., et al. (1975). Pioneer 11 observations of energetic particles in the Jovian magnetosphere. *Science*, **188**, 459–462.

Varotsou, A., Boscher, D., Bourdarie, S., Horne, R. B., Glauert, S. A., and Meredith, N. P. (2005). Simulation of the outer radiation belt electrons near geosynchronous orbit including both radial diffusion and resonant interaction with Whistler-mode chorus waves. *Geophys. Res. Lett.*, **32**, 19106.

Vogt, R. E., et al. (1979a). Voyager 1 – Energetic ions and electrons in the Jovian magnetosphere. *Science*, **204**, 1003–1007.

Vogt, R. E., et al. (1979b). Voyager 2 – Energetic ions and electrons in the Jovian magnetosphere. *Science*, **206**, 984–987.

Williams, D. J., McEntire, R. W., Jaskulek, S., and Wilken, B. (1992). The Galileo Energetic Particles Detector. *Space Sci. Rev.*, **60**, 385–412.

Wilson, R. J., Bagenal, F., Delamere, P. A., Desroche, M., Fleshman, B. L., and Dols, V. (2013). Evidence from radial velocity measurements of a global electric field in Saturn's inner magnetosphere. *J. Geophys. Res.: Space Physics*, **118**, 2122–2132.

Woodfield, E. E., Horne, R. B., Glauert, S. A., Menietti, J. D., and Shprits, Y. Y. (2013). Electron acceleration at Jupiter: input from cyclotron-resonant interaction with whistler-mode chorus waves. *Annales Geophysicae*, **31**, 1619–1630.

Xiao, F., Thorne, R. M., Gurnett, D. A., and Williams, D. J. (2003). Whistler-mode excitation and electron scattering during an interchange event near Io. *Geophys. Res. Lett.*, **30**, 1749.

16

Fields and Waves Influencing Radiation Belt Dynamics—Results from the Van Allen Probes Mission

C. A. Kletzing

■ Abstract

An overview of early results that have been published in the peer-reviewed literature from the electric and magnetic field instruments on the NASA Van Allen Probes mission is presented. New science on a variety of waves from low frequencies—ULF and EMIC waves—to higher frequencies—whistler mode chorus and plasmaspheric hiss are summarized with an emphasis on key result. Additionally, the physics of how these waves interact with and alter the populations of radiation belt particles is discussed.

16.1 Introduction

The Van Allen Probes mission (originally called the Radiation Belt Storm Probes and renamed in November, 2012) consists of two identically instrumented spacecraft in

Waves, Particles, and Storms in Geospace. First Edition. Georgios Balasis, Ioannis A. Daglis, Ian R. Mann.

elliptical orbits around the Earth. Apogee is 5.8 R_E and perigee is 600 km with an orbital period of about nine hours. Complete local time coverage (due to precession of the orbit) occurs with a period of approximately one year and eight months. The spacecraft have the most complete and modern set of instruments ever flown in the radiation belts and are rapidly evolving our understanding of the radiation belts. This chapter focuses on the electric and magnetic field wave and DC field measurements from the Van Allen Probes.

To orient the reader for the results that are presented in what follows, an overview of the key wave modes that are important for wave-particle interactions in the inner magnetosphere is shown in Figure 16.1. Four types of waves are shown by the shaded regions in the figure. Whistler mode plasmaspheric hiss is found inside the plasmasphere at all local times. Whistler mode chorus is found predominantly on the morning side extending from midnight to early afternoon. Radially, chorus is located ouside the plasmasphere and can be observed well beyound the orbit of the Van Allen Probes. Magnetosonic equatorial noise is thought to be primarily a dayside phenomenon, while electromagnetic ion cyclotron (EMIC) waves occur in the evening sector. Figure 16.1 also shows the directions of particle drifts; eastward for electons and westward for ions.

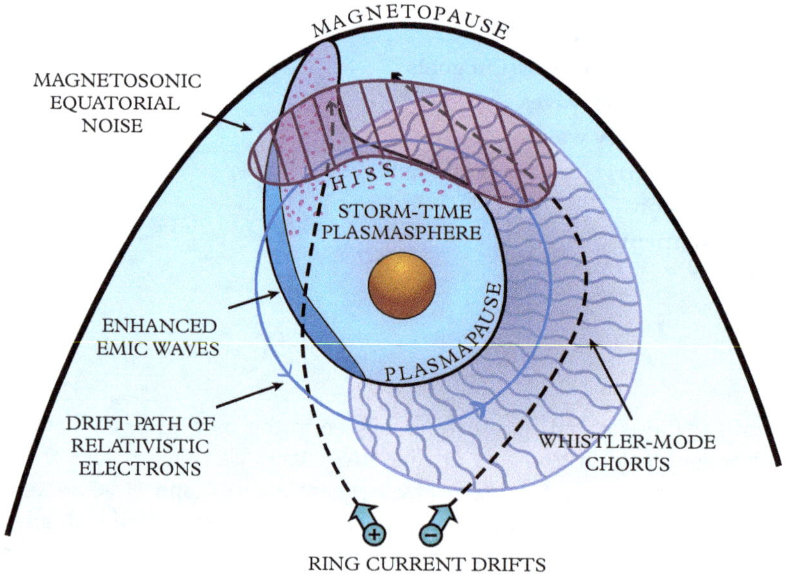

Figure 16.1 *An overview of the key waves important for radiation belt wave-particle interactions. The shaded areas show the approximate locations within the inner magnetosphere at which the waves occur.*

16.2 Instrumentation

There are two sets of instrumentation on the Van Allen Probes for the measurement of both the DC and wave components of electric and magnetic fields. The Electric and Magnetic Field Instrumentation and Integrated Science (EMFISIS) investigation provides both the wave electric and magnetic field measurements as well as the DC magnetic field measurements. The wave instrumentation on the Van Allen Probes consists of hardware to make three-axis measurements of both the electric and magnetic field from 2 Hz to 12 kHz. The AC magnetic field measurements are from a three-axis magnetic search coil (MSC) on a boom which extends 3 m from the spacecraft body. The AC electric field measurements are made from signals provided to the EMFISIS receivers by the Electric Field and Wave (EFW) experiment. These signals are derived from differencing the potential from pairs of spherical probes at the ends of booms both in the spin plane of the spacecraft (100 m separation) and along the spin axis (14 m separation). Both AC electric and AC magnetic field measurements are simultaneously sampled at 35 000 samples/s. Additionally, a single, selectable axis of the electric field is measured from 10 kHz to 500 kHz. The DC magnetic field is measured by a three-axis fluxgate magnetometer (MAG) on a boom extending 3 m from the spacecraft and opposite to the MSC. The DC magnetic field is sampled at 64 vectors/s, providing a frequency range of 0–30 Hz. Further details of the EMFISIS instrumentation can be found in (Kletzing et al., 2013)

The EFW investigation measures the electric field along three orthogonal axes. The EFW measurements include vector electric field measurements from DC to 15 Hz as well as power in wave frequency bands, and higher time resolution measurements of electric fields and potentials relative to the spacecraft body utilizing a large burst storage memory. Data are stored in the memory and then specific intervals are telemetered after examining survey data for interesting events. The experiment also measures the spacecraft potential relative to the plasma which can be used as a proxy measurement for the local plasma density when calibrated to the upper hybrid frequency. Further details can be found in (Wygant et al., 2013).

16.3 Van Allen Probes mission goals

The purpose of the Van Allen Probes mission is to investigate the acceleration. loss, and transport of radiation belt particles. The requirements for the mission break this into three primary science questions:

- Which physical processes produce radiation belt enhancement events?
- What are the dominant mechanisms for relativistic electron loss?
- How do ring current and other geomagnetic processes affect radiation belt behavior?

These questions defined the overall mission design including the needed instrumentation, orbit, and spacecraft design as well as, most importantly, the fundamental science to be fulfilled for mission success.

To further elucidate the science of the Van Allen Probes, these broad goals are broken down into more specific scientific areas. Specific to this report are those aspects of radiation belt physics which involve electric and magnetic field effects on the particle populations. More specifically, the mission requirements for field measurements focus on spatiotemporal properties of radiation belt electric and magnetic fields including frequency ranges, amplitudes, polarizations, wave normal directions, coherence, energy transport, and phase between components. From these parameters, the mission science objectives are to determine types of waves, wave growth rates, diffusion coefficients, wave-particle acceleration, particle pitch angle scattering and loss rates, as well as statistical occurrence rates and properties of both waves and DC electric and magnetic field as a function of position and geomagnetic activity.

16.4 Whistler mode waves

Whistler mode waves, including whistlers, chorus, and plasmaspheric hiss, have long been associated with radiation belt phenomena (Millan and Thorne, 2007; Thorne et al., 2010; Millan and Baker, 2012). Several important advances in our understanding of these waves have come about due to the high quality of the Van Allen Pores measurements. The EMFISIS measurements have provided extensive coverage of whistler mode waves, including a large amount of waveform data in addition to very regular coverage in the survey mode which provides the full spectral matrix enabling a variety of studies of the wave propagation parameters (Santolík et al., 2003).

16.4.1 Chorus emissions

Chorus emissions are thought to be one of the most important local accelerators of outer belt radiation belt electrons (Summers et al., 1998). The Van Allen Probes have enabled the most detailed look at these waves ever flown. Figure 16.2 shows an example of whistler-mode chorus as measured by the EMFISIS magnetic search coil. The lower panel shows the waverform and the upper panel shows the spectrogram of the waveform. To guide the eye, a black line on the spectrogram shows one-half of the electron cyclotraon frequency. Both upper and lower band chorus are visible in Figure 16.2. The lower band shows several discrette "rising tone" chorus elements while the upper band shows a more unstructured emission. The two bands are separated by a gap around one-half the cyclotron frequency.

The fine structure of large amplitude chorus (greater than a few hundred pT) has been investigated using the EMFISIS waveform data to discern the detailed character of rising tone elements. (Santolík et al., 2014). This study found that chorus elements are composed of several subpackets and that the peak amplitudes of the subpackets tends

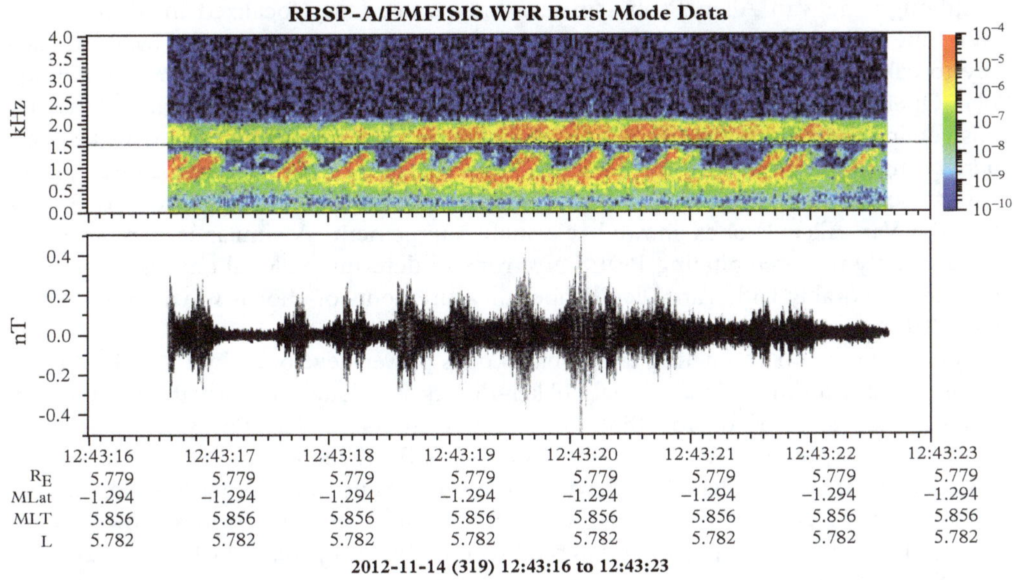

Figure 16.2 *An example of chorus elements as measured by the EMFISIS WFR receiver. The lower panel shows the magnetic field waveform along one axis and the upper panel shows a spectrogram dervied from the waveform. The black line denotes one-half of the electron cyclotron frequency.*

to decrease with rising frequency. The study also found that largest amplitude chorus element subpackets tend to be quasi-field aligned relative to the background magnetic field while the weaker packets turned away from the field-aligned direction.

An intriguing event correlating quasi-periodic bursts of chorus emission with distinct conic-like pitch angle bunching in the 17–26 keV electrons was reported by Fennell et al. (2014). The enhanced electrons fluxes were observed between 75° and 80° in pitch angle while fluxes at lower pitch angles and near 90° were unchanged. Estimates of the resonant energy for the observed chorus frequencies utilizing the local plasma density determined from the upper hybrid line combined with the measured background magnetic field gave a result of 15–35 keV in reasonable agreement with the particle distributions. No waveform burst data was available at the times of the events, so resolution of the question of whether the waves were energizing the particles or vice versa will await further event studies.

The Van Allen Probes mission has also enabled advances in modeling the energization of outer belt electrons by chorus. During a period of relatively quiet geomagnetic conditions in February, 2013, Su et al. (2014) examined outer belt electron enhancement for values of $L > 5$, finding elevated levels of chorus activity along with increased fluxes of relativistic electrons above 1 MeV. Using a Gaussian distribution of power to approximate the chorus levels observed by the RBSP-A satellite, they were able to get reasonable agreement between their two-dimensional Fokker-Planck model and the observed fluxes.

Although the Van Allen Probes measurements are fairly localized in MLT at any given time (due to the nature of their overlapping orbits), they have been used successfully to calibrated a proxy model for chorus using particle data from the low-altitude Polar Operational Environmental Satellites (POES) spacecraft (Li et al., 2013a). By constructing the ratio between precipitating and mirroring particles in the 30–100 keV energy range on each of the multiple POES spacecraft, a proxy for global chorus power is constructed. Comparisons of this proxy with direct measurements of chorus power from the Van Allen Probes showed reasonable agreement. A similar empirical model that uses only the precipitating POES electrons to determine global chorus power also compared favorably with Van Allen Probes measurements of chorus wave power (Chen et al., 2014).

The technique for estimating the global chorus power developed by Li et al. (2013a) was used in a two-dimensional global Fokker-Planck simulation to study the energizaiton of outer belt electrons during the storm of 9 October, 2012. The study compared high-resolution phase space density measurements from the higher energy particle instruments on the Van Allen Probes with the model results for the phase space density (Thorne et al., 2013b). Chorus observations from the Van Allen Probes showed significantly elevated levels of power and agreed with the POES proxies of Li et al. (2013a). Figure 16.3 shows a comparison of the time evolution of the phase space density over time taken form the report. The left hand side shows phase space density from the REPT instrument on the Van Allen Probes (Baker et al., 2013a). The right hand side shows the results from the Fokker-Planck model. As can be seen, the agreement is remarkably good. In the comparison of the detailed pitch angle distributions of phase space density, similarly good agreement was achieved. It is also interesting to note that the use of the measured plasma density instead of a model density (which tends to produce an overestimate) was essential to the excellent agreement in the several MeV energy range. This result has made a compelling case validating the theoretical work that has shown that chorus can lead to rapid acceleration of radiation belt electrons into the relativistic and ultrarelativistic regimes.

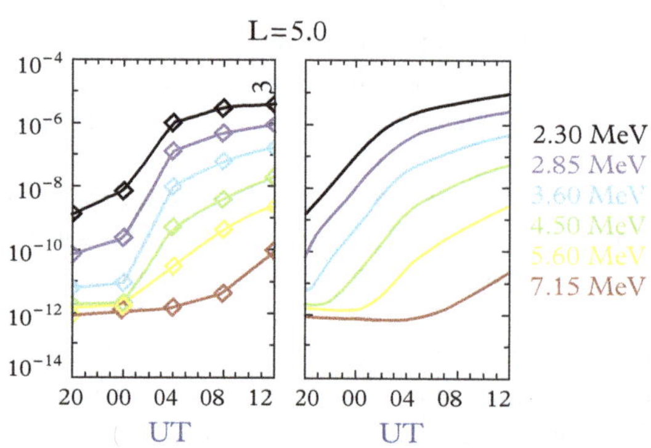

Figure 16.3 *Comparison of phase space density measured by the Van Allen Probes and the results form a model of the effect of chorus on electron acceleration from Thorne et al. (2013b). Phase space density from the REPT instrument is shown on the left and the model results are shown on the right.*
(Thorne et al., 2013b.)

16.4.2 Plasmaspheric hiss

The loss of radiation belt electrons has long been associated with pitch angle scattering by plasmaspheric hiss. Quantitative models of the this loss have been tested by recent Van Allen Probes measurements. After the report of the creation of a third, remnant "storage ring" radiation belt early in the mission (Baker et al., 2013b), data on plasmaspheric hiss was incorporated into a model of loss of relativistic electrons (Thorne et al., 2013a). The measured hiss was used to derive pitch angle diffusion coefficients which were in turn used to compute electron loss rates as a function of energy. The model showed good agreement between the measured decrease in electrons over many days and that predicted based on the hiss observed by the Van Allen probes.

The source of plasmaspheric hiss has been much debated with models suggesting the source is local with some type of non-linear amplification, whistlers from lightning which bounce between hemispheres, and, most recently, chorus that is emitted outside the plasmasphere which, as it propagates away from the equator, refracts and reflects into the plasmasphere (Bortnik et al., 2008). The Van Allen Probes have discovered that local generation is clearly one mechanism that be demonstrated to be a source of plasmaspheric hiss. A study of an observation of extremely low frequency hiss, with frequencies

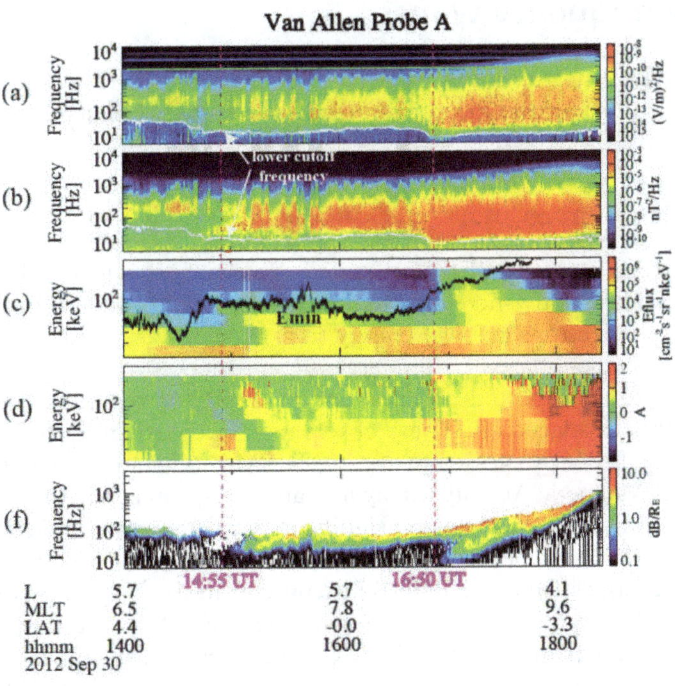

Figure 16.4 *Data from the Van Allen Probes mission showing observations during a period of very low frequency plasmaspheric hiss.*
From Li et al. (2013b)

below 100 Hz, used particle data to calculate the growth rate of these low frequency waves. Figure 16.4, taken from Li et al. (2013b) shows these observations. The top two panels are the electric and magnetic field wave spectrograms, respectively. As can be seen, plasmaspheric hiss with frequencies below 100 Hz begin around 1455 UT. The white line in the two spectrograms traces out the low frequency edge of the hiss spectrum panel (c) shows the omnidirectional electron flux spectrogram from the MagEIS Low Instrumented (Blake et al., 2013). Panel (d) shows the anisotropy as measured by the same instrument. The dashed vertical lines show times when the electron flux was enhanced and the anisotropy increased. The black line in panel (c) represents the minimum electron cyclotron resonance condition calculated from the background magnetic field and the plasma density determined from the upper hybrid line. The bottom panel shows the calculated convective linear growth rate from the electron observations. The upper cutoff in the growth rate frequency corresponds to the minimum energy of the electrons measured by the MagEIS instrument. As can be seen, there is remarkable agreement between the frequencies where the growth rate is positive and the low frequency hiss. This result provides evidence that local generation of plasmaspheric hiss occurs and may be an important mechanism for generating hiss near the plasmapause.

16.5 Low frequency waves

Combining the DC magnetic and electric fields from the EMFISIS and EFW experiments enables studies of the lowest frequency waves with frequencies from a few Hz down to a few mHz.

16.5.1 EMIC waves

Electromagnetic ion cyclotron (EMIC) waves have long been suggested as a mechanism for scattering of ultrarelativistic electrons (> 2 MeV) and have also been suggested as possibly scattering electrons at lower energy (e.g. Meredith et al. 2003).

During the recovery phase of a moderate geomagnetic storm on 11 October, 2012, Usanova et al. (2014) reported observations of EMIC waves in the helium band on the Van Allen Probes spacecraft. These observations were roughly conjugate to ground measurements of EMIC waves from the CARISMA (Canadian Array for Real-time Investigations of Magnetic Activity) array in Canada (Mann et al., 2008). Despite these extensive EMIC wave observations, no significant variation in precipitation of energetic electrons > 0.7 MeV was observed. Similarly, omnidirectional electron flux measurements at these energies from the REPT instrument on the Van Allen Probes show no variation either. However, closer examination of the pitch angle distribution over a longer interval from 9 October to 29 November 2012 revealed that during times of enhanced EMIC wave activity, fluxes near 0° and 180° pitch angle were reduced with no change to fluxes between 45° and 135° in pitch angle. The calculated quasi-linear pitch angle diffusion rates confirmed this result suggesting that the minimal variation in omnidirectional electron fluxes could be explained by a lack of any significant pitch angle diffusion other than near the field-aligned directions.

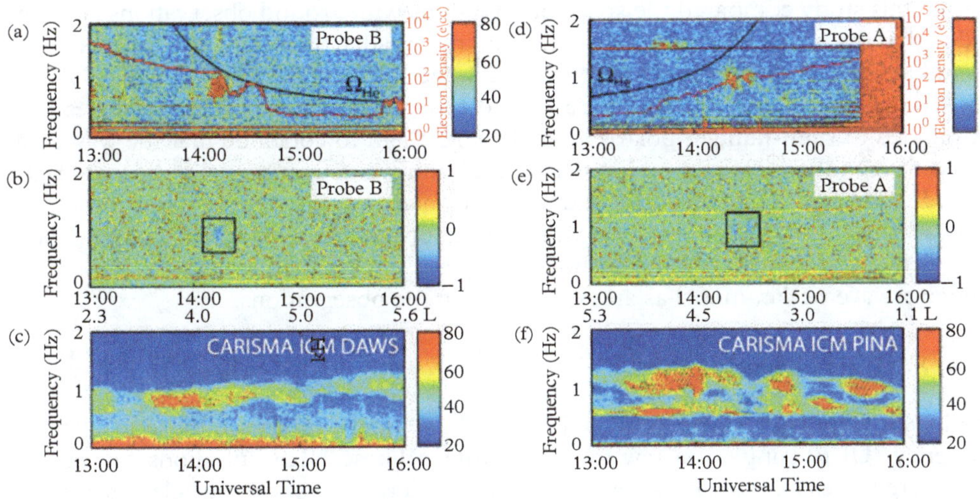

Figure 16.5 *EMIC observations from the Van Allen Probes mission compared with CARISMA array ground observations. The top panels show the EMIC waves observed in space, the bottom panels show the ground signatures extending over several hours, middle panels show the left hand polarization expected in EMIC waves.*
From Mann et al. (2014).

The spatial extent of EMIC waves on the ground and compared with in-situ space measurements were studied for the same 11 October 2012 even by Mann et al. (2014). In Figure 16.5, data from the Van Allen Probe fluxgate magnetometer is compared with data from the CARISMA ground array. The top panels show the observations from RBSP-A and RBSP-B which in both cases show very limited extent of the order of ~0.1–0.4 in L. These waves were observed near the plasmapause where density gradients suggest that the waves can be ducted and thereby amplified by multiple passes through the generation region. The middle panels show the polarization of the in-situ observations, giving the expected left-handed polarization as indicated by the boxed blue regions with values close to −1. The bottom panel shows the data from CARISMA array demonstrating that the same frequencies are observed and that the EMIC observations from the ground extend for many hours, which is equivalent to several hours in MLT. For this event, it was estimated that the ground observations indicated that the EMIC waves covered 18 hours of MLT. This work shows that it can be difficult to estimate the time extent of EMIC waves from satellite data. They also compared ground data and a model of ducted EMIC propagation to show that, although polarization an amplitude are modified in transit to the ionosphere, ground stations can do a good job of determining the temporal extent of EMIC events.

A special case of EMIC waves are the events identified as Pc1 "pearl" pulsations. These are bursts of EMIC waves that create a pearl-like pattern in frequency-time spectrograms. Paulson et al. (2014) reported 14 such events observed by the Van Allen Probes in the first year of operation. In agreement with the study of Mann et al.

(2014), this study compared the space observations with ground observations and concluded that the waves were narrowly extended in the radial direction, but quite extended azimuthally. By analyzing polarization of the waves observed on the Van Allen Probes spacecraft, Paulson et al. (2014) were able to observe the transition of the waves from left-handed to right-handed polarization allowing them to conclude that the waves had passed though the crossover region where left-hand waves can tunnel to become right-handed. Although no direct observations were made of the source region producing the pearl structures, the comparison with ground observations did not show the type of wave element spacing predicted by bouncing wave packet models, thus ruling out the bouncing wave packet model as an explanation of the observations.

16.5.2 ULF waves

The lowest frequency waves observed by the Van Allen Probes are those in ultra-low frequency (ULF) range of a few to tens of mHz. These ULF pulsations can interact with energetic electrons to resonantly exchange energy via wave-particle interactions. One of the early observations from the Van Allen Probes showed exactly this kind of interaction. Claudepierre et al. (2013) found a localized drift resonance between energetic electrons and poloidal mode ULF waves. Figure 16.6 shows the evidence of this interaction taken from Claudepierre et al. (2013). The top panel shows traces of the

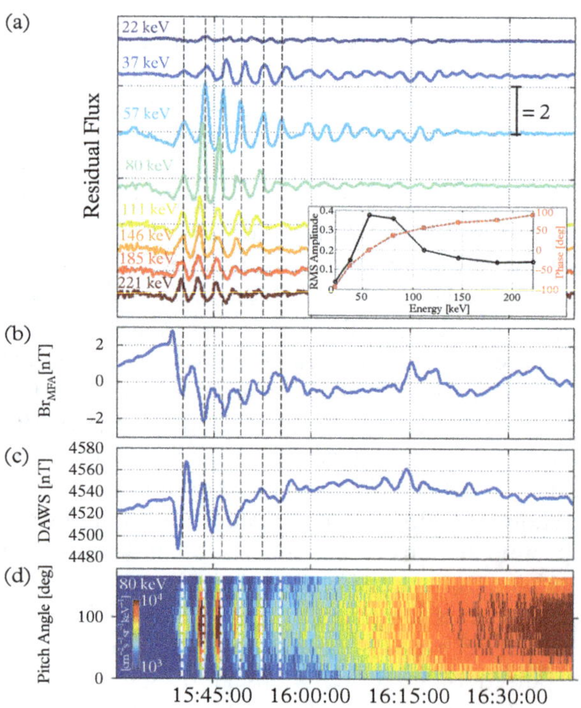

Figure 16.6 *Data from the Van Allen Probes mission showing observations of ULF pulsations exhibiting resonance with electrons in the energy range of 22–221 keV with peak response at 80 keV. From Claudepierre et al. (2013).*

residual electron flux after the background trend has been subtracted. Panel (b) shows the magnetic field measured on the RBSP-B spacecraft. Panel (c) shows data from the ground-based magnetometer from Dawson City which is part of the CARISMA array. Panel (d) shows the pitch angle distribution of the 80 keV channel of the MagEIS instrument. As can be clearly seen, the particle observations show modulation at the same frequency as the both the space-borne and ground magnetometers. The largest amplitude response from the electrons is at 80 keV and is bunched at 90° in pitch angle. The 3 min period of the waves corresponds to the fundamental poloidal mode and matches the modeled drift resonance frequency for ~60 keV electrons, in reasonable agreement with the peak resonance in the measured flux at 80 keV.

ULF drift resonance has also been reported for poloidal waves resonant with ions. Dai et al. (2013) showed that waves in with a wave period of ~84 s at a radial distance of 5 R_E. The waves exhibited a clear signature of a radially directed magnetic field and an azimuthally directed electric field with amplitudes of the order of 2 nT and 10 mV/m, respectively. The waves showed a clear 90° phase difference with the electric field leading the magnetic field. Using a realistic model field line with appropriate ionospheric conductivity, the study identified these as the fundamental mode of a standing poloidal wave. Ion measurements showed resonance with energies the range of 64 to 164 keV, with the peak in response at 88 keV. These 88 keV ions showed zero phase shift with respect to the wave and ions with energies higher and lower showed the expected phase shifts for interactions with the wave. The inferred radial gradient of the phase space density of the ions suggested that the waves were generated by the resonant ions and represent a mechanism for energy loss from the ring current.

16.5.3 Alfvén waves

Large amplitude (>100 mV/m) low frequency kinetic scale Alfvén waves observed by the Van Allen Probes were reported by Chaston et al. (2014). The perpendicular scale size of the waves was estimated to of the order of an ion gyroraidus or less and were associated with plasma injections in the inner magnetosphere. Using the measured background magnetic field and an assumed mass density, the measured E/B ratio of the waves agreed well with the local dispersion relation for kinetic Alfvén waves as a function of frequency. Although the mass density was a fit parameter, the overall number density associated with the mass density agreed well with the number density determined from the upper hybrid resonance. The phase of the electric field relative to magnetic field as a function of frequency showed the oscillatory character on either side of zero phase expected of a broad spectrum of standing kinetic Alfvén waves, further confirming the wave mode identification. The waves are localized near the front of plasma injections into the inner magnetosphere suggesting that the waves are driven by the injection. This energy source coupled with the narrow perpendicular scale size is consistent with large amplitudes observed and suggests that these waves can contribute to ion energization.

16.6 Global electric and magnetic fields

A study of the recovery phase of the 17 March 2013 geomagnetic storm examined plasma redistribution in the dusk and noon sectors arising from erosion flux observed in the ionosphere by DMSP (Defense Meteorological Satellite Program) and at high latitude by the Van Allen Probes. (Foster et al., 2014). By using Van Allen Probes measurements of the large scale convection electric field in concert with density determined from the spacecraft potential, they calculated the convection speed and plasmaspheric erosion flux which reached values of the order of 10^{12} ions $cm^{-2}s^{-1}$. Combining Millstone Hill radar data, the authors determined that the sunward erosion flux at ionospheric heights was slightly more than ten times that determined near the equatorial plane by the Van Allen Probes. This work demonstrated that during this event, there was an enhanced source of cold ionospheric plasma that was transported into the magnetosphere.

Models of the global scale ring current, which impacts the large scale magnetic field, global inductive electric field, and indices such as AE/AL and Dst/SYM-H ,have been developed in concert with Van Allen Probes data. Yu et al. (2014) presented results from the RAM-SCB (Ring current-AtMosphere-Self-Consistent B) model coupled to the BATS-R-US (Block-Adaptive-Tree-Solarwind-Roe-Upwind-Scheme) global MHD code that self–consistently solves for the coupling between plasma and electromagnetic fields during substorm injections. A key improvement in these results was providing two-way coupling between the models such that the RAM-SCB ring current model now provides feedback to the BATS-R-US global MHD code. This allows for realistic substorm injections in which the time-varying magnetic field creates an inductive electric field to create the injection instead of the ad hoc electric fields used in earlier models. The coupled model was used to study the 17 March 2013 storm event. The results showed good agreement between modelled and measured SYM-H and Dst as well as showing reasonable agreement between the magnetic dipolarizations observed on the RBSP-B spacecraft. Modeled and measured ring current particle fluxes measured by the MagEIS instrument on the Van Allen Probes also agreed reasonably well. These results illustrated the importance of the two-way coupling which produces more fine scale structure in the global fields and consequently, a more realistic injection process.

16.7 Conclusion

The initial results from the Van Allen Probes show that the mission has already achieved significant success in addressing its science goals. The electric and magnetic field observations made by the twin spacecraft are among the best ever made in the radiation belts and have contributed in a very significant way to this early success. As the mission completes its primary objectives, it will achieve full MLT coverage enabling a new set of detailed statistical models of wave and global scale field observations. This exciting re-examination of the radiation belts has just begun and promises to bring further surprises and insights into this important region of near-Earth geospace

Acknowledgment

This work was supported by JHU/APL contract no. 921647 under NASA Prime contract No. NAS5-01072.

REFERENCES

Baker, D. N., et al. (2013a). The relativistic electron-proton telescope (rept) instrument on board the radiation belt storm probes (rbsp) spacecraft: Characterization of earth's radiation belt high-energy particle populations. *Space Sci. Rev.*, **179**(1–4), 337–381.

Baker, D. N., et al. (2013b). A long-lived relativistic electron storage ring embedded in earth's outer van allen belt. *Science*, **340**(6129), 186–190.

Blake, J. B., et al. (2013). The magnetic electron ion spectrometer (mageis) instruments aboard the radiation belt storm probes (rbsp) spacecraft. *Space Sci. Rev.*, **179**(1–4), 383–421.

Bortnik, Jacob, Thorne, Richard M., and Meredith, Nigel P. (2008). The unexpected origin of plasmaspheric hiss from discrete chorus emissions. *Nature*, **452**(7183), 62–66.

Chaston, Christopher C., et al. (2014). Observations of kinetic scale field line resonances. *Geophys. Res. Lett.*, **41**(2), 209–215.

Chen, Yue, Reeves, Geoffrey D., Friedel, Reiner H. W., and Cunningham, Gregory S. (2014). Global time- dependent chorus maps from low- earth- orbit electron precipitation and van allen probes data. *Geophys. Res. Lett.*, **41**(3), 755–761.

Claudepierre, S. G., et al. (2013). Van allen probes observation of localized drift resonance between poloidal mode ultra-low frequency waves and 60 kev electrons. *Geophysical Research Letters*, **40**(17), 4491–4497.

Dai, Lei, et al. (2013). Excitation of poloidal standing alfven waves through drift resonance wave-particle interaction. *Geophys. Res. Lett.*, **40**(16), 4127–4132.

Fennell, J. F., et al. (2014). Van allen probes observations of direct wave- particle interactions. *Geophys. Res. Lett.*, **41**(6), 1869–1875.

Foster, J. C., et al. (2014). Storm time observations of plasmasphere erosion flux in the magnetosphere and ionosphere. *Geophys. Res. Lett.*, **41**(3), 762–768.

Kletzing, C. A., et al. (2013). The electric and magnetic field instrument suite and integrated science (emfisis) on rbsp. *Space Sci. Rev.*, **179**(1–4), 127–181.

Li, W., et al. (2013a). Constructing the global distribution of chorus wave intensity using measurements of electrons by the POES satellites and waves by the van allen probes. *Geophys. Res. Lett.*, **40**(17), 4526–4532.

Li, W., et al. (2013b). An unusual enhancement of low-frequency plasmaspheric hiss in the outer plasmasphere associated with substorm-injected electrons. *Geophys. Res. Lett.*, **40**(15), 3798–3803.

Mann, I. R., et al. (2008). The upgraded carisma magnetometer array in the themis era. *Space Sci. Rev.*, **141**(1–4), 413–451.

Mann, I. R., et al. (2014). Spatial localization and ducting of emic waves: Van allen probes and ground- based observations. *Geophys. Res. Lett.*, **41**(3), 785–792.

Meredith, N. P., Thorne, R. M., Horne, R. B., Summers, D., Fraser, B. J., and Anderson, R. R. (2003). Statistical analysis of relativistic electron energies for cyclotron resonance with emic waves observed on crres. *J. of Geophys. Res-Space Physics*, **108**(A6), 1250.

Millan, R. M. and Baker, D. N. (2012). Acceleration of particles to high energies in earth's radiation belts. *Space Sci. Rev.*, **173**(1–4).

Millan, R. M. and Thorne, R. M. (2007). Review of radiation belt relativistic electron losses. *J. of Atmos. and Sol-Terr. Phy.*, **69**(3), 362–377.

Paulson, K. W., Smith, C. W., Lessard, M. R., Engebretson, M. J., Torbert, R. B., and Kletzing, C. A. (2014). In situ observations of pc1 pearl pulsations by the van allen probes. *Geophys. Res. Lett.*, **41**(6), 1823–1829.

Santolík, O., Kletzing, C. A., Kurth, W. S., Hospodarsky, G. B., and Bounds, S. R. (2014). Fine structure of large-amplitude chorus wave packets. *Geophys. Res. Lett.*, **41**(2), 293–299.

Santolík, O., Parrot, M., and Lefeuvre, F. (2003). Singular value decomposition methods for wave propagation analysis. *Radio Sci.*, **38**(1), 1010.

Su, Zhenpeng, et al. (2014). Nonstorm time dynamics of electron radiation belts observed by the van allen probes. *Geophys. Res. Lett.*, **41**(2), 229–235.

Summers, D., Thorne, R. M., and Xiao, F. L. (1998). Relativistic theory of wave-particle resonant diffusion with application to electron acceleration in the magnetosphere. *J. of Geophys. Res-Space Physics*, **103**(A9), 20487–20500.

Thorne, R. M., et al. (2013a). Evolution and slow decay of an unusual narrow ring of relativistic electrons near l similar to 3.2 following the september 2012 magnetic storm. *Geophys. Res. Lett.*, **40**(14), 3507–3511.

Thorne, R. M., et al. (2013b). Rapid local acceleration of relativistic radiation-belt electrons by magnetospheric chorus. *Nature*, **504**(7480), 411–414.

Thorne, Richard M., Ni, Binbin, Tao, Xin, Horne, Richard B., and Meredith, Nigel P. (2010). Scattering by chorus waves as the dominant cause of diffuse auroral precipitation. *Nature*, **467**(7318), 943–946.

Usanova, M. E., et al. (2014). Effect of emic waves on relativistic and ultrarelativistic electron populations: Ground-based and van allen probes observations. *Geophys. Res. Lett.*, **41**(5), 1375–1381.

Wygant, J. R., et al. (2013). The electric field and waves instruments on the radiation belt storm probes mission. *Space Sci. Rev.*, **179**(1–4), 183–220.

Yu, Yiqun, Jordanova, Vania, Welling, Dan, Larsen, Brian, Claudepierre, Seth G., and Kletzing, Craig (2014). The role of ring current particle injections: Global simulations and van allen probes observations during 17 March 2013 storm. *Geophys. Res. Lett.*, **41**(4), 1126–1132.

17

An Overview of Early Results from the Radiation Belt Storm Probes Energetic Particle, Composition, and Thermal Plasma Suite on NASA's Van Allen Probes Mission

Harlan E. Spence, Geoff D. Reeves, and Ramona Kessel

■ Abstract

NASA's Van Allen Probes began its two-year prime science mission phase following its launch into the inner magnetosphere in August 2012. Designed to study and understand radiation belt structure and dynamics "ideally to the point of predictability," the dual-spacecraft Van Allen Probes mission comprises a comprehensive suite of charged particle and fields measurements needed to achieve closure on critical science questions. The Radiation Belt Storm Probes-Energetic Particle, Composition, and Thermal Plasma (RBSP-ECT) suite consists of three primary instrument types that collectively provide clean, robust measurements of the electrons and key ions in the inner magnetosphere, with high energy spectral and pitch angle resolution, spanning energy ranges covering the cold/warm plasmasphere populations, the hot ring current populations, the medium-energy electron "seed" population, as well as the core relativistic and ultra-relativistic radiation

Waves, Particles, and Storms in Geospace. First Edition. Georgios Balasis, Ioannis A. Daglis, Ian R. Mann.

belt populations. In this paper, we provide a high-level summary of the major science accomplishments from approximately the first year of operations of the RBSP-ECT instrument suite. We demonstrate that at the time of this writing, representing approximately half of the prime science mission phase, the Van Allen Probes science team has already met important mission objectives and achieved mission science goals, with results that not only resolve long-standing controversial questions using definitive RBSP-ECT measurements but also reveal fundamental features about the radiation belt that have heretofore gone undiscovered despite 50+ years of prior missions and study.

17.1 Introduction to mission science

The Van Allen Probes (formerly referred to as Radiation Belt Storm Probes) mission is a science-driven mission with top-level requirements [Kessel et al., 2012; Mauk et al., 2012] defined to ultimately serve as metrics against which the science community and NASA may measure mission success. The overarching science objective is "to provide understanding, ideally to the point of predictability, of how populations of relativistic electrons and penetrating ions in space form or change in response to variable inputs of energy from the Sun" [Mauk et al., 2012]. The three Van Allen Probes mission science objectives that in turn drive science mission requirements are to:

1. Discover which processes, singly or in combination, accelerate and transport radiation belt electrons and ions and under what conditions.
2. Understand and quantify the loss of radiation belt electrons and determine the balance between competing acceleration and loss processes.
3. Understand how the radiation belts change in the context of geomagnetic storms.

Minimum mission success requires first achieving a set of threshold measurements on both spacecraft, over time, needed to make progress on the collection of key science goals that are necessary for achieving the three science objectives. These science goals can be grouped into five broad areas. They are: (1) understanding the characteristics and sources of the electron radiation belt; (2) understanding the specific physical candidate processes which lead to radiation belt dynamics; (3) understanding the VLF, ELF, and ULF wave effects that control dynamics; (4) understanding the role of storm time and other global effects, such as shock-related events; and (5) understanding the role of flows and fronts that feed and modify the inner magnetosphere from outer regions. Minimum mission success requires progress in all five areas. Even after only one year of operations, the Van Allen Probes mission had already made at least good progress in all areas, and in some cases, truly significant progress. As the two-year prime mission nears its end, progress continues to accelerate with new results and discoveries. We discuss some of the prime successes in this paper.

17.2 Mission design

To address the aforementioned science objectives and goals, the Van Allen Probes mission required functionally identical, two-point measurements of the charged particle and electromagnetic field environments of the inner magnetosphere. The 50-plus year history of earlier radiation belt studies and open science questions motivated the need for a new mission (see Baker et al. [2013a] for a recent review) to provide definitive estimates of the full phase space density of energetic particles, particularly the electrons that dominate the outer radiation belt. Such a need for science closure drove many mission requirements, including, but not limited to: the energy range, energy resolution, and pitch angle coverage of the charged particle sensors; the need for local measurements (and global estimates) of the magnetic and electric fields, both the DC and the wave components; the desire for the Van Allen Probes spacecraft to orbit near the magnetic equator; and, the need for two spacecraft flying through the same regions at different times in order to resolve temporal structure and its dynamism.

Collectively, these science drivers led to a mission architecture comprising a pair of identically instrumented satellites, each with a comprehensive suite of charged particle and fields instruments. The Van Allen Probes spacecraft explore the radiation belts in near-equatorial (10° inclination), highly elliptical orbits with low-altitude perigees ($\sim$600 km altitude) and apogees of approximately 5.8 R_E. These parameters lead to orbital periods of approximately 9 hours, thus permitting comparatively rapid reencounters with the belts during a single spacecraft's orbit. Furthermore, by design, the orbital parameters of the two spacecraft are carefully tuned to be slightly different such that their relative orbital phase varies continuously (and slowly) with time. When the spacecraft are 180° out of phase, in other words, when the time between successive apogee passages is approximately 4.5 hours, this dual configuration results in passages through the radiation belts every couple of hours. This particular mission design feature provides unprecedented time resolution of belt dynamics near the magnetic equator, allowing us, for example, to help distinguish unambiguously diffusive from direct acceleration processes for the first time.

As noted earlier, the science goals also dictated measurement requirements, which in turn drove the instrument requirements. To achieve science closure, each Van Allen Probes spacecraft has identical instrumentation; Mauk et al. [2012] provide an overview of the instrumentation. The instrumentation includes two suites specifically designed to quantify the steady and time-varying electric and magnetic fields, namely the Electric and Magnetic Field Instrument Suite and Integrated Science or EMFISIS [Kletzing et al., 2013] and the Electric Field and Waves or EFW [Wygant et al., 2013] suite. The mission also includes two charged particle instruments, one called the Radiation Belt Storm Probes Ion Composition Experiment or RBSPICE [Mitchell et al., 2013], designed primarily to characterize the quiet- and storm-time ring current ions, and the second called the Relativistic Proton Spectrometer or RPS [Mazur et al., 2013], designed to measure inner zone protons. Each of these instruments plays key roles in achieving mission science but are not the primary focus of this publication; their contributions are specifically highlighted in other publications.

In this paper, we focus on the last of the Van Allen Probes mission instrument suites known as the Radiation Belt Storm Probes-Energetic Particle, Composition, and Thermal Plasma or RBSP-ECT suite [Spence et al., 2013]. RBSP-ECT contains a complement of particle instruments making the highest quality measurements ever made in the radiation belts and the inner magnetosphere. The coordinated RBSP-ECT particle measurements analyzed in combination with the fields and waves observations (see also Kletzing [2015, this volume] for details of scientific discoveries of the EMFISIS suite), complementary charged particle measurements, and state of-the-art theory and modeling, are providing new understanding on the acceleration, global distribution, and variability of radiation belt electrons and ions. We therefore next provide a very brief overview of the RBSP-ECT instrument suite, before summarizing mission scientific results from an energetic charged particle perspective in the following sections.

The RBSP-ECT suite consists of three highly coordinated and complementary instruments: the Helium Oxygen Proton Electron or HOPE spectrometer [Funsten et al., 2013], the Magnetic Electron Ion Spectrometer or MagEIS [Blake et al., 2013], and the Relativistic Electron Proton Telescope or REPT [Baker et al., 2012]. Collectively these three instrument types cover comprehensively the full electron and ion spectra from ~1 eV (or the spacecraft potential, whichever is higher) to tens of MeV with sufficient energy resolution, pitch angle coverage and resolution, and with composition measurements in a critical energy range up to 50 keV. All three instruments are based on measurement techniques proven in the radiation belts, then optimized to provide unambiguous separation of ions and electrons and clean energy responses even in the presence of extreme penetrating background environments. The instruments were designed to provide complete coverage in energy with overlaps between each sensor type. Each instrument underwent extensive ground characterization and calibration as well as inflight cross-instrument and cross-spacecraft calibration. As with all instruments on the mission, RBSP-ECT definitive data products are provided to the community, with data release versions that reflect improved understanding of the instrument responses and cross-calibration.

17.3 Representative examples of early mission science

A comprehensive description of Van Allen Probes mission accomplishments is well beyond the scope of this paper. So, too, is even a full accounting of mission accomplishments from just the RBSP-ECT suite. Rather, in this paper we provide a high level accounting of early publications and how they map into achieving mission science. One outcome of doing so is to assist readers in becoming acquainted with this broad body of work, distributed over time and different journals, and how they are collectively inter-related to one another and to higher-level mission objectives. Finally, we highlight just a few representative examples of the sorts of publications that have already made significant impact. Many are presently rewriting the textbooks on radiation belt physics, and with a still active mission, many more to come will continue to do so with little doubt. Though admittedly this is an imperfect and incomplete snapshot of progress made to date, only halfway (approximately one year of publications) through the prime science mission, we already have many accomplishments toward realizing each of the five aforementioned mission science goals, and, with the promise of more to follow. In this paper, we next highlight especially the first three science goals for which significant progress has been overwhelmingly demonstrated.

17.3.1 Understanding the characteristics and sources of the electron radiation belt

This first mission science goal possesses several elements encompassing: understanding and quantifying energization, loss, and transport of radiation belt electrons; understanding and quantifying source populations of radiation belt electrons; determining how these two prior aspects lead to spatial-temporal variability of the radiation belt electrons; and finally, demonstrating that this understanding leads to improved predictability of radiation belt models. Substantial progress has been made in each area; several examples are illustrated below.

Importance of plasmapause position Witnessing the creation of an isolated, persistent storage ring of extremely high-energy electrons just inside the plasmapause was an early and significant discovery of the mission [Baker et al., 2013b]. Shortly after the Van Allen Probes science phase began, an event on September 2, 2012, resulted in the formation of this energetic electron torus, effectively a new third belt, which persisted for more than a month before being disrupted in response to the passage of an interplanetary shock past Earth's magnetosphere; Figure 17.1 illustrates the time history of this event (after Figure 3 of Baker et al. [2013b]) as measured by the REPT instruments. Li et al. [2013] reported on the same belt structure and time dependence with independent measurements made at low altitudes onboard the CSSWE CubeSat. This dynamical phenomenon challenged the conventional wisdom regarding the position of the plasmapause relative to the location of electrons in a simplified radiation belt comprising an outer zone electron population and an inner zone. Traditionally, one would expect the outer zone electrons to persist outside of the plasmapause location with the slot region inside. During this event, both the new storage ring feature as well as much of the outer zone electrons persisted inside the high density plasmasphere.

In a related study, Thorne et al. [2013a] explored the physical mechanisms by which this feature formed, how it evolved, and how it eventually decayed. They demonstrated that the position of the plasmapause was key in this evolution, including its control over where the inner edge of the outer zone became depleted during the initial stages of the

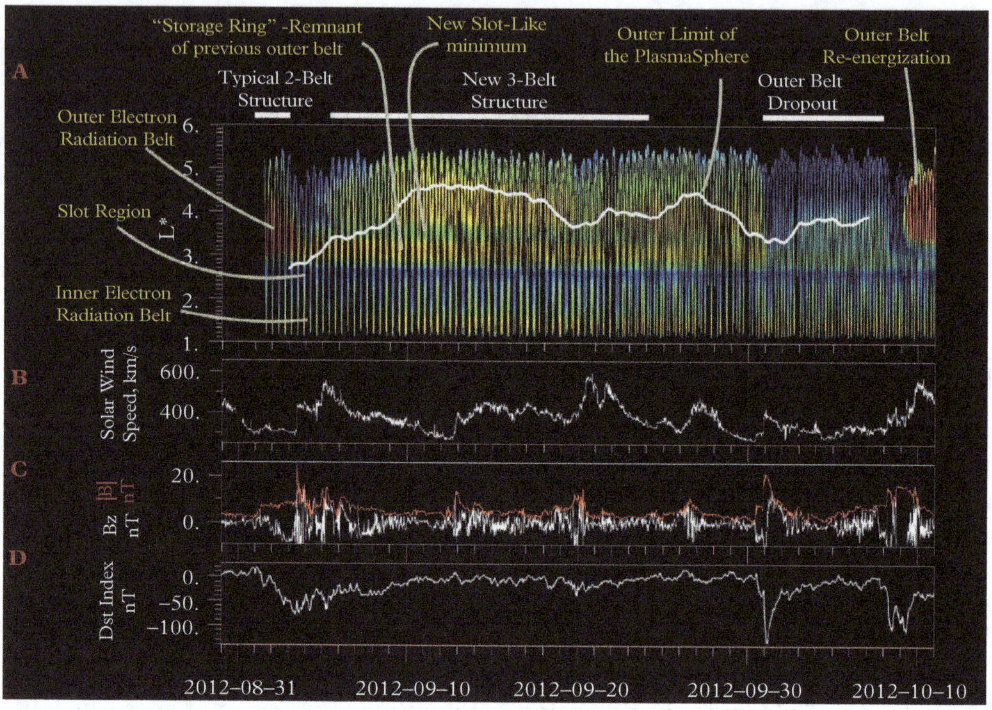

Figure 17.1 *Development of storage ring of highly energetic electrons.*
(Figure 3 from Baker et al., 2013; Reprinted with permission from AAAS.)

event, where the outer belt eventually dynamically reformed beyond the plasmapause during the second phase of the event, and finally the maintenance of the storage ring and reformed outer zone electrons when the plasmapause moved outward for a prolonged period engulfing them both. Being embedded within the plasmasphere meant that the remnant and new belts were largely sheltered from dynamical processes occurring beyond the plasmapause, with the slow final decay of the storage ring the result of pitch angle scattering into the atmospheric loss cone by plasmaspheric hiss, with energy-dependent timescales (~10 to 20 days at >3 MeV and much more rapid at lower energies) consistent with the REPT observations. This study and others (including the related work of Shprits et al. [2013]) underscore the powerful control that the lowest energy plasmas in the inner magnetosphere have over the dynamics of the highest energy electrons in the collocated radiation belts.

An additional study also showed the importance of the plasmapause to radiation belt dynamics. Foster et al. [2014] studied the dynamics of outer zone electrons during an event on March 17, 2013, in association with a large geomagnetic storm. In this event, multi-MeV electrons in the outer zone were first significantly depleted and then rapidly replenished following a substorm dipolarization. The location of the Van

Allen Probes spacecraft near the magnetic equator, the high temporal cadence provided by two spacecraft, and a full accounting of particles and fields allowed this study (among others) to accurately account for losses and gains by tracking phase space density at fixed invariants, an important advance enabled routinely by this mission over previous ones. Foster et al. [2014] showed how the relativistic electrons formed through the presence of a lower energy seed population (see Boyd et al. [2014] for additional insights on this seed population) and intense VLF waves, growing rapidly over the course of just several hours. This study also showed again the critical role of the plasmapause location. Foster et al. [2014] localized the re-energization of electrons to regions outside the plasmapause, after which this newly formed belt was engulfed by the plasmapause; this study reinforces the importance of knowing not only the location of the plasmapause, but also its time evolution in the context of belt forming and belt destroying events. To that end, Goldstein et al. [2014] has demonstrated how well relatively simple models predict the actual time dependence of Van Allen Probes plasmapause encounters; those results provide substantial promise of using the outputs of such models as inputs to other predictive radiation belt models.

Quantifying energization processes and conditions A second major result of the mission to date is the unambiguous observation that relativistic electrons are accelerated to high energies through local resonant interactions with waves deep in the heart of the radiation belts. Reeves et al. [2013] presented a first compelling example of this process in action by studying a belt-forming event that occurred on October 8–9, 2012. As noted earlier, the design of the Van Allen Probes mission and the instrument suites permits for the first time the ability to routinely and robustly express the evolution of radiation belt particles in phase space density, which is needed to differentiate between local acceleration and so-called radial acceleration. A fortuitous arrangement of the two Van Allen Probes spacecraft passing subsequently through the outer zone allowed Reeves et al. [2013] to reconstruct the time evolution of phase space density in a way that unambiguously identified local acceleration as the operative agent for this particular radiation belt enhancement. Figure 17.2 illustrates their key result (taken from Figure 3 of Reeves et al. [2013]). Phase space density profiles were constructed from differential energy spectra measurements made by REPT, in combination with the locally measured magnetic field and with estimates of the adiabatic invariants obtained from global magnetic field models (see Morley et al. [2013] for a further discussion of related issues). Figure 17.2 illustrates the time evolution of phase space density versus L^* (at a particular fixed first and second invariant) from spacecraft RBSP-A (squares) and RBSP-B (circles), color-coded by the time at which each spacecraft passed through $L^* = 4.2$ for each inbound and outbound pass. As noted by Reeves et al. [2013], the rapid rise in phase space density near $L^* = 4.2$ and the slower, delayed increase at higher L^* are characteristic signatures of local acceleration. Such a process produces this characteristic L^* peak in phase space density with positive radial gradients at lower L^* and negative radial gradients at higher L^*. While such a process has been implicated in the past, the Van Allen Probes mission has definitively affirmed this is an

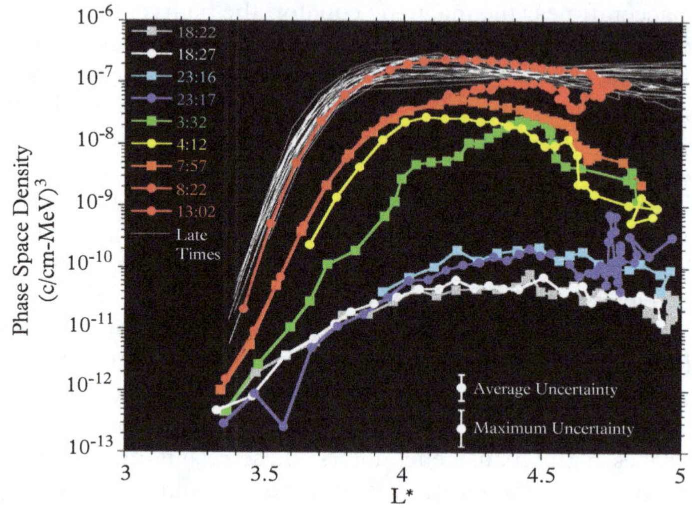

Figure 17.2 *Phase space density profiles for relativistic radiation belt electrons as a function of* L* *at fixed first and second invariant.* *(Figure 3 from Reeves et al., 2013; Reprinted with permission from AAAS.)*

operative process with unprecedented certainty and has propelled us on the way toward quantifying its relative importance.

In a related study, Thorne et al. [2013b] explored the physical mechanism responsible for the local acceleration. They studied the same event from Reeves et al. [2013] and focused on one wave mode, magnetospheric chorus, implicated in prior studies as a potential candidate for resonant interaction with and local acceleration of relativistic electrons. They compared the observed high-resolution electron measurements from RBSP-ECT with those inferred from a two-dimensional, data-driven, time-varying model of global chorus emissions and demonstrated convincingly that scattering of electrons by chorus emissions can be solely responsible for the temporal evolution of the energy and angular distributions.

Su et al. [2014] complemented and reinforced the Thorne et al. [2013b] results by studying a different dynamical radiation belt event, in this instance occurring during non-storm times, a rarely reported phenomenon. A belt-enhancing event occurred on February 21–24, 2013, during a relatively quiet geomagnetic period. Despite the lack of geomagnetic activity, powerful chorus waves were observed nonetheless during this period. As with the previously mentioned study, Su et al. [2014] used a model of chorus waves to approximately reproduce the time evolution of the energy and pitch angle dependence of these newly accelerated, spatially localized electrons. Their study illustrates the importance of chorus-driven local acceleration even during non-storm times. We note that Schiller et al. [2014] studied another event on January 13–14, 2013, showing strong evidence of local acceleration and dramatic enhancements in the outer radiation belt, again during non-storm times.

Clarifying a variety of loss processes In parallel with this rapidly expanding understanding of the processes which create radiation belt enhancements, the Van Allen Probes mission is also allowing us to identify, clarify, and quantify a variety of loss processes. In one such study, Blum et al. [2013] combined contemporary measurements of electron precipitation at low altitudes from the Colorado Student Space Weather Experiment (CSSWE) CubeSat (whose main instrument is derived from the Van Allen Probes REPT instrument) and the Balloon Array for Radiation belt Relativistic Electron Losses (BARREL) [Millan et al., 2013]. They identified conjugate intervals during which the multipoint measurements could be used to quantify the temporal and spatial properties of so-called precipitation bands. These precipitation bands are narrow features that have been implicated in playing a major role in depleting outer zone electrons; however, previous studies have been limited in advancing our understanding of their spatial/temporal properties needed in order to quantify their role. Blum et al. [2013] show that precipitation bands may indeed play a major role in radiation belt losses; their study estimates that ~20 such events could rapidly empty the entire outer belt through precipitation into the atmosphere.

In addition to precipitation loss, another significant loss process for outer zone electrons is drift loss to the magnetopause. Hudson et al. [2014] quantified this mechanism through detailed test-particle simulations of the October 8–9, 2012, event. They used a sophisticated global magnetohydrodynamic model, coupled with a comprehensive ring current model, all driven by solar wind inputs, to perform test particle simulations of electrons in the outer zone. They presented strong evidence, through exploring the energy and pitch-angle dependence of outer zone electron depletions, that magnetopause shadowing played a major role in this particular event. They note that drift loss to the magnetopause results both from the direct inward motion of the magnetopause during compression events, as well as through the production of enhanced ULF waves that drive rapid outward radial diffusion throughout the outer zone. In related work, Shprits et al. [2013] concluded that magnetopause shadowing loss becomes important outside $L \sim 4$, while precipitation due to pitch angle scattering by electromagnetic ion cyclotron waves was the primary source of loss inside of $L \sim 4$. Turner et al. [2014a] used a combination of Van Allen Probes, THEMIS, and GOES measurements during a September 30, 2012, outer zone electron dropout to quantify the depth of the loss owing to magnetopause shadowing. At larger L, the dropout was inferred to be a >90% reduction; for $L^* <\sim 4$ they concluded that another loss process must be operating for the multi-MeV electrons.

Toward improved model predictability The Van Allen Probes mission, as noted above, permits us to explore specific acceleration and loss mechanisms. Because of these advances, we are also able to better quantify the relative importance of energization and loss during specific events needed ultimately to understand the net, global dynamical behavior of the belts. This is exemplified in several studies. For example, Tu et al. [2014] show how outward radial diffusion to the solar-wind driven magnetopause, an event-specific chorus wave model, and a dynamic lower-energy seed electron population, are all critical for robustly modeling the dynamics of the complex October 2012, "double dip" magnetic storm event. Their diffusion model results emphasize the critical role that the comprehensive Van Allen Probes measurements (in energy, pitch angle, and L) play in constraining the various competing mechanisms. Jordanova et al. [2014] used an improved ring current-atmosphere interactions model, including self-consistent magnetic fields to explore its ability to predict the behavior of the lower energy charged particles. Their study identified strengths and weaknesses in the ring current model through comparison of event studies against MagEIS and HOPE ion observations. These studies amply demonstrate how an improved understanding of the individual processes involved in radiation belt dynamics can markedly improve the predictive capability of models. This is still very much a work in progress, though these improved validation studies represent an important and significant step forward.

17.3.2 Understanding the specific physical candidate processes which lead to radiation belt dynamics

This second mission science goal possesses two main elements: distinguishing between candidate processes of acceleration, transport, and loss; and characterizing statistically these processes as a function of solar and magnetospheric input conditions. As with the first science goal, many publications demonstrate substantial progress toward closure. Significant progress toward achieving this goal was already described in the prior section and so is not repeated here. Rather, we next focus on the critical mission goal associated with disentangling the various competing mechanisms at work that sculpt the unique temporal evolution and spatial properties of the belts during any particular event.

Differentiating between competing mechanisms While all studies contribute to this goal to some degree, two studies in particular have given prime attention to the important task of differentiating between competing mechanisms. Baker et al. [2014] explored comprehensively the evolution of the multi-MeV radiation belts during the extended March 2013, interval, an overall period of relative calm during which the belt behaved one way, punctuated by dynamical events which influenced the belts in dramatically different ways. This interval is illustrated in Figure 17.3 (taken from Figure 1 of Baker et al. [2014]) Their study demonstrated the classic signatures both of inward radial diffusive energization (quiet times) as well as abrupt localized acceleration (impulsive times); both competing mechanisms of multi-MeV electron energization are at play in the radiation belts during this interval, often acting concurrently or in rapid succession. Furthermore, these two energization components are also incessantly competing with loss mechanisms, including precipitation, which has both a slow diffusive component and an impulsive one. Even though daunting, studies such as those by Baker et al. [2014] (and enabled by the Van Allen Probes mission) are allowing us to disentangle all of these effects and how their simultaneous evolution determines the overall behavior of the belts.

This effort is no better illustrated than by the work of Turner et al. [2014b]. They focused on the particular period spanning September 30 to October 3, 2012, using a comprehensive constellation of 15 spacecraft, including the twin Van Allen

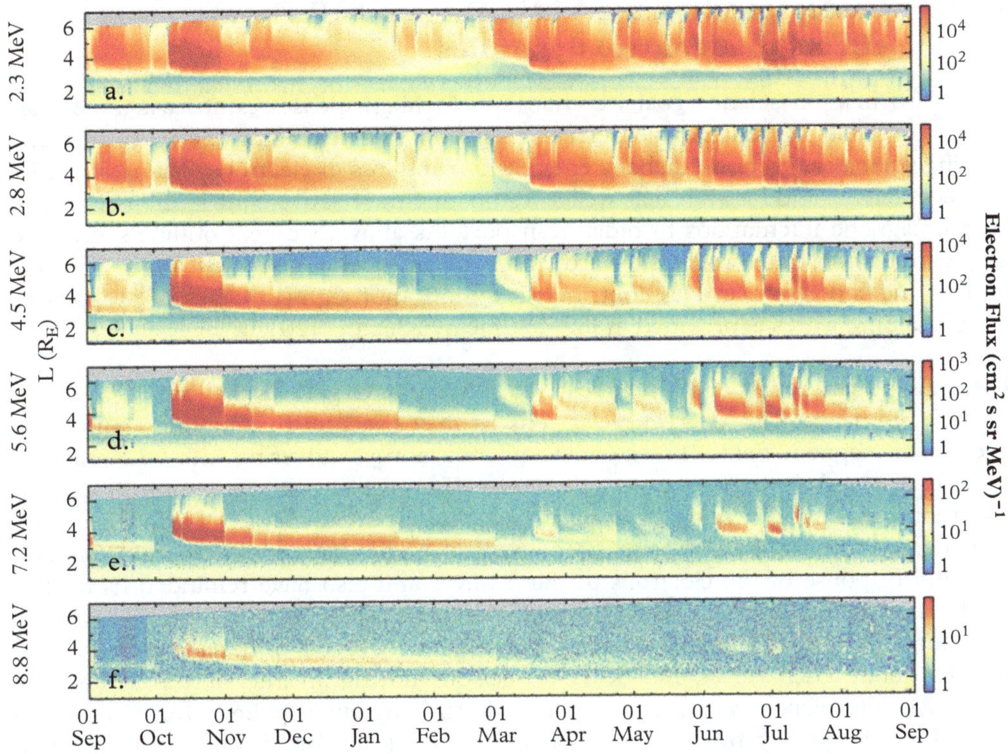

Figure 17.3 *Color-coded fluxes of relativistic electron in an L versus time format in selected energy channels of REPT for the first year of Van Allen Probes operation spanning September 2012 to September 2013. The March 2013 period near the center of this figure is the period during which both gradual and impulsive evolution of the belts are particularly easily differentiated.*
(Figure 1 from Baker et al., 2014; Reprinted with permission from John Wiley and Sons.)

Probes spacecraft, to investigate a powerful geomagnetic storm period and the concomitant response of the radiation belts. From the array of multipoint measurements, they were able to unravel how the various components worked together over time, including key elements such as: enhanced precipitation losses to the atmosphere at $L^* < 4$ consistent with pitch-angle scattering by EMIC waves during the main storm phase; evidence for local acceleration also active during the main phase; early recovery phases corresponding to IMF Bz southward when electron injections and whistler-mode chorus waves led to relativistic electron enhancements; followed ultimately by the IMF Bz turning northward, during which injections, chorus activity, and enhancements ceased. Such studies represent an ultimate test of our understanding of the individual components of a radiation belt event, and are enabled by the unique quality of the Van Allen Probes measurements in combination with a fortuitous constellation of spacecraft needed to fully quantify the connected inner magnetosphere system.

17.3.3 Understanding the VLF, ELF, and ULF wave effects that control dynamics

This third mission science goal we choose to highlight possesses at least three main elements related to radiation belt electron dynamics: determining the types of waves causing adiabatic and non-adiabatic energization and loss of electrons; estimating diffusion coefficients and loss rates; as well as estimating the relative contribution of various electromagnetic fluctuations to radial transport. As above, we next outline several examples to demonstrate progress toward closure of this science goal, with a primary focus on how the energetic particles contributed; a more thorough summary of progress on understanding waves is provided in this volume by Kletzing [2015].

Ultra Low Frequency (ULF) waves ULF waves have periods that can be in bounce or drift resonance with energetic electrons as they execute their motions both along and across magnetic fields. The magnetosphere is always quivering with ULF waves to some extent and during some periods, these ULF wave fields can be significant and coherent. It is during these times, in particular, that the interactions between ULF waves and energetic electrons can be significant. Several studies have explored the oftentimes-beautiful response of the electrons to these waves and also have reminded us of their utility in understanding the system beyond their sheer aesthetics.

Claudepierre et al. [2013] showed evidence for localized drift resonance between polodial mode ULF waves and 60 keV electrons during an interval on October 31, 2012. An interplanetary shock swept past Earth's magnetosphere, which ultimately excited low frequency fundamental poloidal mode oscillations in the magnetosphere. Electron flux variations were observed at the same oscillatory period (~3 minutes) suggesting a drift-resonance mechanism with energies near ~60 keV. The high resolution energetic electron data allowed Claudepierre et al. [2013] to estimate the azimuthal mode number of the interacting wave as well as its localization. Figure 17.4 (taken from Figure 2 of Claudepierre et al. [2013]) illustrates the quality of the observations both in the magnetosphere and on the ground that inform such studies. Their study highlights the importance of localized ULF wave-particle interactions for understanding energetic particle dynamics in the inner magnetosphere. In another study, Dai et al. [2013] identified the generation of a fundamental mode standing poloidal ULF wave excited through drift-resonance interactions of medium energy ions in the inner magnetosphere. Similarly, Hao et al. [2014] explored these same sorts of interactions during a ULF wave event that occurred on April 13, 2013. Together, these studies illustrate the way by which wave particle interactions produce a coupled system in which waves and particles may exchange energy in a complex manner and through a host of different pathways. Finally, Mann et al. [2013] used Van Allen Probes data as a way to illustrate indeed how coherent acceleration of drifting radiation belt electrons may occur through resonant interactions with ULF waves on a planetary scale, whether excited from external drivers such as interplanetary shocks or internally: essentially a process that can be thought of as a geophysical synchrotron.

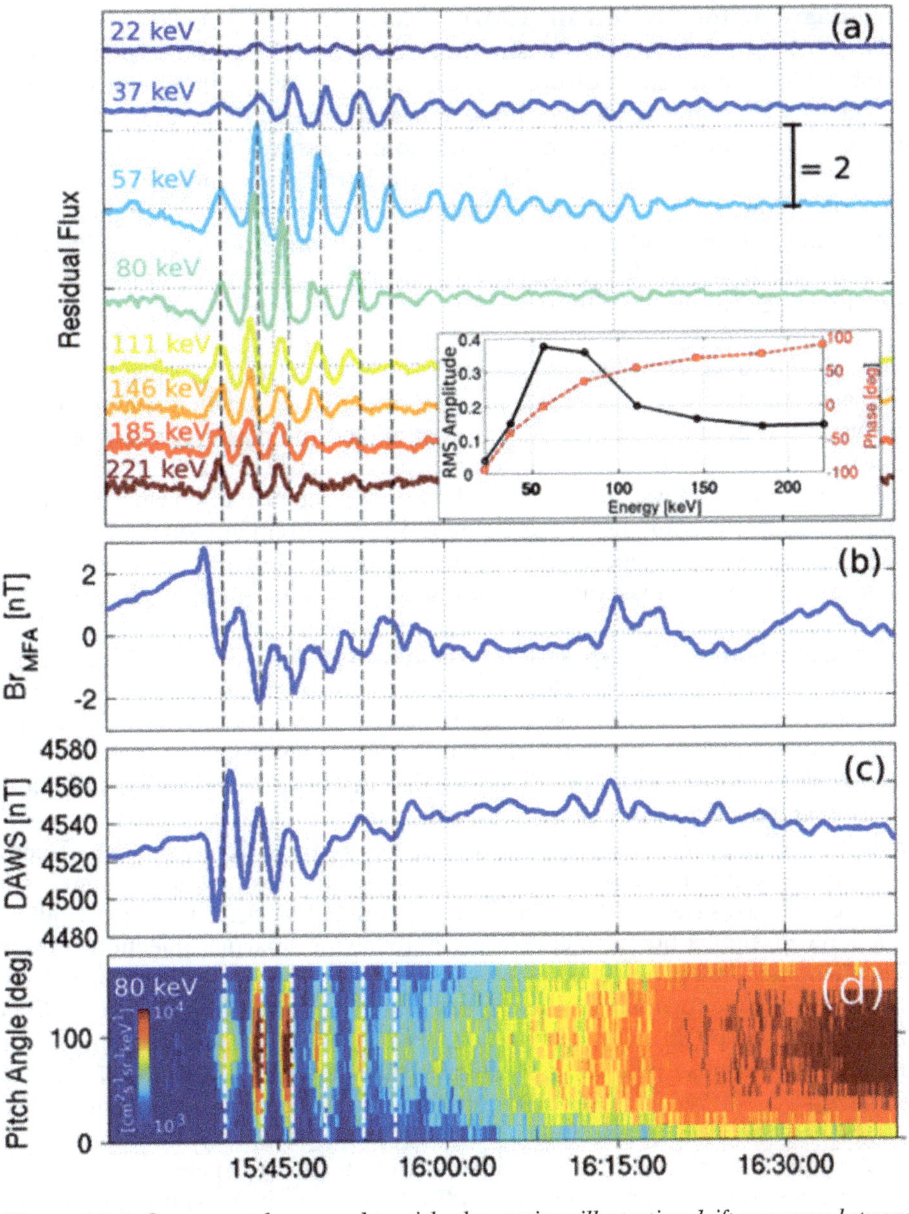

Figure 17.4 *Summary of wave and particle observations illustrating drift resonance between magnetospheric ULF waves and energetic electrons. (a) Residual electron flux (J–J_0) from MagEIS-A. (Figure 2a, inset) The amplitude (solid trace) and J_0 phase (dashed trace) of the residual flux oscillations. (b) The poloidal component of the magnetic field from the EMFISIS-A magnetometer. (c) East-west component of the Earth's magnetic field measured at the Dawson City (CARISMA) ground magnetometer station. (d) Pitch angle spectrogram from the 80 keV channel on MagEIS-A.*

(Figure 2 from Claudepierre et al. [2013]; Reprinted with permission from John Wiley and Sons.)

Electromagnetic Ion Cyclotron (EMIC) waves Usanova et al. [2014] studied the effects of EMIC waves on relativistic and ultra-relativistic electrons during a storm event on October 11, 2012. Their study benefited from the high-resolution electron and EMIC wave data from the Van Allen Probes mission near the magnetic equator, combined with conjugate measurements from ground magnetometer arrays. They used the observed wave power to compute expected radiation belt electron pitch-angle diffusion rates and demonstrated that rapid pitch-angle diffusion is confined to low pitch angles. For the first time, they showed evidence both experimentally and from models of how EMIC waves may trigger ~2–8 MeV electron loss, but confined to pitch angles below ~45° and not affecting the core distribution.

Plasmaspheric hiss Li et al. [2013] explored the complex interplay between electron dynamics and hiss (and chorus) throughout the inner magnetosphere. They used an interval when the twin Van Allen Probes straddled the plasmapause to explore this relationship. Probe-A was located inside the plasmapause where it observed intense hiss emissions; Probe-B was located outside the plasmapause observing chorus waves. Through cross-analysis of these two point wave measurements, including properties of observed injected energetic electrons, they were able to show how unusually low-frequency plasmaspheric hiss was likely amplified in the outer plasmasphere in this case.

Direct comparison of wave-particle interactions In the ideal world, a direct comparison between waves and particles is made difficult by the fact that you only rarely measure the particles at the location where the waves interact with them. Once particles interact with a potentially spatially located wave field, they drift out of that region and are thus spatially dislocated from the waves that modified them. This makes a direct comparison and hence science closure difficult.

Fennell et al. [2014] found and studied an example where such a direct comparison was possible, in this case, associated with chorus wave bursts. Bursts of lower energy electron fluxes (tens of keV) measured by MagEIS occurred quasi-periodically in conjunction with similar bursts of chorus waves following a particle injection. This wave-particle phenomenon was observed by both Van Allen Probes on January 13, 2013, when they were just outside the plasmasphere (see Figure 17.5 taken from Figure 2 of Fennell et al. [2014]). The pitch angle distributions evolved systematically during the bursts, transitioning from those peaked at 90 degree pitch angle to those in which maxima formed at $\alpha = 75°–80°$ while fluxes at 90° and <60° remained nearly unchanged. The expected resonant electron energy, inferred from the plasma density, cyclotron frequency and pitch angle of the peak flux, was consistent with the observed electrons in the tens of keV range. Such wave-particle observations (in this case for chorus) provide new, robust observational tests of our understanding in how and when these interactions take place, which in turn aid ultimately in quantifying their global effects on the radiation belts.

Mozer et al. [2014] presented another important example of direct observations of particle responses to waves. They discovered a powerful connection between nonlinear whistler waves and acceleration signatures in the electrons across the HOPE and MagEIS energy ranges, implicating a two-step process involving first parallel acceleration from

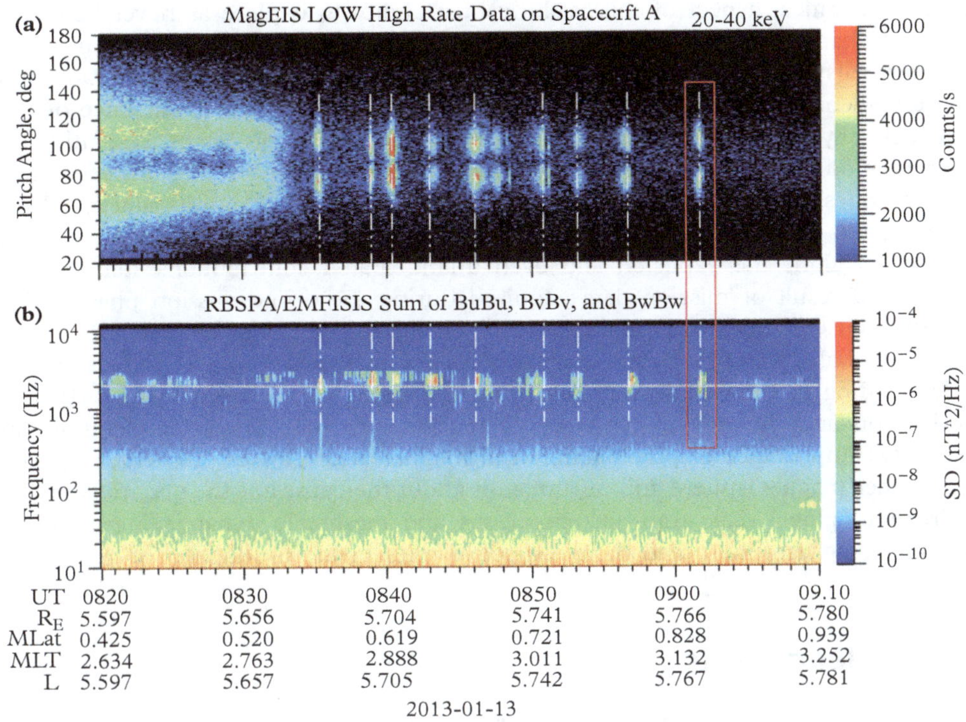

Figure 17.5 *Comparison of electron flux bursts and chorus waves on January 13, 2013. (a) A pitch angle spectrogram of 20–40 keV electrons with flux intensification bursts highlighted with vertical dash-dotted lines. (b) A frequency-time spectrogram from EMFISIS showing bursts of chorus emissions. The gray horizontal line demarks half the electron cyclotron frequency. (Figure 2 of Fennell et al. [2014]; Reprinted with permission from John Wiley and Sons.)*

time-domain electric field structures parallel to the magnetic field at lower energies, then resonance of those accelerated electrons with Doppler-shifted upper band whistler frequencies. This mechanism provides the capacity to locally accelerate electrons with surprisingly high efficiency and on exceptionally short time scales. They used test particle calculations of electrons interacting with the modeled large-amplitude whistler waves to predict successfully the temporal correlations between observed quantities, including, magnetic field compressions, parallel electric field spikes, electron flux, and electron pitch angle distributions.

17.4 Summary

Informed by earlier mission measurements, modeling, and theory, the Van Allen Probes mission design and instrumentation capabilities are allowing for remarkable improvements in temporal and spatial resolution of key parameters near the magnetic equator. A critical aspect of this is the robust ability to explore as never before the complete charged particle phase space density as a function of adiabatic invariants, at two points simultaneously, needed for so much science closure. The advent of the Van Allen Probes mission has thereby introduced a veritable renaissance of radiation belt physics, complete with deeper understanding of processes, resolution and differentiation between competing processes, and discovery of new aspects of the belt dynamics and structures unimagined before launch. The results illustrated only briefly in this paper represent a mere fraction of mission accomplishments achieved to date, let alone the expected and even unanticipated contributions still to be reported as a result of mission data acquired during the prime mission phase (which lasts through October 2015). An extended science mission phase thereafter promises great potential. New prioritized science goals for this next phase include advancing understanding of the relative: (1) roles of local versus global mechanisms of particle acceleration and the role that non-linear mechanisms play in the acceleration processes; (2) importance of precipitation and magnetopause losses of energetic particle in the inner magnetosphere and information about the causes of the precipitation; and (3) roles of global scale transport processes and mesoscale dynamic injections into the inner magnetosphere and their respective roles in the production of geoeffective waves.

Acknowledgments

This research was supported by RBSP-ECT funding under NASA's Prime contract no. NAS5-01072. We acknowledge and thank both the AAAS and John Wiley and Sons for their permission in reprinting the as-noted figures in this paper.

REFERENCES

Baker, D. N., et al. (2012), The relativistic electron-proton telescope (REPT) instrument on board the radiation belt storm probes (RBSP) spacecraft: Characterization of earth's radiation belt high-energy particle populations, *Space Sci. Rev.*, **179**, 337–381, doi:10.1007/s11214-012-9950-9.

Baker, D. N, et al. (2013a), James Van Allen and his namesake NASA mission, *EOS*, 94, 49, 469–470, doi:10.1002/2013EO490001.

Baker, D. N., et al. (2013b), A long-lived relativistic electron storage ring embedded in Earth's outer Van Allen belt, *Science*, **340**, 6129, doi:10.1126/science.1233518.

Baker, D. N., et al. (2014), Gradual diffusion and punctuated phase space density enhancements of highly relativistic electrons: Van Allen Probes observations, *Geophys. Res. Lett.*, **41**, 5, doi:Doi 10.1002/2013gl058942.

Blake, J. B., et al. (2013), The magnetic electron ion spectrometer (MagEIS) instruments aboard the radiation belt storm probes (RBSP) spacecraft, *Space Sci. Rev.*, **179**, 383, doi:10.1007/s11214-013-9991-8.

Blum, L. W., et al. (2013), New conjunctive CubeSat and balloon measurements to quantify rapid energetic electron precipitation, *Geophys. Res. Lett.*, **40**, 5833–5837, doi:10.1002/2013GL058546.

Boyd, A. J., et al. (2014), Quantifying the radiation belt seed population in the 17 March 2013 electron acceleration event, *Geophys. Res. Lett.*, **41**, 7, 2275–2281, doi:Doi 10.1002/2014gl059626.

Claudepierre, S. G., et al. (2013), Van Allen Probes observation of localized drift resonance between poloidal mode ultra-low frequency waves and 60 keV electrons, *Geophys. Res. Lett.*, **40**, 17, 4491–4497, doi:Doi 10.1002/Grl.50901.

Dai, L., et al. (2013), Excitation of poloidal standing Alfvén waves through drift resonance wave-particle interaction, *Geophys. Res. Lett.*, **40**, 16, 4127–4132, doi:10.1002/grl.50800.

Fennell, J. F., et al. (2014), Van Allen Probes observations of direct wave-particle interactions, *Geophys. Res. Lett.*, DOI: 10.1002/2013GL059165.

Foster, J. C., et al. (2014), Prompt energization of relativistic and highly relativistic electrons during a substorm interval: Van Allen Probes observations, *Geophys. Res. Lett.*, **41**, 1, doi:Doi 10.1002/2013gl058438.

Funsten, H. O., et al. (2013), Helium, oxygen, proton, and electron (hope) mass spectrometer for the radiation belt storm probes mission, *Space Sci. Rev.*, **179** (1–4), 423–484, doi: 10.1007/s11214-013-9968-7.

Goldstein, J., et al. (2015), Simulation of Van Allen Probes plasmapause encounters, *J. Geophys. Res.*, **119**, 7464–7484, doi:10.1029/2014JA020252.

Hao, Y., et al. (2014), Interactions of energetic electrons with ULF waves triggered by interplanetary shock: Van Allen Probes observations in the magnetotail, *J. Geophys. Res.*, **119**, 8262–8273, DOI: 10.1002/2014JA020023.

Hudson, M. K., et al. (2014), Simulated magnetopause losses and Van Allen Probe flux dropouts, *Geophys. Res. Lett.*, **41**, 4, 1113–1118, doi:10.1002/2014gl059222.

Jordanova, V. K., et al. (2014), Simulations of inner magnetosphere dynamics with an expanded RAM-SCB model and comparisons with Van Allen Probes observations, *Geophys. Res. Lett.*, **41** (8), 2687–2694, DOI: 10.1002/2014GL059533.

Kessel, R. L., et al. (2012), The radiation belt storm probes (RBSP) and space weather, *Space Sci. Rev.*, **179** (1–4), 531–543, doi:10.1007/s11214-012-9953-6.

Kletzing, C. A., et al. (2013), The electric and magnetic field instrument suite and integrated science (EMFISIS) on RBSP, *Space Sci. Rev.*, **179**, 127–181, doi:10.1007/s11214-013-9993-6.

Kletzing, C.A. (2015), Waves chapter of this volume.

Li, W., et al. (2013), An unusual enhancement of low-frequency plasmaspheric hiss in the other plasmasphere associated with substorm injected electrons, *Geophys. Res. Lett.*, **40**, 3798–3803, doi:10.1002/grl.50787.

Li, X., et al. (2013), First results from CSSWE: Characteristics of relativistic electrons in the near-earth environment during the October 2012, magnetic storms, *J. Geophys. Res.*, **118** (10), 6489–6499, doi:10.1002/2013JA019342.

Mann, I. R., et al. (2013), Discovery of the action of a geophysical synchrotron in the earth's Van Allen radiation belts, *Nat. Commun.*, **4**, 1–6, doi:Artn 2795, doi:10.1038/Ncomms3795.

Mauk, B. H., et al. (2012), Science objectives and rationale for the radiation belt storm probes mission, *Space Sci. Rev.*, **179** (1–4), 3–27, doi:10.1007/s11214-012-9908-y.

Mazur, J. , et al. (2013), The Relativistic Proton Spectrometer (RPS) for the Radiation Belt Storm Probes Mission, *Space Sci. Rev.*, **179** (1), 221–261, doi:10.1007/s11214-012-9926-9.

Millan et al. (2013), The Balloon Array for RBSP Relativistic Electron Losses (BARREL), *Space Sci. Rev.*, **179**, 503–530, doi: 10.1007/s11214-013-9971-z.

Mitchell, D. G., et al. (2013), Radiation Belt Storm Probes Ion Composition Experiment (RBSPICE), *Space Sci. Rev.*, **179** (1–4), 263–308, doi:10.1007/s11214-013-9965-x.

Morley, S. K., et al. (2013), Phase space density matching of relativistic electrons using the Van Allen Probes: REPT results, *Geophys. Res. Lett.*, **40** (18), 4798–4802, doi:Doi 10.1002/Grl.50909.

Mozer, F. S., et al. (2014), Direct observation of radiation-belt electron acceleration from electron-volt energies to megavolts by nonlinear whistlers, *Phys. Rev. Lett.*, **113** (3), 35001-1–35001-5, doi:10.1103/PhysRevLett.113.035001.

Reeves, G. D., et al. (2013), Electron acceleration in the heart of the Van Allen radiation belts, *Science*, **341**, 6149, doi:10.1126/science.1237743.

Schiller, Q., et al. (2014), A nonstorm time enhancement of relativistic electrons in the outer radiation belt, *Geophys. Res. Lett.*, **41** (1), 7–12, doi:10.1002/2013gl058485.

Shprits, Y. Y., et al. (2013), Unusual stable trapping of the ultrarelativistic electrons in the Van Allen radiation belts, *Nat. Phys.*, **9** (11), 699–703, doi:Doi 10.1038/Nphys2760.

Spence, H. E., et al. (2013), Science Goals and Overview of the Energetic Particle, Composition, and Thermal Plasma (ECT) Suite on NASA's Radiation Belt Storm Probes (RBSP) Mission, *Space Sci. Rev.*, **179** (1–4), 311–336, doi: 10.1007/s11214-013-0007-5.

Su, Z. P., et al. (2014), Nonstorm time dynamics of electron radiation belts observed by the Van Allen Probes, *Geophys. Res. Lett.*, **41** (2), 229–235, doi:Doi 10.1002/2013gl058912.

Thorne, R. M., et al. (2013a), Rapid local acceleration of relativistic radiation-belt electrons by magnetospheric chorus, *Nature*, **504**, 7480, doi:10.1038/nature12889.

Thorne, R. M., et al. (2013b), Evolution and slow decay of an unusual narrow ring of relativistic electrons near L$\sim$3.2 following the September 2012, magnetic storm, *Geophys. Res. Lett.*, **40**, 3507–3511, DOI: 10.1002/grl.50627.

Tu, W. C., et al. (2014), Event-specific chorus wave and electron seed population models in DREAM3D using the Van Allen Probes, *Geophys. Res. Lett.*, **41**, 5, 1359–1366, doi:Doi 10.1002/2013gl058819.

Turner, D. L., et al. (2014a), On the cause and extent of outer radiation belt losses during the 30 September 2012 dropout event, *J. Geophys. Res.-Space Physics*, **119**, 3, 1530–1540, doi:Doi 10.1002/2013ja019446.

Turner, D. L., et al. (2014b), Competing source and loss mechanisms due to wave-particle interactions in earth's outer radiation belt during the 30 September to 3 October 2012 geomagnetic storm, *J. Geophys. Res.-Space Physics, 119* (3), 1960–1979, doi:Doi 10.1002/2014ja019770.

Usanova, M. E., et al. (2014), Effect of EMIC waves on relativistic and ultrarelativistic electron populations: Ground-based and Van Allen Probes observations, *Geophys. Res. Lett.*, **41** (5), 1375–1381, doi:Doi 10.1002/2013gl059024.

Wygant, J. R., et al. (2013), The electric field and waves instruments on the radiation belt storm probes mission, *Space Sci. Rev.*, 179 (1–4), 183–220, doi: 10.1007/s11214-013-0013-7.

Index

The manufacturer's authorised representative in the EU for product safety is Oxford University Press España S.A. of el Parque Empresarial San Fernando de Henares, Avenida de Castilla, 2 – 28830 Madrid (www.oup.es/en or product.safety@oup.com). OUP España S.A. also acts as importer into Spain of products made by the manufacturer.

www.ingramcontent.com/pod-product-compliance
Lightning Source LLC
LaVergne TN
LVHW062213200926
848484LV00006B/1386
* 9 7 8 0 1 9 8 7 0 5 2 4 6 *